Lecture Notes in Computer Science 11184

Commenced Publication in 1973
Founding and Former Series Editors:
Gerhard Goos, Juris Hartmanis, and Jan van Leeuwen

More information about this series at http://www.springer.com/series/7407

Raffaele Cerulli · Andrea Raiconi
Stefan Voß (Eds.)

Computational Logistics

9th International Conference, ICCL 2018
Vietri sul Mare, Italy, October 1–3, 2018
Proceedings

 Springer

Editors
Raffaele Cerulli ⓘD
University of Salerno
Fisciano, Italy

Andrea Raiconi ⓘD
University of Salerno
Fisciano, Italy

Stefan Voß ⓘD
Institute of Information Systems
University of Hamburg
Hamburg, Germany

ISSN 0302-9743   ISSN 1611-3349   (electronic)
Lecture Notes in Computer Science
ISBN 978-3-030-00897-0   ISBN 978-3-030-00898-7   (eBook)
https://doi.org/10.1007/978-3-030-00898-7

Library of Congress Control Number: 2018954767

LNCS Sublibrary: SL1 – Theoretical Computer Science and General Issues

This Springer imprint is published by the registered company Springer Nature Switzerland AG
The registered company address is: Gewerbestrasse 11, 6330 Cham, Switzerland

Preface

While digital transformation and digital innovation are spreading the word and penetrating almost every business, logistics is still omnipresent. If we do not move freight or people, we move bits and bytes, or both or even everything together. We claim that the data is there, the technology is there, and we need to get our ideas and solution concepts into running systems. Whether we are emphasizing new buzzwords like mobility-as-a-service or synchromodality, or whether we are still using existing phrases like stowage planning or service network design, computational logistics is here to support these processes and make things better and better. We are solving classical combinatorial optimization problems related to, for instance, pickup and delivery, we are thinking about already established concepts of recent years such as e-mobility, and we are considering new concepts like autonomous vessels. But in all cases we appreciate the connection to computational tools for solving complex problems in logistics and supply chain management as well as public transport.

The International Conference on Computational Logistics (ICCL) is a forum where recent advances on the topic are presented and discussed. This volume offers a selection of 32 peer-reviewed papers out of well over 70 contributions submitted to the 9th International Conference on Computational Logistics (ICCL 2018), held in Vietri sul Mare, Italy, during October 1–3, 2018. The papers show various directions of importance in computational logistics, classified into five topic areas reflecting the interest of researchers and practitioners in this field. The papers in this volume are grouped according to the following parts:

1. *Maritime Shipping and Routing*:
   As a major mode in freight transportation we see considerable work in the area of maritime shipping. The ICCL has been strong in addressing issues of maritime shipping as it is also observed here. The papers in this area address various problems arising, among others, in maritime inventory routing, fleet deployment, as well as the upcoming topic of autonomous shipping. Moreover, inland waterways are addressed, too.
2. *Container Handling and Container Terminals*:
   Moving from the waterside into ports, in this part we have papers focussing on the relocation of containers within the yard area of container terminals, stowage planning, vessel capacity, as well as related key performance indicators.
3. *Vehicle Routing and Multi-modal Transportation*:
   Being a classical area of research, vehicle routing problems still need quite a bit of attention, especially when they become richer and include more and more practical constraints. Additionally, concomitant factors such as those of pricing services, as well as sensing vehicle coverage, etc., have received more attention recently and are considered here.

4. *Network Design and Scheduling*:
   Network design may refer to any mode of transportation and relates to freight as
   well as public transport. The papers in this part span demand-responsive transport
   on a large scale as well as on a small scale. Scheduling papers deal with crew
   scheduling in maritime shipping as well as a rich machine scheduling problem.
5. *Selected Topics in Logistics Oriented Combinatorial Optimization*:
   In this part we have included papers treating warehouse operations, sourcing in
   health care, e-mobility, as well as one of the classical combinatorial optimization
   problems that are still in need of better algorithmic developments, that is, the
   quadratic assignment problem.

The ICCL 2018 was the ninth edition of this conference series, following the earlier
ones held in Shanghai, China (2010, 2012), Hamburg, Germany (2011), Copenhagen,
Denmark (2013), Valparaiso, Chile (2014), Delft, The Netherlands (2015), Lisbon,
Portugal (2016), and Southampton, UK (2017).

The editors thank all the authors for their contributions and the reviewers for their
invaluable support and feedback. We trust that the present volume supports the continued advances within computational logistics and inspires all participants and readers
to its fullest extent.

October 2018

Raffaele Cerulli
Andrea Raiconi
Stefan Voß

Organization

Program Committee

Panagiotis Angeloudis	Imperial College London, UK
Claudia Archetti	University of Brescia, Italy
Buyang Cao	Tongji University, China
Francesco Carrabs	University of Salerno, Italy
Carmine Cerrone	University of Molise, Italy
Raffaele Cerulli	University of Salerno, Italy
Stefano Coniglio	University of Southampton, UK
Francesco Corman	ETH Zurich, Switzerland
Ciriaco D'Ambrosio	University of Salerno, Italy
René De Koster	Erasmus University Rotterdam, The Netherlands
Karl Doerner	University of Vienna, Austria
Jan Fabian Ehmke	Otto-von-Guericke University Magdeburg, Germany
Kjetil Fagerholt	Norwegian University of Science and Technology, Norway
Yingjie Fan	Erasmus University, Rotterdam, The Netherlands
Paola Festa	University of Naples Federico II, Italy
Monica Gentili	University of Louisville, USA
Rosa Guadalupe Gonzalez Ramirez	Universidad de Los Andes, Chile
Luis Gouveia	University of Lisbon, Portugal
Peter Greistorfer	Karl-Franzens-Universität Graz, Austria
Irina Gribkovskaia	Molde University College, Norway
Hans-Otto Günther	TU Berlin, Germany
Richard F. Hartl	University of Vienna, Austria
Leonard Heilig	University of Hamburg, Germany
Sin C. Ho	Chinese University of Hong Kong, SAR China
Rune Møller Jensen	Information Technology University of Copenhagen, Denmark
Herbert Kopfer	University of Bremen, Germany
Eduardo Lalla-Ruiz	University of Hamburg, Germany
Jasmine Siu Lee Lam	Nanyang Technological University, China
Gilbert Laporte	HEC Montréal and GERAD, Canada
Janny Leung	Chinese University of Hong Kong, SAR China
Ridha Mahjoub	LAMSADE, University Paris-Dauphine, France
Vittorio Maniezzo	University of Bologna, Italy
Rudy Negenborn	Delft University of Technology, The Netherlands
Dario Pacino	Technical University of Denmark, Denmark
Julia Pahl	University of Southern Denmark, Denmark

Ana Paias	University of Lisbon, Portugal
Andrea Raiconi	University of Salerno, Italy
Ruben Ruiz	Universitat Politècnica de València, Spain
Mario Ruthmair	University of Vienna, Austria
Simona Sacone	University of Genoa, Italy
Juan José Salazar González	Universidad de La Laguna, Spain
Frederik Schulte	University of Hamburg, Germany
L. Douglas Smith	University of Missouri-St. Louis, USA
M. Grazia Speranza	University of Brescia, Italy
Sven Spieckermann	Simplan AG, Germany
Lori Tavasszy	Delft University of Technology, The Netherlands
Kevin Tierney	Bielefeld University, Germany
Walter Ukovich	University of Trieste, Italy
Juan G. Villegas R.	Universidad de Antioquia, Colombia
Stefan Voss	University of Hamburg, Germany
David Woodruff	University of California, Davis, USA
Tsz Leung Yip	Hong Kong Polytechnic University, SAR China
Shiyuan Zheng	Shanghai Maritime University, China
Bülent Çatay	Sabanci University, Turkey

Additional Reviewers

Ajspur, Mai	Molenbruch, Yves
Ambrosino, Daniela	Pacciarelli, Dario
Cacchiani, Valentina	Pentangelo, Rosa
Christensen, Jonas	Saibene, Giorgio
Koza, David Franz	Sammarra, Marcello
Kuhlemann, Stefan	Stecca, Giuseppe
Larsen, Rie	Sterle, Claudio
Laureana, Federica	Szeto, Wai Yuen
Miranda, Pedro	Tanaka, Shunji

Contents

Vehicle Routing and Multi-modal Transportation

Network Design and Scheduling

Selected Topics in Logistics Oriented Combinatorial Optimization

Maritime Shipping and Routing

Applying a Relax-and-Fix Approach to a Fixed Charge Network Flow Model of a Maritime Inventory Routing Problem

Marcelo W. Friske[(✉)] and Luciana S. Buriol

Departamento de Informática, Universidade Federal do Rio Grande do Sul,
Porto Alegre, Brazil
{mwfriske,buriol}@inf.ufrgs.br

Abstract. This work presents a Relax-and-Fix algorithm to solve a class of single product Maritime Inventory Routing Problems. The problem consists of routing and scheduling a heterogeneous fleet of vessels to supply a set of ports, respecting lower and upper limits of inventory at production and consumption ports, along with a time horizon. A fixed charge network flow is used to model the problem, and valid inequalities are incorporated into the formulation, providing tight bounds and enabling the Relax-and-Fix algorithm to obtain good solutions in reasonable processing times. Three MIP-based local search procedures are proposed for improving solutions. Tests performed on a set of benchmark instances from the literature show that the solution approach can be effective for solving the problem.

Keywords: Maritime Inventory Routing Problem · Fixed Charge
Network Flow · Relax-and-Fix · MIP-Based Local Search

1 Introduction

The Maritime Inventory Routing Problem (MIRP) is a combinatorial problem where one has to manage the voyage of vessels and the inventory of ports along a finite planning horizon. This problem consists of routing and scheduling a heterogeneous fleet of vessels for transporting one or more products from production to consumption ports. Each port has a storage capacity and a production or consumption rate that may vary along the planning horizon. A vessel must arrive at a production port to load a certain quantity of product before the port becomes full, and similarly, it must deliver the product to a consumption port before inventory becomes empty. Vessels can differ between them by capacity, operating costs, and traveling time between each port. The objective usually aims to minimize the traveling and operational costs. Most of the MIRP discussed in the literature are particular variations based on real scenarios, where different assumptions and side constraints are considered.

As the nature of the problem is very complex, and maritime transportation involves substantial costs, it is crucial to use optimization techniques for

© Springer Nature Switzerland AG 2018
R. Cerulli et al. (Eds.): ICCL 2018, LNCS 11184, pp. 3–16, 2018.
https://doi.org/10.1007/978-3-030-00898-7_1

obtaining better schedules of vessels. A literature review of works involving optimization in maritime transportation can be found in [4–6]. They describe three major maritime areas in which optimization can be applied: tramp, liner, and industrial shipping. Also, the problems can be classified into three planning levels: strategic, tactical, and operational. MIRP can be viewed as an industrial and tactical problem, as the cargo owner also controls the vessels and the planning horizon is greater than one month. In [12], a specific review of MIRP models and solution methods is presented, and a core model for the problem with additional features and side constraints is proposed. The paper introduces a benchmark library for the problem called MIRPLIB [1].

This work considers the MIRP model proposed by [12]. The problem considers a single product that needs to be transported from loading ports to discharging ports. The ports are grouped geographically in production and consumption regions. Each port has a variable production (or consumption) rate along the planning horizon, which is discretized and each time-period corresponds to one day. The objective function of the problem is to maximize the revenue obtained per delivered product at discharging ports minus the cost involving the transportation of the product, and some penalizations that will be explained.

The model proposed by [12] is based on a simple time-space network, which is also used in different articles of MIRPs and Inventory Routing Problems, a similar planning problem where land vehicles are considered instead of vessels [8,14,15]. However, this structure produces weak bounds, and solving even small problems can be hard for a state-of-the-art mathematical solver.

The work of [2] proposed a discrete time fixed charge network flow model (FCNF) to obtain a better formulation for a MIRP. Also, new valid inequalities generalized from the lot-sizing problem were proposed with branching priorities. The proposed model was capable of proving tight bounds and obtaining optimal solutions faster than the standard time-space network. The works of [3,9] compared the use of FCNF discrete time and continuous time formulations for modeling MIRPs. Continuous time formulations have the advantage of having a smaller number of variables and constraints than discrete models. Thus, they tend to be faster in solving mainly large instances. On the other hand, discrete time formulations can model situations such as variable production and consumption rates, which is usually considered constant in continuous time models. Also, they can provide better linear relaxation when using valid inequalities.

For solving the core MIRP of [12] with some features, [11] proposed an iterative algorithm with two phases. They consider a practical assumption that one vessel must visit at most two ports of the same type and geographical region sequentially. From this, a special time network is built to incorporate the assumptions indirectly. The model is improved with valid inequalities, and an improvement phase with a MIP-based local search is applied. The proposed approach provided high-quality solutions in reasonable computational time.

Using the time-space network, [7] proposed a Relax-and-Fix (R&F) algorithm for solving the same MIRP of [11]. Additional constraints were considered to improve the efficiency of the algorithm, and four MIP-based local searches were proposed. The approach used was able to obtain good solutions, including two new best-known values. However, for the hard (large) instances, the proposed algorithm cannot achieve feasible solutions.

This work aims to extend the work of [7], applying the Relax-and-Fix framework for solving the FCNF model. In this case, we use the problem described in [12], modeling it as the FCNF similar to [2]. The R&F is a matheuristic that consists in dividing the problem into subproblems, solving them iteratively by relaxing and fixing subsets of variables to obtain a feasible solution for the problem. For more explanation about R&F, we recommend the book of [13]. We also implemented valid inequalities based on knapsack sets, and MIP-based local searches were used for improving the solution quality. The objective of this work is to verify if a relatively simple algorithmic approach such as R&F can obtain good solutions for a hard problem, without using problem-specific algorithms, as proposed by [11].

The remainder of this work is organized as follows. Section 2 describes the problem and presents the FCNF formulation with the valid inequalities. Section 3 describes the R&F and MIP-based local search procedures. Computational experiments are presented in Sect. 5. Finally, Sect. 6 presents the conclusions and future works.

2 Problem Formulation

In this section, we present the FCNF model for the MIRP, that is based on the model described in [12]. This model considers a deep-sea MIRP, where the traveling times are much larger than the operating times. Each vessel must visit different ports such that production ports do not reach their maximum capacity, and discharge ports do not achieve an empty inventory. The discharging ports pay different revenue values according to the quantity discharged by vessels.

In some cases, it is not possible to supply loading and discharging demands due to the limited fleet size and the large traveling times between ports. In this case, it is possible to consider the inventory as soft constraints, where the excess or lack of the product can be sold to or bought from simplified spot markets. However, the quantity sold to or bought from each port is limited.

In the FCNF, the loading of vessels is modeled as a commodity that flows along the nodes, represented by port-times. Let V be the set of vessels, J the set of ports, and T the set of time periods, with $T = |T|$. Ports are split into subsets J^P for production or loading ports, and J^C for consuming or discharging ports. Ports are grouped in production regions R^P and discharging regions R^C.

A port-time $(i, t), i \in J, t \in T$ represents a possible operation of a vessel in port i at time t. Source node $o(v)$ represents the starting point of vessel v, and sink node $d(v)$ is the end of vessel route. Binary variable x_{ijvt} defines if a vessel v travels from port i to port j, departing at time period t. Parameter T_{ijv} denotes the traveling time between port i and j by vessel v. When vessel v arrives at specific node (i, t), it can wait for one time-period (binary variable w_{it}^v is set to 1), or it can start to operate (binary variable o_{ivt}^A is set to 1). When vessel v starts operating, continuous variable f_{it}^v represents the amount discharged (or loaded) at node (i, t). The consumption rate of port i in each time period t is denoted by D_{it}, and s_{it} is the inventory of port i in time t. Parameter s_i^0 corresponds to the initial inventory of port i. Continuous variable α_{it} represents the amount of product bought from the simplified spot market. After started to operate, a vessel can continue to operate in the same port, using variable $o_{ivt}^B = 1$, or it can leave the port, traveling to another port or the sink node using a traveling arc x_{ijvt}.

Figure 1 illustrates the FCNF model for a discharging port $i \in \mathcal{J}^P$ and one vessel $v \in \mathcal{V}$, with $T = 4$. A possible route of the vessel is highlighted.

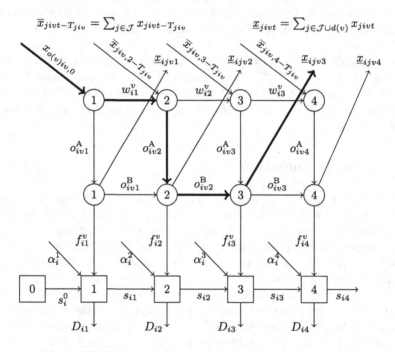

Fig. 1. FCNF for a discharging port $i \in \mathcal{J}$ and vessel $v \in \mathcal{V}$.

In Fig. 1 each port-time (i, t) is divided into three layers. The top layer coordinates the traveling of vessels between ports, including dummy source and sink nodes. The middle layer coordinates the operation of vessel at the port, and the bottom layer represents the transfer of product between vessel v and port i.

In the FCNF model, when a vessel arrives at a port, it must operate (discharging or loading) before departing to another destination. Also, after ending the operation, it is not allowed to wait at the same port. These constraints are not considered in the original formulation of [12]. However, they are implicit and useful for a good solution in a real scenario.

The MIRP model of [12] is formulated as a FCNF model as follows:

$$
\max \quad
\begin{aligned}
&\sum_{i \in \mathcal{J}^c} \sum_{t \in \mathcal{T}} \sum_{v \in \mathcal{V}} R_{it} f_{it}^v - \sum_{v \in \mathcal{V}} \sum_{i \in \mathcal{J} \cup \{o(v)\}} \sum_{j \in \mathcal{J} \cup \{d(v)\}} \sum_{t \in \mathcal{T}} C_{ijv} x_{ijvt} \\
&- \sum_{i \in \mathcal{J}} \sum_{t \in \mathcal{T}} \sum_{v \in \mathcal{V}} (t\epsilon_z) o_{ivt} - \sum_{i \in \mathcal{J}} \sum_{t \in \mathcal{T}} P_{it} \alpha_{it}
\end{aligned}
\tag{1a}
$$

s.t.

$$
x_{o(v)j_v^0 v0} = 1, \ \forall v \in \mathcal{V}, \tag{1b}
$$

$$
\sum_{i \in \mathcal{J}} \sum_{t \in \mathcal{T}} x_{id(v)vt} = 1, \ \forall v \in \mathcal{V}, \tag{1c}
$$

$$
\sum_{j \in \mathcal{J} \cup \{o(v)\}} x_{jiv,t-T_{jiv}} + w_{iv,t-1} = w_{ivt} + o_{ivt}^{\mathrm{A}}, \ \forall v \in \mathcal{V}, i \in \mathcal{J}, t \in \mathcal{T}, \tag{1d}
$$

$$
o_{ivt}^{\mathrm{A}} + o_{iv,t-1}^{\mathrm{B}} = o_{ivt}^{\mathrm{B}} + \sum_{j \in \mathcal{J} \cup \{d(v)\}} x_{ijvt}, \ \forall v \in \mathcal{V}, i \in \mathcal{J}, t \in \mathcal{T}, \tag{1e}
$$

$$
o_{ivt}^{\mathrm{A}} + o_{iv,t-1}^{\mathrm{B}} = o_{ivt}, \ \forall v \in \mathcal{V}, i \in \mathcal{J}, t \in \mathcal{T}, \tag{1f}
$$

$$
\sum_{j \in \mathcal{J} \cup \{o(v)\}} f_{jiv,t-T_{jiv}}^{\mathrm{X}} + f_{iv,t-1}^{\mathrm{W}} = f_{ivt}^{\mathrm{W}} + f_{ivt}^{\mathrm{OA}}, \ \forall v \in \mathcal{V}, i \in \mathcal{J}, t \in \mathcal{T}, \tag{1g}
$$

$$
f_{ivt}^{\mathrm{OA}} + f_{iv,t-1}^{\mathrm{OB}} + \Delta_i f_{it}^v = f_{ivt}^{\mathrm{OB}} + \sum_{j \in \mathcal{J} \cup \{d(v)\}} f_{ijvt}^{\mathrm{X}}, \ \forall v \in \mathcal{V}, i \in \mathcal{J}, t \in \mathcal{T}, \tag{1h}
$$

$$
f_{o(v)j_v^0 v0}^{\mathrm{X}} = L_v^0, \ \forall v \in \mathcal{V}, \tag{1i}
$$

$$
f_{ijvt}^{\mathrm{X}} \geq Q^v x_{ijvt}, \ \forall v \in \mathcal{V}, t \in \mathcal{T}, i \in \mathcal{J}^P, j \in \mathcal{J}^D \cup \{d(v)\}, \tag{1j}
$$

$$
f_{ijvt}^{\mathrm{X}} \leq Q^v (1 - x_{ijvt}), \ \forall v \in \mathcal{V}, t \in \mathcal{T}, i \in \mathcal{J}^D, j \in \mathcal{J}^P \cup \{d(v)\}, \tag{1k}
$$

$$
\sum_{v \in \mathcal{V}} o_{ivt} \leq B_i, \ \forall i \in \mathcal{J}, t \in \mathcal{T}, \tag{1l}
$$

$$
s_{it} = s_{i,t-1} + \Delta_j (D_{it} - \sum_{v \in \mathcal{V}} f_{it}^v - \alpha_{it}), \ \forall i \in \mathcal{J}, t \in \mathcal{T}, \tag{1m}
$$

$$
\sum_{t \in \mathcal{T}} \alpha_{it} \leq \alpha_i^{max}, \ \forall i \in \mathcal{J}, \tag{1n}
$$

$$
0 \leq \alpha_{it} \leq \alpha_{it}^{max}, \ \forall i \in \mathcal{J}, t \in \mathcal{T}, \tag{1o}
$$

$$
S_i^{min} \leq s_{it} \leq S_i^{max}, \ \forall i \in \mathcal{J}, t \in \mathcal{T}, \tag{1p}
$$

$$
s_{i,0} = S_i^0, \ \forall i \in \mathcal{J}, \tag{1q}
$$

$$
0 \leq f_{ijvt}^{\mathrm{X}} \leq Q^v x_{ijvt}, \forall v \in \mathcal{V}, i \in \mathcal{J} \cup \{o(v)\}, j \in \mathcal{J} \cup \{d(v)\}, t \in \mathcal{T}, \tag{1r}
$$

$$
0 \leq f_{ivt}^{\mathrm{OA}} \leq Q^v o_{ivt}^{\mathrm{A}}, \forall v \in \mathcal{V}, i \in \mathcal{J}, t \in \mathcal{T}, \tag{1s}
$$

$$
0 \leq f_{ivt}^{\mathrm{W}} \leq Q^v w_{ivt}, \forall v \in \mathcal{V}, i \in \mathcal{J}, t \in \mathcal{T}, \tag{1t}
$$

$$
0 \leq f_{ivt}^{\mathrm{OB}} \leq Q^v o_{ivt}^{\mathrm{B}}, \forall v \in \mathcal{V}, i \in \mathcal{J}, t \in \mathcal{T}, \tag{1u}
$$

$$
F_i^{\min} o_{ivt} \leq f_{it}^v \leq F_i^{\max} o_{ivt}, \ v \in \mathcal{V}, \forall i \in \mathcal{J}, t \in \mathcal{T}, \tag{1v}
$$

$$
w_{ivt}, o_{ivt}, o_{ivt}^{\mathrm{A}}, o_{ivt}^{\mathrm{B}} \in \{0,1\}, \ v \in \mathcal{V}, \forall i \in \mathcal{J}, t \in \mathcal{T}, \tag{1w}
$$

$$
x_{ijvt} \in \{0,1\}, \ \forall v \in \mathcal{V}, i \in \mathcal{J} \cup \{o(v)\}, j \in \mathcal{J} \cup \{d(v)\}, t \in \mathcal{T}. \tag{1x}
$$

Objective function (1a) maximizes the revenue R_{it} of the unloaded product at discharging ports, subtracting arc costs C_{ijv} used by each vessel. The third term is an additional value that induces vessels to operate as soon and as few times as possible. Variable o_{vit} is equal to one when vessel v operates at port i in time t. The penalization value P_{it} for using spot markets is accounted in the last term of the equation.

Constraints (1b) fix the source arc of each vessel v. It assumes that vessel v travels from the source node $o(v)$ to its initial port j_v^0, departing from time period 0. The traveling time between the source node and the initial port is the first time t_v^0 in which vessel v becomes available. Constraints (1c) impose that all vessels end its route, reaching the sink node $d(v)$. Constraints (1d) and (1e) are the flow balance of each vessel along the nodes. Constraints (1f) define that if a vessel is operating, it started to operate in the current time period, or it continues to operate from the previous time period.

Constraints (1g)-(1h) represent inventory balance of each vessel, where $\Delta_i = 1$ if $i \in \mathcal{J}^P$, and $\Delta_i = -1$ if $i \in \mathcal{J}^C$. Variable f_{ijvt}^X is the load on board of vessel v before traveling from port i to port j in time period t, f_{ivt}^W is the load on board of vessel v before waiting at port i in time t, f_{ivt}^{OA} is the load on board of vessel v before starting to operate at port i in time t, and f_{ivt}^{OB} is the load on board of vessel v before continuing to operate at port i in time t. Constraints (1i) define the initial inventory of each vessel, where parameter L_v^0 corresponds to the initial inventory.

Side constraints (1j) and (1k) impose that each vessel must depart from a loading region to a discharging region at full capacity and must depart from a discharging region to a loading region empty. Constraints (1l) limit to B_i (number of berths) the number of vessels that can operate simultaneously at a node. The inventory balance at ports is defined by (1m). Constraints (1n) and (1o) limit the daily amount and the cumulative amount of product that can be bought from or sold to the simplified spot market, respectively. Constraints (1p) impose the inventory limit of each port to be between the lower limit S_i^{min} and upper limit S_i^{max}, and (1q) define the initial inventory for each port. Constraints (1r)–(1u) limit the flow of vessel v to the capacity Q_v along the arcs, while (1v) impose a minimum F_i^{min} and maximum F_i^{max} amount of product that can be loaded or discharged by vessel v when operating in each port i and time period t. Finally, (1w) and (1x) define the scope of binary variables.

2.1   Valid Inequalities

In order to improve the bounds of FCNF model relaxation, valid inequalities based on well-known knapsack sets are included. They are derived from the inequalities proposed in [2], which detailed step-by-step the building of each constraint. We present here just the final valid inequalities that were added a priori to the model.

Let T be a time interval such that $T = [l, k] \subseteq \mathcal{T}$. Also let $\overline{K} = \max\{Q_v : v \in V\}$. The knapsack inequalities for a loading port i are:

$$\sum_{v \in V} \left(o_{ivk}^B + \sum_{t \in T} \sum_{j \in \mathcal{J} \cup \{d(v)\}} x_{ijvt} \right) \geq$$
$$\left\lceil \frac{\sum_{t \in T} D_{ij} + S_i^{\min} - S_i^{\max}}{\overline{K}} \right\rceil, \tag{2}$$

$$\sum_{v \in V} \sum_{t \in T} o_{ivt} \geq \left\lceil \frac{\sum_{t \in T} D_{ij} + S_i^{\min} - S_i^{\max}}{F_i^{\max}} \right\rceil \tag{3}$$

Inequalities (2) impose a minimum number of vessel departures, while (3) impose a minimum number of loadings at port i.

For discharging ports, we can consider three inequalities sets:

$$\sum_{v \in V} \left(\sum_{j \in \mathcal{J} \cup \{o(v)\}} \sum_{t \in T} x_{jiv,t-T_{jiv}} + w_{iv,l-1} + o_{iv,l-1}^B \right) \geq$$
$$\left\lceil \frac{\sum_{t \in T} D_{it} - S_i^{\max} + S_i^{\min}}{\overline{K}} \right\rceil \tag{4}$$

$$\sum_{v \in V} \left(\sum_{t \in T} o_{ivt}^A + o_{iv,l-1}^B \right) \geq \left\lceil \frac{\sum_{t \in T} D_{it} - S_i^{\max} + S_i^{\min}}{\overline{K}} \right\rceil \tag{5}$$

$$\sum_{v \in V} \sum_{t \in T} o_{ivt} \geq \left\lceil \frac{\sum_{t \in T} D_{it} - S_i^{\max} + S_i^{\min}}{F_i^{\max}} \right\rceil \tag{6}$$

Inequalities (4) impose a minimum number of arrivals at discharging port i, while (5) impose a minimum number of starting of operations. Finally, (6) impose a minimum number of operations at port i in the interval $[l, k]$.

3   The Relax-and-Fix Algorithm

The proposed R&F is built in a rolling horizon fashion, i.e., the subproblems are defined according to the time horizon, which is divided into n intervals. Figure 2 illustrates the first, second, and last iterations of the R&F.

At the first iteration, all binary variables belonging to the *Integer block* are restricted to be integral. In the *Relaxed block* the integrality constraints on binary variables are relaxed. Variables and constraints that belong to the *End block* are omitted from the model for a while with the objective of reducing the size of the problem at the first iterations. A MIP solver is then used to solve the current problem. At the second iteration, the blocks "move forward": binary variables from the integer interval are fixed to their corresponding solution value, belonging now to the *Fixed block* (original continuous variables of the problem are kept unfixed), integrality constraints are re-introduced to a part of the relaxed block,

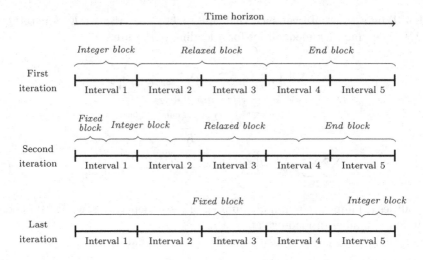

Fig. 2. Relax-and-Fix algorithm

and a part of the model that was omitted is now considered in the model as relaxed block. The problem is then solved again by the MIP solver. The algorithm continues iterating until all intervals have been removed from the end-block and integrality constraints are reintroduced to the variables of all intervals. At this point, a solution to the original problem is obtained.

In the Relax-and-Fix strategy, solving each interval up to optimality does not necessarily leads to an optimal solution for the original problem. In this case, we use MIP relative GAP and time limit as stopping criteria in each iteration, as suggested in [17], for accelerating the search. Initially, the MIP relative GAP is set to a positive value, which is linearly decreased along the iterations such that in the last iteration, the MIP relative GAP is set to 0.1%.

According to Fig. 2, there is an overlap between the blocks at each iteration This strategy is used to avoid infeasible solutions after some part of the model was fixed and integrality constraints are re-introduced to relaxed binary variables [13]. However, for the MIRP even using the overlap, port-time inventory bounds can be violated. It occurs when no vessel can reach a port at specific times due to the previously fixed routing decisions and the spot markets variables are not sufficient to avoid lack or surplus of inventory. To handle this issue, we introduce nonnegative auxiliary variables $\beta_{jt}, \theta_{jt}, j \in \mathcal{J}, t \in \mathcal{T}$. They transform the port-time inventory capacity in soft constraints, where the variables β_{jt}, θ_{jt} work as an unlimited spot market either for buying and selling product amounts, being highly penalized in the objective function. Equation (1m) is reformulated as follows:

$$s_{it} = s_{i,t-1} + \Delta_j(D_{it} - \sum_{v \in \mathcal{V}} f_{it}^v - \alpha_{it} - \beta_{it} + \theta_{it}), \quad \forall i \in \mathcal{J}, t \in \mathcal{T} \qquad (7)$$

Note that the auxiliary variables are used to prevent the solver from stopping prematurely, but if some of these variables are positive at the end of R&F, the solution for the original problem remains infeasible.

4   Improvement Phase

After the R&F algorithm obtained a solution, an improvement phase starts. It is used for improving the quality of solution and in some cases for removing infeasibilities due to the use of auxiliary variables β_{jt} and θ_{jt}.

In the proposed improvement phase, we use three MIP-based local search strategies. MIP-based local search is a technique that combines heuristic and mathematical programming, being used in several works, including MIRPs [11, 15]. It consists of first fixing all integer variables with the current solution values. Systematically a subset of these variables is unfixed, and a solver is used to solve the subproblem. Then, the variables are fixed with the newly obtained values, and a new subset is selected to be optimized. All continuous variables are kept free to be optimized in all iterations. The definition of each subset defines different strategies for MIP-based local search. The proposed strategies are:

1. **Time Intervals:** Consists of dividing the time horizon into m intervals, as in the R&F. At each iteration, the integer variables of one interval are unfixed. After optimized, these variables are fixed to the newly obtained values. This procedure is repeated iteratively until no improvement is achieved by optimizing at least one interval in m iterations, or before a time limit is reached. This strategy is similar to one presented in [10], for a Berth Allocation Problem with Time-dependent Limitations, where the POPMUSIC (Partial Optimization Metaheuristic Under Special Intensifications and Conditions) is applied by dividing the planning horizon into intervals, but selecting two intervals for optimizing at each iteration. One interval is selected at random (seed part), and the second by a distance function from the seed part, in this case, the adjacent interval.

2. **Vessels Pairs:** Similar to the strategy of [8] which explores the neighborhood between two vessels, this procedure consists in iteratively selecting a pair of vessels to be optimized. Let v_1 and v_2 be the vessels selected to be optimized in an iteration. Then, all binary variables indexed by $v \in \mathcal{V} : v = \{v_1, v_2\}$ are unfixed, and the solver is started. The vessel pairs are selected at random with no repetitions. The algorithm runs until no improvement is achieved for $\binom{|\mathcal{V}|}{2}$ iterations or within the stipulated time limit.

3. **Improving Vessels and Time Intervals:** This improvement approach is a combination of the two previous methods. The time horizon is divided into m intervals, allowing one interval to be optimized at a time. Also, all integer variables corresponding to one vessel are allowed to be optimized per iteration. After optimizing a solution, all integer variables of this vessel are fixed to the new values, except those belonging to the interval which is being

optimized. Then, a next vessel with the same time interval is optimized. The algorithm iterates between all time intervals and all vessels, $m|\mathcal{V}|$ steps in a complete iteration. The search stops when no improvement is achieved in one complete iteration or if the time limit is reached.

According to preliminary tests, we decided to define the order of running each MIP-based local search strategy as {3,2,1}, which demonstrated better objective values.

5   Computational Results

This section presents the results obtained by solving the Maritime Inventory Routing Problem with the solution approach described in Sects. 3 and 4. As in [11], we solved the model as a minimization problem, turning negative the objective function (1a) for comparison purposes. The algorithms were implemented using CPLEX 12.5 C++ API and compiled with the optimization parameter $-O3$. Experiments were carried out on an Intel Core i7-3632QM computer running at 2.2 GHz on a single core, with 8 GB RAM.

5.1   Tested Instances

For testing the algorithms, we use the "Group 1" instances available in the MIRPLIB [1]. The instances name present their characteristics. For example, instance "LR2_22_DR2_22_VC3_V10a" means that there exists 2 loading regions (LR), and in each region there is two loading ports, two discharging regions (DR), each of them with two ports, three vessel classes (VC), and a total of ten available vessels (V), at least one for each vessel class. The letter at the end of the name is used for differentiating instances with the same size. Each instance was tested with time horizons of 45 and 60 time-periods, where each time-period corresponds to one day.

5.2   Parametrization

Some parameters should be defined for the proposed algorithm. Although the tested instances have different sizes and there are differences in their characteristics, we defined a more general parametrization to be tested. Table 1 presents the parameters and the tested values.

For the MIP-based local search procedures, parameter "Total time limit (s)" is divided equally for each strategy. If the first strategy ends before the time limit, the remaining time is equally divided for the other two strategies and so on. The improvement phase uses the default value of the CPLEX solver for the MIP relative GAP.

The valid inequalities were added to the model a priori for all combinations of $\mathcal{T}$, such that T includes either the first time period or the last time period, i.e. $T = \{1, \ldots, t\}$, or $T = \{t, \ldots, T\}, t \in \mathcal{T}$.

Table 1. Parameters and values used on the computational results

	Parameter	Value	
Relax-and-Fix	Number of time periods per interval (T/n)	15	
	Size of model (in time periods) at the first iteration	45	Instances prefix "LR1_1"
		30	Instances prefix "LR1_2" or "LR2"
	Overlap (%)	50	
	Time limit (s) per iteration	600	Instances prefix "LR1" or "LR2_11"
		1200	Instances prefix "LR2_22"
	Initial MIP GAP (%)	5	
Local Search	Time limit (s) per iteration	120	
	Total time limit (s)	7200	
	Number of time intervals m	3	

5.3   Main Results

Table 2 presents the main results obtained by our solution approach. Columns "Relax-and-Fix" and "Local Search" present the results obtained by the algorithms described in Sects. 3 and 4, respectively. Columns "Time(s)" present the CPU time in seconds to obtain the objective value presented in columns "Obj". Column "Total Time(s)" presents the total computation time in seconds required for our algorithm. We compared our results with the best-known values obtained by [7,11], which are presented in column "BKV". The BKVs obtained by [7] are marked with an "*" in the Obj column. The CPU times were normalized using the *PassMark Software*.[1] Column "GAP(%)" presents the relative deviation ($\frac{Obj-BKV}{-BKV}$) * 100, where Obj corresponds to the objective value computed by our algorithm, while BKV corresponds to the best-known value.

As observed in Table 2, our algorithm is on average 25% faster than the BKV algorithm, although needing more processing time for some instances. Just for one instance no feasible solution could be found, where column GAP(%) is marked with "-". For two instances the same objective value of BKV was found ($GAP(\%) = 0.0$). Further, our algorithm could find four new best-known values (highlighted in bold), two of them in less processing time than the time reported in [11]. Except for three instances in which the deviation GAP from the BKV was greater than 10%, the average GAP deviation was 0.5% for $T = 45$ and 3.73% for $T = 60$, which can be considered very satisfactory.

The improvement phase was capable of removing the infeasibility of solutions provided by the Relax-and-Fix for six instances. This is observed in Table 2 when the value in column "Obj" of the Relax-and-Fix is positive for the respective instance, while for the Local Search the "Obj" column value is negative. Excluding these cases, the improvement on R&F solution was on average 3.13%,

---

[1] http://www.cpubenchmark.net/.

Table 2. MIRP results using Relax-and-Fix and improvement phase algorithms

Instance	Relax-and-Fix		Local Search		Total time(s)	GAP(%)	BKV	
	Time(s)	Obj	Time(s)	Obj			Time(s)	Obj
T = 45								
LR1_1_DR1_3_VC1_V7a	101	−13,271	375	−13,272	476	0.0	190	−13,272
LR1_1_DR1_4_VC3_V11a	912	−10,537	1,879	−11,009	2,792	2.1	1,436	*-11,243
LR1_1_DR1_4_VC3_V12a	1,421	−10,492	2,721	−10,709	4,142	0.2	8,566	−10,732
LR1_1_DR1_4_VC3_V12b	1,234	−9,023	2,469	−9,028	3,703	0.6	1,768	*−9,085
LR1_1_DR1_4_VC3_V8a	296	−5,060	2,039	−5,060	2,335	0.9	4,943	−5,106
LR1_1_DR1_4_VC3_V9a	728	−6,921	492	**−6,921**	1,220	**−0.4**	549	−6,891
LR1_2_DR1_3_VC2_V6a	729	−10,455	1,652	−10,717	2,380	3.7	6,396	−11,134
LR1_2_DR1_3_VC3_V8a	296	−10,658	906	−11,889	1,202	1.0	8,188	−12,010
LR2_11_DR2_22_VC2_V6a	225	149,598	1,775	−8,510	2,000	12.4	8,740	−9,718
LR2_11_DR2_33_VC4_V11a	2,465	22,761	2,875	−11,651	5,340	16.9	9,559	−14,017
LR2_11_DR2_33_VC5_V12a	1,418	−18,083	3,232	−18,395	4,650	0.2	9,582	−18,423
LR2_22_DR2_22_VC3_V10a	1,742	−23,905	4,148	**−24,855**	5,890	**−0.3**	9,359	−24,789
LR2_22_DR3_333_VC4_V14a	5,242	−20,013	5,043	−21,925	10,285	0.1	10,088	−21,952
LR2_22_DR3_333_VC4_V17a	5,748	−21,253	5,045	**−22,294**	10,793	**−2.7**	10,218	−21,713
AVG	1,611		2,475		4,086	2.5	6,399	
T = 60								
LR1_1_DR1_3_VC1_V7a	732	−16,673	2,190	−16,675	2,922	0.0	476	−16,675
LR1_1_DR1_4_VC3_V11a	2,211	−11,249	4,806	−12,155	7,018	8.3	7,657	−13,257
LR1_1_DR1_4_VC3_V12a	1,348	−10,907	4,419	−11,022	5,768	0.2	8,566	−11,040
LR1_1_DR1_4_VC3_V12b	2,122	−9,785	4,735	−9,830	6,857	2.2	9,342	−10,053
LR1_1_DR1_4_VC3_V8a	1,684	−4,410	2,829	−4,502	4,513	13.3	8,245	−5,191
LR1_1_DR1_4_VC3_V9a	2,032	−7,278	4,494	−7,328	6,526	3.0	8,886	−7,552
LR1_2_DR1_3_VC2_V6a	1,192	−12,406	3,162	−12,843	4,354	5.8	8,902	−13,631
LR1_2_DR1_3_VC3_V8a	1,114	−13,817	3,752	−14,152	4,866	3.4	8,613	−14,652
LR2_11_DR2_22_VC2_V6a	420	223,498	4,373	42,126	4,793	-	9,014	−12,655
LR2_11_DR2_33_VC4_V11a	3,552	39,783	5,043	−14,379	8,595	6.6	9,592	−15,387
LR2_11_DR2_33_VC5_V12a	2,453	59,566	3,882	**−22,948**	6,335	**−1.0**	9,656	−22,730
LR2_22_DR2_22_VC3_V10a	4,143	−5,356	5,043	−31,598	9,186	3.2	9,441	−32,627
LR2_22_DR3_333_VC4_V14a	7,669	−23,307	5,047	−25,069	12,716	6.7	10,234	−26,873
LR2_22_DR3_333_VC4_V17a	9,558	6,874	5,461	−25,236	15,020	6.5	10,312	−27,000
AVG	2,874		4,231		7,105	4.5	8,495	

although consuming approximately 60% of the processing time. Also, we can observe that in some instances no improvement was obtained, and just the R&F could find a new best-known solution.

Previous tests were performed using a small number of time periods per interval for Relax-and-Fix (9 time periods for instances with *T* = 45 and 10 time periods for instances with *T* = 60). Although not presented here, these results showed that for small instances, the approach can be faster and still provide good solutions, sometimes equals to the BKVs. However, in large instances, the approach can become very myopic, and the solution quality decreases considerably.

Comparing the results with the previous work [7], the use of the FCNF formulation in the R&F framework provided a good improvement in the quality of the solutions. Although the parameters values tested are different, in [7] no feasible solutions were found for eight instances, while in this work no feasible solution was found just for one instance. Considering the instances where a feasible solution was found, the average GAP from the best-known solution was 4.9% for the work of [7], while in this work the average GAP was 3.4%. The work of [7] provided results with less average processing time (3,861 s against 5,596 s of this work). However, the values of the time parameters were smaller than those tested in this work.

6 Conclusion and Future Work

This work presented a Relax-and-Fix algorithm for solving a class of Maritime Inventory Routing Problems. We modeled the problem using a fixed charge network flow model, such as described by [2], and also implemented valid inequalities based on knapsack sets. Results were carried out on instances of the MIRPLIB. Our solution approach is relatively simple and general if compared to the specialized algorithm of [11]. Nevertheless, it can find good solutions in some cases with less processing time, including new best-known solutions for four instances. Results have demonstrated that the use of a tighter formulation for the problem can be useful in the performance of the solution approach. As future work, we intend to improve the FCNF formulation adding more valid inequalities for tightening the formulation. Also, we will specialize the MIP-based local search procedures by using the POPMUSIC [16] concepts for defining the subproblems to be optimized, guiding the search through the subproblems that have been successfully optimized.

Acknowledgment. The present work was carried out with the support of CNPq, National Council of Scientific and Technological Development - Brazil and the support of FAPERGS, Foundation for Research Support of the State of Rio Grande do Sul.

References

1. Maritime inventory routing problem library (MIRPLIB). https://mirplib.scl.gatech.edu/. Accessed 05 Apr 2017
2. Agra, A., Andersson, H., Christiansen, M., Wolsey, L.: A maritime inventory routing problem: discrete time formulations and valid inequalities. Networks **62**(4), 297–314 (2013). https://doi.org/10.1002/net.21518
3. Agra, A., Christiansen, M., Delgado, A.: Discrete time and continuous timeformulations for a short sea inventory routing problem. Optim. Eng. **18**(1), 269–297 (2016). https://doi.org/10.1007/s11081-016-9319-0
4. Christiansen, M., Fagerholt, K.: Ship routing and scheduling in industrial and tramp shipping. In: Toth, P., Vigo, D. (eds.) Vehicle Routing: Problems, Methods, and Applications, Chap. 13, 2nd edn., pp. 381–408. No. Vlcc, Society for Industrial and Applied Mathematics, Philadelphia (2014). https://doi.org/10.1137/1.9781611973594.ch13
5. Christiansen, M., Fagerholt, K., Nygreen, B., Ronen, D.: Ship routing and scheduling in the new millennium. Eur. J. Oper. Res. **228**(3), 467–483 (2013). https://doi.org/10.1016/j.ejor.2012.12.002

6. Christiansen, M., Fagerholt, K., Ronen, D.: Ship routing and scheduling: status and perspectives. Transp. Sci. **38**(1), 1–18 (2004). https://doi.org/10.1287/trsc.1030.0036

7. Friske, M.W., Buriol, L.S.: A relax-and-fix algorithm for a maritime inventory routing problem. Computational Logistics. LNCS, vol. 10572, pp. 270–284. Springer, Cham (2017). https://doi.org/10.1007/978-3-319-68496-3_18

8. Goel, V., Furman, K.C., Song, J.H., El-Bakry, A.S.: Large neighborhood search for LNG inventory routing. J. Heuristics **18**(6), 821–848 (2012). https://doi.org/10.1007/s10732-012-9206-6

9. Jiang, Y., Grossmann, I.E.: Alternative mixed-integer linear programming models of a maritime inventory routing problem. Comput. Chem. Eng. **77**, 147–161 (2015). https://doi.org/10.1016/j.compchemeng.2015.03.005

10. Lalla-Ruiz, E., Voß, S., Expósito-Izquierdo, C., Melián-Batista, B., Moreno-Vega, J.M.: A popmusic-based approach for the berth allocation problem under time-dependent limitations. Ann. Oper. Res. **253**(2), 871–897 (2017). https://doi.org/10.1007/s10479-015-2055-6

11. Papageorgiou, D.J., Keha, A.B., Nemhauser, G.L., Sokol, J.: Two-stage decomposition algorithms for single product maritime inventory routing. INFORMS J. Comput. **26**(4), 825–847 (2014). https://doi.org/10.1287/ijoc.2014.0601

12. Papageorgiou, D.J., Nemhauser, G.L., Sokol, J., Cheon, M.S., Keha, A.B.: MIRPLib - a library of maritime inventory routing problem instances: survey, core model, and benchmark results. Eur. J. Oper. Res. **235**(2), 350–366 (2014). https://doi.org/10.1016/j.ejor.2013.12.013

13. Pochet, Y., Wolsey, L.A.: Production Planning by Mixed Integer Programming. Springer Science & Business Media, New York (2006). https://doi.org/10.1007/0-387-33477-7

14. Savelsbergh, M., Song, J.H.: An optimization algorithm for the inventory routing problem with continuous moves. Comput. Oper. Res. **35**(7), 2266–2282 (2008). https://doi.org/10.1016/j.cor.2006.10.020

15. Song, J.H., Furman, K.C.: A maritime inventory routing problem: practical approach. Comput. Oper. Res. **40**(3), 657–665 (2013). https://doi.org/10.1016/j.cor.2010.10.031

16. Taillard, É.D., Voß, S.: Popmusic-partial optimization metaheuristic under special intensification conditions. Essays and surveys in metaheuristics, pp. 613–629. Springer, Boston (2002). https://doi.org/10.1007/978-1-4615-1507-4_27

17. Uggen, K.T., Fodstad, M., Nørstebø, V.S.: Using and extending fix-and-relax to solve maritime inventory routing problems. TOP **21**(2), 355–377 (2013). https://doi.org/10.1007/s11750-011-0174-z

Offshore Supply Planning in a Rolling Time Horizon

Eirik Fernández Cuesta[1], Henrik Andersson[1], and Kjetil Fagerholt[1,2(✉)]

[1] Department of Industrial Economics and Technology Management,
Norwegian University of Science and Technology, Trondheim, Norway
kjetil.fagerholt@ntnu.no
[2] SINTEF Ocean, Trondheim, Norway

Abstract. This paper presents a real transportation problem stemming from offshore oil and gas logistics and shows how optimization models used in a rolling horizon simulation framework can be very valuable to assess and improve the operation's performance. With this aim, we study how the Order Selection Problem (OSP), a problem that helps the logistics provider decide which orders to carry to and from the platforms and which to postpone, and the *Vessel Routing Problem with Selective Pickups and Deliveries* (VRPSPD), that in addition to the order selection also routes the vessels carrying the orders, can be used in a practical planning setting. To quantify and justify the benefits of using the VRPSPD and OSP models in a real planning situation, an industry case based on real data was simulated in a rolling horizon framework and solved for an entire year. In addition to the traditional cost metric used, the focus of this paper lies on the implication these have on other important aspects that are often neglected in traditional optimization models; regularity and level of service. Several strategies for overbooking and postponing orders were also evaluated with respect to their cost, regularity, and level of service.

Keywords: Vehicle routing · Pickup and delivery · Rolling horizon
Simulation · Offshore supply

1 Introduction

The upstream offshore oil and gas supply logistics deals with the transportation of equipment and supplies used at offshore platforms. The main bulk of this transport work is performed by supply vessels. Several studies have shown that using advanced optimization methods can yield large benefits in the planning and organizing of offshore supply logistics, see for example Gribkovskaia et al. [8], Halvorsen-Weare et al. [9], Fernández-Cuesta et al. [4], and Norlund et al. [11]. On the other hand, improvements resulting from the use of optimization tools sometimes result in frequent re-design of established sailing patterns or the relaxing of the regularity of the platform visits and other requirements that can

© Springer Nature Switzerland AG 2018
R. Cerulli et al. (Eds.): ICCL 2018, LNCS 11184, pp. 17–31, 2018.
https://doi.org/10.1007/978-3-030-00898-7_2

be important for the planners. The challenge when designing optimization models is therefore often to model the problems in such way that improvements can be made, but at the same time keep changes on a level that can be realistically implemented.

Figure 1 shows the layout for an offshore oil and gas logistic network off the Brazilian coast, which is the case studied in this paper. The platforms place new order requests throughout the year. The requests include both pickup orders destined from the platforms back to the base and delivery orders destined out to the platforms from an onshore base. The resulting planning problem faced by the logistics provider consists of transporting the requested orders to and from the platforms so that these can produce oil and gas without any disturbances.

For practical purposes, the transportation is organized into *scheduled departures* that are repeated on a weekly basis throughout the year, see the examples in Fig. 1. Each scheduled departure is associated with a voyage termed *regular voyage* that has a predetermined starting time and follows a predetermined route. The transportation of the orders is performed by a supply vessel sailing the regular voyage along the predetermined route while delivering and picking up orders. If the supply vessel performing the regular voyage has insufficient capacity to carry all orders for the platforms visited on a scheduled departure, these are normally transferred to the order pool of the subsequent scheduled departure that visits the corresponding platform. For every scheduled departure the planner therefore has an Order Selection Problem (OSP) where it has to

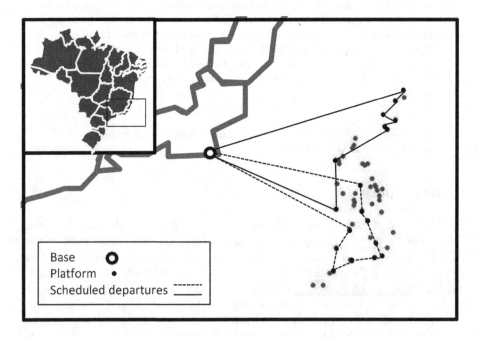

Fig. 1. Offshore supply layout with two scheduled departures.

decide which orders to carry and which to postpone for later. The exception to this is that some orders are *urgent* and should not be postponed. In this case, an *express voyage* is requested to carry the order instead. An express voyage requires a crew working outside their regular working hours.

The overall planning problem can therefore be seen as a number of consecutive OSPs that consist of selecting the subset of available orders to carry and to postpone. Even though each OSP can be considered as a static problem, the overall problem is dynamic because the decision of which orders to carry and postpone influences the consecutive problems. This interdependency between the different OSPs for the scheduled departures is in this paper modeled in a rolling horizon simulation framework where the information about the available order requests is revealed as time passes. Subsequent departures are dependent on the new order arrivals as well as the previous decisions made. Early implementations of a rolling horizon approach can be found in Sethi and Sorger [12]. General literature and classification on the combination of simulation and optimization can be found in Fu [5] and Gosavi [6].

The problem of order selection for offshore supply logistics in a rolling horizon framework was formulated by Andersson et al. [1]. However, it is clear that sailing the same historical route on the regular voyage week after week can be suboptimal. Instead, routing and order selection could be performed jointly. The *Vessel Routing Problem with Selective Pickups and Deliveries* (VRPSPD) is a pickup and delivery problem (PDP) with optional pickup and delivery orders where both the order selection and routing of the regular voyage is decided after the available orders become known. The VRPSPD was introduced by Fernández-Cuesta et al. [4] and belongs to a class of general *Pickup and Delivery Problems* (PDPs), see for example the classification schemes suggested by Berbeglia et al. [3] and Battarra et al. [2]. The class of PDPs to which the VRPSPD is most closely related to is the *One-to-Many-to-One Vehicle Pickup and Delivery Problem* (1-M-1 PDP), which is described in Gribkovskaia and Laporte [7]. The expression 1-M-1 refers to the fact that all supplies destined to the set of customers originate from the depot and all orders picked up at the customer locations must be returned to the depot. There exist also a few practical applications for the 1-M-1 PDP in the upstream offshore petroleum industry, see for example Gribkovskaia et al. [8] and Fernández-Cuesta et al. [4].

The purpose of this paper is to show through a rolling horizon simulation study how the practical planning of a real problem arising from offshore oil and gas logistics can benefit from solving the VRPSPD and OSP. Furthermore, we estimate the consequences this will have on platform visiting regularity and level of service in terms of the number of platform calls. An additional purpose is to test different strategies for selecting and postponing orders and to analyse their impact on the transportation system.

The remainder of this paper is organized as follows: Sect. 2 describes the transportation problem faced by the company and how the OSP and VRPSPD relate to this. Section 3 presents the mathematical models that correspond to the current industry practice (OSP) and the VRPSPD. Section 4 presents the case study on which the VRPSPD has been tested, followed by conclusions in Sect. 5.

2   Problem Description

The planning problem faced by the logistics provider originates from the necessity to transport supplies in the form of maintenance and production equipment (delivery orders) to offshore platforms and collect waste and redundant or depleted equipment (pickup orders) destined for the onshore base. The pickup and delivery orders are placed throughout the course of the year by the platforms. These orders cannot be split into smaller orders. The order arrival process for a platform is illustrated in Fig. 2. Every time a new order arrives it is added to the order pool from which the orders to be carried are selected. The order pool for a given departure consists of all the pickup and delivery orders originating from or destined for one of the platforms scheduled to be visited on that departure. In addition, some order requests are *urgent*. It is the platforms who decide whether an order is urgent or not. Urgent orders cannot be postponed. Therefore, if the regular voyage is unable to carry all the urgent orders, an express voyage is requested to carry the remaining ones. A regular order can be postponed to the next departure for that platform, but becomes urgent after it has been postponed once.

All materials are stored in a warehouse at a separate location away from the onshore base from which all orders are transported to the platforms and where the fleet is located. There is limited storage space at the harbor front of the base, so orders planned on a scheduled departure are transported to the

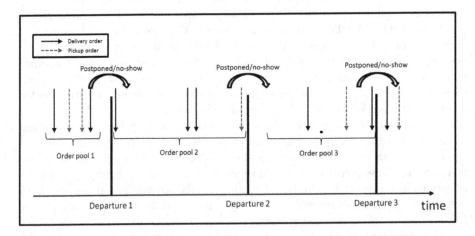

Fig. 2. The order arrivals are simulated in a rolling horizon where arriving orders are added to the order pool as time increases and carried at each departure point while missing (no-show) and postponed orders are transferred to the order pool of the following schedule visiting the same platform.

harbor front and directly onto the vessel sailing the voyage. Due to congestion issues between the warehouse and the harbor front and issues with orders not being available at the warehouse as expected, some delivery orders may not be available for the departure in which they are planned. An order that does not show up at the harbor front at the determined departure time is termed a *no-show*. No-shows are transferred to the following departure just as the postponed ones. This means that an urgent order will be transferred to the order pool of the following departure if it becomes a no-show.

Based on the orders available in the order pool for a given departure, the orders to carry and the ones to postpone (if any) have to be decided based on their importance. The orders are therefore sorted according to their utility. In general, the utility for an urgent order is higher than for regular orders and higher for delivery orders than for pickup orders. The latter is because the imbalance between the quantity of delivery and pickup orders makes the former more constraining on the vessel capacity. The decision of which orders to carry is made by either solving the OSP or the VRPSPD.

Prior to every departure an initial set of orders to carry is decided and a request to transport the orders from the warehouse to the harbor front is placed. At this point, whether an order is a no-show is unknown. At the point of loading the orders on the vessel, the no-shows become known. To reduce the negative effects of the no-show orders, an overbooking policy can be implemented. This basically means that more orders are planned on the regular voyage than the vessel's capacity. If the initial plan reveals that there is insufficient capacity on the regular voyage to carry all the urgent orders, an additional express voyage has to be requested. The order selection for the express voyage is the same as for the regular voyage but with the reduced order pool consisting of the orders left behind by the regular voyage.

For the transportation of the orders the planner has a homogeneous fleet of specialized supply vessels at its disposal. The fleet is chartered on an annual basis so the charter costs are considered sunk. In the case study considered here, a schedule of 13 departures (routes/voyages) with a weekly regularity has been designed by the company a priori using historic knowledge about the expected demand of the platforms. There are 52 platforms that are serviced in total. Most platforms are visited twice per week, although some are serviced only once per week. Each scheduled departure is carried out by one of the available vessels on a regular voyage and visits each platform in a predetermined subset of the platforms exactly once according to the schedule. Under the current industry planning practice, the only way a regular voyage will deviate from the historical route is when a platform does not have any pickup or delivery orders. In this case the platform is skipped and the voyage continues to the next platform along the route.

An alternative to using the historic routes is to route the vessel solving the VRPSPD. Then, the route of the regular voyage will change from one scheduled departure to the next depending on the set of orders available. Based on the number of working hours of the crew, each regular voyage has a four day limit on its duration.

Whenever the vessel sailing the regular voyage has insufficient capacity to carry all the urgent orders or if there are overbooked orders left at the harbor front an express voyage has to be requested to carry the surplus. However, using an express voyage is expensive as it entails using a crew outside their ordinary working hours. Since the crew requires additional time to assemble, the express voyage is planned (routed) sequentially after the regular voyage using the VRP-SPD. The express voyage has a two day maximum duration. For both regular and express voyages, the vessels require half a day turnaround time in the base before being ready to sail a new voyage.

3   Mathematical Formulations

In this section we formulate the mathematical models for the two versions of the planning problem. The OSP, which assumes given routes, is presented in Sect. 3.1. This is used to represent the current planning practice and is a variation of the model presented by Andersson et al. [1]. The VRPSPD, which considers the integrated routing and order selection problem is presented in Sect. 3.2 and is a variation of the model presented in Fernández-Cuesta et al. [4].

3.1   Order Selection Problem

The current practice is modeled through an order selection model that determines which orders from the order pool to service on the next given departure (route) and which to postpone and transfer to the subsequent order pools. Let $\mathcal{N}$ be the set of all nodes corresponding to platforms to be visited on a given scheduled departure and $\mathcal{O}^{Pool}$ be the set of all non-splitable pickup and delivery orders available in the order pool of the scheduled departure. Nodes that are scheduled for a visit but have no pickup or delivery orders are removed from $\mathcal{N}$. In addition, let o be an order in the set $\mathcal{O}_i$ for node/platform $i \in \mathcal{N}$. There may be several delivery or pickup orders at each node.

The binary variable u_o is 1 if order o is carried, and 0 if it is postponed. Let the time it takes to pickup or deliver o be T_o and its size be S_o. In the OSP we let S_o be negative for delivery orders and positive for pickup orders. The capacity of the vessel is given by Q. Let the variable l_i represent the total load on the vessel when leaving node i. In addition, let T^r be the given sailing time for the scheduled departure/route after the nodes without orders have been removed. T^r includes the time for visiting the platforms (i.e. mooring etc.), but excludes the time for loading/unloading. Let T^L be the maximum allowed scheduled time for the departure. Let $n(t)$ be a function that returns the order in which the nodes are visited, so that $n(t)$ is the tth node that is visited according to the schedule. The set $\mathcal{O}^{Post}$ represents the subset of the orders in $\mathcal{O}^{Pool}$ that are postponed in a given solution. Finally, we define the objective function $U(\mathcal{O}^{Post})$

to represent the loss in utility postponed orders $\mathcal{O}^{Post}$. This yields the following compact order selection model, denoted the OSP.

$$\text{Minimize} \quad U(\mathcal{O}^{Post}) \tag{1}$$

subject to

$$l_i \leq Q \qquad i \in \mathcal{N} \tag{2}$$

$$l_0 = \sum_{i \in \mathcal{N}} \sum_{o \in \mathcal{O}_i \mid S_o < 0} -S_o u_o \tag{3}$$

$$l_{n(t)} - l_{n(t-1)} = \sum_{o \in \mathcal{O}_{n(t)}} S_o u_o \qquad t = 1, .., |\mathcal{N}| \tag{4}$$

$$T^r + \sum_{o \in \mathcal{O}_i} T_o u_o \leq T^L \tag{5}$$

$$u_o \in \{0,1\} \qquad i \in \mathcal{N}, o \in \mathcal{O}_i \tag{6}$$

$$l_i \geq 0 \qquad i \in \mathcal{N}, \tag{7}$$

The objective (1) minimizes the loss in utility of the postponed orders. Constraints (2) ensure that the capacity of the vessel is not violated, whereas constraints (3) and (4) control the start load and load continuity, respectively. The total duration is controlled by constraint (5). The domain of the variables are defined through constraints (6) and (7).

3.2   Vessel Routing Problem with Selective Pickups and Deliveries

The VRPSPD is an extension to the OSP where routing is included. For this we let $\mathcal{N} = \{\mathcal{N}^P, \mathcal{N}^D\}$ be the set of all nodes with orders on a given scheduled departure where $\mathcal{N}^P$ is the set of pickup locations and $\mathcal{N}^D$ is the set of delivery locations. Note that this means that a node no longer corresponds to a platform as in the OSP. Instead, each platform in the set of all platforms $\mathcal{P}$ is therefore represented by at most two nodes corresponding to the pickup and delivery node, respectively. Let sets $\mathcal{O}^{Pool}$ and $\mathcal{O}_i$ be as defined for the OSP. The vessel starts in the depot node 0 and ends in the depot node $|\mathcal{N}| + 1$.

Let v_i be a binary variable that controls whether node i is visited, and let variable l_i from the OSP be split into l_i^P and l_i^D for the pickup and delivery loads on the vessel immediately after leaving i, respectively. This eliminates the need for subtour elimination constraints (see for example Hoff et al. [10]). The binary variable u_o is 1 if order o is carried, and 0 otherwise, like before. The size of order o is given by S_o as in the OSP, although the delivery quantities are no longer defined as negative. Furthermore, let the set of arcs in the network be $\mathcal{A}$. This set consists of the arcs between all nodes in the network, except for the ones from the pickup to the delivery nodes for the same platform. For the arc between nodes i and j, the travel time is given as T_{ij} and the cost of traversing it as C_{ij}. The binary variable x_{ij} is 1 if the arc from i to j is used, and 0 otherwise. The time it takes to visit a node consists of a fixed time T^F (for mooring etc.)

and a variable time for loading or unloading each unit of order o denoted (as before) by T_o. The duration of each voyage is limited by T^L as for the OSP. Then VRPSPD can be formulated as

$$\text{Minimize} \sum_{(i,j)\in\mathcal{A}} C_{ij}x_{ij} + U(\mathcal{O}^{Post}) \tag{8}$$

subject to

$$\sum_{(0,j)\in\mathcal{A}} x_{0j} = 1 \tag{9}$$

$$\sum_{(i,|\mathcal{N}|+1)\in\mathcal{A}} x_{i,|\mathcal{N}|+1} = 1 \tag{10}$$

$$\sum_{j\in\mathcal{N}} x_{ij} = v_i \qquad i \in \mathcal{N} \tag{11}$$

$$\sum_{i\in\mathcal{N}} x_{ij} = v_j \qquad j \in \mathcal{N} \tag{12}$$

$$v_i \geq u_o \qquad i \in \mathcal{N}, o \in \mathcal{O}_i \tag{13}$$

$$0 \leq l_i^P + l_i^D \leq Q \qquad i \in \mathcal{N} \tag{14}$$

$$l_0^D = \sum_{i\in\mathcal{N}^D} \sum_{o\in\mathcal{O}_i} S_o u_o \tag{15}$$

$$l_j^D \geq l_i^D - \sum_{o\in\mathcal{O}_j} S_o u_o - Q(1 - x_{ij}) \qquad i \in \mathcal{N}, j \in \mathcal{N}^D \tag{16}$$

$$l_j^D \geq l_i^D - Q(1 - x_{ij}) \qquad i \in \mathcal{N}, j \in \mathcal{N}^P \tag{17}$$

$$l_j^P \geq l_i^P + \sum_{o\in\mathcal{O}_j} S_o u_o - Q(1 - x_{ij}) \qquad i \in \mathcal{N}, j \in \mathcal{N}^P \tag{18}$$

$$l_j^P \geq l_i^P - Q(1 - x_{ij}) \qquad i \in \mathcal{N}, j \in \mathcal{N}^D \tag{19}$$

$$\sum_{(i,j)\in\mathcal{A}} T_{ij}x_{ij} + \sum_{i\in\mathcal{N}} T^F v_i + \sum_{i\in\mathcal{N}} \sum_{o\in\mathcal{O}_i} T_o u_o \leq T^L \tag{20}$$

$$x_{ij} \in \{0,1\} \qquad (i,j) \in \mathcal{A} \tag{21}$$

$$l_i^P \geq 0 \qquad i \in \mathcal{N} \tag{22}$$

$$l_i^D \geq 0 \qquad i \in \mathcal{N} \tag{23}$$

$$u_o \in \{0,1\} \qquad i \in \mathcal{N}, o \in \mathcal{O}_i \tag{24}$$

$$v_i \in \{0,1\} \qquad i \in \mathcal{N} \tag{25}$$

The objective function (8) minimizes the travel cost plus the loss in utility for not handling orders. Constraints (9)–(10) ensure that the vessel starts and ends at the depot, whereas constraints (11) and (12) express that all visited nodes have one inbound and one outbound arc, respectively. Constraints (13) state that only orders from visited nodes can be handled. The capacity of the vessels is respected through constraints (14). The start delivery loads on the vessel is controlled by constraints (15), whereas constraints (16) and (18) together with constraints (17) and (19) ensure continuity in the pickup and delivery load aboard the vessel, respectively. Finally, constraint (20) limits the total travel time of the vessel, and constraints (21)–(25) define the domain of the variables.

It can be noted that by fixing variables v_i and x_{ij} corresponding to the scheduled departures (routes), the VRPSPD and OSP become equivalent.

4 Computational Study

The purpose of the computational study is to solve the VRPSPD and compare the results with a benchmark provided by the current industry practice represented by solving the OSP in a rolling horizon simulation framework corresponding to a full year of logistics operations. An additional purpose is to evaluate and analyze other strategies for improving the transportation system. Section 4.1 describes the case study, while Sect. 4.2 presents and discusses the computational results.

4.1 Case Study Setup

The case study is based on historical data from the Brazilian offshore oil and gas industry. The case company serves 52 platforms and other production/drilling units. A homogeneous fleet of eight supply vessels sailing 13 scheduled weekly departures is available during the course of a year. A sketch showing the layout of the platforms and two of the scheduled departures was shown in Fig. 1. 47 of the 52 platforms are visited twice per week whereas the remaining five are visited once per week. There are 98 platform calls per week and in total there are 676 scheduled departures in a year. There are roughly 8 000 delivery and 5 000 pickup orders totalling 350 000 m^2 of cargo to be transported over that period. On average there are 380 m^2 of delivery orders and 150 m^2 of pickup orders on each scheduled departures. The vessels in the fleet all have a capacity of 620 m^2. The charter cost of the fleet is based on a regular charter contract in the offshore supply industry and is considered sunk over the planning period. The operational costs for the routing are calculated based on an estimate of the bunker price, the fuel consumption at the service speed for the vessel and an estimate of the crew costs.

Order Arrival. Each platform is modeled with its own independent order arrival process as a homogeneous Poisson process. The inter-arrival time between two orders to a platform i is given by $(\lambda_i^{delivery})^{-1}$ and $(\lambda_i^{pickup})^{-1}$ for pickup and delivery orders, respectively. The inter-arrival time is estimated from historical data and the order sizes are calculated so that the total expected load for a given platform is proportional to the number of times the platform is visited. On average each platform that is visited twice per week has 3.2 delivery and 2.1

pickup requests per week. Pickup and delivery order requests have a 10% and 50% chance of being *urgent*, respectively. In addition, every delivery order has a 25% chance of being a no-show (to mimic the real operation). Both the no-show and urgent probabilities are approximated based on the data provided by the case company. The order arrival process for the entire year is drawn a priori and is therefore identical across all the runs.

Order Selection. Based on the orders available in the order pool for a given departure, the orders to carry and the orders to postpone are decided, either by solving the OSP or the VRPSPD. To represent the utility of carrying the orders each order o is associated with an artificial penalty C_o for not being carried so that

$$U(\mathcal{O}^{Post}) = \sum_{i \in \mathcal{N}} \sum_{o \in \mathcal{O}_i} C_o(1 - u_o) \tag{26}$$

For the VRPSPD, the artificial penalty for not carrying an order is set higher than the operational cost $\overline{C}$ of visiting and returning from the most distant platform. $\overline{C}$ is set to zero for the OSP. C_o also scales with the size of the order compared to $\overline{S}$, which is the size of the largest order available. For every departure t an order is postponed the penalty is increased by a factor 4 so that

$$C_o(t) = F_o^{Base} \cdot 4^t \left(\overline{C} + \frac{S_o}{\overline{S}} \right) \tag{27}$$

where the factor F_o^{Base} is

$$F_o^{Base} = \begin{cases} 1 & o \in O^P \\ 2 & o \in O^D \\ 4 & o \in O^{Urgent} \end{cases}$$

All express voyages are routed with the VRPSPD using the same penalties as for the regular voyage.

Strategies and Setup. To evaluate the transportation system, several different strategies and setups have been tested. These are referred to as *settings*. The basic setting that corresponds to the current practice is termed **Base**. In this, the no-show probability is set to 25% for delivery orders and there is no overbooking policy in place. The order penalties are set so that the vessels are filled up as much as possible. To see if the vessel utilization can be improved, several overbooking settings (**OB**) ranging from 5% to 25% have also been tested. Since the 25% no-show rate leads to reduced vessel capacity utilization, the setting **Ideal** using a no-show probability of 0% is implemented to quantify the consequences. Lastly, we have also tested the **Opportunistic** setting where we allow postponing regular orders by two departures instead of one before they become urgent. In addition, the factor F_o^{Base} is replaced by a new factor F_o^{Opp} that is defined

$$F_o^{Opp} = \begin{cases} 0.25 & o \in O^P \\ 1 & o \in O^D \\ 4 & o \in O^{Urgent} \end{cases}$$

Opportunistic is run with 0% overbooking and 25% no-show as in the **Base** setting. Either setting can be solved using both the OSP and the VRPSPD.

4.2   Case Results

The simulation results regarding delays and costs for the different settings run over a 360 day rolling horizon are summarized in Table 1. Delays are measured in the number of departures an order is postponed or weighted with the size (in m^2) of the order. The results for **Base**-OSP are given in absolute numbers. All other results are in percent using the **Base**-OSP as comparison except the results where **Base**-OSP is zero. These results are given in absolute numbers. Note that the first and last week (26 departures) have been cleaned from the results to allow the rolling horizon to be in a steady state. The level of service (LoS) is measured in number of platform calls during the period. Regularity is calculated for each of the 98 weekly departures for each platform and is measured as the difference in hours between the actual visiting time and the average visiting time from the start of the scheduled departure. The total regularity is then averaged across all platforms. In the following we discuss each of the settings to evaluate the VRPSPD and the transportation system. All VRPSPD runs were performed with Gurobi 6.0 running on a 2.4 GHZ 4-core computer with 16 GB of RAM.

VRPSPD vs. OSP. By comparing the results for the **Base** setting in Table 1 for the OSP and VRPSPD models we see that it is possible to reduce the total cost by improving the routing when solving the VRPSPD by 8.3%. Part of these savings stem from the reduction in express sailings by 21.7% (from 23 to 17). We also see that the number of one–departure delays stays roughly the same (−0.3%) whereas the number of two-departure delays has increased by 24.6%. This seems substantial but is an increase from 118 to 147 orders out of a total of 2 049 delayed orders. Since two-departure delays are not planned but rather only happen by chance if a previously postponed order happens to be a no-show the two-departure delay increase can be considered insignificant. Note that because of the no-shows, there are also 1 015 urgent one-departure delays. As expected, the primary advantage of using the historical routes is that they lead to more regular visits to the platforms as can be seen from the regularity in Table 1. In **Base**-OSP the average visiting deviation was five hours, whereas this average was increased by +178% to over 13 h for the VRPSPD. On the other hand, the level of service (LoS) in the form of total number of platform calls is slightly reduced (−0.7%) to 4 659 from 4 699 annual calls for the VRPSPD. Considering that the initial annual schedule has 4 900 platform calls it is hard to argue that the use of the VRPSPD would have a notable adverse impact on the service level.

Overbooking. In Table 1 the results from overbooking from 5% to 25% are presented for both the OSP and VRPSPD. The total cost and delay (# delayed departures · size) is also presented in Fig. 3. From this we see that it is beneficial to plan with overbooking up until a certain point. Because there is no way of storing the surplus overbooked orders, these will require an additional express voyage. The lowest total cost for the system was achieved by overbooking with 5% for both the OSP and the VRPSPD. From Fig. 3 it is clear that if minimizing

Table 1. Results for settings **Base**, **OB** (Overbooking 5%–25%), **Ideal**, for both the OSP and the VRPSPD. **Opportunistic** (Opp.) setting for VRPSPD only. Results from one-year run using **Base**-OSP as the baseline [±%]. Results where baseline is 0 are shown in absolute numbers. 1-dpt = one departure delay etc., NS = no-show, OB = overbooking, LoS = Level of Service [# platform calls]

| | | | OSP | | | | | | | VRPSPD | | | | | | | |
| | | | | Overbooking [%] | | | | | | | Overbooking [%] | | | | | | |
			Base	5	10	15	20	25	Ideal	Base	5	10	15	20	25	Ideal	Opp.
Delays	Regular	# of 1-dpt	1 326	-3.3	-10.9	-15.1	-18.6	-20.0	-89.2	-0.3	-4.4	-11.3	-15.1	-17.9	-19.8	-89.7	3.8
		# of 2-dpt	118	-13.6	-38.1	-60.2	-84.7	-87.3	-100.0	24.6	-6.8	-37.3	-67.8	-80.5	-90.7	-100.0	122.9
		# of 3-dpt	0	0	0	0	0	0	0	0	0	0	0	0	0	0	16
		Sum	1 444	-4.2	-13.1	-18.8	-24.0	-25.5	-90.1	1.7	-4.6	-13.4	-19.4	-23.0	-25.6	-90.6	14.6
	Urgent	# of 1-dpt	1 015	0.0	0.0	0.0	0.0	0.0	-100.0	0.0	0.0	0.0	0.0	0.0	0.0	-100.0	0.0
		# of 2-dpt	0	0	0	0	0	0	0	0	0	0	0	0	0	0	0
		Sum	1 015	0.0	0.0	0.0	0.0	0.0	-100.0	0.0	0.0	0.0	0.0	0.0	0.0	-100.0	0.0
	Total	# delays	2 459	-2.4	-7.7	-11.0	-14.1	-15.0	-94.2	1.0	-2.7	-7.9	-11.4	-13.5	-15.0	-94.5	8.6
		delay dpt	2 577	-2.9	-9.1	-13.3	-17.3	-18.3	-94.5	2.1	-2.9	-9.2	-14.0	-16.6	-18.5	-94.7	15.1
		Size [m2]	77 736	-2.5	-7.6	-11.1	-14.1	-15.0	-94.3	1.0	-2.8	-7.9	-11.4	-13.4	-15.0	-94.6	8.4
		Size-dpt	81 437	-2.9	-9.0	-13.3	-17.4	-18.3	-94.6	2.1	-3.1	-9.2	-14.0	-16.4	-18.5	-94.8	15.2
Costs [MUSD]	Regular	Operational	66.5	-0.1	-0.1	-0.2	-0.2	-0.2	-0.5	-12.7	-12.7	-12.6	-12.7	-12.7	-12.6	-12.8	-13.3
		Charter	43.2	0.0	0.0	0.0	0.0	0.0	0.0	0.0	0.0	0.0	0.0	0.0	0.0	0.0	0.0
		Total	109.7	-0.1	-0.1	-0.1	-0.1	-0.1	-0.3	-7.7	-7.7	-7.6	-7.7	-7.7	-7.7	-7.7	-8.0
	Express	Fuel	2.0	-44.7	-36.6	-37.0	1.1	4.0	-96.0	-22.2	-49.7	-39.9	-38.1	-5.6	2.6	-96.0	-66.1
		Crew	2.6	-44.4	-36.1	-36.4	4.3	6.4	-95.9	-22.1	-49.2	-38.6	-36.1	-3.0	5.3	-95.9	-65.8
		Total	4.6	-44.5	-36.3	-36.7	2.9	5.4	-95.9	-22.1	-49.4	-39.2	-36.9	-4.1	4.1	-95.9	-65.9
		TOTAL	114.3	-1.9	-1.5	-1.6	0.0	0.1	-4.2	-8.3	-9.4	-8.9	-8.8	-7.5	-7.2	-11.3	-10.4
Departures		Scheduled	650	0.0	0.0	0.0	0.0	0.0	0.0	0.0	0.0	0.0	0.0	0.0	0.0	0.0	0.0
		Express	23	-43.5	-34.8	-34.8	13.0	13.0	-95.7	-21.7	-47.8	-34.8	-30.4	4.3	13.0	-95.7	-65.2
		# with OB	0	4	9	12	25	25	0	0	2	10	12	22	25	0	0
		# with NS	613	0.2	0.2	0.0	0.0	0.0	-100.0	0.0	0.3	0.2	0.2	0.0	0.0	-100.0	0.2
Orders		# of OB	0	4	12	17	31	37	0	0	2	13	15	29	38	0	0
		# of NS	2 049	0.0	0.0	0.0	0.0	0.0	-100.0	0.0	0.0	0.0	0.0	0.0	0.0	-100.0	0.0
		# carried	13 218	0.0	0.0	0.0	0.0	0.0	0.0	0.0	0.0	0.0	0.0	0.0	0.0	0.0	0.0
Vessels		# used	8	0.0	0.0	0.0	0.0	0.0	-12.5	0.0	0.0	-12.5	-12.5	-12.5	-12.5	-12.5	-12.5
		Utilization	58.9	-1.1	-0.9	-1.0	-0.2	-0.2	-2.8	0.0	-6.5	-6.7	-6.5	-5.9	-5.7	-8.2	-7.4
		LoS	4 699	-1.4	-1.7	-1.9	-1.6	-1.7	-3.7	-0.7	-1.6	-2.0	-2.1	-2.0	-1.9	-3.9	-2.4
		Regularity	5.0	0.0	0.0	0.0	1.2	1.2	1.0	178	176	176	176	178	178	180	174

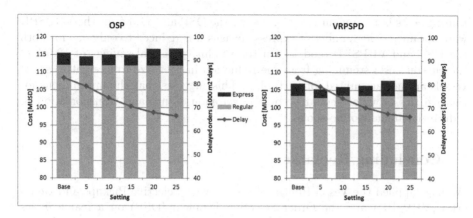

Fig. 3. Comparison of the cost and delays with varying levels of overbooking for the OSP and the VRPSPD.

the number of order delays is the only concern of the logistics provider, it is better to overbook as much as possible. For the OSP at 25% overbooking there is a total of 2 091 delay orders. When considering that 2 049 of these are no-shows it means that only 42 orders were postponed voluntarily. However, 26 express voyages would be required in the 25% overbooking setting.

Relaxing Delay Requirements. In the **Opportunistic** setting we measure the consequence of allowing regular orders to be postponed twice and use the penalty factor F_o^{Opp} instead of F^{Base}. The results are presented in Table 1 in the **Opportunistic** setting for the VRPSPD. Note that we do not run the **Opportunistic** setting for the OSP as there are limited gains from postponing orders if the platforms are visited anyway. The results show that this setting would reduce the cost by 10.4% with respect to the **Base**-OSP. This corresponds to 2.3% of savings compared with the **Base**-VRPSPD. This is achieved by increasing the number of delays and thereby reducing the routing cost and also reducing the number of express voyages to seven. We also note that this setting has more than double the number of two-departure delays (+122.9%) and 16 three-departure delays due to no-shows. Note also that in addition to the savings in operational costs, an additional 12.5% savings in charter costs could be obtained because the number of vessels used is reduced from eight to seven. Both the regularity and level of service is comparable with the other VRPSPD settings.

Cost of No-Shows. The presence of no-shows in the logistic system is an important cost driver as it leads to reduced vessel capacity utilization on the voyages. From the **Ideal** setting in Table 1 the obtainable benefits by removing no-shows in the system can be seen. The results show that it would be possible to remove all urgent delays and reduce the number of regular one-departure delays by around 89% for both the OSP and the VRPSPSD. In addition, savings in routing costs in the magnitude of 4.2% for the OSP and 11.3% for the VRPSPD were obtained. The number of express voyages was reduced to only one for both the OSP and VRPSPD models. On top of this, potential savings of 12.5% of the charter costs stemming from the reduction in fleet size from eight to seven vessels are not accounted for in Table 1. This suggests that the initial focus for the logistics provider should be on the warehouse and land transportation to reduce the amount of no-shows.

5   Concluding Remarks

We have applied the Vessel Routing Problem with Selective Pickups and Deliveries (VRPSPD) and the Order Selection Problem (OSP) on a real logistics planning problem from offshore oil and gas supply logistics and suggested some strategic improvements. This was comprehensively tested in a simulated rolling horizon over a full year of operation based on historical data. The settings tested showed that savings of over 8% would be attainable by leaving the fixed route policy without having a large impact on the level of service provided, and still maintain the original scheduled departures and corresponding platform visits. However, doing this will lead to an decrease in the regularity as seen from the platforms. In addition, strategies for overbooking were found to be useful in reducing the overall cost up on to a certain point where the trade-off between the increased vessel utilization on the regular voyages was offset by the increase in express voyages. Relaxing the delay requirements was also analyzed and found to have a savings potential although at the cost of increased number of order delays. Lastly, an estimate of the cost of no-shows in the system was provided and potential savings of over 4% under the current planning and over 11% when using the VRPSPD could be attainable by improving the logistics and information systems.

Aknowledgements. This research is funded and carried out as part of the Center for Integrated Operations in the Petroleum Industry. This support is gratefully acknowledged.

References

1. Andersson, H., Cuesta, E.F., Fagerholt, K., Gausel, N.T., Hagen, M.R.: Order management in the offshore oil and gas industry. In: Corman, F., Voß, S., Negenborn, R.R. (eds.) ICCL 2015. LNCS, vol. 9335, pp. 648–657. Springer, Cham (2015). https://doi.org/10.1007/978-3-319-24264-4_44
2. Battarra, M., Cordeau, J.-F., Iori, M.: Pickup-and-delivery problems for goods transportation. In: Toth, P., Vigo, D. (eds.) Vehicle Routing: Problems, Methods, and Applications, vol. 18, pp. 161–181. MOS-SIAM Series on Optimization, Philadelphia (2014)
3. Berbeglia, G., Cordeau, J.-F., Gribkovskaia, I., Laporte, G.: Static pickup and delivery problems: a classification scheme and survey. TOP 15(1), 1–31 (2007)

4. Fernández-Cuesta, E., Andersson, H., Fagerholt, K., Laporte, G.: Vessel routing with selective pickups and deliveries: an application to the supply of offshore oil platforms. Comput. Oper. Res. **79**, 140–147 (2017)
5. Fu, M.C.: Optimization for simulation: theory vs. practice. INFORMS J. Comput. **14**(3), 192–215 (2002)
6. Gosavi, A.: Simulation-based optimization: an overview. In: Gosavi, A. (ed.) Simulation-Based Optimization, pp. 29–35. Springer, Boston (2015). https://doi.org/10.1007/978-1-4899-7491-4_3
7. Gribkovskaia, I., Laporte, G.: One-to-many-to-one single vehicle pickup and delivery problems. In: Golden, B.L., Raghavan, S., Wasil, E.A. (eds.) The Vehicle Routing Problem: Latest Advances and New Challenges. Operations Research/Computer Science Interfaces, vol. 43, pp. 359–377. Springer, New York (2008). https://doi.org/10.1007/978-0-387-77778-8_16
8. Gribkovskaia, I., Laporte, G., Schlopak, A.: A tabu search heuristic for a routing problem arising in the servicing of oil and gas offshore platforms. J. Oper. Res. Soc. **59**, 1449–1459 (2006)
9. Halvorsen-Weare, E., Fagerholt, K., Nonås, L.M., Asbjørnslett, B.E.: Optimal fleet composition and periodic routing of offshore supply vessels. Eur. J. Oper. Res. **223**(2), 508–517 (2012)
10. Hoff, A., Gribkovskaia, I., Laporte, G., Løkketangen, A.: Lasso solution strategies for the vehicle routing problem with pickups and deliveries. Eur. J. Oper. Res. **192**(3), 755–766 (2009)
11. Norlund, E.K., Gribkovskaia, I., Laporte, G.: Supply vessel planning under cost, environment and robustness considerations. Omega **57**, 271–281 (2015)
12. Sethi, S., Sorger, G.: A theory of rolling horizon decision making. Ann. Oper. Res. **29**(1), 387–415 (1991)

Maritime Fleet Deployment with Speed Optimization and Voyage Separation Requirements

Venke Borander[1], Anders Straume[1], Bo Dong[1(✉)], Kjetil Fagerholt[1,2], and Xin Wang[1]

[1] Norwegian University of Science and Technology, Trondheim, Norway
{venkeb,andstrau}@stud.ntnu.no,
{bo.dong,kjetil.fagerholt,xinwan}@ntnu.no
[2] SINTEF Ocean, Trondheim, Norway

Abstract. A shipping company operates a heterogeneous fleet of ships to service a given number of voyages on a number of trade routes over the planning horizon. Each ship has a predefined speed range within which it can sail. Fuel consumption, and hence fuel cost, significantly depends on the chosen speed. Furthermore, the shipping company makes Contracts of Affreightments with the shippers stating that the voyages on each trade route should be fairly evenly spread. This leads to the maritime fleet deployment problem with speed optimization and voyage separation requirements. We propose two formulations for this problem, i.e. one arc flow and one path flow model. The non-linear relationship for fuel consumption as a function of ship speed is linearized by choosing discrete speed points and linear combinations of these. Computational results show that the path flow model performs better than the arc flow model and that incorporating speed decisions in the fleet deployment gives better solutions and more planning flexibility.

Keywords: Maritime fleet deployment · Speed optimization
Voyage separation requirements

1 Introduction

Maritime transportation is the main distribution network for international trade and has a key role in today's globalized world. According to the International Maritime Organization (IMO), 90% of all transported goods across borders worldwide is transported by the shipping industry, corresponding to approximately 10 billion tons in 2015 [9]. Even though the global demand has steadily been increasing over decades, there has been a tendency of overcapacity in the fleet since the financial crisis around 2009 [9]. In 2015, the shipping industry, with the exception of tankers, suffered from historic low levels of freight rates and weak earnings. As a result, the margins are pushed down. For an industry

R. Cerulli et al. (Eds.): ICCL 2018, LNCS 11184, pp. 32–46, 2018.
https://doi.org/10.1007/978-3-030-00898-7_3

that has high investment and operational costs, the quest for profitable operations is of higher importance than ever. One of the main targets in order to achieve this is to utilize the fleet capacity at all times and reduce ballast sailing (i.e. sailing without payload) to a minimum. Proper planning of maritime routes and schedules is therefore important.

In this paper, we extend the problem studied by Norstad et al. [5]. They considered a real maritime fleet deployment problem with voyage separation constraints for a shipping company operating in the open hatch dry bulk segment. The voyage separation constraints arise from contracts with the shippers which require that the trade routes are serviced regularly and that consecutive voyages along each trade are sufficiently separated in time. The main task in this problem is to assign available ships to the voyages on the different trade routes, such as to utilize the fleet in an optimal manner. Two models, an a priori path generation method and an arc flow method, were presented in [5], where the path flow model performed best. Vilhelmsen et al. [10] developed a Branch-and-Price procedure for the problem studied in [5]. They used a dynamic programming algorithm and a modified time window branching scheme, and found solutions that were at least as good as those by Norstad et al. [5] in shorter time.

Within other transportation modes, several examples of voyage separation requirements and time dependencies between routes can be found, though in a different context than ours. In vehicle routing, Reinhardt et al. [8] consider a dial-a-ride problem for airport passengers with complicating synchronization constraints. Dohn et al. [4] also consider synchronization and precedence constraints in two compact formulations of the vehicle routing problem with time windows. Dantzig–Wolfe decompositions of these formulations are presented and four different master problem formulations are proposed.

Most of the models found in the maritime transportation literature assume fixed and known speeds for the ships, either as implicit input or explicit input [7]. This is also the case in [5, 10]. However, in reality fuel consumption, and hence sailing costs, is strongly dependent on speed. Therefore, incorporating speed in ship routing and scheduling can yield significant improvements in profits for the shipping company [6]. In addition, fuel consumption influences the emissions of Greenhouse Gas (GHG). Many papers assume that daily fuel consumption is a cubic function of ship speed. Andersson et al. [2] use a linear combination of predefined discrete speed alternatives and interpolation in order to provide the desired fuel consumption as a piecewise linear function of speed. It should be noted that this problem differs from most other problems where speed optimization have been incorporated, including [2, 6], in that we cannot optimize speed locally in each route due to the voyage separation requirements, resulting in inter-dependency among the ship routes.

Based on these findings, we extend the models [5, 10] by integrating speed decisions along the different sailing legs, and we denote it the Maritime Fleet Deployment problem with Speed Optimization and Voyage Separation requirements (MFDSOVS). Our main contributions are to propose two models for the MFDSOVS, i.e. an arc flow and a path flow model, both which are extended based on [5] by integrating speed decisions. A number of realistic instances based on data from a shipping company are used to test the performance of the models and the effect of incorporating speed optimization on the solution quality, and it is shown that the path flow model performs better than the arc flow model.

2   Problem Description

We will now give a description of the MFDSOVS. Section 2.1 describes the fleet deployment part of the problem, which basically consists of assigning voyages to ships in the fleet (and implicitly ship routes). Section 2.2 describes the speed optimization part of the problem. Section 2.3 describes the voyage separation requirements before we end the section by summarizing the MFDSOVS.

2.1   Fleet Deployment

The fleet deployment problem can be described as a tactical planning problem of assigning ships from a heterogeneous fleet to voyages on different trade routes efficiently in terms of costs and service. A trade route is a predefined, typically intercontinental, sailing route from an origin region (including one or more ports) to a destination region (including one or more ports). Figure 1 shows intercontinental trade routes. A voyage is a sailing along a trade route. The number of voyages to be serviced along each trade may vary according to some frequency requirements. The trades can be separated into two types; contractual (mandatory) and optional trades. The shipping company seeks to maximize its profit by servicing voyages on optional trades while satisfying all contractual voyages on the contractual trades. If the company's own fleet is not capable to carry all contractual voyages, additional spot ships are chartered to serve contractual voyages. It is assumed that the ballast sailing costs associated with chartering a spot ship is included in the charter costs. The ships usually serve several voyages

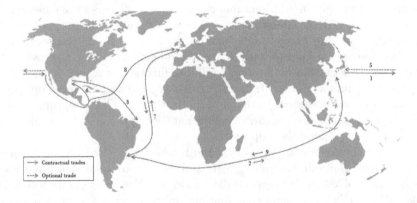

Fig. 1. Contractual and optional trade routes

in a sequence within the planning horizon. To start the next voyage, a ship might have to sail in ballast from the end of its previous voyage to re-position itself.

Each voyage has a predefined time window within which the voyage must start, instead of a fixed start-up time as is common in container shipping [3], which provide some flexibility for the shipping company.

2.2 Fuel Consumption and Speed

The operational costs of a fleet depend heavily on fuel consumption, which is also an environmental concern. Thus, optimizing sailing speeds along the ships' routes should be integrated with the fleet deployment. Fuel consumption is typically a cubic (quadratic) and convex function of speed per time (distance) unit.

Speed optimization means to adjust the sailing speed to seek higher profits. Increasing the speed increases the total available fleet capacity, which can in some cases be cheaper than chartering in spot ships to service contractual voyages. Increased fleet capacity also enables the possibility for the company to service optional voyages, which leads to additional revenue. On the other hand, higher sailing speeds incurs higher sailing costs. Therefore, it is not straight forward to find the optimal sailing speeds along all sailing legs that a ship performs during the planning horizon. In addition, a ship's fuel consumption also depends on the load onboard the ship as illustrated in Fig. 2.

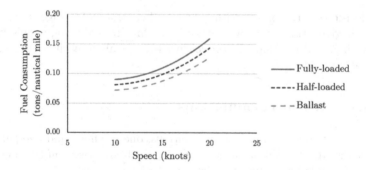

Fig. 2. Fuel consumption curves for ballast, half-loaded, and fully-loaded sailing [11].

2.3 Voyage Separation Requirements

Shippers enter Contracts of Affreightment (CoAs) with the shipping company. The most important part of the CoAs is where the cargo is heading, the amount transported, at what time and the freight rate. A commonly used term in CoAs, regarding the frequency and timing of voyages on a trade route, is "fairly evenly spread". This means that consecutive voyages on the same trade should be suf-ficiently spread in time. This introduces voyage separation requirements to the MFDSOVS. Norstad et al. [5] shows an example of the spread of voyages on a trade with or without voyage separation requirements as in Fig. 3, which clearly shows that without voyage separation constraints some of the consecutive voyages start very close in time to each other, which would possibly be in conflict with the "fairly evenly spread" terms that are stated in the CoAs.

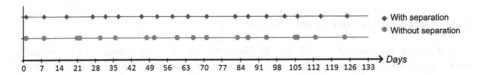

Fig. 3. Starting days for voyages on a trade with or without voyage separation [5].

2.4 Problem Summary

The objective in the MFDSOVS problem is to maximize profit, i.e. total freight income minus the sum of operation costs of ships in fleet and the charter costs for spot ships. The decisions to be made are: (1) the ship routes (i.e. which ship should perform which voyages and in what sequence), (2) the ships' sailing speeds for each sailing leg along their routes, (3) the start time for each voyage, (4) which optional voyages to sail, and (5) which voyages should be serviced by spot ships.

The decisions must comply with (1) that all contractual voyages are serviced within their given time windows, either by a ship from the company's fleet or by a spot ship, and (2) that all consecutive voyages along each trade route are fairly evenly spread.

3 Mathematical Formulations

In this section, two mathematical formulations, one arc flow, and one path flow model, for the MFDSOVS problem are given. Both are based on the ones from [5], though extended with speed optimization.

3.1 Arc Flow Model

Notation. Let $\mathcal{V}$ be the set of heterogeneous ships in the fleet of the shipping company, indexed by v. The ships have individual starting positions and maintenance schedules, and should therefore be treated individually, as treating them as a group could lead to infeasible solutions. We use the same approach as Andersson et al. [2] for handling the non-linear relationship between speed and fuel consumption/cost and sailing times, where we choose a number of discrete speed alternatives from the non-linear function (see Fig. 2) and allocate weights to these speed points. The set $\mathcal{S}$ is an ordered set containing all available discrete speed points, from minimum to maximum speed, indexed by s.

The set $\mathcal{R}$ denotes the set of trade routes operated by the company, indexed by r. $\mathcal{R}_v$ is a subset of $\mathcal{R}$ for which trade routes ship v can carry out. Let the set $\mathcal{I}_r = \{1, 2, 3, ..., n_r\}$ be the set of voyages on trade route r, where n_r is the number of voyages on trade route r that has to be serviced over the planning horizon. The set of voyages is indexed by i.

The given problem can be formulated on a directed graph $\mathcal{G} = (\mathcal{N}, \mathcal{A})$, where $\mathcal{N}$ denotes nodes, and $\mathcal{A}$ represents the set of arcs. The set $\mathcal{N}$ consists of four different kinds of nodes: Origin nodes, destination nodes, voyage nodes and maintenance nodes. The set $\mathcal{N}_v \subseteq \mathcal{N}$ consists of the nodes that ship v can visit. For each ship v, its origin node $o(v)$ in set $\mathcal{N}$ represents the initial position and its destination node $d(v)$ in set $\mathcal{N}$ corresponds to an artificial destination which does not exist physically. Each voyage i on trade route r is given by a voyage node (r, i). There are two types of voyage nodes (contracted voyage nodes and optional voyage nodes), which consists of two disjoint subsets of $\mathcal{N}$. The set $\mathcal{N}^C$ represents the contracted voyages that the shipping company must service, while the set $\mathcal{N}^O$ represents the optional voyages. The set $\mathcal{N}_v^M$ is the set of maintenance nodes for ship v, indexed by (r, i) like voyage nodes. For each ship v without any maintenance requirements during the planning period, the set $\mathcal{N}_v^M$ will be empty. If ship v is due for maintenance, it is assumed to visit exactly one maintenance node during the planning period. The set $\mathcal{A}$ includes all arcs. The set $\mathcal{A}_v \subseteq \mathcal{A}$ consists of the arcs that can be traversed by ship v. The arc $((r, i), (q, j))$ corresponds to sailing ballast directly from the end of voyage or maintenance node (r, i) to the start of voyage or maintenance node (q, j). The arcs sailing directly from the origin node of ship v to voyage or maintenance node (r, i), $((o(v)), (r, i))$, and the arcs travelling directly from the voyage or maintenance node (r, i) to the destination node of ship v, $((r, i), (d(v)))$, are also included in $\mathcal{A}$. The set $\mathcal{A}_v$ consists of the arcs $(o(v), d(v))$ such that the ship v sails directly from its starting node $o(v)$ to the ending node $d(v)$, i.e. the ship is idle over the planning horizon.

Let T_{vriqjs}^B be the time ship v takes to sail ballast from the last unloading port of voyage (r, i) to the first loading port of voyage (q, j), or in other words sailing the arc $((r, i), (q, j))$, with speed alternative s. The corresponding cost is C_{vriqjs}^B. The time it takes to sail ballast from the starting position to start position of voyage (r, i) with speed alternative s is $T_{vo(v)ris}^B$, and the corresponding cost is $C_{vo(v)ris}$. The time it takes to sail voyage (r, i) with speed alternative s is denoted by T_{vris}, which corresponds to sailing time between all ports on a trade route plus the operation time at all port calls. The corresponding cost is C_{vris}, which mainly consists of fuel costs. The estimated freight income minus the port costs, for sailing voyage (r, i) is R_{ri}. C_{ri}^S is the cost of chartering a ship from the spot market to service voyage (r, i). Each voyage has to start at its first port within a given time window, $[E_{ri}, L_{ri}]$. The parameter E_{ri} is the earliest time for starting voyage i on trade r, while L_{ri} is the latest time for starting the voyage. Let $E_{o(v)}$ be the earliest time ship v can depart from its initial starting position. B_r represents the minimum accepted time interval between two consecutive voyages on trade r.

Let variable x_{vriqj} be a binary variable, which is 1 if ship v sails directly from node (r, i) to node (q, j), otherwise 0. The binary flow variable $x_{vo(v)ri}$ is 1 if ship v travels directly from it initial position to node (r, i), otherwise 0. Let variable $x_{vrid(v)}$ equal 1 if (r, i) is the last node ship v services, and 0 otherwise. Similarly, variable $x_{o(v)d(v)}$ is 1 if ship v is idle, and 0 otherwise. Let variable u^S_{ri} be 1 if voyage i on trade r is carried out by a spot ship, and 0 otherwise. The start time of voyage i on trade r is defined by the variable t_{ri}. Let variable w_{vris} be the weight of speed alternative s for ship v sailing voyage (r, i). Let variable w^B_{vriqjs} be the weight of speed alternative s for ship v sailing ballast from the last unloading port of voyage (r, i) to the first loading port of voyage (q, j). Let variable $w^B_{vo(v)ris}$ be the weight of speed alternative s for ship v sailing ballast from its initial position $o(v)$ to the first loading port of the voyage (r, i). The weights of the speed alternatives should sum up to 1 if an arc is serviced by that ship, and 0 otherwise. The maritime fleet deployment problem with speed optimization and voyage separation requirements can be formulated as follows:

Objective Function. The objective function (1) maximizes the total profit by summing the profits from servicing the voyages by ships in the fleet (the estimated freight income minus the voyage costs minus ballast sailing costs) and the spot ships (the estimated freight income minus the voyage costs).

$$\max \sum_{v \in V} \sum_{r \in \mathcal{R}_v} \sum_{i \in \mathcal{I}_r} \left[\sum_{s \in S} (R_{ri} - C_{vris}) w_{vris} - \sum_{q \in \mathcal{R}_v} \sum_{j \in \mathcal{I}_q} \sum_{s \in S} C^B_{vriqjs} w^B_{vriqjs} \right.$$
$$\left. - \sum_{s \in S} C^B_{vo(v)ris} w^B_{vo(v)ris} \right] + \sum_{r \in \mathcal{R}} \sum_{i \in \mathcal{I}_r} (R_{ri} - C^S_{ri}) u^S_{ri} \qquad (1)$$

Service Constraints. Constraints (2) represent that each contracted voyage should be serviced exactly once by either a ship in the fleet or a spot ship. Constraints (3) state that each optional voyage can be serviced at most once by a ship in the fleet. Constraints (4) ensure that all required maintenance for ships in the fleet are performed.

$$\sum_{v \in V_r} \left[\sum_{q \in \mathcal{R}_v} \sum_{j \in \mathcal{I}_q} x_{vriqj} + x_{vrid(v)} \right] + u^S_{ri} = 1, \quad (r, i) \in \mathcal{N}^C \qquad (2)$$

$$\sum_{v \in V_r} \left[\sum_{q \in \mathcal{R}_v} \sum_{j \in \mathcal{I}_q} x_{vriqj} + x_{vrid(v)} \right] \leq 1, \quad (r, i) \in \mathcal{N}^O \qquad (3)$$

$$\sum_{q \in \mathcal{R}_v} \sum_{j \in \mathcal{I}_q} x_{vriqj} + x_{vrid(v)} = 1, \quad v \in V, (r, i) \in \mathcal{N}^M_v \qquad (4)$$

Network Flow Constraints. Constraints (5)–(7) ensure network flow for each ship. Constraints (5) state that a ship must either be idle or leave its starting position to a node (r, i), while constraints (7) state that a ship must either be idle or arrive at its ending position from a node (r, i). Constraints (6) ensure that each voyage starts in an origin node, that every visited voyage or maintenance node is also exited, and that each voyage ends up in a destination node. Constraints (8)–(10) describe the relation between the flow variables and the speed weighting variables for initial ballast sailing, ballast sailing and voyage sailing, respectively.

$$x_{vo(v)d(v)} + \sum_{r \in \mathcal{R}_v} \sum_{i \in \mathcal{I}_r} x_{vo(v)ri} = 1, \quad v \in \mathcal{V} \tag{5}$$

$$x_{vrid(v)} + \sum_{q \in \mathcal{R}_v} \sum_{j \in \mathcal{I}_q} x_{vriqj} - \sum_{q \in \mathcal{R}_v} \sum_{j \in \mathcal{I}_q} x_{vqjri} - x_{o(v)ri} = 0,$$
$$v \in \mathcal{V}, r \in \mathcal{R}_v, i \in \mathcal{I}_r \tag{6}$$

$$x_{vo(v)d(v)} + \sum_{r \in \mathcal{R}_v} \sum_{i \in \mathcal{I}_r} x_{vrid(v)} = 1, \quad v \in \mathcal{V} \tag{7}$$

$$x_{vo(v)ri} - \sum_{s \in \mathcal{S}} w_{vo(v)ris}^B = 0, \quad v \in \mathcal{V}, r \in \mathcal{R}_v, i \in \mathcal{I}_r \tag{8}$$

$$x_{vriqj} - \sum_{s \in \mathcal{S}} w_{vriqjs}^B = 0, \quad v \in \mathcal{V}, ((r,i),(q,j)) \in \mathcal{A}_v \tag{9}$$

$$x_{vrid(v)} + \sum_{q \in \mathcal{R}_v} \sum_{j \in \mathcal{I}_q} x_{vriqj} - \sum_{s \in \mathcal{S}} w_{vris} = 0, \quad v \in \mathcal{V}, r \in \mathcal{R}_v, i \in \mathcal{I}_r \tag{10}$$

Time Constraints. Constraints (11) state that time spent sailing ballast from the initial position of ship v to its first voyage (r, i) does not exceed the latest start time of voyage (r, i). Constraints (12) ensure the time spent on voyage (r, i) and ballast sailing to the start of voyage (q, j) does not exceed the latest start time of voyage (q, j). Constraints (13) secure that time window for each voyage is not violated. Constraints (11) and (12) have been linearized by applying the big-M method.

$$E_{o(v)} + \sum_{s \in \mathcal{S}} T_{vo(v)ris}^B w_{vo(v)ris}^B - t_{ri} - E_{o(v)}(1 - x_{vo(v)ri}) \leq 0,$$
$$v \in \mathcal{V}, r \in \mathcal{R}_v, i \in \mathcal{I}_r \tag{11}$$

$$t_{ri} + \sum_{s \in \mathcal{S}} (T_{vris} w_{vris} + T_{vriqjs}^B w_{vriqjs}^B) - t_{qj} - L_{ri}(1 - x_{vriqj}) \leq 0,$$
$$v \in \mathcal{V}, ((r,i),(q,j)) \in \mathcal{A}_v \tag{12}$$

$$E_{ri} \leq t_{ri} \leq L_{ri}, \quad r \in \mathcal{R}, i \in \mathcal{I}_r \tag{13}$$

Voyage Separation Constraints. Constraints (14) take care of the minimum accepted time between two consecutive voyages on a trade route.

$$t_{r,i+1} - t_{ri} \geq B_r, \quad r \in \mathcal{R}, i \in \mathcal{I}_r \setminus \{n_r\} \tag{14}$$

Binary and Non-negativity Constraints.

$$x_{vo(v)d(v)} \in \{0,1\}, \quad v \in \mathcal{V} \tag{15}$$

$$x_{vo(v)ri}, x_{vrid(v)} \in \{0,1\}, \quad v \in \mathcal{V}, r \in \mathcal{R}_v, i \in \mathcal{I}_r \tag{16}$$

$$x_{vriqj} \in \{0,1\}, \quad v \in \mathcal{V}, ((r,i),(q,j)) \in \mathcal{A}_v \tag{17}$$

$$w^B_{vo(v)ris}, w_{vris} \in [0,1], \quad v \in \mathcal{V}, r \in \mathcal{R}_v, i \in \mathcal{I}_r, s \in \mathcal{S} \tag{18}$$

$$w^B_{vriqjs} \in [0,1], \quad v \in \mathcal{V}, ((r,i),(q,j)) \in \mathcal{A}_v, s \in \mathcal{S} \tag{19}$$

$$t_{ri} \geq 0, \quad r \in \mathcal{R}_v, i \in \mathcal{I}_r \tag{20}$$

$$u^S_{ri}, \in \{0,1\} \quad r \in \mathcal{R}_v, i \in \mathcal{I}_r \tag{21}$$

3.2 Path Flow Model

Notation. Some of the notation presented for the arc flow model is still valid for the path flow model. Only new notation for the path flow model is presented in this section. $\mathcal{P}_v$ represents the set of all feasible paths for ship v. $\mathcal{P}_{vriqj}$ is a subset of $\mathcal{P}_v$ including all paths where ship v travels directly from voyage i on trade route r to voyage j on trade route q. $\mathcal{P}_{vri}$ is a subset of $\mathcal{P}_v$, which contains all paths where ship v services voyage i on trade route r. Another subset of $\mathcal{P}_v$, $\mathcal{P}_{vo(v)ri}$, which contains all paths where ship v sails directly from its initial position to voyage i on trade route r as its first voyage.

Let E_{vpri} be a the earliest service start time for ship v at voyage i on trade route r for a path p.

Let variable z_{vp} be a binary variable, which equals 1 if ship v sails path p, and 0 otherwise. Let t_{vri} be a variable that sets the start time of voyage i on trade route r with ship v. Variable t^S_{ri} applies when a spot ship starts sailing voyage i on trade route r. A path flow model describing the fleet deployment problem with speed optimization and voyage separation constraints can be described as follows.

Objective Function. The objective function (22) aims to maximize profit by finding the optional speed on the paths.

$$\max \sum_{v \in \mathcal{V}} \sum_{r \in \mathcal{R}_v} \sum_{i \in \mathcal{I}_r} \left[\sum_{s \in \mathcal{S}} (R_{ri} - C_{vris}) w_{vris} - \sum_{q \in \mathcal{R}_v} \sum_{j \in \mathcal{I}_r} \sum_{s \in \mathcal{S}} C^B_{vriqjs} w^B_{vriqjs} \right.$$

$$\left. - \sum_{s \in \mathcal{S}} C^B_{vo(v)ris} w^B_{vo(v)ris} \right] + \sum_{r \in \mathcal{R}} \sum_{i \in \mathcal{I}_r} (R_{ri} - C^S_{ri}) u^S_{ri} \tag{22}$$

Service Constraints. Constraints (23) ensure that all contractual voyages are carried out exactly once, either by a ship within the fleet or by a spot ship, where constraints (24) ensure that the optional voyages may be carried out at most once by a ship within the fleet. All ships have to be assigned to exactly one path, as in constraints (25).

$$\sum_{v\in\mathcal{V}_r}\sum_{p\in\mathcal{P}_{vri}} z_{vp} + u_{ri}^S = 1, \quad (r,i)\in\mathcal{N}^C \tag{23}$$

$$\sum_{v\in\mathcal{V}_r}\sum_{p\in\mathcal{P}_{vri}} z_{vp} \le 1, \quad (r,i)\in\mathcal{N}^O \tag{24}$$

$$\sum_{p\in\mathcal{P}_v} z_{vp} = 1, \quad v\in\mathcal{V} \tag{25}$$

Network Flow Constraints. Constraints (26)–(28) ensure that the speed weighting variables for each ship on a path can take non-zero values only when the ship sails that path.

$$\sum_{s\in\mathcal{S}} w_{vris} = \sum_{p\in\mathcal{P}_{vri}} z_{vp}, \quad v\in\mathcal{V}, r\in\mathcal{R}_v, i\in\mathcal{I}_r \tag{26}$$

$$\sum_{s\in\mathcal{S}} w_{vo(v)ris}^B = \sum_{p\in\mathcal{P}_{vo(v)ri}} z_{vp}, \quad v\in\mathcal{V}, r\in\mathcal{R}_v, i\in\mathcal{I}_r \tag{27}$$

$$\sum_{s\in\mathcal{S}} w_{vriqjs}^B = \sum_{p\in\mathcal{P}_{vriqj}} z_{vp}, \quad v\in\mathcal{V}, r\in\mathcal{R}_v, i\in\mathcal{I}_r, q\in\mathcal{R}_v, j\in\mathcal{I}_q \tag{28}$$

Time Constraints. Constraints (29) say that the start time for a voyage has to be within the time window. The same goes for the start time for a voyage by spot ships as in constraints (30). Constraints (31) ensure that a ship can not start a voyage before it has sailed ballast from its origin position to the start point of the voyage. Likewise, constraints (32) ensure that a ship can not start a voyage before it has completed the previous voyage and sailed ballast to the start of the next voyage.

$$\sum_{p\in\mathcal{P}_{vri}} E_{vpri} z_{vp} \le t_{vri} \le \sum_{p\in\mathcal{P}_{vri}} L_{ri} z_{vp}, \quad v\in\mathcal{V}, r\in\mathcal{R}_v, i\in\mathcal{I}_r \tag{29}$$

$$E_{ri} u_{ri}^S \le t_{ri}^S \le L_{ri} u_{ri}^S, \quad r\in\mathcal{R}, i\in\mathcal{I}_r \tag{30}$$

$$\sum_{s\in\mathcal{S}} \left(T_{vo(v)ris}^B + E_{o(v)}\right) w_{vo(v)ris}^B \le t_{vri}, \quad v\in\mathcal{V}, r\in\mathcal{R}_v, i\in\mathcal{I}_r \tag{31}$$

$$t_{vri} + \sum_{s\in\mathcal{S}} \left(T_{vris} w_{vris} + T_{vriqjs}^B w_{vriqjs}^B + (L_{ri} + T_{vri,1}) w_{vriqjs}^B\right)$$

$$- L_{ri} - T_{vri,1} - t_{vqj} \le 0, \quad v\in\mathcal{V}, r\in\mathcal{R}_v, i\in\mathcal{I}_r, q\in\mathcal{R}_v, j\in\mathcal{I}_q \tag{32}$$

Voyage Separation Constraints. Constraints (33) show the voyage separation constraints, which ensures a minimum time spread between two consecutive voyages on the same trade route.

$$B_r + \sum_{v\in\mathcal{V}} t_{vri} + t_{ri}^S - \sum_{v\in\mathcal{V}} t_{vr,i+1} - t_{r,i+1}^S \le 0, \quad r\in\mathcal{R}, i\in\mathcal{I}_r\setminus\{n_r\} \tag{33}$$

Binary and Non-negativity Constraints.

$$z_{vp} \in \{0,1\}, \quad v \in \mathcal{V}, p \in \mathcal{P}_v \tag{34}$$

$$u_{ri}^S \in \{0,1\}, \quad r \in \mathcal{R}, i \in \mathcal{I}_r \tag{35}$$

$$w_{vris}, w_{vo(v)ris}^B \in [0,1], \quad v \in \mathcal{V}, r \in \mathcal{R}_v, i \in \mathcal{I}_r, s \in \mathcal{S} \tag{36}$$

$$w_{vriqjs}^B \in [0,1], \quad v \in \mathcal{V}, r \in \mathcal{R}_v, i \in \mathcal{I}_r, q \in \mathcal{R}_v, j \in \mathcal{I}_q, s \in \mathcal{S} \tag{37}$$

$$t_{ri}^S \geq 0, \quad r \in \mathcal{R}_v, i \in \mathcal{I}_r \tag{38}$$

$$t_{vri} \geq 0, \quad v \in \mathcal{V}, r \in \mathcal{R}_v, i \in \mathcal{I}_r \tag{39}$$

4   Computational Study

The mathematical models presented in Sect. 3 have been implemented in Mosel and solved using Xpress 31.01.09. All computational tests are performed on a HP Elitedesk computer with Intel Core i7-7700 CPU (4 × 3.60 GHz) and 32 GB RAM running on Windows 10.

4.1   Test Instances

The instances are based on data from the case shipping company as in [5] and shown in Table 1. The instances are divided into four sets, with three (six), five (ten), seven (14) and nine (18) trades (ships), respectively. All four sets have also been divided into planning horizons of 60, 90 and 120 days, totaling 12 instances. For example, the nine trade routes instances shown in Fig. 1 correspond to the largest set, set 4 in Table 1. Instance sets 1–3 are reduced versions of instance set 4 where some trade routes and ships have been removed. The fifth column shows the number of voyages that should be serviced for each instance. The numbers represent the contractual voyages out of the total number of voyages (including optional voyages). The optional voyages are organized as one trade that consists of optional voyages only. All feasible paths for each instance are generated using Matlab code as input to the path flow model in Sect. 3.2, and are shown in the last column in Table 1. The fuel cost is set to 388 USD/ton, which is the global average for the first quarter of 2018 for the 20 largest ports in the world [1].

4.2   Comparison of the Arc Flow and Path Flow Models

The 12 test instances in Table 1 have been solved by both the arc flow and the path flow models using three speed points (i.e. the minimum, maximum and the middle speed points). The results of these comparisons are shown in Table 2. The columns Time report the computational times in seconds. Here, we have allocated a maximum running time of one hour (i.e. 3600 s). The columns Obj val. show the objective values found by the two models. The columns Gap show the gap in percentage between the best integer solution and the best bound found after the time limit. The columns LP Rel. show the LP relaxation.

Table 1. Summary of test instances

Set	Instance	Ships	Trades	Voyages	Planning days	Paths
1	1	6	3	11/11	60	159
	2	6	3	15/15	90	299
	3	6	3	20/20	120	985
2	4	10	5	13/15	60	364
	5	10	5	18/21	90	823
	6	10	5	24/28	120	3277
3	7	14	7	24/26	60	1886
	8	14	7	34/37	90	8711
	9	14	7	46/50	120	69776
4	10	18	9	30/32	60	3073
	11	18	9	44/47	90	16199
	12	18	9	59/63	120	138292

From the results in Table 2 we see that for the smallest problem instances (1, 2 and 4) there are no significant differences in performance between these two models. They both find the optimal solutions to these instances in little computational time. For the larger problem instances, however, there is a tendency that the path flow model is faster than the arc flow model (14.05% improvement in time). For large test instances 8–12, both the arc flow and the path flow models are not able to prove optimality within the time limit with an average of 17.25% and 3.46% gaps, respectively.

Comparing the average performance of the two models, we see that the path flow models gives 11.5% improvement in solution quality with reduced solution times compared to the arc flow model. The LP relaxation achieves a 4.1% improvement for the path flow compared to the arc flow model. The conclusion from this comparison is that using the path flow model gives better (or equally good) solutions for all large instances due to smaller average gaps, better LP relaxation, and less computational times. Therefore, only the path flow model is used in Sect. 4.3 for analyzing the speed optimization in detail.

4.3   Comparison of Using Different Speed Points for Linearization

All test instances in Table 1 with planning horizon of 120 days have been solved using the path flow model with one, two, and three speed points, respectively. When solving with one speed point (without speed optimization), we have used the maximum speed, as this was shown to give better solutions compared to planning with only the medium (or service) speed of the ships [6]. The results are shown in Table 3. For a fair comparison, it should be noted that the column Profit show the best solutions found by the model after a posteriori speed optimization (using 10 points), and will therefore slightly deviate from Obj. value in Table 2

Table 2. Comparison of arc flow and path flow models using three speed points.

Instance	Arc flow model				Path flow model			
	Time	Obj. val.	Gap	LP Rel.	Time	Obj. val.	Gap	LP Rel.
1	0.6	13,837	0.00%	14,742	0.1	13,837	0.00%	13,949
2	1.7	17,350	0.00%	19,874	1.2	17,350	0.00%	17,965
3	80.6	22,223	0.00%	26,123	10.6	22,223	0.00%	23,308
4	1.3	17,456	0.00%	18,035	0.3	17,456	0.00%	17,555
5	1857.1	22,949	0.00%	24,485	13.0	22,949	0.00%	23,845
6	3600.0	28,141	11.77%	32,018	3600.0	28,795	3.99%	31,090
7	3600.0	24,995	4.72%	26,186	1704.3	25,339	0.00%	25,835
8	3600.0	31,579	18.62%	37,752	3600.0	33,934	4.56%	35,967
9	3600.0	40,751	23.10%	50,587	3600.0	42,227	12.62%	47,621
10	3600.0	29,435	5.20%	31,610	3600.0	30,288	0.14%	30,621
11	3600.0	34,419	28.61%	44,466	3600.0	40,755	6.43%	43,465
12	3600.0	27,510	114.99%	59,340	3600.0	51,161	13.76%	58,223
Average	2261.8	25,887	17.25%	32,101	1944.1	28,860	3.46%	30,787

Table 3. Comparison of different number of speed points.

Instance	1 speed point (max)				2 speed points (min/max)				3 speed points (min/avg/max)			
	Profit	Gap	# Spot	Time	Profit	Gap	# Spot	Time	Profit	Gap	# Spot	Time
3	20,175	0.00%	3	0.1	22,246	0.00%	3	11.1	23,308	0.00%	3	10.6
6	27,666	0.00%	1	0.2	28,555	5.56%	3	3600.0	28,852	3.99%	2	3600.0
9	41,207	0.00%	0	211.4	42,367	13.61%	7	3600.0	42,359	12.62%	6	3600.0
12	51,173	0.00%	0	994.3	50,781	16.16%	1	3600.0	51,553	13.76%	3	3600.0
Average	35,055	0.00%	1	301.5	35,987	8.83%	4	2702.8	36,518	7.59%	4	2702.7

for three speed points. The columns Gap show the gap between the best integer solutions and the best bounds found after the 3600 s time limit. The columns #Spot show the number of voyages performed by spot ships in the problem instance, while the columns Time report the computational time in seconds.

Comparing the average solutions from the different number of speed points, we can see that using three speed points gives the best solution quality, though with longer solution times. A larger number of speed points might provide better results with longer solution times. As a compromise, three speed points are used in our study. The one speed point instances are all solved to optimality within the maximum time limit. The average number of voyages performed by spot ships with two and three speed points are higher than that with one speed point, especially for the medium and large instances (instances 9 and 12), which implies that integrating speed optimization in a fleet deployment problem not only achieves better profits, but also gives much more planning flexibility for shipping companies.

4.4   Path Reduction Heuristics

It was shown in the previous sections that the gaps are large for the largest instances with long planning horizons. We have therefore tested three simple path reduction rules on instances 9 and 12. In the first, we remove all paths/routes that have higher percentage ballast sailing than a threshold level. In the next two, which will be used during the path generation, we remove paths with any ballast seiling leg and consecutive waiting time (assuming maximum speed) above specified threshold levels. We show results for the following four combinations of applying these rules in Table 4 (Max percentage ballast sailing − Max length ballast sailing in nautical miles − Max consecutive waiting days): (A) 45% - 10.500 - 20, (B) 35% - 10.500 - 20, (C) 40% - 10.500 - 10, and (D) 30% - 10.000 - 10.

Table 4. Effect of heuristic combinations

Comb.	Instance 9			Instance 12			Average	
	Paths	Obj. val.	Impr.	Paths	Obj. val.	Impr.	Obj. val.	Impr.
None	69776	42,227'	-	138392	51,161'	-	46,694'	-
A	19603	42,716'	1.2%	42338	51,841'	1.3%	47,279'	1.3%
B	8127	43,075'	2.0%	20431	52,098'	1.8%	47,587'	1.9%
C	7542	43,208'	2.3%	18294	52,305'	2.2%	47,757'	2.3%
D	2365	42,113'	−0.3%	6956	52,740'	3.1%	47,427'	1.6%

Table 4 shows both the number of paths and the solution improvement compared to the results without any path reduction rules. We see that the number of paths are significantly reduced and that we are able to obtain improved solutions to both instances 9 and 12, except for combination D on instance 9, where we obviously lose at least one of the optimal paths when using the path reduction heuristic.

5   Concluding Remarks

We have extended a previously studied problem [5,10] by incorporating speed optimization. This gives the Maritime Fleet Deployment problem with Speed Optimization and Voyage Separation requirements (MFDSOVS). Two formulations for this problem, one arc flow formulation and one path flow formulation, are proposed in this paper. The non-linear relationship for fuel consumption as a function of ship speed is linearized by choosing discrete speed points and linear combinations of these. Computational results show that the path flow model is faster and generate better results than the arc flow model. Furthermore, we show that speed in the fleet deployment results in not only better profits, but also gives much more planning flexibility for shipping companies by having more voyages taken by spot ships. Finally, a priori path reduction heuristics are tested to solve the large instances more efficiently.

References

1. World bunker prices (2018). https://shipandbunker.com/prices
2. Andersson, H., Fagerholt, K., Hobbesland, K.: Integrated maritime fleet deployment and speed optimization: case study from roro shipping. Comput. Oper. Res. **55**, 233–240 (2015)
3. Brouer, B.D., Alvarez, J.F., Plum, C.E.M., Pisinger, D., Sigurd, M.M.: A base integer programming model and benchmark suite for liner-shipping network design. Transp. Sci. **48**(2), 281–312 (2014)
4. Dohn, A., Rasmussen, M.S., Larsen, J.: The vehicle routing problem with time windows and temporal dependencies. Networks **58**(4), 273–289 (2011)
5. Norstad, I., Fagerholt, K., Hvattum, L.M., Arnulf, H.S., Bjørkli, A.: Maritime fleet deployment with voyage separation requirements. Flex. Serv. Manuf. J. **27**(2–3), 180–199 (2015)
6. Norstad, I., Fagerholt, K., Laporte, G.: Tramp ship routing and scheduling with speed optimization. Transp. Res. Part C Emerg. Technol. **19**(5), 853–865 (2011)
7. Psaraftis, H.N., Kontovas, C.A.: Ship speed optimization: concepts, models and combined speed-routing scenarios. Transp. Res. Part C Emerg. Technol. **44**, 52–69 (2014)
8. Reinhardt, L.B., Clausen, T., Pisinger, D.: Synchronized dial-a-ride transportation of disabled passengers at airports. Europ. J. Oper. Res. **225**(1), 106–117 (2013)
9. UNCTAD: Review of Maritime Transportation 2016. UN Publications (2016)
10. Vilhelmsen, C., Lusby, R.M., Larsen, J.: Tramp ship routing and scheduling with voyage separation requirements. OR Spectr. **39**(4), 913–943 (2017)
11. Wang, X., Norstad, I., Fagerholt, K., Christiansen, M.: Tramp ship routing and scheduling: effects of market-based measures on CO_2 reduction. In: Sustainable Shipping: A Cross Disciplinary View, Chap. 8, Springer (2018, forthcoming)

A Novel Method for Solving Collision Avoidance Problem in Multiple Ships Encounter Situations

Shijie Li[1(✉)], Jialun Liu[2(✉)], Xiaohua Cao[1], and Yu Zhang[1]

[1] School of Logistics Engineering, Wuhan University of Technology,
No.1178 Heping Road, Wuhan, People's Republic of China
{lishijie,sanli}@whut.edu.cn, tomm_cao@163.com
[2] National Engineering Research Center for Water Transport Safety
(Ministry of Science and Technology), Wuhan, People's Republic of China
jialunliu@whut.edu.cn

Abstract. With the emergence and development of larger and faster ships, and the increased maritime traffic, situations in which one ship must take actions to avoid collisions with multiple ships are also likely to increase, which makes anti-collision decision making more complicated. This paper proposes a novel method for solving collision avoidance problem in multiple ships encounter situations, consisting of three phases: firstly, predictions of each ship's potential trajectories with different rudder angles are made considering ship dynamics; secondly, each ship evaluates the collision risk it has with the other encountering ships based on trajectory prediction; thirdly, a distributed constraint optimization strategy is adopted to assist ships in making efficient anti-collision decisions, which concerns determining the optimal rudder angles alterations for each ship to avoid collisions. This method takes into account both the maneuverability of ships and the reciprocally-affected nature of multiple ships' anti-collision decisions. Simulation experiments of 7-ships encounter situations are carried out to evaluate the effectiveness of the proposed method. Experimental results show that the proposed method could provide ships with optimal rudder angle alteration decisions to avoid collisions in a safe and efficient way.

1 Introduction

With the emergence and development of larger and faster ships, and the increased maritime traffic, situations in which one ship must take actions to avoid collisions with multiple ships are also likely to increase, which make anti-collision decision making more complicated. While many advanced assistant systems such as GPS, ARPA, AIS and ECDIS have been developed and installed on ships, collision accidents still happen every now and then. This causes great loss of lives and property, and brings negative impacts to the maritime environment. According to many investigations on maritime accidents, a large majority of maritime accidents are caused by or related to human mistakes [3]. To reduce human error

© Springer Nature Switzerland AG 2018
R. Cerulli et al. (Eds.): ICCL 2018, LNCS 11184, pp. 47–66, 2018.
https://doi.org/10.1007/978-3-030-00898-7_4

and improve maritime safety, it is important to enhance navigational intelligence and autonomy.

There are several ways to prevent ship collisions, such as lookouts, radar, and VHF radio. In early years, the 1972 International Regulations for Preventing Collision at Sea (COLREGs) proposed by the International Maritime Organization (IMO) is supposed to be obeyed by all ships. The COLREGs describes potential collision scenarios between two encountering ships and provide a set of guidelines for safe maneuvering at sea. However, it is stated in a general way and does not guarantee the efficiency of anti-collision operations. Therefore, a number of maritime collision avoidance methods such as ship domain, fuzzy theory, evolutionary algorithms, and real-time control algorithms have been proposed, the literature reviews can be found in [4,12]. These methods work well in one-to-one and one-to-many situations, in which each ship determines its own course based on the assumption that surrounding ships keep their sailing states unchanged. For many-to-many situations, few methods have been suggested in the literature. Therefore, this paper mainly discusses the most recent and relevant research work.

Regarding multi-ship encounter situations, Distributed Local Search Algorithm (DLSA) and Distributed Tabu Search Algorithm (DTSA) have been adopted in [5]. The drawback of these algorithms is that it takes a relatively large number of messages for the ships to coordinate their actions. Therefore, the authors extend their work by introducing Distributed Stochastic Search Algorithm (DSSA), which allows each ship to change her intention in a stochastic manner immediately after receiving all of the intentions from the target ships [6]. The authors in [4] presents a ship collision avoidance system based on model predictive control. A finite set of alternative control behaviors are generated by varying offsets to the guidance course angle commanded to the autopilot, and changes to the propulsion command ranging from nominal speed to full reverse. Using simulated ship trajectories predictions, each alternative control behavior is evaluated regarding its compliance with the COLREGs and associated collision risks to find the optimal control behavior.

A decision support system for ship collision avoidance is proposed for Istanbul Strait in [9]. The system uses manually controlled and reciprocally passing ships' data to train artificial neural networks (ANN), with the aim to make predictions of ships' future locations three minutes in advance. If collision risks exist, warnings would be sent to the Vessel Traffic Services (VTS) center and to the ships' personnels. A two-step trajectory planning procedure for ship collision avoidance is introduced in [3]. Firstly, the trajectory generation phase uses A* algorithm to generate initial collision avoidance trajectories for each ship, based on the prediction of ship's future motions, without considering the information of other ships. Secondly, trajectory negotiation incorporates the algorithm in [14] to find a global optimal solution for all ships. The authors in [11] develops a deterministic collision avoidance path planning algorithm to provide collision-free path for all involved ships. The algorithm plans navigation paths for all encountering ships in a cooperative mode, from a multi-ship perspective.

When multiple ships encounter with one another, the maneuvering decisions for avoiding potential collisions depend on many factors such as speed, course, relative position and maneuverability of the encountering ships. These decisions are highly-related, as a small change in the course of one ship may affect the future decisions of the other ships, and vice versa [6]. To deal with such complex nature of multiple ship collision avoidance problem, the dynamic relations among ships in dense traffic and the impacts of ship maneuverability should both be considered. In [13], a ship maneuverability-based collision avoidance support system in close-quarters situations is presented. It includes a mathematical model and a control mechanism of ship maneuvering, as well as a dynamic calculation model of collision avoidance parameters. However, it is still from the a single ship perspective and does not consider the other ships simultaneously.

This paper proposes a novel method for solving collision avoidance problem in multiple ships encountering situations, which considers both ship maneuverability and reciprocally-affected anti-collision decisions of multiple ships. The proposed method consists of three phases: firstly, predictions regarding each ship's potential trajectories with different rudder angles are made based on ship dynamics, within a pre-defined prediction time; secondly, each ship evaluates the potential collision risks that are incurred by its predicted trajectories and the other encountering ships' predicted trajectories; thirdly, a distributed constraint optimization strategy is adopted to determine the most efficient rudder angle alterations for each ship to prevent collisions. This paper models ships as individual parties who can communicate and coordinate their anti-collision operations with one another, to find the most efficient anti-collision operations. It is assumed that each ship can exchange information regarding each ship's position, course and speed with the other ships to cooperatively make decisions on the routes to avoid collisions.

This rest of the paper is organized as follows. Section 2 introduces the structure of the proposed method. Section 3 gives details regarding the models and solution steps. Preliminary results are presented in Sect. 4. Conclusions and future work are given in Sect. 5.

2 Structure of the Proposed Method

According to Rule 8 of COLREGs, ship course alteration is generally more effective than ship speed alteration as the it is faster to take effects, and it is easier to be observed both visually and on radar [13]. Therefore, it is important to make sure that ships change their courses at suitable times, and that the changing should not increase the collision risks. In practice, a ship's rudder angle affects its rudder forces and moments, thereby leads to changes in the ship's course. As a result, rudder angle alteration is considered as the main anti-collision operation in this paper.

As it is inefficient for a ship to constantly change its course to avoid collisions, this paper divides continuous time into a series of discrete time slots, and each ship makes decisions regarding rudder angle alteration during each time slot.

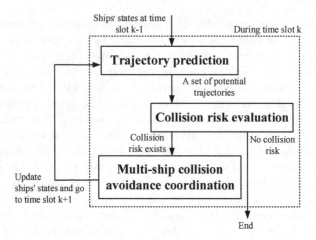

Fig. 1. Structure of the proposed method.

The structure of the proposed method is shown in Fig. 1. At time slot k, based on the encountering ships' states at the end of time slot $k-1$, including its coordinates, speed, course, and rudder angle, an optimization procedure is carried out. This procedure determines for each ship the most suitable rudder angle it should take within this time slot k. Then the encountering ships' states will be updated again with the optimal rudder angles, after which the same procedure starts again at the beginning of the next time slot $k+1$ with the updated ships' states at the end of time slot k. The optimization procedure during each time slot consists of three phases: firstly, each ship's trajectory is predicted based on its dynamic equations, given a set of candidate rudder angles; secondly, each ship evaluates its potential collision risks with all the encountering ships according to the predicted trajectories: if no collision risk exists, no coordination on their rudder angle alterations is required; otherwise, in the third phase, the ships need to coordinate their decisions on rudder angle alterations with one another, and search for globally optimal solutions. It is an iterative process at the end of which all the potential collision risks will be eliminated. The overall optimization objective is find the most efficient anti-collision maneuvering operations for all encountering ships, which is are series of rudder angle alterations in a set of discrete time slots.

3   Models and Solution Steps

This section firstly introduces the parameters and decision variables that are used in this paper and then gives details on the optimization procedure.

3.1   Decision Variables and Parameters

The parameters that are used is given in Table 1. It is assumed that information regarding each ship $i \in V$'s current speed (u_i, v_i), initial rudder angle $\delta_i(0)$ and

Table 1. Relevant parameters.

Symbols	Definitions
V	A set of encountering ships
K	A set of discrete time slots
N^{unit}	The length of a discrete time slot in continuous time
$T^{\text{prediction}}$	The length of trajectory prediction time
C^{MSD}	The minimum safe distance that each two ships should keep to avoid collision
(u_i, v_i)	Forward speed u and sway speed v of ship i
$\delta_i(0)$	Initial rudder angle of ship i
$(x_i(0), y_i(0))$	Initial coordinates of ship i

coordinates $(x_i(0), y_i(0))$ are available to other ships. Parameter K represents a set of discrete time slots. Parameter N^{unit} represents the length of a discrete time slot k in continuous time. Parameter $T^{\text{prediction}}$ determines the length of

Table 2. Decision variables.

Symbols	Definitions
$\delta_i(k)$	The rudder angle that ship i takes during time slot k
$D_i(k)$	The domain of variable $\delta_i(k)$ during time slot k
$(x_i(t), y_i(t))$	The coordinates of ship i on x axis and y axis at continuous time t
$C_{ij}^{\text{dCPA}}(\delta_i(k), \delta_j(k))$	The distance between ship i's current position and its closest point of approach (CPA) with ship j during time slot k, if they take rudder angles $\delta_i(k)$ and $\delta_j(k)$, respectively
$C_{ij}^{\text{tCPA}}(\delta_i(k), \delta_j(k))$	The traveling time from ship i's position to its CPA with ship j during time slot k, if they take rudder angles $\delta_i(k)$ and $\delta_j(k)$, respectively
$C_{ij}^{\text{distance}}(\delta_i(k), \delta_j(k))$	The nearest distance between ships i and j during time slot k, if they take rudder angles $\delta_i(k)$ and $\delta_j(k)$, respectively
$C_i^{\text{closest}}(\delta_i(k))$	The distance between ship i's current position to the nearest CPA with the other ships during time slot k, if it takes rudder angle $\delta_i(k)$
$T_i^{\text{closest}}(\delta_i(k))$	The traveling time from ship i's position to the nearest CPA with the other ships during time slot k, if it takes rudder angle $\delta_i(k)$
$N_i^{\text{ship}}(\delta_i(k))$	The number of ships that ship i has collision risk with during time slot k, if it takes rudder angle $\delta_i(k)$

each ship's trajectory prediction time. Parameter C^{MSD} represents the minimum safe distance that each two ships should keep to avoid potential collisions.

Table 2 presents the decision variables in this paper. The main decision each ship makes at each time slot is to choose a suitable rudder angle to change its course. Variable $\delta_i(k)$ represents the current rudder angle that ship i takes during time slot k. Considering typical ship maneuvering, this paper assumes that the rudder angle ranges from $+30°$ on the port side to $+30°$ on the starboard side ($\pm 30°$) in steps of $10°$. Therefore, variable $\delta_i(k) \in D_i(k) = \{\pm 30°, \pm 20°, \pm 10°, 0°\}$. Variables $x_i(t)$ and $y_i(t)$ represents the coordinates of ship i's trajectory on x and y axes. During time slot k, when ship i takes different values for its rudder angle $\delta_i(k)$, trajectories within time $T^{\mathrm{prediction}}$ are constructed based on ship dynamic equations. Variables $C_{ij}^{\mathrm{dCPA}}, C_{ij}^{\mathrm{tCPA}}, C_{ij}^{\mathrm{distance}}$ and C_{ij}^{time} are used to evaluate the potential collision risks between any two ships i and j, when they take different rudder angles. To evaluate the cumulative collision risk of ship i's each candidate rudder angle, variables $C_i^{\mathrm{closest}}, T_i^{\mathrm{closest}}$ and N_i^{ship} are introduced.

3.2   Phase I: Trajectory Prediction

The basic steps of trajectory prediction for each ship is given in Algorithm 1. During each time slot, trajectories of each ship with different rudder angles are calculated based on ship dynamic equations as follows:

$$\begin{cases} (m+m_x)\dot{u} - (m+m_y)vr - x_G mr^2 = & X_H + X_P + X_R \\ (m+m_y)\dot{v} + (m+m_x)ur + x_G m\dot{r} = & Y_H + Y_P + Y_R \\ (I_z + x_G^2 m + J_z)\dot{r} + x_G m(\dot{v} + ur) = & N_H + N_P + N_R \end{cases} \tag{1}$$

where, subscripts H, P, R represents the hull, the propeller and the rudder; m, m_x and m_y are ship mass, added mass in x axis, and added mass in y axis; I_z and J_z are moment of inertia and added moment of inertia around the z axis, u and v are ship longitudinal and lateral speed, r is ship yaw rate around midship, and the dot notation of u, v and r represents the derivative of each parameter. For more details regarding the ship dynamic equations, we refer readers to the empirical ship model in [8].

Given different rudder angles, the hydrodynamic force X_R due to rudder acting on midship in x direction is determined, thereby the forward speed u and acceleration $\dot{u}$ in x-axis, as well as sway speed v and acceleration $\dot{v}$ in y-axis are also determined. According to ship motion variables (u, v) and $(\dot{u}, \dot{v})$, coordinates $(x(t), y(t))$ of the ship on x-axis and y-axis at time t can be calculated. A series of ship coordinates during a continuous time constitute the ship's trajectory. It is noted that during each time slot k, trajectories are predicted with a time length of $T^{\mathrm{prediction}}$ in Phase I, which are the basis of the optimization procedure that will be carried out in the next phases. To be exact, the predicted trajectory consists of two parts: the first part is formulated when each ship keeps the chosen rudder angle for N^{unit} (the length of time slot k in continuous time), and the

---

Algorithm 1. Basic steps of ship i's trajectory prediction

1: **while** $1 \leq k \leq K^{\max}$, for each time slot k
2:   **for** each ship $i \in V$, it takes a value for its course $\delta_i(k)$ from variable domain $D_i(k) = \{\pm 30°, \pm 20°, \pm 10°, 0°\}$ **do**
3:      calculate the trajectories with different rudder angles with a prediction period of $T^{\mathrm{prediction}}$, based on ship i's maneuvering model in (1),
4:   **return** a set of trajectories $(x_i(t), y_i(t))$ $(k-1 \leq t \leq k$ in continuous time) of ship i, for each $\delta_i(k) \in D_i(k)$.

---

second part is formulated when it switches back to its original rudder angle and keeps it with a time length of $(T^{\mathrm{prediction}} - N^{\mathrm{unit}})$.

3.3   Phase II: Collision Risk Evaluation

The basic steps of collision risk evaluation is shown in Algorithm 2. Each ship evaluates its potential collision risks with the other encountering ships according to its potential trajectories. In each time slot $k \in K$, for each ship $i \in V$, firstly it checks if collision risk exists between it and any other ship $j \in V$ by calculating their relative distance (line 1–2). If the distance between ships i and j is larger than the minimum safe distance and increases with time, it implies that these

---

Algorithm 2. Basic steps of collision risk evaluation

Require: A set of potential courses and the corresponding predicted trajectories of ship i as calculated in Algorithm 1
1: **while** $1 \leq k \leq K^{\max}$, in each time slot k
2:   **for** each ship $i \in V$ **do** evaluate its collision risk with any other ship $j \in V$ according to their trajectories $(x_i(t), y_i(t))$ and $(x_j(t), y_j(t))$ $(t \in [k-1, k])$:
3:      **if** the distance between ships i and j, $\sqrt{|x_i(t) - x_j(t)|^2 + |y_i(t) - y_j(t)|^2} \geq T^{\mathrm{MSD}}$ and increases with time, no collision risk exists;
4:      **otherwise**, collision risk exists, **do** find the closest point of approach (x_i^*, y_i^*), and the distance C_{ij}^{dCPA} and time C_{ij}^{tCPA} to the closest point during time slot k, when ships i and j take rudder angles $\delta_i(k) \in D_i(k)$ and $\delta_j(k) \in D_j(k)$, respectively:
$$C_{ij}^{\mathrm{dCPA}}(\delta_i(k), \delta_j(k)) = \sqrt{|x_i(k) - x_i^*|^2 + |y_i(k) - y_i^*|^2}$$
$$C_{ij}^{\mathrm{tCPA}}(\delta_i(k), \delta_j(k)) = \frac{C_{ij}^{\mathrm{dCPA}}(\delta_i(k), \delta_j(k))}{(u_i, v_i)}$$
$$C_{ij}^{\mathrm{distance}}(\delta_i(k), \delta_j(k)) = \sqrt{|x_j^* - x_i^*|^2 + |y_j^* - y_i^*|^2}$$
5:   **then** calculate for each ship i's distance and time to its nearest CPA during time slot, and the number of ships that ship i has potential collision risk with, when it takes the rudder angle $\delta_i(k) \in D_i(k)$:
$$C_i^{\mathrm{closest}} = \min_{j \in V} C_{ij}^{\mathrm{dCPA}}(\delta_i(k), \delta_j(k))$$
$$T_i^{\mathrm{closest}}(\delta_i(k)) = \min_{j \in V} C_{ij}^{\mathrm{tCPA}}(\delta_i(k), \delta_j(k))$$
$$N_i^{\mathrm{ship}}(\delta_i(k)) = N_i^{\mathrm{ship}}(\delta_i(k-1)) + 1$$
6: **return** $C_{ij}^{\mathrm{distance}}$ between any two ships $i, j \in V$ when they take each of the rudder angles in $D_i(k)$ and $D_j(k)$, and $C_i^{\mathrm{closest}}, T_i^{\mathrm{closest}}$, and N_i^{ship} of each ship $i \in V$ when it takes each rudder angle in $D_i(k)$

---

two ships are departing from each other and that no collision risk exist (line 3). Otherwise, it means ships i and j are getting closer to each other, and collision risk exists. Therefore, it is important to calculate the time and distance from ship i's current position to its closest point of approach with ship j, as well as the closest distance between them, according to their trajectories within this time slot. In other words, when ships i and j take rudder angles $\delta_i(k) \in D_i(k)$ and $\delta_j(k) \in D_j(k)$ during time slot k, calculate the values of C^{dCPA}, C^{tCPA} and $C_{ij}^{distance}$ (line 4). It is assumed that the speeds and courses of ships i and j unchanged in the calculation. After that, collision risks that may have been caused by each ship i's each candidate rudder angle $\delta_i(k) \in D_i(k)$ also need to be determined. Consequently, the shortest time and distance from ship i's current position to the nearest CPA with the other ships, and the number of ships that ship i has collision risks with are calculated (line 5). At the end of collision risk evaluation in time slot k, values to variables $C_{ij}^{distance}$ regarding each pair of rudder angles $(\delta_i(k), \delta_j(k))$, and values to variables $C_i^{closest}, T_i^{closest}$, and N_i^{ship} regarding each $\delta_i(k)$ are returned (line 6). These values will be exploited in Phase *III* to construct utility functions.

3.4   Phase *III*: Multi-ship Collision Avoidance Coordination

Phase *III* formulates a multi-ship collision avoidance coordination model based on distributed constraint optimization (DCOP). This paper adopts the DCOP formalism as defined in [10]. A DCOP is represented by a triple $\langle \mathcal{A}, \mathcal{COP}, \mathcal{R}^{ia} \rangle$, where:

- $\mathcal{A} = \{A_1, \ldots, A_M\}$ is a set of M agents;
- $\mathcal{COP} = \{COP_1, \ldots, COP_M\}$ is a set of disjoint, local Constraint Optimization Problems (COPs); COP_m is called the local sub-problem of agent A_m; COP_i is defined by a triple $\langle \mathcal{X}_m, \mathcal{D}_m, \mathcal{R}_i \rangle$, where $\mathcal{X}_m = \{X_{m1}, \ldots, X_{m|\mathcal{X}_m|}\}$ is a set of $|\mathcal{X}_m|$ variables that belong to A_m; $\mathcal{D}_m = \{d_{m1}, \ldots, d_{m|\mathcal{X}_m|}\}$ is a set of finite variable domains of the variables in $\mathcal{X}_m$; $\mathcal{R}_i = \{r_{m1}, \ldots, r_{m|\mathcal{R}_m|}\}$ is a set of $|\mathcal{R}_m|$ utility functions, where each utility function $r_{m|\mathcal{R}_m|}$ is with scope $\mathcal{X}_m$, $r_{m|\mathcal{R}_m|} : d_{m1} \times \cdots \times d_{m|\mathcal{X}_m|} \rightarrow \mathbb{R} \cup \{-\infty\}$. The utility functions are used to represent objectives, as well as both hard and soft constraints. For hard constraints, the value of the utility function is 0 if the constraint is satisfied; otherwise the value is $-\infty$. For soft constraints, for different combinations of the values for variables, different values will be assigned to the utility functions.
- $\mathcal{R}^{ia} = \{r_1^{ia}, \ldots, r_{|\mathcal{R}^{ia}|}^{ia}\}$ is a set of so-called inter-agent utility functions defined over variables of multiple agents. Each $r_l^{ia} : scope(r_l^{ia}) \rightarrow \mathbb{R}$ expresses the utility for a joint decision obtained by the agents that have variables involved in r_l^{ia}. The agents that have variables can decide on the values of these variables involved in r_l^{ia} and are called "responsible" for r_l^{ia}. Inter-agent utility functions are considered known to all agents involved, i.e., those agents of which the local variables are part of the inter-agent utility function.

The objective of the agents solving a DCOP is to find the assignment to all variables such that the sum of values of all utility functions (representing the objectives, hard and soft constraints) are minimized. So, the agents determine:

$$X^* = \arg\min \sum_{m=1}^{M} \left(\sum_{v=1}^{|\mathcal{R}_m|} r_{mv}(X_{m1}, \ldots, X_{i|\mathcal{X}_m|}) \right) + \sum_{l=1}^{|\mathcal{R}^{ia}|} r_l^{ia}$$

This paper considers ships as individual agents, with the aim to minimize the sum of values to utility functions that reflect the collision risks among the ship agents, when they choose different rudder angles. Therefore, each ship agent i owns rudder angle variable $\delta_i(k)$. The collision risks of each ship agent i and each two ship agents i and j are represented via two types of utility functions, including individual utility functions for each ship agent, and inter-agent utility functions for each two ship agents. These utility functions are constructed based on the values calculated in Phase II.

Utility Functions. When collision risk exists, ships change their rudder angles to avoid collisions. Each ship has many preferences regarding its rudder angle selection, these preferences are presented via utility functions $r_{i1}, r_{i2}, r_{i3}, r_{i4}$ and r_{ij}^{inter}, as concluded in Table 3.

Table 3. Utility functions and ships' preferences

Utility function	Relevant variables	Preference
r_{i1}	$\delta_i(k)$	Each ship prefers smaller rudder angle alterations
r_{i2}	N_i^{ship}	Each ship does not like to change rudder angles frequently
r_{i3}	C_i^{closest}	Each ship prefers to keep larger distances with other ships
r_{i4}	T_i^{closest}	Each ship prefers to switch back to its original course as soon as possible
r_{ij}^{inter}	C_{ij}^{distance}	Minimum safe distance must be kept between any two ships

Greater changes in a ship's rudder angles leads to large deviations from its current course, which also implies that it takes more efforts time-wise for the ship to switch back to its initial course. Therefore, smaller rudder angle changes are more preferable for ships. This is considered via utility function r_{i1}:

$$r_{i1} = \begin{cases} U_i^1(k) : \text{if} \quad \delta_i(k) = \beta \\ 0 : \quad \text{otherwise} \end{cases} \qquad \forall i \in V, \forall d_{ik} \in D_i(k) \qquad (2)$$

The values of $U_i^1(k) = 0$ when ship i takes rudder angle $0°$, as it does not need to change its course; as smaller rudder angle alteration is preferable, $U_i^1(k) = 1$ when ship i takes rudder angle $\pm 10°$, $U_i^1(k) = 2$ when ship i takes rudder angle $\pm 20°$, and that $U_i^1(k) = 3$ when ship i takes rudder angle $\pm 30°$.

If one ship has collision risks with several ships, and that these ships do not have collision risks with one another, it will be more efficient if this ship adjusts its own rudder angle accordingly so that the other ships can keep their original courses and do not need to take actions in avoiding collisions. Therefore, rudder angles that cause collision risk with less number of ships are more preferable. In other words, for ship i, rudder angles with smaller N_i^{ship} values will be given priorities. This is considered via utility function r_{i2}:

$$r_{i2} = \begin{cases} U_i^2(k) : \text{if} \quad \delta_i(k) = \beta & \forall i \in V, \forall \beta \in D_{ik} \\ +\infty : \quad \text{otherwise} \end{cases} \tag{3}$$

The values of $U_i^2(k)$ is determined by arranging the values of N_i^{ship} of each rudder angle $\delta_i(k)$ in ascending order: $U_i^2(k) = 1$ if $\delta_i(k)$ causes to the smallest N_i^{ship}; $U_i^2(k) = 2$ if $\delta_i(k)$ causes to the 2nd-smallest N_i^{ship}; $U_i^2(k) = 3$ if $\delta_i(k)$ causes to the 3rd-smallest N_i^{ship}, and so forth.

To guarantee safety, ships prefer to keep larger distances with the other ships. Therefore, rudder angles that ensure larger inter-ship distances are more preferable. In other words, for ship i, rudder angles with larger C_i^{closest} values will be given high priorities. This is represented via utility function r_{i3}:

$$r_{i3} = \begin{cases} U_i^3(k) : \text{if} \quad \delta_i(k) = \beta & \forall i \in V, \forall \beta \in D_i(k) \\ +\infty : \quad \text{otherwise} \end{cases} \tag{4}$$

The values of $U_i^3(k)$ is determined by arranging the values of C_i^{closest} of each rudder angle $\delta_i(k)$ in descending order: $U_i^3(k) = 1$ if $\delta_i(k)$ leads to the largest C_i^{closest}; $U_i^3(k) = 2$ if $\delta_i(k)$ leads to the 2nd-largest C_i^{closest}, and so forth.

In addition, ships also prefer to encounter the other ships with collision risks in a shorter time, as it means ships can go across the other ships and switch back to its original course in a shorter time. In other words, for ship i, rudder angles with smaller T_i^{closest} values will be given high priorities. This is represented via utility function r_{i4}:

$$r_{i4} = \begin{cases} U_i^4(k) : \text{if} \quad \delta_i(k) = \beta & \forall i \in V, \forall \beta \in D_i(k) \\ +\infty : \quad \text{otherwise} \end{cases} \tag{5}$$

The values of $U_i^4(k)$ is determined by arranging the values of T_i^{closest} of each rudder angle $\delta_i(k)$ in ascending order: $U_i^4(k) = 1$ if $\delta_i(k)$ leads to the smallest T_i^{closest}; $U_i^4(k) = 2$ if $\delta_i(k)$ leads to the 2nd-smallest T_i^{closest}, and so forth.

Besides considering each ship's preferences regarding rudder angle alterations, it is also important to take into account the collision risks between any two ships. Therefore, inter-agent utility function r_{ij}^{inter} is introduced.

$$r_{ij}^{\text{inter}} = \begin{cases} U_{ij}^{\text{inter}-\text{ship}}(k)\colon \text{if } \delta_i(k) = \alpha, \delta_j(k) = \beta \,\forall i, j \in V, \forall \alpha \in D_i(k), \forall \beta \in D_j(k), \\ +\infty : \text{otherwise} \end{cases}$$

$$(6)$$

The values of $U_{ij}^{\text{inter}-\text{ship}}(k)$ are determined in the following ways:

- Firstly, if the distance between ships i and j with rudder angles $\delta_i(k)$ and $\delta_j(k)$ during time slot k is smaller than the minimum safe distance C^{MSD}, it is unsafe for these two ships to take these rudder angles. Therefore, the values of $U_{ij}^{\text{inter}-\text{ship}}(k) = +\infty$.
- Secondly, if the distance between ships i and j decreases during time slot k, it means they are approaching each other. It is reasonable to claim that rudder angles $\delta_i(k)$ and $\delta_j(k)$ that lead to larger distances between them would be preferable. Therefore, the values of $C_{ij}^{\text{distance}}(\delta_i(k), \delta_j(k))$ are arranged in descending order: $U_{ij}^{\text{inter}-\text{ship}}(k) = 1$ if the corresponding rudder angles lead to the largest C_{ij}^{distance} value; $U_{ij}^{\text{inter}-\text{ship}}(k) = 2$ if the corresponding rudder angles lead to the 2nd-largest C_{ij}^{distance} value, and so forth.

Optimization Objective. This paper aims to find the most efficient anti-collision operations for multiple encountering ships, which concerns determining the optimal rudder angle each ship chooses at each time slot k. For the formulated DCOP at time slot k, the optimization objective is to minimize the sum of utility values in utility functions $r_{i1}, r_{i2}, r_{i3}, r_{i4}$, and inter-agent utility function r_{ij}^{inter}, which is defined as follows:

$$\min \left(\omega_0 \sum_{i \in V} \sum_{j \in V, j \neq i} r_{ij}^{\text{inter}} + \omega_1 \sum_{i \in V} r_{i1} + \omega_2 \sum_{i \in V} r_{i2} + \omega_3 \sum_{i \in V} r_{i3} + \omega_4 \sum_{i \in V} r_{i4} \right).$$

Parameter ω_0, ω_1, ω_2, ω_3, and ω_4 are the weights assigned to each utility function. By introducing the weights we can find the balance of whether to satisfying ship's preferences regarding smaller rudder angle alteration (ω_0), or minimize the number of ships with collision risks (ω_1), or maximize the distances between any two ships (ω_2), or maximize the sum of time each ship spends in sailing to its nearest CPA (ω_3), while making sure that minimum safe distance is kept between any two ships (ω_4).

Solution Methods. Once a problem has been modeled as a DCOP, a solution method is required to solve it. DCOP solution algorithms can be categorized as complete and incomplete algorithms. Complete algorithms are guaranteed to find optimal solutions, if they exist. Complete algorithms typically do an exhaustive search over the problem space, while incomplete algorithms usually use local search methods to find locally optimal solutions. This paper incorporates the traditional complete DCOP algorithm, SynchBB (Synchronous Branch and Bound) [2] to solve the formulated multi-ship collision avoidance coordination problem. SynchBB is a straightforward distributed adaptation of the centralized branch-and-bound mechanism, which guides the search through a heuristic applied over the optimization function.

Algorithm 3 shows the solution process of SynchBB. Firstly, all variables and agents are arranged along a total order with the priority $\delta_i \succ \cdots \succ \delta_j$. The message passing starts with the highest priority variable δ_i, the corresponding agent of which sends a so-called single Current Partial Assignment (CPA) message that includes the value assignment to δ_i and the current associated utility value to the next agent(line 6–8). Each agent that receives the CPA extends it by including a value assignment to its own variable, as well as the utility value it has because of the utility function is has with other variable assignments

---

Algorithm 3. Basic steps of multi-ship coordination based on SynchBB

Require: a fixed, known, linear ordering of variables $\delta_1 \succ \delta_m \succ \cdots \succ x^n$

1: //Join all utility functions involving δ_m and only previous variables
2: $u^m(\delta_m, \cdot) \leftarrow \wedge_{u_{mi} \in \{u' \in \mathcal{U} | \delta_m \in scope(u') \wedge \forall y \in scope(u'), y \succ \delta_m\}} u(\delta_m, \cdot)$
3: $D'_m \leftarrow D_m$ // a copy of δ_m's domain D_m
4: $\bar{u}^m \leftarrow 0$ // utility value of the CPA up to and including δ_{m-1}
5: $u^* \leftarrow \infty$ // utility value of the best solution found so far
6: **if** $m = 1$ **then** $\delta_1 \leftarrow$ first $\delta_1^* \in D'_1$ such that $u^1(\delta_1^*) < \infty$
7:     **if** there exists such a δ_1^* **then** send message (CPA, (δ_1^*), $u^1(\delta_1^*)$) to δ_2
8:     **else** broadcast messages INFEASIBLE
9: **for** each received message M **do**
10:     **if** $M = (UB, (\delta_1^*, \ldots, \delta_n^*), u)$ **then**
11:         $u^* \leftarrow u$ **and** record $(\delta_1^*, \ldots, \delta_n^*, u)$ as the best solution found so far
12:         **continue**
13:     **if** $M = (CPA, (\delta_1^*, \ldots, \delta_{m-1}^*), u)$ **then**
14:         $D'_m \leftarrow D_m$ **and** $(\delta_1, \ldots, \delta_{m-1}, u) \leftarrow (\delta_1^*, \ldots, \delta_{m-1}^*)$ **and** $\bar{u}^m \leftarrow u$
15:     **else if** M=(BACK) **then** $D'_m \leftarrow D'_m \setminus \{\delta_m^*\}$
16: // Look for a (better) value for δ_m
17: $\delta_m \leftarrow$ first $\delta_m^* \in D'_m$ such that $\bar{u}_i^m + u_i^m(\delta_m^*, \cdot) > u^*$
18:     **if** there exists such a δ_m^* **then**
19:         **if** $m = n$ **then** $\delta_n \leftarrow \delta_n^* = \arg\min_{\delta_{n'} \in D'_n} \{u^n(\delta_{n'}, \cdot)\}$, $u^* \leftarrow \bar{u}^n + u_I^n(\delta_n^*, \cdot)$
20:         Record $(\delta_1^*, \ldots, \delta_n^*)$ as the best solution found so far
21:         Broadcast message $M = (UB, (\delta_1^*, \ldots, \delta_n^*), u^*)$
22:             **if** $D'_n = 0$ **then** broadcast message TERMINATE
23:             **else** send message BACK to δ_{n-1}
24:         **else** send message (CPA, $(\delta_1^*, \ldots, \delta_n^*)$, $\bar{u}^m + u^m(\delta_m^*, \cdot)$) to δ_{m+1}
25:     **else**
26:         **if** $m = 1$ **then** broadcast message TERMINATE
27:         **else** send message BACK to δ_{m-1}

---

appearing in the received CPA (line 13–14). Whenever a CPA reaches a new full assignment at the last agent (line 19), the accumulated utility value of the CPA is the utility value of the full variable assignment. This utility value will then be broadcast to all other agents, and each agent can use this utility value as an upper bound (UB). When the utility value of a new CPA exceeds utility value of the currently known upper bound (line 19–21), it will be broadcast to all agents as the new upper bound (line 10–12). Recursively, each agent holding CPA then checks whether its CPA accumulated utility value is larger than the upper bound. If this is true (line 17), it means the old previous variable assignment is sub-optimal, and the agent will assigns the next value in the domain of its variable instead of the current value, and send it again to the next agent (line 24) and checks again. An agent encountering an empty domain of values (when all values have been used) erases its assignment (and its utility value) and sends the CPA back to the previous agent (line 22–23). When the domain of the first agent is exhausted, the last discovered full assignment is reported as the optimal solution (line 26–27).

At the end of time slot k, with the optimal rudder angle δ_i of each ship i obtained from Phase *III*, ships' coordinates are updated based on the ship dynamic equations, which will be considered as initial ship states in the multi-ship collision avoidance problem at time slot $k + 1$. The above-mentioned three steps are carried out in an iterative way, until a series of rudder angle alterations at different time slots have been found for each ship, to make sure that all the encountering ships can pass one another safely and efficiently.

4   Preliminary Results

Simulation experiments are carried out to assess and analyze the effectiveness of the proposed method. This section first presents a typical example of 7-ships encounter situation and solved results of different weights assigned to utility functions r_{i1}, r_{i2}, r_{i3} and r_{i4}, and gives an overall comparison of different anti-collision operations.

4.1   Experimental Settings

Our tests are performed on an Intel Core i7-7500 CPU with 8GB RAM running Windows 10. The proposed method is implemented in MATLAB. The SynchBB algorithm is implemented in the latest version of the FRODO2 toolbox [7]. This paper selects KVLCC2 tanker as a sample ship and adopts KLVCC2 ship parameters [1] in the ship dynamic equations of Phase *I*. The trajectory prediction time $T^{\text{prediction}} = 300\,\text{s}$. The length of each discrete time slot, $N^{\text{unit}} = 60\,\text{s}$. We set up 10 scenarios in which 7 homogeneous ships are encountering, with different courses and coordinates. The minimum safe distance that each two ships should keep is set as 150 m.

Key Performance Indicators (KPIs) 1,2 and 3 are used to evaluate the efficiency of the anti-collision operations generated by the proposed method.

- KPI 1: completion time of collision avoidance per ship, which is defined as the time when all collision risks of a ship have been eliminated, in other words, the time after which a ship can switch back to its original course.
- KPI 2: distance traveled for completing collision avoidance, which is defined as the distance a ship needs to travel to avoid potential collisions with the other ships.
- KPI 3: number of rudder angle alterations, which reflects how many times a ship has to changes it rudder angle to avoid potential collisions with the other ships.

KPIs 1 and 2 reflect the costs for each ship to avoid collisions time-wise, and KPI 3 reflect the costs from a ship maneuvering perspective. As the values of KPIs 1–3 for each ship may be different, this paper uses averages values of all the encountering ships.

To investigate the impacts of ships' different preferences, represented by r_{I1}, r_{i2}, r_{i3} and r_{i4}, on the performance of their anti-collision operations, in each scenario, the proposed optimization procedure is carried out with different optimization objectives, i.e., considering different combinations of utility functions as the optimization function. Five types of optimization objectives are considered in each scenario, including:

1. $Obj_1 = \min \left(\sum_{i \in V} \sum_{j \in V, j \neq i} r_{ij}^{\text{inter}} + \sum_{i \in V} r_{i1} \right)$;
2. $Obj_2 = \min \left(\sum_{i \in V} \sum_{j \in V, j \neq i} r_{ij}^{\text{inter}} + \sum_{i \in V} r_{i2} \right)$;
3. $Obj_3 = \min \left(\sum_{i \in V} \sum_{j \in V, j \neq i} r_{ij}^{\text{inter}} + \sum_{i \in V} r_{i3} \right)$;
4. $Obj_4 = \min \left(\sum_{i \in V} \sum_{j \in V, j \neq i} r_{ij}^{\text{inter}} + \sum_{i \in V} r_{i4} \right)$;
5. $Obj_5 = \min \left(\sum_{i \in V} \sum_{j \in V, j \neq i} r_{ij}^{\text{inter}} + \sum_{i \in V} r_{i1} + \sum_{i \in V} r_{i2} + \sum_{i \in V} r_{i3} + \sum_{i \in V} r_{i4} \right)$.

In these optimization objectives, inter-agent utility function r_{ij}^{inter} is always included. This is because it ensures that the ships will keep the minimum safe distance, otherwise the corresponding rudder angles may lead ship collisions. It is a hard constraint that must be respected in the optimization procedure.

4.2 An Example of Simulated Ships' Trajectories to Avoid Collisions

Figure 2 presents an typical example of multi-ships encountering situation, which gives the ships' predicted trajectories when they keep their original courses unchanged. The initial coordinates of ships are marked with ∗. It can be seen that collision risks exist among Ships 1, 2 and 4, between Ships 3 and 7, as well as between Ships 5 and 6.

Figures 3, 4, 5, 6 and 7 are the simulated trajectories of ships with the optimized rudder angle alterations considering Obj_1, Obj_2, Obj_3, Obj_4 and Obj_5 as the optimization objectives in Phase *III*, respectively. The initial coordinates of each ship is marked with ∗, and the point at which each ship can switch back to its original course is marked with ×. It is noticed that the lengths of some ships' trajectories are not equal, this is because the collision risks of the ships are eliminated at different times. Figure 8 shows the distances between the involved

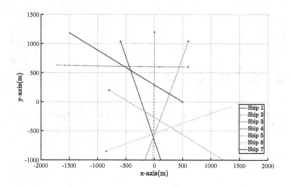

Fig. 2. Simulated ships' trajectories with original courses.

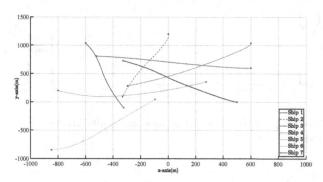

Fig. 3. Simulated ships' trajectories of rudder angle alterations solved with utility functions r_{i1} and r_{ij}^{inter}.

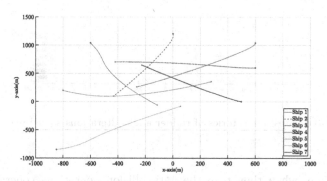

Fig. 4. Simulated ships' trajectories of rudder angle alterations solved with utility functions r_{i2} and r_{ij}^{inter}.

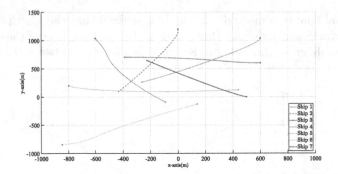

Fig. 5. Simulated ships' trajectories of rudder angle alterations solved with utility functions r_{i3} and r_{ij}^{inter}.

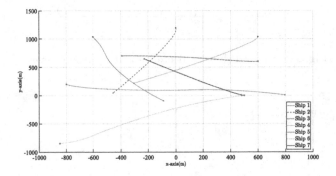

Fig. 6. Simulated ships' trajectories of rudder angle alterations solved with utility functions r_{i4} and r_{ij}^{inter}.

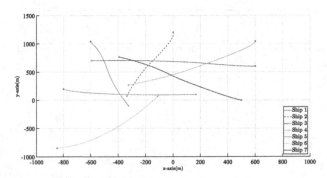

Fig. 7. Simulated ships' trajectories of rudder angle alterations solved with all utility functions.

ships over time, when they take the anti-collision operations generated with Obj_5. In these figures, all the ships can keep minimum safe distance with the other ships and are able to avoid the potential collisions with different rudder

Table 4. KPI 1: The time that each ship takes to avoid potential collisions (/minutes)

	Ship 1	Ship 2	Ship 3	Ship 4	Ship 5	Ship 6	Ship 7
Obj_1	13.20	13.20	10.58	15.71	12.51	12.51	15.71
Obj_2	9.98	9.98	11.24	8.87	12.29	13.80	13.80
Obj_3	9.98	9.98	11.98	16.00	11.98	14.49	16.00
Obj_4	9.98	9.98	14.70	15.45	15.80	15.80	15.45
Obj_5	13.31	13.31	12.49	15.71	12.49	10.94	15.71

Table 5. KPI 2: The distance that each ship travels to avoid collisions (/km)

	Ship 1	Ship 2	Ship 3	Ship 4	Ship 5	Ship 6	Ship 7
Obj_1	3.037	3.015	2.157	3.821	2.515	2.756	3.764
Obj_2	1.933	1.952	2.374	1.579	2.575	3.184	3.148
Obj_3	1.933	1.952	2.622	3.921	2.590	3.433	3.878
Obj_4	1.933	1.952	3.388	3.792	3.851	3.828	3.574
Obj_5	3.076	3.053	2.795	3.821	2.780	2.201	3.764

angle alterations. According to the experimental results, KPIs 1, 2 and 3 regarding each optimization objective are given in Tables 4, 5 and 6.

It can be seen from Table 4 that: Ships 1 and 2 can switch back to its original course in a shorter time with Obj_2, Obj_3, and Obj_4; Ships 3 can switch back to its original course in a shorter time with Obj_1; Ship 4 can switch back to its original course in a shorter time with Obj_2; Ship 5 can switch back to its original course in a shorter time with Obj_3; Ship 6 can switch back to its original course in a shorter time with Obj_5; Ship 7 can switch back to its original course in a shorter time with Obj_2.

In Table 5, Ships 1 and 2 can travel shorter distances to avoid collisions with Obj_2, Obj_3, and Obj_4; Ships 3 can stravel shorter distances to avoid collisions with Obj_1; Ship 4 can travel shorter distances to avoid collisions with Obj_2; Ship 5 can travel shorter distances to avoid collisions with Obj_1; Ship 6 can travel shorter distances to avoid collisions with Obj_5; Ship 7 can travel shorter distances to avoid collisions with Obj_2.

It can be found in Table 6 how many times each ship needs to change its rudder angle at least, with different optimization objectives: Ship 1 is able to avoid collisions with fewest rudder angle alterations with Obj_1, Obj_2, Obj_3, and Obj_4; Ship 2 is able to avoid collisions with fewest rudder angle alterations with Obj_2, Obj_3, and Obj_4; Ship 3 is able to avoid collisions with fewest rudder angle alterations with Obj_2, Obj_3, and Obj_5; Ship 4 is able to avoid collisions with fewest rudder angle alterations with Obj_4; Ship 5 is able to avoid collisions with fewest rudder angle alterations with Obj_4 and Obj_5; Ship 6 is able to avoid collisions with fewest rudder angle alterations with Obj_1, Obj_3, Obj_4 and Obj_5; Ship 7 is able to avoid collisions with fewest rudder angle alterations with Obj_3.

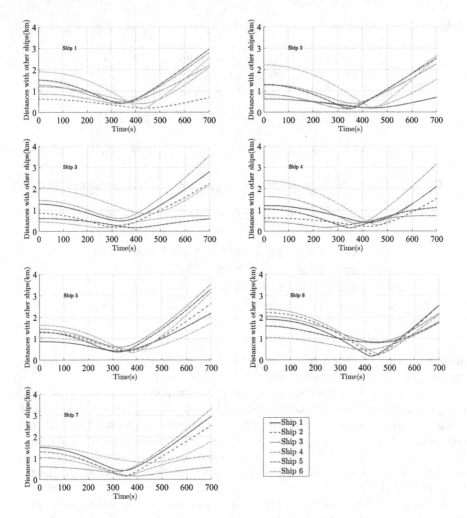

Fig. 8. Relative distances between different ships over time.

Table 6. KPI 3: Number of rudder angle alterations to avoid collisions

	Ship 1	Ship 2	Ship 3	Ship 4	Ship 5	Ship 6	Ship 7
Obj_1	2	2	3	2	3	3	3
Obj_2	2	1	2	2	5	5	3
Obj_3	2	1	2	2	3	3	2
Obj_4	2	1	4	1	2	3	3
Obj_5	3	2	2	2	2	3	3

4.3 Overall Comparison of Different Anti-collision Operations

Tables 4, 5 and 6 present the experimental results of a typical example, in which the most efficient solution for each ship is generated with different optimization objectives. To evaluate the overall performance of anti-collision operations, more experiments with different scenarios are required.

Table 7 concludes the performance of the generated anti-collision operations considering different optimization objectives in the 10 scenarios. The values of Relative KPI 1[1], Relatvie KPI 2[2] and Relative KPI 3[3] are the ratio of KPI values between each objective function Obj_i and the benchmark objective function Obj_5 in the 10 scenarios, in average. Optimization Obj_5 is used as the benchmark because it includes all the utility functions.

Table 7. Performance comparison of the generated anti-collision operations considering different optimization objectives

	Obj_1	Obj_2	Obj_3	Obj_4	Obj_5
Relative KPI 1	100.63%	98.01%	95.26%	92.05%	100%
Relative KPI 2	101.61%	102.70%	96.48%	88.56%	100%
Relative KPI 3	108.26%	116.86%	115.98%	115.49%	100%

It can be seen from Table 7 that ships are able to avoid potential collisions with fewest rudder angle alterations with Obj_5. While ships are able to avoid collisions in a a shorter time and travels shorter distances with Obj_4, they need to change their rudder angles more frequently. In general, comparing with other optimization objectives, Obj_5 leads to relatively better anti-collision plans, as it performs better regarding all KPIs.

5 Conclusions and Future Work

This paper proposes a new method for solving the multi-ship collision avoidance problem, which considers both the ship dynamics and the inter-related characteristic of the anti-collision decision making of multiple ships. It could assist ships in determining the rudder angle alterations to prevent collisions. This could increase the safety and reliability of a ship's automated navigation, reduce the psychological and physical burden of ship operators, and reduce the occurrence of ship collisions.

To enhance the applicability of the proposed method, further research is required. Firstly, the rudder angle alteration is only optimal in each time slot.

---

[1] Relative KPI 1 = $\dfrac{\text{Values of KPI 1 with } Obj_i \text{ as the optimization objective}}{\text{Values of KPI 1 with } Obj_5 \text{ as the optimization objective}}$.

[2] Relative KPI 2 = $\dfrac{\text{Values of KPI 2 with } Obj_i \text{ as the optimization objective}}{\text{Values of KPI 2 with } Obj_5 \text{ as the optimization objective}}$.

[3] Relative KPI 3 = $\dfrac{\text{Values of KPI 3 with } Obj_i \text{ as the optimization objective}}{\text{Values of KPI 3 with } Obj_5 \text{ as the optimization objective}}$.

It is possible that the solution is not optimal during the whole time horizon. Therefore, further research will be carried out. Secondly, this paper adopts a DCOP method to solve the multi-ship collision avoidance coordination problem, it would be also interesting to compare the performance of this method with other promising methods in the future. Last but not least, real-world data and further simulation experiments are required, in order to enhance the applicability of the proposed method, and to investigate the impacts of ships' preferences on the efficiency of their anti-collision decisions in a systematic way.

References

1. MOERI Tanker KVLCC2: Geometry and test conditions. http://www.simman2008.dk/KVLCC/KVLCC2/tanker2.html. Accessed 1 June 2018
2. Hirayama, K., Yokoo, M.: Distributed partial constraint satisfaction problem. In: Smolka, G. (ed.) CP 1997. LNCS, vol. 1330, pp. 222–236. Springer, Heidelberg (1997). https://doi.org/10.1007/BFb0017442
3. Hornauer, S., Hahn, A., Blaich, M., Reuter, J.: Trajectory planning with negotiation for maritime collision avoidance. TransNav Int. J. Mar. Navig. Saf. Sea Transp. **9**(3), 335–341 (2015)
4. Johansen, T.A., Perez, T., Cristofaro, A.: Ship collision avoidance and colregs compliance using simulation-based control behavior selection with predictive hazard assessment. IEEE Trans. Intell. Transp. Syst. **PP**(99), 1–16 (2016)
5. Kim, D.G., Hirayama, K., Okimoto, T.: Ship collision avoidance by distributed tabu search. Transnav Int. J. Mar. Navig. Saf. Sea Transp. **9**(1), 23–29 (2015)
6. Kim, D., Hirayama, K., Okimoto, T.: Distributed stochastic search algorithm for multi-ship encounter situations. J. Navig. **70**(4), 1–20 (2017)
7. Léauté, T., Ottens, B., Szymanek, R.: FRODO 2.0: an open-source framework for distributed constraint optimization. In: Proceedings of the IJCAI 2009 Distributed Constraint Reasoning Workshop (DCR 2009), Pasadena, California, USA, 13 July 2009, pp. 160–164 (2009). https://frodo-ai.tech
8. Liu, J., Hekkenberg, R., Quadvlieg, F., Hopman, H., Zhao, B.: An integrated empirical manoeuvring model for inland vessels. Ocean Eng. **137**, 287–308 (2017)
9. Perera, L.P., Ferrari, V., Santos, F.P., Hinostroza, M.A., Soares, C.G.: Experimental evaluations on ship autonomous navigation and collision avoidance by intelligent guidance. IEEE J. Oceanic Eng. **40**(2), 374–387 (2015)
10. Petcu, A.: A class of algorithms for distributed constraint optimization. Ph.D. thesis, École Polytechnique Fédérale de Lausanne, Lausanne, Switzerland (2009)
11. Tam, C.K., Bucknall, R.: Cooperative path planning algorithm for marine surface vessels. Ocean Eng. **57**(2), 25–33 (2013)
12. Tam, C.K., Bucknall, R., Greig, A.: Review of collision avoidance and path planning methods for ships in close range encounters. J. Navig. **62**(3), 455–476 (2009)
13. Wang, X., Liu, Z., Cai, Y.: The ship maneuverability based collision avoidance dynamic support system in close-quarters situation. Ocean Eng. **146**, 486–497 (2017)
14. Waslander, S.L.: Multi-agent systems design for aerospace applications. Ph.D. thesis, Stanford University (2007)

Stimulating Inland Waterway Transport Between Seaports and the Hinterland from a Coordination Perspective

Shijie Li[1](✉), Rudy R. Negenborn[2], and Jialun Liu[3](✉)

[1] School of Logistics Engineering, Wuhan University of Technology,
No. 1178 Heping Road, Wuhan, People's Republic of China
lishijie@whut.edu.cn
[2] Department of Maritime and Transport Technology,
Delft University of Technology, Mekelweg 2, 2628 CD Delft, The Netherlands
r.r.negenborn@tudelft.nl
[3] National Engineering Research Center for Water Transport Safety
(Ministry of Science and Technology), Wuhan, People's Republic of China
jialunliu@whut.edu.cn

Abstract. With the trend towards less-polluting and sustainable transport solutions, the European Commission aims to strengthen the competitive position of inland waterway transport, and to facilitate its integration into synchromodal logistic chains. To stimulate inland waterway transport, it is essential to ensure smooth containers transshipments from seaports to hinterland and vice versa. Currently, inland vessels usually spend unnecessary long times in the port area due to insufficient terminal and quay planning with respect to the sailing schedules of the vessels. Coordination among multiple vessel operators and multiple terminal operators is required in order to improve the efficiency and reliability of inland waterway transport within the port. For this, four recently proposed classes of coordination strategies from our earlier work are reviewed. Two levels of cooperativeness, including partially-cooperative and fully-cooperative, as well as two types of interaction, including single-level and multi-levels are considered. The proposed coordination strategies are compared and evaluated from a methodological perspective and from an information needs perspective. Our results provide insights for vessel and terminal operators in the ways in which they can cooperate with each other: vessel operators can decide to what extent they would like to coordinate their actions based on the information requirements of each coordination strategy; terminal operators can estimate information that should be made available during different coordination phases. Moreover, our results also provide insights for policy makers or practitioners to determine the most suitable coordination strategy under different circumstances.

© Springer Nature Switzerland AG 2018
R. Cerulli et al. (Eds.): ICCL 2018, LNCS 11184, pp. 67–85, 2018.
https://doi.org/10.1007/978-3-030-00898-7_5

1   Introduction

For centuries, transport systems have been developed for moving cargo and passengers from one location to another. Transport systems used to be based on roads in the beginning. Later on, alternative modes started being developed and used, including transport over water, over rail, and through the air [12]. Compared to other transport modes, transport over water ensures a higher level of safety, less CO_2 emission per ton, and the capability of handling large volumes of cargo without congestion [4]. With the trend towards less-polluting and sustainable transport solutions, the European Commission aims to strengthen the competitive position of transport over water, especially inland waterway transport, and to facilitate its integration into synchromodal logistic chains [3,4]. Using the potential of inland waterway transport could significantly contribute to achieve the "EU2020" Strategy and the EU transport policy targets of the European Commission [1]. In addition, to alleviate the congestion on roads and railways, as well as reduce pollutant emissions, the Dutch government also aims for an increase in the proportion of transport over water [6].

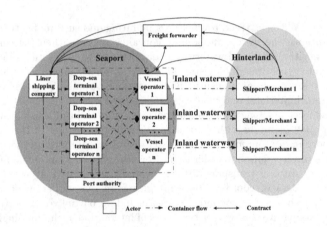

Fig. 1. Actors and correlations in hinterland transport chain and seaport (adapted from [5]).

Figure 1 describes the actors and correlations between the hinterland transport chain and seaport. The relations of these actors mainly depends on the contracts they have with one another. As can be seen, liner shipping companies always have contractual relations with terminals. They always make agreements about the transshipment of containers from a sea-going vessel to a subsequent hinterland transport modality (truck, train or inland container vessel) and vice versa. The vessel operators have contractual relations with the carriers (in carrier haulage) or the shippers/merchants (in merchant haulage). Meanwhile, no contractual relation exists between terminal operators and inland container vessel operators, or among multiple container terminals, or among multiple vessel operators. This implies that these actors cannot charge each other if these agreements are not carried out satisfactorily, which affects the efficiency of inland vessel operations in the port area. It is commonly recognized in literature that the inland vessels spend an unnecessarily long time in the port, and that the planning of their schedules at terminals is insufficient [2,5,7,13]. This has a negative influence on the total cost of inland vessel services and undermines the competitiveness of inland waterway transport [7].

To stimulate the use of inland waterway transport, it is important to ensure that the containers are transported from the seaport to shippers in the hinterland efficiently. More specifically, to significantly reduce waiting time and turnaround times in the seaports and enable a higher capacity utilization of inland vessels. For this, four classes of coordination strategies are proposed based on our earlier work in [8,9,11]. These strategies are compared and evaluated from a methodological perspective and an information needs perspective, which gives insights for vessel operators and terminal operators on the ways in which they can cooperate with each other, as well as on the required information and communication schemes to implement these coordination strategies.

The outline of this paper is as follows. Section 2 gives the problem description. Section 3 presents four classes of coordination strategies and discusses the models and solution methods to implement these strategies. In Sect. 4, a comparative study of the four coordination strategies is given, from a methodological perspective and an information needs perspective. Section 5 completes the paper with conclusions and future perspectives.

2   Problem Description

Every time an inland container vessel enters the port, it calls at many different terminals spread over the port area. Since many inland container vessels call at the same terminal, congestion and waiting times are inevitable [7]. In addition, when a delay at a terminal happens, the vessel's agreed time window at the next terminal will be missed. Vessel operators have to anticipate such events by inserting large margins when planning their visits to terminals, otherwise the reliability of the transport service might be undermined [7]. At a first glance, it seems that the nature of inland vessels coordination problem is similar to many other traditional planning problems, such as vehicle routing problems (VRP) and ship routing and scheduling problems (SRSP). However, it has unique characteristics. The most important difference between our coordination problem and other traditional planning problems lies in the way the waiting time of an inland vessel at an (un)loading location (terminal) is considered.

1. In inland vessels coordination, the fact that each container terminal can only serve a limited number of vessels causes the vessels with later arrival times to wait or go to other terminals. Meanwhile, VRP does not consider the capacity of (un)loading locations for handling vehicles. Although the capacity constraint of terminals also exists in SRSP, it is unlikely to cause the waiting

of sea-going vessels. This is because sea-going vessels have made appointments with terminals for (un)loading operations a long time in advance and because they always have priorities over inland vessels at terminals. In SRSP, the waiting time caused by other vessels is not common and therefore is often not considered in literature.

2. In inland vessels coordination, if a vessel operator decides to wait until being handled, the loading and unloading operations of the vessel only start after the vessels that arrived earlier at the terminal have been handled. This implies that inland vessels are allowed to miss the scheduled time window at terminals, as long as they are willing to wait. This also implies that the waiting time may be caused by the other vessels. Meanwhile, for the vehicles in VRP and SRSP, they must start within a given time window, otherwise the generated route is no longer feasible.

3. The vehicles in VRP and the sea-going vessels in SRSP are actually cooperating with each other to minimize the total costs, as they are all planned and scheduled within one liner organization. Meanwhile, the inland vessels are owned by different parties, and they are usually in competitive relations and therefore are not necessarily willing to cooperate with each other.

Due to these differences, the existing methods for the traditional planning problems are therefore not directly applicable in our problem, and innovative coordination strategies need to be developed.

3   Proposed Coordination Strategies

This section identifies levels of cooperativeness and interactions in the multiple-vessel-multiple-terminal coordination problem, and presents the corresponding models and solution methods.

3.1   Levels of Cooperativeness and Interactions

The main physical elements involved in the inland vessel transport in the port area are multiple vessels and multiple terminals. Although one terminal operator can operate more than one terminal and one vessel operator can operate more than one vessel, without loss of generality, it is assumed that every terminal operator operates exactly one terminal and one vessel operator operates exactly one vessel. Vessel operators communicate with terminal operators only to make appointments for planning loading and unloading operations and do not communicate with other vessel operators.

In practice, inland vessel operators are in competitive positions, which means they are non-cooperative. Therefore, it is important to investigate in what way the inland vessel operators can be motivated to cooperate and share information with one another. For this, two levels of cooperativeness are considered, including partially-cooperative and fully-cooperative: partially-cooperative means that vessel operators only share limited information with other vessel operators; fully-cooperative means that vessel operators are willing to share all information.

Information exchange is the fundamental basis of coordination, and it is therefore critical to determine suitable schemes. The concept of a multi-agent system (MAS) is adopted [15], which considers an agent as a computer system that is capable of independent action on behalf of its user or owner, and considers a multi-agent system as consisting of a number of agents that interact with each other, typically via the exchange of messages. This paper assumes that for each physical element (a terminal or an inland vessel), there is a local agent that controls the physical element's operations and sends information to the other agents. Two types of interactions are considered: single-level interaction refers to a situation in which the information exchange among different agents occurs at same level, either in a distributed or centralized way; multi-level interactions refer to a situation in which the information exchange happens at two levels, in a multi-stage way.

Based on these categories, four classes of coordination strategies can be established, as shown in Figs. 2, 3, 4 and 5, respectively:

- Coordination strategy 1: partially-cooperative coordination with single-level interaction. Each agent shares limited information with the other agents and the coordination is carried out in a distributed way, with all agents in equal positions;
- Coordination strategy 2: fully-cooperative coordination with single-level interaction. All local agents send all necessary information to the overall coordinator, and the coordinator searches for globally optimal solutions;
- Coordination strategy 3: partially-cooperative coordination with multi-level interactions. Each agent firstly solves its local optimization problem for each physical element at Coordination Level 1, and shares limited information to the overall coordinator at Coordination Level 2, after which the coordinator searches for globally optimal solutions for all agents;
- Coordination strategy 4: fully-cooperative coordination with multi-level interactions. All local agents send their information to the overall coordinator at Coordination Level 1 to determine globally optimal solutions, the solutions will be sent to and evaluated by the local agents at Coordination Level 2, based on the information sent back from the agents at Level 2 and the termination criteria, the coordinator decides if the coordination continues.

Coordination Strategy 1 in Fig. 2 has the advantage that vessel agents only reveal information partly to the other agents, which ensures certain degrees of information privacy. Meanwhile, it also has the disadvantage that the overall coordination process could be slower than the centralized Coordination Strategy 2 as in Fig. 3, due to a considerable amount of information exchange between terminal agents and vessel agents caused by its distributed coordination scheme. With the increase of problem sizes, the information exchange in a distributed coordination scheme also increases substantially. This implies that for large-scale problems a distributed coordination strategy would cost longer time to find appropriate solutions. Therefore, Coordination Strategy 2 may get quicker solutions than Coordination Strategy 1, at the cost of information privacy. However, it is difficult for both Coordination Strategies 1 and 2 to get solutions

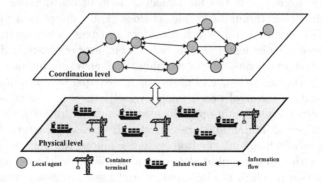

Fig. 2. Structure of coordination Strategy 1

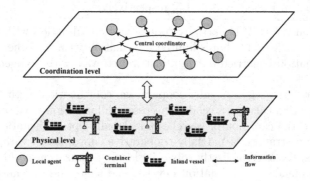

Fig. 3. Structure of coordination Strategy 2

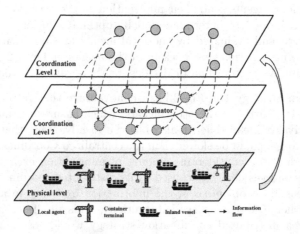

Fig. 4. Structure of coordination Strategy 3

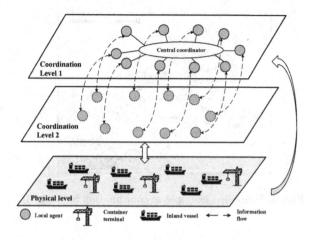

Fig. 5. Structure of coordination Strategy 4

for a large coordination problem with a reasonable amount of time [9,10], as all the constraints are considered in a large optimization problem in a single-level. Therefore, Coordination Strategy 3 in Fig. 4 and Coordination Strategy 4 in Fig. 5 are designed to solve the coordination problem in larger sizes, with a two-level structure in which approximate methods can be incorporated.

3.2   Models and Solution Methods

After establishing coordination strategies, models and the corresponding solution methods are required. This paper mainly uses constraint programming (CP) techniques instead of mathematical programming (MP) techniques for several reasons. Firstly, an MP model typically consists of several linear or non-linear equalities, while a coordination problem usually involves many equalities and logical conditions. In addition, CP has been rarely applied to this type of coordination problems, research on application of CP techniques for solving this would be a novel contribution to literature provide insights for those who have met similar coordination problems.

Partially-Cooperative Coordination with Single-Level Interaction. As the vessel operators and terminal operators have different preferences and interests, it is natural to model their interactions in a distributed way. Moreover, they are also conservative in information sharing. A distributed and exact method would be preferable for these operators.

The interactions among agents in Coordination Strategy 1 are modeled based on a distributed version of the constraint programming problem, namely a distributed constraint optimization problem (DCOP). In a DCOP, the knowledge of variables and constraints is distributed among several agents. These agents

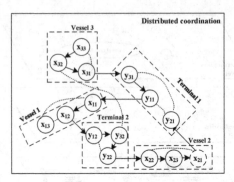

Fig. 6. Simplified example in coordination Strategy 1

jointly make decisions on values of variables so as to minimize the sum of constraint costs, or to maximize the sum of utility values [14]. A DCOP consists of a set of agents, variables and constraints that reflect the costs/utilities of assignments to variables. Control of values of variables in DCOPs is also distributed, with agents only able to assign values to variables that they own. Furthermore, agents are assumed to know only the constraints involving variables that they own.

Each vessel/terminal is considered as an individual agent that owns a set of variables, and each variable is exclusively controlled by the corresponding agent. Agents need to communicate with each other through message exchange to find optimal solutions. It is commonly assumed that agents can only communicate with agents that hold variables constrained with their own variables. In DCOP algorithms, a variable constitutes a variable node. These variable nodes are connected based on different structures, for example, a linear ordering of nodes, or a depth-first-search structure that connects all the nodes. Different DCOP algorithms define how the variables assignments and the corresponding utility values are passed from one node to another, as well as from one agent to other agents. Figure 6 gives an simplified example that includes three vessel agents and two terminal agents, and in which the variables are connected with a linear ordering. As can be seen, the message passing starts with variable node x_{33} in vessel agent 3, and then the information regarding its variable assignments and the associated utility values are passed to the next connected variable node x_{31}, until all the variable nodes have been reached. For details of different DCOP algorithm we refer the readers to our work in [9].

Therefore, the DCOP-based methods do not require a central controller to receive and send information from/to all the agents. This also means that a central coordinator is not required to coordinate the rotations of different vessel agents. As each agent only knows the variable assignments and utility values from the agents with whom it shares inter-agent utility functions, each agent does not know these types of information from the agents with whom it does not share any inter-agent utility function. For example, vessel agent 2 in Fig. 6

does not know the variable assignments of variable node x_{13} of vessel agent 1. Therefore, it guarantees to some extent the privacy of vessel or terminal agents, as each agent does not necessarily need to reveal all the information to all the other agents. This means that the vessel agents can be coordinated in a partially-cooperative way.

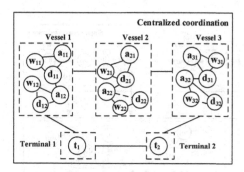

Fig. 7. Simplified example in coordination Strategy 2

Fully-Cooperative Coordination with Single-Level Interaction. In Coordination Strategy 2, all the vessel agents send information to a central coordinator to decide their schedules. Although these variables are owned by different agents, their values are determined by the central coordinator. All constraints are considered in a large constraint programming problem, in which commercial solvers are used by the central coordinator to search for solutions. Figure 7 gives a simplified example that includes three vessel agents and two terminal agents with a centralized scheme. Here, variables t_1 and t_2 are considered as known information that reflects the current status of terminals 1 and 2. Once the solutions have been found, the coordinator send the solutions of the variables for arrival times $a_{11}, a_{12}, a_{21}, a_{22}$ and a_{31}, a_{31}, departure times d_{11}, d_{12}, d_{21}, d_{22} and d_{31}, d_{31}, and the waiting times w_{11}, w_{12}, w_{21}, w_{22} and w_{31}, w_{31} back to vessels and terminals for implementation. Therefore, unlike distributed coordination, the information exchange in centralized coordination only happens twice, firstly, the vessel agents send information to the central coordinator, then the coordinator sends the determined solutions to the agents. Consequently, the communication cost of this coordination strategy is far less than in Coordination Strategy 1.

Partially-Cooperative Coordination with Multi-level Interactions. To consider the privacy issue and enlarge the problem size that can be solved simultaneously, Coordination Strategy 3 is proposed. Each vessel first decides on locally optimal solution in the first level and then shares part of the information with other vessels in the second level. Therefore, Coordination Level 1 includes multiple unconnected local problems of different vessels, each vessel has its local optimization problem based on mixed-integer programming, with the aim to minimize its round-trip time for loading and unloading the required number of containers at each terminal in the port. Constraints on vessel capacity, required number of containers to be transported, as well as time constraints for arrival and departure are considered at this level, while the constraints of inter-vessels relations are considered at Level 2.

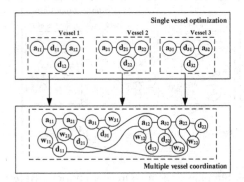

Fig. 8. Simplified example in coordination Strategy 3

Figure 8 gives a simplified example of a small-sized problem that includes three vessel agents and 3 terminal agents. Terminal agents are not shown in this figure because terminal-related constraints are not considered in Level 1, instead, they are considered in the large coordination problem in Level 2. Variables a_{11}, d_{11}, a_{12} and d_{12} formulate a local optimization problem of vessel agent 1. After determining optimal solutions of this problems using a commercial solver, not only optimal solution but also a set of feasible but not optimal solutions are kept. Those solutions are all possible schedules for the vessel agent 1, and the set of possible solutions is referred as a solution pool. This is because locally optimal schedules of some vessels may be conflicted with one another at certain terminals. If one vessel 1 takes the priority, the vessel 2 and vessel 3 have to wait or adjust their schedules accordingly. This could cause domino effects that increase the total round-trip time. Consequently, Coordination Level 2 formulates a multiple vessel coordination problem based on constraint programming, in which all 3 vessels are considered simultaneously. The central coordinator in Level 2 searches for better schedules for each vessel after considering the impacts of other vessels. Depending on the size of the multiple vessel coordination problem in Level 2, commercial solvers and heuristic methods are applicable for the central coordinator. A heuristic method based on large neighborhood search can be found in our work in [11]. Once the solutions for arrival times $a_{11}, a_{12}, a_{21}, a_{22}$ and a_{31}, a_{31}, departure times $d_{11}, d_{12}, d_{21}, d_{22}$ and d_{31}, d_{31}, and the waiting times $w_{11}, w_{12}, w_{21}, w_{22}$ and w_{31}, w_{31} have been found by the central coordinator, vessels will implement them immediately and the central coordinator will not send any information back to vessel agents 1, 2 and 3 for evaluation or re-calculation.

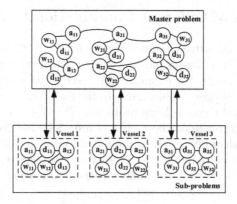

Fig. 9. Simplified example in coordination Strategy 4

Fully-Cooperative Coordination with Multi-level Interactions. Coordination Strategy 4 considers the original problem as consisting of a master problem in Coordination Level 1 and several subproblems in Coordination Level 2. Figure 9 gives an simplified example that includes three vessel agents and two terminals. Similar to Coordination Strategy 3, terminal agents are not shown in this figure because the impact of terminal-related constraints are considered in a large problem of Level 2. As can be seen, the variables from different agents are considered together in a large master problem. The master problem is formulated as a constraint optimization problem, in which the vessels are planned by the central coordinator before they enter the port area with the current information of vessels and terminals. The central coordinator decides on the sequences of vessel visits to terminals in the master problem of Coordination Level 1 using approximate methods such as heuristics.

The subproblems of Level 2 is formulated as constraint satisfaction problems, in which the vessel agents evaluates the visiting sequences that are determined in Level 1, represented as $x_{11}^*, x_{12}^*, x_{21}^*, x_{22}^*$ and x_{31}^*, x_{31}^*. Based on the values of these variables, the vessel agents of Level 2 calculates the arrival times $a_{11}, a_{12}, a_{21}, a_{22}$ and a_{31}, a_{31} and departure times $d_{11}, d_{12}, d_{21}, d_{22}$ and d_{31}, d_{31}, and the waiting times $w_{11}, w_{12}, w_{21}, w_{22}$ and w_{31}, w_{31} at different terminals. Based on the values of the waiting times, constraints on the waiting time variables $w_{11}, w_{12}, w_{21}, w_{22}$ and w_{31}, w_{31} and the visiting sequence variables $x_{11}, x_{12}, x_{21}, x_{22}$ and x_{31}, x_{31} of vessels in the master problem are derived. These constraints are sent from Level 2 to Level 1 and are added to the master problem in the next iteration, in order to exclude variable assignments that can be no better than the best variable assignments of the previous solution. After that, the master problem in Level 1 is re-solved again with the newly added constraints to find a better solution. When the solutions are finally obtained when a pacific termination criteria has been met (time or solution quality), the final determined vessel schedules will be sent from the central coordinator to the local agents and then to the physical level, in the waterborne transport system for implementation.

4   Comparison of Different Coordination Strategies

This paper takes a qualitative perspective on comparing the proposed coordination strategies. This is because each coordination strategy applies to a specific problem size with a range, for example, the distributed coordination strategy mainly applies to small-sized problem, due to the high communication costs, and that it cannot be used to solve large problems. Therefore, it is difficult to test those coordination strategies on the same benchmark system from a quantitative perspective. The readers interested in the individual quantitative performances of these strategies is referred in [9–11]. This section presents a comparison of the proposed coordination strategies by analyzing the information exchange involved in their coordination processes, as well as the models that are used to implement these strategies.

We categorize the information exchange into two types, viz. vessel-related and terminal-related information, as shown in Table 1. Table 2 concludes the characteristics of the proposed coordination strategies: before coordination refers to the information that is known or required by the vessel and terminal agents before the coordination process; during coordination refers to the information that is exchanged among the agents, from agents to the coordinator and from the coordinator to agents during the coordination process starts; after coordination refers to the final information received by vessel and terminal agents, and will be used by the waterborne transport system for implementation; frequency of information exchange refers to the way in which agents and central coordinator communicate with each other.

It can be seen from Table 2 that Coordination Strategy 1 requires less information compared to the other strategies. This strategy considers discrete time in the model, therefore continuous time-related information such as T_3 is not considered. In Coordination Strategies 3 and 4, vessel information V_8 and V_9 in Level 2 is obtained after the coordination process in Level 1 have completed. In these two strategies, terminal-related information is only required on Level 2.

During coordination, the information exchange in Coordination Strategy 1 mainly happens among agents, as there is no central coordinator, while the information exchange in Coordination Strategies 2, 3 and 4 mainly happens between the central coordinator and the agents, without direct information exchange among vessel agents.

The information received after coordination is the same in all strategies: the visiting sequences to terminals and the planned arrival times at terminals, as well as the handling sequences for vessels in each terminal. On the other hand, the information privacy of the four coordination strategies differs: each agent in Coordination Strategy 1 only knows about the current variable assignments and the corresponding utility values of its directly connected agents and the constraints that involve its owned variables; each agent in both Coordination Strategies 2 and 4 knows about all the variable domains and all the constraints; each agent in Coordination Strategy 3 knows the variable a limited set of variable assignments of the other agents, but does not know about the corresponding round-trip times or waiting times of their variable assignments, and that all the

Table 1. Vessel/terminal-related information

Symbols	Descriptions	Explanatory notes	Agents
V_1	The set of terminals to visit	Known information	Vessel
V_2	Number of containers to load/unload at each terminal	Known information	Vessel
V_3	Traveling time between terminals	Known information	Vessel
V_4	Vessel capacity	Known information	Vessel
V_5	Initial number of containers on-board	Known information	Vessel
V_6	Preferences regarding visiting sequences to terminals	Known information	Vessel
V_7	Utility values of each visiting sequence	Obtained during coordination	Vessel
V_8	Currently chosen visiting sequence to terminals	During coordination	Vessel
V_9	A set of candidate solutions (visiting sequences)	Obtained during coordination	Vessel
V_{10}	Round-trip time caused by the currently chosen solution	Obtained during coordination	Vessel
V_{11}	Waiting time caused by the currently chosen solution	Obtained during coordination	Vessel
V_{12}	Final determined visiting sequence	Final solution	Vessel
V_{13}	Final determined candidate solution	Final solution	Vessel
V_{14}	Final determined arrival time at each terminal	Final solution	Vessel
T_1	Terminal capacity for handling vessels	Known information	Terminal
T_2	The latest departure time of vessels that are currently being served at the terminal	Obtained during coordination	Terminal
T_3	Closing and opening times of terminals	Known information	Terminal
T_4	Service time window of sea-going vessels	Known information	Terminal
T_5	Final determined handling sequences of vessels in each terminal	Final solution	Terminal

Table 2. Information exchange analysis

Involved information		Strategy 1	Strategy 2	Strategy 3	Strategy 4
Before coordination	Vessel	V_1, V_2, V_3, V_6	V_1, V_2, V_3, V_4, V_5	V_1, V_2, V_3, V_4, V_5 (Level1); V_9 (Level2)	V_1, V_2, V_3, V_4, V_5 (Level1); V_8 (Level2)
	Terminal	T_1, T_3	T_1, T_2, T_3, T_4	T_1, T_2, T_3, T_4 (Level2)	T_1, T_2, T_3, T_4 (Level2)
During coordination	Agent-to-agent	V_7, V_8	-	-	-
	Agent-to-coordinator	-	$V_1, V_2, V_3, V_4, V_5, T_1, T_2, T_3, T_4$	$V_9, V_{10}, T_1, T_2, T_3, T_4$	V_{11}
	Coordinator-to-agent	-	$V_{12}, V_{13}, V_{14}, T_5$	V_8	V_8
After coordination	Vessel	V_{12}, V_{14}	V_{12}, V_{14}	V_{12}, V_{13}, V_{14}	V_{12}, V_{14}
	Terminal	T_5	T_5	T_5	T_5
Information privacy	Variable domains	Each agent only knows the current variable assignments of directly connected agents	Each agent knows all variable domains of all other agents	Each agent only knows a limited set of variable assignments of the other agents but does not know which one it finally chooses	Each agent knows all variable domains of all other agents
	Constraints	Each agent only knows about the constraints which involve its owned variables	Each agent knows all the constraints	Each agent only knows about constraints related with terminals	Each agent knows all the constraints
Frequency of information exchange		Iterative process among an ordering of agents	From local agents to central coordinator	From Level 1 to Level 2	Iterative process between Level 1 and Level 2

terminal-related constraints are known to each agent. This implies that Coordination Strategies 1 and 3 ensure relatively better information privacy.

Moreover, the frequency of information exchange in the four coordination strategies also differs: the distributed scheme in strategy 1 requires an iterative process in which the message passing starts with a pacific agent, then the information is accumulated from one agent to the next agent until all agents have been reached and then the last agent sends back other types of messages to the next agent. The sequence of the message passing depends on how the agents are connected (linear ordering, DFS structure, ...), i.e., which DCOP algorithm is used; the centralized schemes in both Strategy 2 and Strategy 3 only require one-time only information exchange from local agents in Level 1 to the central coordinator in Level 2, respectively; the coordination scheme in Strategy 4 requires an iterative process in which the central coordinator in Level 1 sends information to the agents in Level 2 for evaluation, and the agents in Level 2 send back information to Level 1, after which the coordinator in Level 1 starts searching for solutions again, information exchange between the two levels continues until a pre-determined termination criteria is met. Comparing with other strategies, Coordination Strategy 1 has the highest communication requirements, as it involves an iterative information exchange among agents. Although the information exchange process in Coordination Strategy 4 is also iterative, it happens between two levels instead of among multiple agents. Therefore, it does not have as high communication requirements as in Coordination Strategy 1.

In order to describe the relations among multiple vessel operators and multiple terminal operators, detailed model formulations are presented in our earlier works [9–11]. This section focuses on the analysis of the differences among these models, as shown in Table 4, and the definitions of types of constraints is given in Table 3.

It can be seen from Table 4 that all the decision variables are the same, as the primary goal of the inland vessels coordination problem is to determine the most efficient way in which these vessels visit multiple terminals, to be more pacific, to determine the visiting sequences and arrival times at terminals. Although the decision variable in Strategy 1 seems to be different from the decision variables in Strategies 2, 3 and 4, the set of time slots at which each vessel visits each terminal actually determines the visiting sequence to all the terminals. The arrival time of each vessel at each terminal is also determined by its visiting sequences, as the traveling time, loading and unloading time are considered as know in this paper.

The distributed model in Coordination Strategy 1 is based on discrete time slots, with the aim to maximize the sum of vessels' utility values. Different utility values represent different vessels' preferences for visiting terminals at pacific time slots, maximizing the sum of utility values means satisfying the preferences of all the agents as much as possible. The models in Coordination Strategies 2, 3 and 4 are formulated based on continuous time, with the same objective to minimize the sum of vessels' round-trip times.

Table 3. Definitions of different types of constraints

Symbols	Definitions	Agents
C_1	Utility function that describes the preference of each vessel for visiting a particular terminal during different time slots, higher value represents higher preferences	Vessel
C_2	The number of containers on a vessel cannot exceed its capacity	Vessel
C_3	Time-related constraints regarding the relations among the sequence variables, arrival time and departure time variables	Vessel
C_4	The arrival times of a set of vessels that arrive at the same terminal on will be ranked on a first-come-first-served basis	Vessel
C_5	Each terminal can handle limited number of vessels simultaneously, without exceeding its capacities	Terminal
C_6	Terminal-related constraints regarding the possible start and end times of each terminal for handling each vessel	Terminal
C_7	The vessels cannot be handled during the closing time of each terminal	Terminal
C_8	Priority of sea vessels: inland vessels cannot be handle at the quays that have been reserved for handling sea vessels	Terminal
C_9	Waiting time of each vessel is determined by the difference between the arrival time of that vessel and the earliest possible starting time of the terminal it visits	Vessel; terminal
C_{10}	Loading time of each vessel is determined by	Vessel; terminal
C_{11}	Visiting sequences of the candidate solution that the vessel chooses	Vessel

The types of constraints in Coordination Strategy 1 are fewer than the other three coordination strategies. This is partly because the increase on the number of constraints would substantially increase the communication costs, thereby the problem sizes that Coordination Strategy 1 can solve are limited. In our preliminary results in [9], it can solve the problem in which 3–6 vessels enter the port area, and each vessel visits 3 terminals. The types of constraints that are considered in Coordination Strategies 2, 3 and 4 are the same, but they are considered at different levels. In Coordination Strategy 2, the constraints are considered altogether in a large optimization problem, while in Strategies 3 and 4 the relatively complicated constraints such as C_4, C_5, C_6, C_7, C_8 and C_9 are considered in Level 2. Coordination strategy 4 can solve the largest problem sizes, as it solves a optimization problem only at Level 1, Level 2 consists of multiple satisfaction problems and requires less computational efforts.

Table 4. Model formulation analysis

		Strategy 1	Strategy 2	Strategy 3	Strategy 4
Time horizon		Discrete	Continuous	Continuous	Continuous
Decision variables	Level 1	Time slot at which each vessel visits each terminal	Visiting sequence of each vessel to terminals	Visiting sequence of each vessel to terminals	Visiting sequence of each vessel to terminals
	Level 2	-	-	The rotation plan that each vessel chooses	The waiting time of the chosen visiting sequence
Optimization Objective	Level 1	Maximizing the sum of vessels' utility values	Minimizing the sum of vessels' round-trip times	Minimizing the round-trip time of each vessel	Minimizing the sum of vessels' round-trip times
	Level 2	-	-	Minimizing the sum of vessels' round-trip times	-
Constraints	Level 1	$C_1, C_3, C_5, C_9, C_{10}$	$C_2, C_3, C_4, C_5, C_6, C_7, C_8, C_9, C_{10}$	C_2, C_3, C_{10}	$C_2, C_3, C_4, C_6, C_{10}$
	Level 2	-	-	$C_4, C_5, C_6, C_7, C_8, C_9, C_{11}$	$C_3, C_4, C_5, C_6, C_7, C_8, C_9$
Problem sizes	No. of upcoming vessels	3–6	8	8	8–16
	No. of terminal visits per vessel	3	6–8	6–8	8

To conclude, from an information exchange perspective, Coordination Strategies 1 and 3 ensures better information privacy, in which Strategy 1 has higher requirements on communication costs, as the agents exchange information in an iterative way. Coordination Strategies 2 and 4 consider fully-cooperative vessel and terminal operators, in which Strategy 2 has higher requirements on communication costs, as the central coordinator requires more information from the local agents than in Strategy 4. From a methodological perspective, Coordination Strategy 1 solves small sized-problem, Coordination Strategies 2 and 3 are able solve medium-sized problem, and Coordination Strategy 4 can solve large problem. Coordination strategies 2, 3 and 4 also consider more practical constraints in the models than Coordination Strategy 1.

5   Conclusions and Future Directions

To ensure efficient inland waterway transport between seaports and hinterland, this paper proposed four classes of coordination strategies to improve the coordination among multiple vessels and multiple terminals in the port area based on our earlier work. A comparative study of these strategies is given from an information perspective and a methodological perspective, which provide insights for practitioners on which types of information are required in each coordination strategy. Considering the four coordination strategies, it can be concluded that: Coordination Strategy 1 mainly applies to a small-sized problem that requires distributed problem-solving with information privacy and solution optimality concerns; Coordination Strategy 2 is most suitable for a medium-sized problem with fully-cooperative vessel and terminal operators, and with optimality requirements on vessel schedules; Coordination Strategy 3 is most suitable for a medium-sized problem with information privacy concerns; Coordination Strategy 4 is most suitable for a large problem with fully-cooperative vessel and terminal operators.

To enhance the applicability of the proposed coordination strategies, further research is required. A more detailed simulation study is important. If reliable historical data from certain seaports can be obtained, the proposed models can be validated. In addition, to use the proposed coordination strategies for practical operations, firstly, an information platform for exchanging information among vessel operators and terminal operators is required; secondly, decision support software also needs to be developed and installed on each vessel in order to send and receive messages to/from the information platform. Moreover, in the process of information exchange, information loss or delay may happen. This could affect the execution of the proposed methods. Therefore, it is important to investigate how to improve the tolerance of the coordination methods for information loss or delay. By implementing the proposed methods on a more practical simulation platform or system, the effectiveness of the proposed coordination methods can be verified.

References

1. European Commission: Strategic Research Agenda for Inland Waterway Transport. Technical report, European Commission (2016)
2. Douma, A.M.: Aligning the operations of barges and terminals through distributed planning. Ph.D. thesis, University of Twente, Enschede, The Netherlands, December 2008
3. European Commission: White paper on transport. Roadmap to a single European Transport Area-towards a competitive and resource efficient transport system (2011)
4. European Commission: Inland waterways (2016). http://ec.europa.eu/transport/modes/inland/index.html
5. Van der Horst, M.R., Langen, P.W.: Coordination in hinterland transport chains: a major challenge for the seaport community. In: Haralambides, H.E. (ed.) Port Management, pp. 57–83. Palgrave Macmillan, London (2015)
6. Kolkman, J.: Binnenvaart en containerlogistiek. Technical report, Ministerie van Verkeer en Waterstaat (2009)
7. Konings, R., Kreutzberger, E., Maraš, V.: Major considerations in developing a hub-and-spoke network to improve the cost performance of container barge transport in the hinterland: the case of the port of Rotterdam. J. Transp. Geogr. **29**, 63–73 (2013)
8. Li, S., Negenborn, R.R., Lodewijks, G.: Approach integrating mixed-integer programming and constraint programming for planning rotations of inland vessels in a large seaport. Transp. Res. Record J. Transp. Res. Board **2549**, 1–8 (2016)
9. Li, S., Negenborn, R.R., Lodewijks, G.: Distributed constraint optimization for addressing vessel rotation planning problems. Eng. Appl. Artif. Intell. **48**, 159–172 (2016)
10. Li, S., Negenborn, R.R., Lodewijks, G.: Closed-loop coordination of inland vessels operations in large seaports using hybrid logic-based benders decomposition. Transp. Res. Part E Logist. Transp. Rev. **97**, 1–21 (2017)
11. Li, S., Negenborn, R.R., Lodewijks, G.: Planning inter-terminal transport for inland vessels in large seaports using a two-phase approach. Comput. Ind. Eng. **106**, 41–57 (2017)
12. Negenborn, R.R., Ocampo-Martinez, C.: Perspectives on transport *of* water versus transport over water. In: Ocampo-Martinez, C., Negenborn, R.R. (eds.) Transport of Water versus Transport over Water. ORSIS, vol. 58, pp. 1–10. Springer, Cham (2015). https://doi.org/10.1007/978-3-319-16133-4_1
13. Nextlogic: Chain optimimalisation container barging (2012). http://www.nextlogic.nl/
14. Weiss, G.: Multi-agent Systems, 2nd edn. MIT Press, Cambridge (2013)
15. Wooldridge, M.: An Introduction to Multi-agent Systems. Wiley, New York (2009)

Autonomous Surface Vessels in Ports: Applications, Technologies and Port Infrastructures

Abhilash Devaraju, Linying Chen$^{(\boxtimes)}$, and Rudy R. Negenborn

Department of Maritime and Transport Technology,
Delft University of Technology, Delft, The Netherlands
{L. Chen-2, R. R. Negenborn}@tudelft.nl

Abstract. Autonomous Surface Vessels (ASVs) have various applications in the maritime sector. However, considerable challenges need to be met before integrating the applications in the current environment. Moreover, infrastructures in ports and waterways need to be upgraded to satisfy the requirements for ASV applications. To have an insight into the potential of ASVs in ports, we analyze the applications of autonomous vessels and the impacts that these applications have on port infrastructures. Future scenarios of the application of ASVs in ports are determined based on the analysis of the following three aspects: ASV applications, ASV technology development, and port infrastructure development. To indicate the development of ASV technology and port infrastructures, the Technology Readiness Level (TRL) is employed. Eleven scenarios of ASVs in ports have been identified based on the analysis.

Keywords: Autonomous surface vessel · Port infrastructures
Technical scenarios

1 Introduction

Autonomous shipping could have a big impact on the operations at the ports and could contribute to the improvement of efficiency, safety, and sustainability. However, the development is not at its final stage since there are a lot of challenges which need to be addressed before the vessels become fully operational. Ports and vessels are synonymous to one another and hence it is important to investigate the required port infrastructure if the aim is to make the autonomous vessels fully operational in the coming years.

In literature, there is a gap in research about potential applications and infrastructure requirements for the applications of ASVs. In this paper, we focus on the question "What scenarios are foreseen considering autonomous shipping in ports?". We analyze the applications and technology development of ASVs, as well as port infrastructure development, by reviewing related studies. All these factors in association with technical maturity and an approximate timeline of implementation results in the formulation of ASV scenarios, which can be used by both the academic and the industrial sector to understand the various possibilities involved with ASVs.

© Springer Nature Switzerland AG 2018
R. Cerulli et al. (Eds.): ICCL 2018, LNCS 11184, pp. 86–105, 2018.
https://doi.org/10.1007/978-3-030-00898-7_6

This paper is organized as follows. Section 2 provides the framework for formulating ASV scenarios. Section 3 describes the various applications involved with shipping. Sections 4 and 5 provide the details about the development of ship technology and port infrastructure. Technology Readiness Level (TRL) is employed to indicate the maturity of the technology. In Sect. 6, we utilize all the above-mentioned information to formulate ASV scenarios in ports. Section 7 provides the conclusions.

2 Research Framework

To formulate ASV scenarios, it is important to understand the factors involved. Figure 1 shows the transport process of an inland vessel in a port. The process consists of a sequence of subprocesses [18]. During these process, vessels need the service provides by the ports.

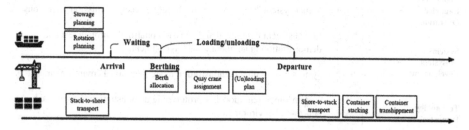

Fig. 1. The transport process of a vessel in a port [18].

To begin with, a vessel makes appointments with operators present at the terminal to determine the time for handling of the containers. A decision is made after the exchange of information between the vessel operator and the terminal. This process involves a lot of communication between the vessel and the terminal. The introduction of an ASV will enhance this process as real-time information can be obtained and subsequent terminal allocations can be decided upon instantly. Subsequently, operations, such as berth allocation, need to be considered. This involves mooring a vessel and it depends on various factors, including vessel length, cargo, loading and unloading operations, availability of quay cranes and berthing and crane requirements of other vessels [18]. ASVs in association with improved port infrastructure, that is, automatic mooring capability will hugely influence the operations at the terminal. After berthing, the next operation under concern is quay crane assignment. Quay crane assignment depends on the accessibility and availability of cranes at the berth. The presence of automated quay cranes at the port in association with ASVs will significantly improve handling operations at the terminal.

Therefore, to formulate ASV scenarios not only need to consider potential ASV applications and ASV technology development, but also port infrastructure development. ASVs can perform the applications of the already existing conventional vessels, such as cargo shipping. Furthermore, it is also capable of additional applications, such as firefighting, scientific research. Ship technology primarily refers to the control strategy applied and the numerous auxiliary requirements, such as sensors and internet technology. ASVs can either be controlled remotely or complete autonomy. There can also be a mix of remote and automated control strategies. Port infrastructure becomes synonymous with ship technology because the ports should be able to accommodate ASVs to achieve full functionality.

To indicate the maturity of the technology, Technology Readiness Level (TRL) is used. TRL is an indicator used to assess the level of a concerned technology. The various levels of technology readiness are defined contrarily by each organization, namely, the U.S. Department of Defense, NASA, ESA, European Commission, Oil & Gas Industry and so on. For this research, the TRL as defined by NASA has been considered, as seen in Fig. 2. The TRL ranges from level 1 to level 9 with level 9 indicating the technology ready to be implemented [34].

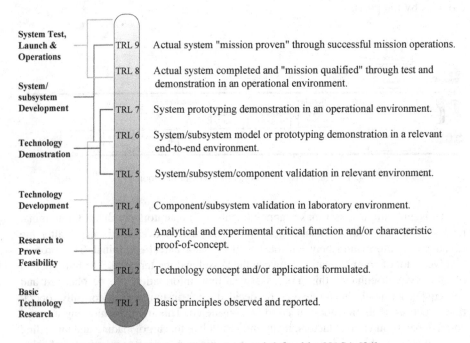

Fig. 2. Technology Readiness Level defined by NASA [34].

3 Applications of ASVs in Ports

In the literature, many ASV projects have been successfully realized, and as many are still under development. In [32], an overview of potential applications of ASVs is provided by analyzing existing ASV projects and related literature. In this paper, we focus on the projects related to applications in Ports, see Table 1.

In 2015, Rolls-Royce together with a host of other companies started an initiative called Advanced Autonomous Waterborne Applications (AAWA). The main objective of this program was to develop the technological, safety, legal and economic aspects of

Table 1. Overview of ASV projects related to applications in Ports.

No.	Name	Company/University	Description	CA	Co	SA	Sa	MR	Ref.
				Challenges*					
1	Rolls Royce ASV 2020	Rolls Royce	Remote controlled vessel with reduced crew	X	X	X	X	X	[30]
2	Rolls Royce ASV 2025		Remote controlled unmanned coastal vessel						
3	Rolls Royce ASV 2030		Remote controlled unmanned ocean-going ship						
4	Rolls Royce ASV 2035		Autonomous unmanned ocean-going ship						
5	MUNIN	MUNIN Consortium – 8 partners	Verify and develop a concept for autonomous ships	X		X		X	[23, 38]
6	RAmora 2400	Robert Allan Ltd	Tug-boat designed primarily for ship assist and berthing operations		X			X	[29, 35]
7	Venus USV	ST Electronics	Unmanned surface vessel with three variants for military operations		X				[41, 42]
8	Mayflower Autonomous (MARS)	Mayflower Autonomous Ship Ltd. (MAS) – 3 partners	An autonomous vessel capable of conducting scientific research					X	[20, 33]
9	The ReVolt	DNV GL	Unmanned, zero emission short-sea vessel	X	X			X	[2, 8, 10]
10	ACTUV	Leidos Maritime Solutions and DARPA	An unmanned surface vessel primarily for military operations			X			[3, 9]
11	Common USV	Textron	Unmanned surface vessel deployed by the US Navy for military operations				X		[12, 24]
12	Svitzer Hermod	Rolls-Royce and Svitzer	A remote-controlled commercial vessel					X	[40]
13	C-Enduro	Autonomous Surface Vehicle Ltd	It is an endurance unmanned vessel ideal for long-term remote data collection		X			X	[39]
14	C-Worker		An offshore unmanned vessel designed for oil and gas operations					X	
15	C-Sweep		It is an endurance unmanned vessel for the purpose of minesweeping missions				X		
16	C-Target 3		Lightweight and fast-moving target drone				X		
17	C-Target 6		Fast and powerful target drone				X		

(continued)

Table 1. (*continued*)

No.	Name	Company/University	Description	Challenges*					Ref.
				CA	Co	SA	Sa	MR	
18	C-Target 9		Highly versatile target craft				X		
19	C-Target 13		Fast and highly realistic target drone				X		
20	C-Cat 4		Multipurpose work USV for sampling and so on		X				
21	C-Stat		Mobile buoy system for when anchoring is difficult	X					
22	Messin	University of Rostock	It is used as a carrier for measuring devices in the field of marine research		X			X	[5]
23	DELFIM	Instituto Superior Tecnico (IST)	Developed to attain automatic marine data and to act as an acoustic relay between submerged craft and support vessel	X		X		X	[1]
24	Yara Birkeland	Yara International and Kongsberg Group	Electric, autonomous and zero emission container ship					X	[4, 16]
25	Hronn	BOURBON and Kongsberg	Light-duty, offshore, utility ship servicing the scientific world					X	[16]
26	PILOT-E Autonomous Ferry	Kongsberg and PILOT-E	Autonomous, zero-emission state-of-the-art ferry			X		X	[16]

* CA: Control Algorithm; Co: Connectivity; SA: Situation Awareness; Sa: Safety; MR: Maritime Regulations.

ASVs [13, 14]. DNV GL, apart from supporting the initiative, also went on to develop an unmanned short-sea vessel called The ReVolt for the purpose of inland marine cargo transport [2, 10].

MUNIN is a project undertaken to verify and develop a concept for autonomous vessels and it was done by using a dry bulk container [19, 38]. Yara Birkeland is a Norwegian container ship which is set to be the world's first electric, autonomous, zero-emission ship. It will initially operate as a manned vessel, followed by being remotely operated in 2019 and it aims to achieve complete autonomy by 2020 [4, 16]. Robert Allan Ltd was one of the leading companies to develop autonomous tugboats. Their flagship vessel, RAmora 2400, is a versatile Tele-Operated Workboat (also referred to as a TOWBoT) [29, 35]. Svitzer Hermod is an autonomous tug designed by Robert Allan. The boat was built in 2016 and is equipped with Rolls-Royce's dynamic positioning system, which is vital for a remotely controlled vessel [36].

Autonomous Surface Vehicles Ltd is a company that deals with autonomous marine vehicle systems. One such vessel designed by this company is called the C-Worker, which is a robust unmanned vessel designed for oil and gas operations in dangerous marine environments [39]. The C-Enduro is an unmanned surface vessel with high endurance capabilities [39]. C-Cat 4 is also developed by Autonomous Surface Vehicles Ltd. It is capable of being remotely controlled as well as being autonomously controlled. It can also be fitted with a diesel engine for longer operations [39].

MESSIN is an autonomous unmanned vessel developed by the University of Rostock. The vessel is expected to have a catamaran hull with glass fiber-reinforced plastic, electrical rudder propulsion system, a variable platform for different measurement technologies, automatic trajectory control, a telemetry system for data transfer and control and a hybrid power supply [5]. DARPA has developed an autonomous unmanned vessel with high endurance capable of operating for months with a payload. The project is referred to as Anti-Submarine Warfare Continuous Trail Unmanned Vessel (ACTUV) and is expected to demonstrate various military applications for the Navy [3]. One of the C-Target range of ASVs developed by Autonomous Surface Vehicles Ltd., C-Target 13, is a reliable and extremely quick target drone. It is equipped with radar, thermal and visual sensors and is capable of either manual or remote operations [39]. Singapore Technologies (ST) Electronics is in the process of developing an ASV, called the Venus USV. The vessel is expected to be equipped with a fast radar tracker, vision-based obstacle or target detection and multi-sensor fusion. The autonomy for maneuvering includes waypoint navigation, course and speed navigation, collision detection and collision avoidance [41].

Textron Systems Advanced Systems and AAI Unmanned Aircraft Systems in association with Maritime Applied Physics Corp (MAPC) have developed a fleet-class Common Unmanned Surface Vessel (CUSV). The CUSV is remotely controlled and it has high endurance levels as it can execute missions for more than 24 h in harsh environments [24].

To sum up the potential applications in ports, 4 categories are identified, as shown in Fig. 3 and detailed below.

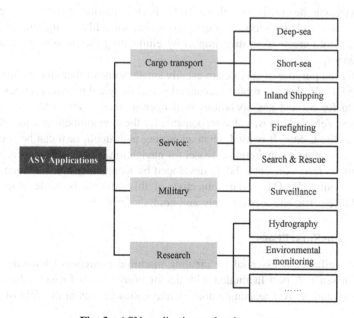

Fig. 3. ASV applications related to ports

3.1   Cargo Transport

Cargo shipping is the most significant of all applications since transportation through waterways is the cheapest mode of transport when compared to land and air. According to the transport distances, cargo transport can be divided into deep-sea shipping, short-sea shipping, and inland shipping. Sustainable and efficient operations are the major drivers of development in the field of automation. The disruptive technology in recent years has also been one of the major reasons for the development of autonomous vessels in the shipping sector.

Various projects have been developed or are in the conceptual stage in the field of cargo shipping. Rolls-Royce has been leading the line in terms of R&D in this sector as it believes that an intelligent shipping era is on the horizon. Hence, Rolls-Royce in association with Advanced Autonomous Waterborne Applications (AAWA) initiative laid out a timeline for the development of ASVs. They aim to start with remotely operated vessels supported by a reduced crew and go on to achieve a completely autonomous ocean-going vessel by 2035. This includes 4 concepts, namely, Rolls-Royce 2020, 2025, 2030 and 2035 [30].

The ReVolt and Yara Birkeland are two other projects in the cargo shipping sector. The ReVolt is short-sea cargo vessel being developed to operate between Oslo and Trondheim with a capacity of 100 TEUs [2]. Yara Birkeland is a Norwegian container ship which is set to be the world's first electric, autonomous, zero-emission ship with a capacity of 150 shipping containers. It is being developed to act as a medium of transport for fertilizer between three ports in southern Norway [16]. MUNIN is another project in the cargo shipping sector, specifically for dry bulk transport [23, 38].

3.2   Maritime Service

Service vessels can be classified as those which assist in marine operations. The advent of autonomous vessels for services in dangerous situations, like firefighting and search and rescue, would create a positive impact by eliminating the possibility of the crew being in danger.

ASVs for the purpose of service are usually small. Some of them have already been implemented. C-Worker is a robust unmanned vessel designed by Autonomous Surface Vehicles Ltd for oil and gas operations in dangerous marine environments. Autonomous Surface Vehicles Ltd was also responsible for the development of a mobile buoy, referred to as the C-Stat. It is used when anchoring is difficult, or it can be used when the costs of operating a conventional vessel for navigation is extremely expensive [39]. An autonomous ferry concept is being developed by Kongsberg in association with an organization called PILOT-E. The objective of this concept is to develop a zero-emission, full electric and a completely autonomous ferry [16].

3.3   Maritime Surveillance

Maritime surveillance is essential for creating maritime awareness ('knowing what is happening at sea'). A port facilitated with autonomous vessels for surveillance has a significant advantage over its competitors. Autonomous vessels in the field of military have a massive impact on the human factor since in most of the cases no humans or a limited number of humans are involved. This would mean a complete or a partial reduction in the number of fatalities and a reduction in costs associated with it.

DARPA has developed an autonomous unmanned vessel with high endurance capable of operating for months with a payload. The project is referred to as Anti-Submarine Warfare Continuous Trail Unmanned Vessel (ACTUV) and is expected to demonstrate various military applications for the Navy. Some of the applications include intelligence, surveillance, reconnaissance for the port authority or for the navy of the country [9]. Autonomous Surface Vehicles Ltd developed a few vessels for military applications and these include the C-Sweep, which is equipped to handle minesweeping missions. It is also capable of deploying Autonomous Underwater Vehicles (AUVs) or Remote Operated Vehicles (ROVs) as an auxiliary application. They also developed a range of C-Target naval target drones to assist the naval personnel train for various scenarios as well as gunnery training, weapons testing and so on. The range of C-Target includes C-Target 3, C-Target 6, C-Target 9 and C-Target 13 [39]. Singapore Technologies (ST) Electronics is in the process of developing an unmanned surface vessel, called the Venus USV. The main application of the Venus USV is in the Republic of Singapore's Navy for autonomous countermining measures [41]. Textron Systems Advanced Systems and AAI Unmanned Aircraft Systems in association with Maritime Applied Physics Corp (MAPC) have developed a fleet-class Common Unmanned Surface Vessel (CUSV). The CUSV can be deployed from another vessel to carry out mine warfare, anti-submarine warfare, intelligence operations, surveillance and reconnaissance, anti-surface warfare as well as launch and recovery operations [24].

3.4 Maritime Research

Autonomous vessels in the field of maritime research are essential since most research is done in dangerous environments. Therefore, reduced crew or completely unmanned research vessels capable of collecting the required data and monitoring the surroundings can have a positive impact in the academic sector.

Autonomous Surface Vehicles Ltd have been responsible for two vessels, one being the C-Enduro USV and the other being the C-Cat 4. The C-Enduro is an unmanned surface vessel with high endurance capabilities and hence it is ideal for remote data collection over a certain period. It is able to conduct activities which include, oceanographic data collection, environmental monitoring and so on for a period of up to 3 months. C-Cat 4 finds its applications in the field of sampling, hydrography and environmental monitoring [39]. MESSIN is an autonomous unmanned vessel developed by the University of Rostock. MESSIN can operate in shallow waters and assist in various applications, such as depth-profile measurement, customer-specific measurements, oceanographic data collection, cartography and other scientific applications [5].

BOURBON, a leading marine offshore services company in association with Automated Ships Ltd and Kongsberg will develop the world's first autonomous, fully-automated and cost-efficient prototype vessel, referred to as Hronn, for offshore operations. The vessel assists in offshore energy, scientific or hydrographic applications and offshore fish farming industries. The various applications capabilities of the vessel include surveying, unmanned vessel launch or deployment, recovery and monitoring and delivery to offshore installations [16]. DELFIM is another autonomous vessel, developed by Institute for Systems and Robotics (ISR) and Instituto Superior Tecnico (IST). Its applications include automatic marine data collection and to serve as an acoustic relay between a submerged craft and a support vessel [1].

4 Development of Ship Technology

The rapid increase in transportation requirements over the years and the need for sustainability in every sector led to the development of ASVs in the marine sector. The idea of ASVs appealed to the industry in various ways, such as reduced operating cost, reduced energy consumption, safer since humans are not involved, and higher productivity [17]. This section of the paper provides an overview of numerous ASV projects that have been developed or are in the conceptual stage of development and the technology being used to realize autonomy. These projects have been tabulated in Table 1, where CA, CT, SA and MR are abbreviated as control algorithm, connectivity, situational awareness and maritime regulations respectively. This section will also address the challenges facing the full functionality of ASVs.

4.1 ASV Technologies

To facilitate an autonomous future, many challenges should be overcome. In Table 1, five significant challenge areas have been identified for ASVs. These include control algorithm, connectivity, situational awareness, safety and maritime regulations, that is, both technical and legal challenges. In this paper, we focus on the challenges from the technical perspective related to following three aspects:

- **Situation awareness:** advanced sensors to get accurate data and algorithm to analyze the data.
- **Communication and cooperation:** methods to enhance vessel to vessel and vessel to infrastructure communication, and algorithms for cooperation among the agents.
- **Computational logistic technologies and control strategies:** better scheduling, guidance and control of the motion systems.

Situation awareness is the key to achieving reliable operations. To provide sufficient information to the vessel about its surroundings accurately, an input from multiple sensors should be available and it needs to be analyzed. Hence, it is important to achieve a reliable sensor fusion which is quite a challenge. To plan an effective collision-free path for an autonomous vessel through numerous moving and static obstacles is also a challenge which needs to be addressed. Weather conditions have a huge impact on the reliability of these controls as well [15]. The projects for which situational awareness is a significant issue are, the Rolls-Royce projects, MUNIN, ACTUV, DELFIM and the PILOT-E autonomous ferry.

Computational logistic technologies and control strategies are the most important challenges that need to be addressed. With the information provided by situation awareness methods, optimization models and algorithms are need both for global planning for executing complex tasks, e.g., for finding the most efficient scheduling and collision avoidance. Furthermore, the ASVs that are not completely autonomous need to be appropriately programmed so that they are remotely controlled from a control center when necessary and when the vessels are in crucial regions, such as, the port or the terminal area [14]. For the fully autonomous vessels, to control the dynamic of vessels moving towards desired position is a fundamental step.

Connectivity is another significant challenge with autonomous vessels since every information from the vessel needs to be monitored, regardless of it being completely autonomous or remotely controlled. The Rolls-Royce vessels, Venus USV, The ReVolt, C-Enduro, C-Cat 4 and MESSIN are the projects involved wherein connectivity needs to be addressed and developed further. This falls under the category of communication and cooperation.

Researchers think of methods to meet the challenges, such as equipping existing vessels with additional sensors, additional communication methods, and remote-control systems. The overview of existing ASV projects and literature provides a track from existing vessels to a remote-controlled vessel with reduced crews, unmanned remote-controlled vessel, and at the end, fully autonomous vessel. As the environment is an important aspect that needs to be concerned, zero-emission vessels have been mentioned frequently in existing literature. Figure 4 shows this trend.

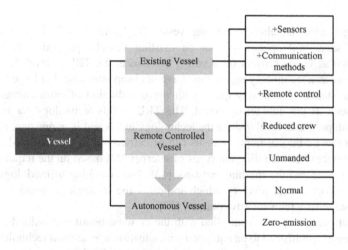

Fig. 4. ASV development trends

4.2 TRL of ASV Technologies

As an indicator of technology maturity, TRL is an important factor in formulating and therefore classifying the ASV scenarios. It is vital to classify them based on their TRL to have a better understanding of the proposed scenarios for ASVs and this has been tabulated in Table 2.

Table 2. TRL of ASV technologies.

TRL	Vessel							
	Existing Vessel				Remote-controlled Vessel		Autonomous Vessel	
	Existing vessel	Additional sensors	Additional communication methods	Remote control	Reduced crew	Unmanned	Normal	Zero-emission
	Technology Development							
	–	Visual sensors	WiFi/WiMAX, 3G/4G, VSAT, InMarsat	Situational Awareness (SA)	SA, Maintenance planning	Control strategy	Control strategy	Control strategy
9	x		x (SS & IS)					
8								
7		x		x	x (SS & IS)			
6			x (DS)		x (DS)			
5								
4						x (SS & IS)		
3						x (DS)		
2							x	x (SS & IS)
1								

NB: SS— Short-sea Shipping; IS – Inland Shipping; DS – Deep-sea Shipping

To begin with, an already existing vessel is classified as TRL 9 since it has undergone successful operations. For an existing vessel equipped with additional sensors, such as visual sensors for berthing assistance, the TRL is level 7. Although, visual sensors as a technology exists, they have not been operational in large vessels for berthing assistance. The next stage is with the introduction of extra communication methods, such as the internet on board. The TRL of this technology for inland and short-sea shipping is at level 9 since the technology exists and is in practice. Regarding deep-sea, a few additional improvements can be carried out in terms of InMarsat [11], switching devices between different types of internet [22] based on the requirements at that point in time and the meshing network [43]. The switching methodology and the meshing network are at level 6, which indicates the technology being tested in a prototype stage in a relevant environment [34].

A vessel being remotely controlled with the crew on board is classified as TRL 7, since the vessel would need to be equipped with situational awareness technology which has been demonstrated in an operational environment [6]. Further, a remotely controlled vessel with reduced crew is considered separately for short-sea, inland shipping, and deep-sea shipping. This is considered to be at TRL 7 and 6, respectively, due to the maintenance planning that would be required onboard. Also, Rolls-Royce expects a vessel of this type to be operational by 2020 [17]. A completely unmanned remotely controlled vessel for short-sea and inland shipping is at a TRL 4 since the technological components are being tested now, whereas, an unmanned remotely controlled vessel for deep-sea operations is at Level 3 since research and development for this sector is still at an early stage. With regard to this type of vessel, Rolls-Royce has provided a timeline of 2025 for the short-sea sector and 2030 for deep-sea sector [17].

The penultimate technology considers a completely unmanned autonomous vessel, which is at TRL 2. A completely autonomous deep-sea vessel is expected to set sail in 2035 [17]. Finally, a completely unmanned zero-emission autonomous vessel is at TRL 2. A good indication of this type of vessel is the Yara Birkeland, which is a zero-emission short-sea vessel and it is set to be operational by 2019 [4].

5 Development of Port Infrastructure

This section describes port infrastructures in detail as well as their corresponding TRL. In this paper, we classify the port infrastructure into 4 categories: information service, navigation assistance, and terminal service, see Fig. 5. A shore control center is also considered for future remotely controlled vessels.

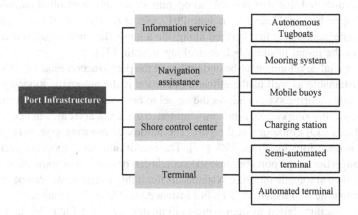

Fig. 5. Port Infrastructure.

5.1 Information Service

The improvements at the port in terms of information are made with the objective being better connectivity between the ship and the shore as this is significant for unmanned shipping. To achieve this, the internet at the port side should be improved considerably. The port should be capable of accommodating the InMarsat technology, switching technology for the internet to achieve efficient operations and the meshing network [11, 22, 42]. For remote controlled or complete autonomy on board a vessel, the port should be equally advanced since the vessel would require continuous or occasional monitoring. Therefore, an advanced shore control center is a requirement at the port side. This would ensure complete monitoring of the vessels, including, efficient path planning operations to achieve maximum efficiency.

5.2 Navigation Assistance

Autonomous tugboats, mooring systems, mobile buoys and charging stations fall under the category of navigation assistance when dealing with port infrastructure. A tugboat is designed to pull vessels that are stranded, oil platforms and barges or huge vessels that are not supposed to move in narrow waterways. In future autonomous ports, tugboats also can help the movements of manned vessels. Moreover, an autonomous tugboat can be equipped to operate in dangerous environments or fight fires for extended periods without causing any risk to humans [29]. The RAmora has been primarily designed for ship assist and berthing operations. Due to its various attributes such as high maneuverability, simple, safe and its ability to operate in hazardous conditions the RAmora has got additional applications such as firefighting, rescue operations, oil spill operations, tanker pull back operations, etc. [35]. Svitzer Hermod is another autonomous tug designed by Robert Allan. Svitzer in association with Rolls-Royce demonstrated that it is possible to operate a remotely controlled unmanned tug in Copenhagen harbor, Denmark in June 2017 [36, 39]. The vessel being controlled from a remote base was at first berthed alongside a quay, then undocked, tuned it 360°, piloted it to the initial position before docking it again [37].

Another vital development to be undertaken at the port to accommodate ASVs and to achieve sustainable as well as the efficient operation is the automatic mooring system. The automatic mooring system allows the vessel to be moored without the use of ropes. The automatic mooring system is not only limited to the quick and efficient operation, but it also reduces CO_2 emissions at the port. The automatic mooring system is known to reduce CO_2 emissions by about 75% [27]. The use of automatic mooring system also reduces congestion at the ports because the vessels can be moored in about 30 s. Another advantage of an automatic mooring system is its flexibility as any vessel can be moored at any berth regardless of its length [21]. For instance, the MoorMaster automatic mooring system is a vacuum-based system which eliminates the need for ropes and handles vessels in a few seconds. It uses controllers which react to the vessel movement and counteract it by absorbing them. The vessels are moored via remote control and are handled by one operator only. The variables involved are monitored continuously in real time and the system is able to moor and release vessels in seconds [21].

Furthermore, the use of mobile buoys would assist in ship navigation and add to the safety of the vessel since they maintain their expected position with the help of an automatic controller [39]. Implementation of a Laser Docking System and the Laser Ranging System (LDR–LRS) at the port will help considerably in berthing operations with the Automatic Identification System (AIS) onboard [26]. Finally, with heading towards zero-emission ASVs, it is crucial for the port to be able to handle such vessels in the future. Cavotec offers a product referred to as the Automatic Plug-in System (APS), which helps in the charging of a vessel at the port. It can be done without any human intervention with a few added advantages namely, short charging time, minimal infrastructure cost and minimal maintenance costs. The APS and the MoorMaster ensure mooring and charging port connections in less than 30 s [7].

5.3 Shore Control Center

One of the most significant developments required at the port side to realize ASVs is a shore control center. A shore control center can be referred to a monitoring center present at the port side capable of monitoring the vessels in real time and hence capable of making the important decisions. A shore control center can prove to be extremely useful as small defects on the vessel can be fixed remotely and hence it improves productivity by reducing downtimes. A conceptual model for a shore control center has been proposed by Rolls-Royce [31].

5.4 Terminal

With the idea being a gradual development of the port infrastructure, the terminal should improve from an already existing semi-automated terminal equipped with semi-automated quay cranes, AGVs and so on to a completely automated terminal. A completely automated terminal is expected to be equipped with automatic quay cranes with automatic loading and unloading of cargo at the quayside, automatic stacking equipment and other significant equipment involved with ship handling. Qingdao New Qianwan Container Terminal (QQCTN) in China demonstrated the automated quay cranes for the first time on May 11, 2017. Laser scanners and various positioning sensors are used to handle the containers at the port. The use of automated quay cranes reduced the number of workers required for cargo handling of one vessel at the port from 60 to 9. Hence, the terminal's labor costs reduced by 70% and the efficiency improved by 30% [28].

5.5 TRL of Port Infrastructures

Similarly, TRL is defined to indicate the development of various port infrastructures in Table 3. An already existing port infrastructure is considered which is at TRL 9. For

Table 3. TRL of port infrastructures.

TRL	Port infrastructure								
	Existing infrastructure	Information service	Navigation assistance				Shore control center	Terminal	
			Autonomous tugboat	Mooring system	Mobile buoys	Charging station		Semi-automated	Automated
	Technology Development (examples)								
	–	LDR, LRS & Internet	Ramora 2400, Svitzer Hermod	MoorMaster by Cavotec	C-Stat	Automatic Plug-in System by Cavotec	–	Quay cranes, Stacking operations	Autonomous quay cranes, Stacking operations
9	x	x		x		x		x	
8		x							
7		x	x		x		x		
6		x							
5									
4									
3									x
2									x
1									

berthing assistance, laser docking systems (LDR) and laser ranging systems (LRS) are at TRL 8 or 9 [25]. This can be followed by an improvement at the shore side connectivity with the introduction of better communication technologies. This is at level 9 if the improvement is the only addition of existing but, expensive internet technologies, namely the InMarsat. However, with the introduction of advanced technology such as the merging network the TRL is at 6. Also, with the introduction of remote controlled vessels, a shore control center becomes vital for control and monitoring of information from the shore side. Automatic mooring system and a charging station are classified as TRL 9, with the existence of fully operational equipment developed by Cavotec [7, 21].

6   Scenarios of ASVs in Ports

Based on the research conducted, there are various possibilities regarding ship technology and ASVs. Various additions are necessary at the port infrastructure development level to accommodate the ASVs and achieve efficient operations. This section introduces ASV scenarios which would include the ship technology, applications involved and the port infrastructure.

With respect to the analysis of ASV applications and the technology development of vessels and port infrastructures, 11 different scenarios are designed for the implementation of ASV in ports, see Table 4. The scenarios range from scenario 0 which refers to the current situation to scenario 10 which refers to the most advanced scenario for ASVs. Each of the defined scenarios has been classified approximately based on their TRL which corresponds to another approximation of the timeline of implementation, which is as follows, 0–10 years is categorized as short-term, 10–25 years is classified as mid-term and 25 years onwards is classified as long-term implementation. TRL 7–9 is classified as short-term, TRL 3–6 is classified as mid-term and TRL 1–3 is classified as long-term implementation scenarios.

Scenario 0 is an indication of the current scenario, an existing vessel in combination with existing port infrastructures. It is classified as TRL 9 and it is a baseline for other scenarios which have been formulated.

Scenario 1 describes the scenario the existing vessel being equipped with additional sensors and berthing assistance at the quayside. The vessels in this scenario are for cargo applications. Implementation of additional sensors would reduce the handling time on the quayside and hence improve productivity. To realize Scenario 1, a few additions would be required on the port, such as LDR–LRS which helps the vessel during berthing operations in coordination with the AIS. Scenario 1 can be cumulatively classified as TRL 8–9, since the LDR-LRS system has been successfully implemented in the Port of Koper, Slovenia [26].

The next scenario features the introduction of an improved communication system on board. The introduction of an improved system would result in better connectivity and monitoring between the ship and the shore. This implementation has been parted based on the involved applications, namely, short-sea, inland shipping and deep-sea shipping. At this stage, the task of voyage scheduling, path planning and collision alarm could all be done by computers. More information is needed to support situation

Table 4. Overview of ASV Scenarios.

Scenario	Description		Applications				TRL
	Vessel	Port	Cargo	Service	Military	Research	
0	Existing Vessel	Existing infrastructure					9
1	Additional Sensors	Navigation Assistance (LDR-LRS)	•				8–9
2	Additional communication methods	Information & Navigation Assistance (Mobile Buoys)	• SS & IS		•	•	8–9
3	Additional communication methods	Information & Navigation Assistance (Mobile Buoys)	• DS				7–9
4	Remotely controlled with crews	Shore control center	•				7–8
5	Remotely controlled with reduced crews	Terminal (Semi-automated)	• SS & IS		•	•	5–7
6	Remotely controlled with reduced crews	Navigation Assistance (Autonomous Tugboat Automatic Mooring)	• DS			•	4–6
7	Unmanned remotely controlled vessel	Navigation Assistance (Autonomous Tugboat Automatic Mooring)	• SS & IS	•	•	•	3–5
8	Unmanned remotely controlled vessel	Navigation Assistance (Autonomous Tugboat Automatic Mooring)	• DS	•		•	3–4
9	Unmanned Autonomous Vessel	Terminal (Automated)	•	•		•	1–3
10	Unmanned Autonomous Zero-emission Vessel	Terminal (Automated) & Charging Station	•	•		•	1–2

NB: SS – Short-sea Shipping; IS – Inland Shipping; DS – Deep-sea Shipping

awareness and better scheduling algorithms. Introducing extra source of information, such as mobile buoys which collecting and sending information to vessels could improve navigation assistance. Moreover, optimization models and algorithms are needed for planning and executing complex logistics tasks, e.g., for finding the most efficient scheduling. The scenario with reference to the short-sea and inland shipping applications is classified as TRL 8–9. The Scenario 3 which involves deep-sea shipping is classified as TRL 7–9 since it also involves the implementation of switching and meshing technology to optimize operations at the vessel and the port level [22, 42].

Scenario 4, called remotely controlled vessel with the crew on board and shore control center, has an approximate TRL 7–8. The vessel should be equipped with remote control technology, and the crews are the back-up for key operations (like berthing), dealing with unexpected situation and maintenance. For remote control, a shore control center is needed to facilitate the vessels. The first four scenarios are classified as short-term implementations based on their TRLs. For these scenarios, the crews are still the core of the control of the vessels. Technologies for situation awareness, scheduling algorithms and control strategies are used to support the crews to make better decisions.

Scenario 5 to Scenario 8 are remotely controlled vessels, which fall under the category of mid-term implementation. Scenario 5 is when the vessel is remotely controlled with the reduced crew on board. This scenario is for short-sea and inland shipping applications with the port being equipped with a semi-automated terminal. The cooperation between port and vessel becomes closer. They have to communicate frequently, and a good scheduling method is needed to avoid congestion at the port. Since the vessel function with reduced crew, maintenance strategies would need to be formulated for both the vessel and the port. Scenario 6 extends the applications to deep-sea shipping. Since sea-going vessels usually have poor maneuverability, autonomous tugboats and automatic mooring system are introduced to assist in handling operations at the port. Scenario 5 has an approximate TRL of 5–7, whereas Scenario 6 has an approximate TRL of 4–6. Then, Scenario 7 further reduces the crew on board. It refers to the implementation of an unmanned remotely controlled vessel for short-sea and inland shipping applications. The unmanned deep-sea shipping is included in Scenario 8. Scenario 7 has an approximate TRL of 3–5, whereas scenario 8 has an approximate TRL of 3-4.

The last two scenarios are classified as long-term implementations since they are of TRL 1-3 and 1–2 respectively. Scenario 9 refers to a completely autonomous and unmanned vessel for all the involved applications and hence this requires an advanced situational awareness technology, control strategy as well as modified maritime laws for operational purposes. The port should be completely automated to handle these vessels, which includes automatic quay cranes, stacking equipment and so on. The final scenario includes a completely autonomous unmanned and zero-emission vessel to improve the sustainability of operations being carried out. These two scenarios are still in a conceptual stage and on paper since a large amount of technological as well as jurisdictional advancements must be made before their implementation.

The formulation of ASV scenarios helps in the understanding of probable steps to be undertaken to realize autonomous shipping. The above 11 scenarios have been formulated after careful analysis of the development of technology and port infras-tructure with the assistance of TRL. For instance, one of the earlier steps towards autonomous shipping would be to improve the technology onboard the vessel by installing additional sensors which can be followed by reduced crew on board with the ability to remotely control the vessel. Final few steps towards autonomous shipping would involve, autonomous control strategies onboard the vessel as well as an advanced terminal that can accommodate and assist these ASVs. All these develop-ments are firstly realized in short sea and inland shipping as the challenges facing these two application areas are relatively few when compared to deep sea shipping.

7 Conclusions

Autonomous ship technologies are needed to allow vessels to operate autonomously, both individually, as well as in groups: (a) accurate sensors and data fusion for situation awareness; (b) computational logistic technologies and control strategies for scheduling and path planning and motion control are particularly important for deciding on what action to take according to information received; (c) enhancing the vessel to vessel and vessel to infrastructure communication and cooperation provides significant assistance for safe and efficient navigation. At the same time, infrastructures in ports need to be upgraded to meet the requirements for autonomous vessel applications, ranging from information service, navigation assistance, to future shore control center and fully autonomous terminal.

By analyzing the development of ship and port infrastructure technologies (using the Technology Readiness Level (TRL) concept), 11 future scenarios of autonomous vessels have been identified. Embedding smart ship equipment into existing vessels is the most likely first step towards obtaining fully autonomous ships, followed by remotely controlled vessels with reduced crews. From the perspective of applied areas, autonomous vessels are more likely to be implemented in inland and short sea shipping than deep sea shipping. Firstly, shore-based sensors and control centers can help to improve the accuracy of information. Secondly, inland and short sea ships are relatively small. Their maneuverability is better and control technology is more mature. Thirdly, the connectivity between vessels can be guaranteed for inland and short-sea ships. Moreover, dealing with the legal aspects of the local application is easier with fewer stakeholders. Last but not the least, maintenance is a big technical challenge that needs to be considered. For inland and short-sea shipping maintenance may be less of an issue due to shorter sailing times and easier accessibility.

Acknowledgment. This research is partially supported by SmartPort project 'TET-SP: Autonomous shipping in the Port of Rotterdam' (2017) and the China Scholarship Council under Grant 201406950041.

References

1. Alves, J.: Vehicle and mission control of the DELFIM autonomous surface craft. In: 14th Mediterranean Conference on Control and Automation (2006)
2. Anton Tvete, H., Engelhardsten, O.: DNV GL's research within autonomous systems: Nor-shipping workshop – Professional insight on unmanned ships. http://www.unmannedship. org/munin/wp-content/uploads/2015/06/MUNIN-Workshop-1-2-DNVGL-DNV-GL%E2% 80%99s-research-within-Autonomous-Systems.pdf. Accessed 10 Sept 2017
3. Barton, T.: The DARPA ACTUV program. In: Presentation, Autonomous Ship Technology Symposium (2016)
4. Bates Ramirez, V.: Singularity Hub: The World's First Autonomous Ship Will Set Sail in 2018. https://singularityhub.com/2017/07/30/the-worlds-first-autonomous-ship-will-set-sail-in-2018/. Accessed 18 Sept 2017

5. Buch, T., Kurowski, M.: MESSIN – An autonomously operating unmanned surface vehicle (2013). https://www.innomar.com/wssa2013/wssa2013-P03-Buch.pdf. Accessed 09 Sept 2017

6. Burmeister, H.C.: Autonomous navigation results from the MUNIN testbed. In: Presentation, Autonomous Ship Technology Symposium (2016)

7. Cavotec:　E-Charging.　http://www.cavotec.com/zh/your-applications/ports-maritime/e-charging. Accessed 20 Oct 2017

8. Chopra, K.: Marine Insight: What are Tug Boats? (2015) http://www.marineinsight.com/types-of-ships/what-are-tug-boats/ Accessed 12 Sept 2017

9. DARPA: ACTUV Unmanned Vessel Helps TALONS Take Flight in Successful Joint Test (2016). https://www.darpa.mil/news-events/2016-10-24. Accessed 15 Sept 2017

10. DNV GL: The ReVolt, a new inspirational ship concept. https://www.dnvgl.com/technology-innovation/revolt/index.html. Accessed 09 Sept 2017

11. FleetBroadband - Inmarsat. https://www.inmarsat.com/service/fleetbroadband. Accessed 26 Oct 2017

12. Fleet-class Common Unmanned Surface Vessel (CUSV). http://www.textronsystems.com/sites/default/files/resourcefiles/TS_US_CUSV_Datasheet.pdf. Accessed 26 Oct 2017

13. Jha, S.K.: Emerging technologies: impact on shipbuilding. Marit. Affairs J. Natl. Marit. Found. India **12**(2), 78–88 (2016)

14. Jokioinen, E.: Advanced Autonomous Waterborne Applications (AAWA) initiative. Presentation, Autonomous Ship Technology Symposium (2016)

15. Jokioinen, E. Remote and Autonomous Ships – The next steps. http://www.rolls-royce.com/~/media/Files/R/Rolls-Royce/documents/customers/marine/ship-intel/aawa-whitepaper-210616.pdf. Accessed 19 Sept 2017

16. Kongsberg: Autonomous Shipping. https://www.km.kongsberg.com/ks/web/nokbg0240.nsf/AllWeb/597733F8A1B8C640C12580AC0049C134?OpenDocument. Accessed 22 Sept 2017

17. Levander, O.: The journey towards unmanned ships in the ship intelligence era. In: Presentation, Autonomous Ship Technology Symposium (2016)

18. Li, S.: Coordinated Planning of Inland Vessels for Large Seaports. TU Delft, The Netherlands (2016)

19. Llyod's Register.: Cyber-enabled ships: ShipRight procedure – autonomous ships, 1st edn. (2016)

20. Mayflower Autonomous Ship. http://www.mayflowerautoship.com/. Accessed 16 Sept 2017

21. MoorMaster: Automated Mooring Systems. http://www.cavotec.com/uploads/2017/05/31/flyercavotec-moormaster11042017ld.pdf. Accessed 20 Oct 2017

22. Mu, L., Prinz, A.: Delay-oriented data traffic migration in maritime mobile communication environments. In: Proceedings of Fourth International Conference on Ubiquitous and Future Networks (ICUFN), pp. 417–422. Phuket (2012)

23. Munin's Objectives and Impact. http://www.unmanned-ship.org/munin/about/munins-objectives/. Accessed 09 Sept 2017

24. Naval-technology.com: Fleet-class Common Unmanned Surface Vessel (CUSV), USA. http://www.naval-technology.com/projects/fleet-class-common-unmanned-surface-vessel-cusv/. Accessed 15 Sept 2017

25. Opensea.pro: Dry bulk market – Shall we trade short sea or deep sea. from https://opensea.pro/blog/short-sea-vs-deep-sea. Accessed 05 Oct 2017

26. Perkovic, M., Gucma, M., Luin, B., Gucma, L., Brcko, T.: Accommodating larger container vessels using an integrated laser system for approach and berthing. Microprocess. Microsyst. **52**, 106–116 (2017)

27. Piris, A.O., Díaz-Ruiz-Navamuel, E., Pérez-Labajos, C.A., Chaveli, J.O.: Reduction of CO_2 emissions with automatic mooring systems. The case of the port of Santander. Atmospheric. Pollut. Res. **9**(1), 76–83 (2017)

28. Port Technology: Asia Enters Fully Automated Terminal Era. https://www.porttechnology. org/news/asia_enters_fully_automated_terminal_era. Accessed 05 Oct 2017

29. Robert Allan Ltd.: Revolutionary RAmora brings Tele-operated capability to ship handling. http://ral.ca/2015/09/18/revolutionary-ramora-brings-tele-operated-capability-to-ship-handling/. Accessed 20 Sept 2017

30. Rolls-Royce: Autonomous Ships, The next step. http://www.rolls-royce.com/~/media/Files/ R/Rolls-Royce/documents/customers/marine/ship-intel/rr-ship-intel-aawa-8pg.pdf. Accessed 08 Sept 2017

31. Rolls-Royce: Rolls-Royce reveals future shore control centre. https://www.rolls-royce.com/ media/our-stories/press-releases/2016/pr-2016-03-22-rr-reveals-future-shore-control-centre. aspx. Accessed 15 July 2018

32. Schiaretti, M., Chen, L., Negenborn, R.R.: Survey on autonomous surface vessels: Part II - categorization of 60 prototypes and future applications. In: Proceedings of the 8th International Conference on Computational Logistics (ICCL 2017), Southampton, UK, pp. 234–252 (2017)

33. Mayflower Autonomous Research Ship (MARS), UK. http://www.ship-technology.com/ projects/mayflower-autonomous-research-ship-mars/. Accessed 16 Sept 2017

34. NASA: Technology Readiness Level (2012). https://www.nasa.gov/directorates/heo/scan/ engineering/technology/txt_accordion1.html. Accessed 08 Sept 2017

35. Stapleton, R.: Revolutionary RAmora brings tele-operated capability to ship-handling tugs. In: Presentation, Autonomous Ship Technology Symposium (2016)

36. The Engineer: Rolls-Royce and Svitzer demonstrate world's first remotely operated commercial vessel (2017). https://www.theengineer.co.uk/rolls-royce-and-svitzer-demonstrate-worlds-first-remotely-operated-commercial-vessel/. Accessed 20 Sept 2017

37. The MUNIN Consortium. http://www.unmanned-ship.org/munin/partner/. Accessed 09 Sept 2017

38. UST: Unmanned Marine Systems. http://www.unmannedsystemstechnology.com/company/ autonomous-surface-vehicles-ltd/. Accessed 17 Sept 2017

39. Wingrove, M.: Tug Technology & Business: Svitzer and Rolls-Royce demonstrate remote control tug operations. http://www.tugtechnologyandbusiness.com/news/view,svitzer-and-rollsroyce-demonstrate-remote-control-tug-operations_48281.htm. Accessed 20 Sept 2017

40. Wong, K.: Jane's 360: Imdex 2017: ST Electronics Venus 16 USV development enters final lap. http://www.janes.com/article/70489/imdex-2017-st-electronics-venus-16-usv-develop ment-enters-the-final-lap. Accessed 11 Sept 2017

41. Wróbel, K., Montewka, J., Kujala, P.: Towards the assessment of potential impact of unmanned vessels on maritime transportation safety. Reliab. Eng. Syst. Safety **165**, 155–169 (2017)

42. Yue Kwok Wai, A.: ST Electronics: Unmanned Surface Vehicle (2016). http://www.mpa. gov.sg/web/wcm/connect/www/01146343-81f0-4dec-8ccd-ef33c82e2240/Presentation+-+Andrew+Yue.pdf?MOD=AJPERES. Accessed 19 Sept 2017

43. Zhou, M., Harada, H.: Cognitive maritime wireless mesh/ad hoc networks. J. Netw. Comput. Appl. **35**(2), 518–526 (2012)

Survey on Short-Term Technology Developments and Readiness Levels for Autonomous Shipping

Laurien E. van Cappelle, Linying Chen[(✉)], and Rudy R. Negenborn

Department of Maritime and Transport Technology,
Delft University of Technology, Delft, The Netherlands
{L.chen-2,R.R.Negenborn}@tudelft.nl

Abstract. Recently, Autonomous Surface Vessels (ASVs) have attracted a lot of attention. Developing a fully autonomous vessel is challenging. Existing research provides a track from existing manned vessels to a remote-controlled vessel with reduced crews, an unmanned remote-controlled vessel, and at the end, a fully autonomous vessel. The first step is to equip existing vessels to realize autonomous sailing. In this paper, we focus on the technologies that make existing vessels "smarter". A categorization of technologies is provided based on the basic architecture of ASV: Navigation, Guidance, Control and Hardware. An overview of the technology developments in each category is presented. The Technology Readiness Level (TRL) is applied to indicate whether these technologies could become commercial in the short term.

Keywords: Autonomous surface vessel · Technology readiness level
Short-term technology development · Review

1 Introduction

Autonomous Surface Vessels (ASVs) have attracted a lot of attention. In [44], an overview of existing ASV projects has been provided. It shows the track from existing maned vessels to remote-controlled vessels with reduced crews, unmanned remote-controlled vessels, and at the end, fully autonomous vessels. Existing papers mostly focus on the last two steps, such as [27,44]. They usually assume the vessels are newly built. However, the number of merchant fleet in the world now is more than 90,000 [52]. Discarding existing vessels is unrealistic and leads to a great waste. Moreover, developing a newly built fully autonomous vessel is a challenging and calls for massive investment. In comparison, to equip existing vessel to realize autonomous sailing is more economical and practical.

In this paper, we focus on the first step, making existing vessels "smarter". The aim is to answer the question "how the vessel technology is going to change in the next 5 to 10 years". An overview of the technologies related to autonomous shipping is provided. We use the Technology Readiness Level (TRL) to indicate

© Springer Nature Switzerland AG 2018
R. Cerulli et al. (Eds.): ICCL 2018, LNCS 11184, pp. 106–123, 2018.
https://doi.org/10.1007/978-3-030-00898-7_7

the maturity of the technologies, i.e., whether these technologies will become commercial in the short term.

This paper is organized as follows: Sect. 2 provides the categorization of ASV technologies and the indicator of technology maturity; an overview of the technology developments with corresponding TRL is presented in Sect. 3; Sect. 4 provides the concluding remarks of this paper.

2 ASV Technologies and Technology Readiness Level

In this section, we classify the technologies related to ASvs into different categories according to their functions. TRL is introduced as an indicator of technology maturity.

2.1 ASV Technologies

An ASV needs different parts to perform different functions. In [7,9,27,43], different categorizations of the subsystems of a typical ASV are provided. Generally, the basic subsystems that are needed for autonomous navigation include 4 parts, as shown in Fig. 1: Navigation, Guidance, Control, and Hardware.

The **Navigation** system of the vessel provides its own states and surrounding information for the decision makers. The **Sensor Fusion** is a software-based

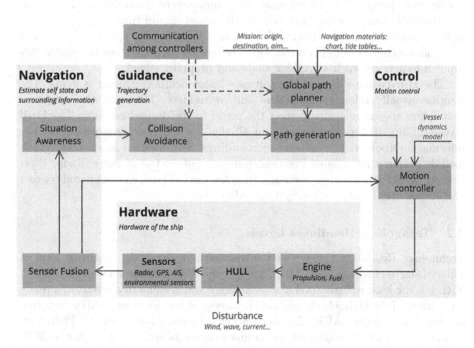

Fig. 1. Subsystems of a ASV [43]

system that combines the information from different sensors to create a visualization of the real world. To create a complete real-life representation of the surroundings is still challenging. Therefore, nowadays, the representation is used as a support system for the Officer On Watch (OOW). Another important function of the navigation subsystem is **Situation Awareness (SA)**. SA involves being aware of what is happening in the vicinity to understand how information, events, and one's own actions will impact goals and objectives, both immediately and in the near future [1]. One example of SA is recognizing objects with collision risks from the picture created by the Sensor Fusion. On current manned vessels, SA is usually done by OOW. Lacking or inadequate SA has been identified as one of the primary factors in accidents attributed to human error.

The **Guidance** system deals with the questions "when will the vessel arrive at which place through which path". The final result is an optimal collision-free path that a vessel should follow. The **Global path planner** uses optimization models and algorithms to make schedules and find the most efficient path for executing the schedules. With the information provided by the Navigation system, the **Collision Avoidance (CA)** block updates the global path to avoid obstacles if necessary. **Communication** among vessels and infrastructures can help to negotiate and cooperate with others and to make better decisions [10]. For existing vessel, communication is usually done by OOW using radio (Very High Frequency, VHF) and (mobile) phones.

With the path decided by the Guidance system, the task of **Motion Controller** is to process this information into commands to the actuators. For example, the path generator could state that the vessel should turn right to an angle of 30° compared to the current position and increase the speed with 1 knots to avoid another vessel. Then the software-based control system translates this input into actions, such as rudder angle and propeller speed.

The **Hardware** supports the software-based decision-making systems. The **Engine** usually refers to propellers and/or rudders. These are the actuators that follow the command and steer the vessel to the desired position. The **Hull** gives stability to the vessel and hold all the components. **Sensors** collect the information about the vessel and surroundings. Sensors used in existing vessels include (differential) Global Position System ((D)GPS), Automatic Radar Plotting Aid (ARPA), Visionary sensors, Internal navigation, environmental sensors and Automatic Identification System (AIS), etc.

2.2   Technology Readiness Levels

Technology Readiness Levels (TRLs) are used to assess the maturity level of a particular technology [34]. There are 9 technology readiness levels, see Fig. 2. TRL 1 is the least mature level of a technology and is equal to a report of a really basic idea, while TRL 9 is the highest, a successful mission operation of an actual system. For example, ARPA has been applied to vessels for decades. Therefore, it is at TRL 9. On the contrary, as an innovative concept, the Waterborne AGV [61] is at TRL 2. In this paper, TRLs are used to indicate which technologies will be applied in reality in the short term.

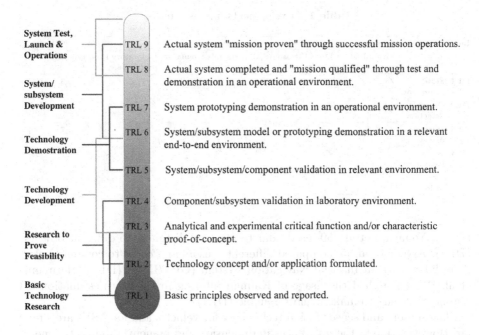

Fig. 2. Technology readiness levels [34]

3   Technology Developments

In this section, the technology developments towards autonomous sailing in the four subsystems and their TRLs are presented. A discussion on which developments could become commercial within 5 to 10 years is provided at the end.

For easier expression, each technology is labeled as 'D$s.n$', where s indicates the subsystem that the technology belongs to (1 means Navigation, 2 means Guidance, 3 means Control, 4 means Hardware, 5 means others), n is the ranking of the technology according to TRL. For example, D1.1 means the Navigation technology at the lowest TRL of the considered technologies.

3.1   Navigation

Sensor Fusion. Sensor fusion aims at using available information from different resources to create a representation of the real world. Table 1 gives the results of sensor fusion technologies that have been used or mentioned in existing research.

The sensor fusion technologies for cars are already at a high level. Tesla was already able to achieve a good sensor fusion by making use of ultrasonic, radar and visual cameras (D1.5) [48]. This was tested in an operational environment near a windfarm. A visual camera made a 2D image to fill in the missing information. This technology was implemented and tested on a golf cart. Wolken [55] created a sensor fusion by using a vessel-lidar system that is able to measure wind

Table 1. Developments in sensor fusion

Label	Description	Sensors					Concept				TRL	Ref.
		Visual	Radar/ ARPA	Lidar	Ultrasonic	AIS	Principle	Simulation	Laboratory prototype	Prototype		
D1.1	State estimation	✓	✓				✓				3	[18]
D1.2	Obstacle detection			✓	✓				✓		4	[3,5]
D1.3	Obstacle detection	✓	✓						✓		7	[59]
D1.4	Wind measuring		✓							✓	7	[55]
D1.5	Sensor fusion	✓	✓	✓					✓		7	[48]

(D1.4). Youngam et al. [59] used lidar to measure distances to different objects (D1.3). Asvade et al. [5] compared different sensors and chose to combine Velodyne lidar with an Internal Navigation System (GPS/IMU) (D1.2). Hermann et al. [18] researched the usage of Kalman filtering in the state estimation of autonomous cars to make it more reliable (D1.1).

The sensors and sensor fusion technology for vehicles and for ASVs are similar. However, due to the differences in dimension and maneuverability, the measuring range, accuracy of sensors and the results of sensor fusion for ASVs are different from those for vehicles. In present, the most representative technology for ASVs is AIS, combining the information from a positioning system such as a GPS receiver, with other electronic navigation sensors, such as a gyrocompass or rate of turn indicator. AIS is intended to assist the OOW to supervise the state of the vessel, detect obstacles and allow maritime authorities to track and monitor vessel movements [3].

Situation Awareness. Understanding what is happening in the surrounding area is essential for the controllers taking actions. Maq et al. [28] proposed an obstacle detection method by making use of the Fuzzy C-Means (FCM) (D1.10). This method was able to create an artificial intelligence that classifies and identifies objects seen on the ARPA. In case of a real vessel, this method was able to recognize a vessel with an accuracy of 91.3%; for a noise, bank or channel target, the accuracy was between 82.6% and 91%. The system detects in real-time obstacles with a range up to 175m. Hermann et al. [18] tested a radar and visual based obstacle detection system successfully on an autonomous vehicle for speeds up to 30 m/s (D1.9). Yalcin et al. [58] proposed an obstacle and road detection for autonomous cars by only using lidar (D1.8). In the system developed by [20], HiCASS, the information from AIS, ARPA and Electronic Chart Display Information System (ECDIS) are combined to create an SA that can view up to 50 km (D1.7). Rødseth and Burmeister [39] stated that good obstacle detection and avoidance can reduce the number of accidents by providing a decision support for the OOW (D1.6) (Table 2).

Table 2. Developments in situation awareness

Label	Description	Autonomy level			Concept				TRL	Ref.
		Supporting	Autonomous with supervision	Autonomous	Principle	Simulation	Laboratory prototype	Prototype		
D1.6	Situation awareness	✓			✓				1	[39]
D1.7	Multiple sensors for SA		✓		✓				3	[20]
D1.8	Lidar for SA		✓		✓				4	[58]
D1.9	Radar for SA		✓				✓		5	[18]
D1.10	ARPA for obstacle detection	✓						✓	7	[28]

According to these studies, human is the core of SA on existing vessels. SA technologies mostly play a supporting role for the OOW. More efforts should be made to apply them for fully autonomous vessels.

3.2 Guidance

Global Path Planning. The optimization of the global path can greatly improve the efficiency of transport. The development of Computational Logistics technology provides a great support for ship scheduling and path planning. [11,12] provided detailed review research on ship routing and scheduling. Typically, an optimization problem is formulated to find the most efficient scheduling for the transport of goods. Regarding path planning, based on [7,13,27,46], existing methods can be classified into three categories, Line-of-Sight (LOS), Potential field methods, Heuristic search algorithms and Evolutionary algorithms.

LOS is a successful guidance technique that is widely employed today, particularly in missile guidance technology [7] (D2.4). The idea behind LOS guidance is that if the vessel converges to a constant LOS heading angle directly between the vessel and target, it eventually converges to the target position. The disadvantage of LOS guidance is potential overshoot caused by reducing the cross-track error due to environmental disturbances [33].

Potential field methods (D2.3) take known obstacles into consideration by building a representation of the environment by potential gradients. Potential Field methods are first proposed by [23] for mobile robots. [57] implemented the potential field method for automatic ship navigation. It shows that the method is effective for ships involved in a complex traffic situation.

Heuristic search algorithms indicate those grid-searching techniques with associated heuristic cost functions (D2.2). A feasible, near-optimal path is found without performing an exhaustive search, as with uninformed (or blind) graph searching algorithms such as Breadth-first or Depth-first searches [7]. Among the group of heuristic search algorithms, A* and its extensions are commonly used to determine the path from an origin to a destination for land-based vehicles [42,53]. A comparison between A* and its extensions, i.e., A* with Post-smoothing, Theta*, and A* on Visibility Graphs, is shown in [9].

Evolutionary algorithms are increasingly employed in the design of path planners inspired by the behavior of biological systems (D2.1) [27]. [15] introduces a solution to the problem of planning for marine vehicles based on Ant Colony Optimization. However, when constraints such as obstacles, dynamic limits, and mission constraints must be satisfied, the method can be time-consuming.

The global path planning system is software-based, and the Computational Logistics technologies are relatively mature. Thus, the above-mentioned methods are all with high TRL levels (8–9).

Collision Avoidance. Ship collision is one major threat to navigation safety. Research on CA is dedicated to finding methods to detect collision dangers as early as possible and to find proper collision-free solutions. Table 3 provides several examples of current collision avoidance technologies. It also provide the autonomy level of the CA system using these technologies: does the OOW take actions (Supporting)? does the CA process need supervision of the OOW? is the CA process done autonomously?

The methods mentioned in existing research can be roughly divided into two types. One is the indicator-based. Some indicators have been defined to help to determine the collision risks and the actions should be taken, such as Distance at Closest Point of Approach (DCPA) and Time to the Closest Point of Approach (TCPA). Wang et al. [54] proposed a dynamic CA system that calculates the DCPA and TCPA (D2.8). Lazarowska et al. [25] proposed a concept where a new Decision Support System uses a Trajectory Base Algorithm (TBA) (D2.6). Tsou et al. A system proposed in [51] which based on AIS and ECDIS data shows a Predicted Area of Danger (PAD) for the vessel at that moment (D2.5). The OOW can choose actions to avoid the marked area.

The other one is rule-based. The rules of the road specify the types of maneuvers that should be taken in situations where there is a risk of collision. The most widely used is the Convention on the International Regulations for Preventing Collisions at Sea (COLREGS) set by the International Maritime Organization (IMO). Hyundai [20] developed the CA system named HiCASS. This system can analyze the locations of the objects and avoid them with respect to COLREGs (D2.12). The system was tested on a 13,800 TEU LNG carrier. The research of Hu et al. In [22], a CA system was proposed using an MPC method to predict the trajectories of (moving) obstacles. Then, own vessel avoids it with compliance with the rules (D2.10). Xu et al. [56] used the danger immune algorithm to find a set of operation instructions obeying the COLREGs (D2.9). Zhang et al. [60] designed a CA strategy where vessels work in cooperation to avoid collisions with respect to the COLREGs (D2.7).

There are methods combined the two type of approaches. For example, [19] used the DCPA and TCPA to detect a possible collision, and then generated new paths which compliance with the COLREGs (D2.11).

Table 3. Developments in collision avoidance

| Label | Description | COLREGs | CA indicator | Predictive? | Autonomy level | | | | Concept | | | | | TRL | Ref. |
					Supporting	Autonomous with supervision	Autonomous	Principle	Simulation	Laboratory prototype	Prototype			
D2.5	Plot PAD to create CA	✓	✓		✓				✓				3	[51]
D2.6	TBA for CA		✓			✓			✓				3	[25]
D2.7	Distributed over vessels	✓		✓			✓		✓				5	[60]
D2.8	Real-time DCPA and TCPA		✓		✓				✓				5	[54]
D2.9	Collision avoidance	✓				✓			✓				6	[56]
D2.10	Collision avoidance by MPC	✓		✓			✓		✓				6	[22]
D2.11	Collision avoidance	✓	✓				✓		✓				6	[19]
D2.12	Collision avoidance	✓				✓						✓	7	[20]

In the literature, many methods have been proposed for CA. However, there are challenges. Most CA methods rely on the prediction of trajectories of own ship or obstacles. However, due to the environmental disturbances and inaccurate ship motion models, the precision of predictions is not always sufficient.

Communication. Vessel-to-Vessel (V2V) and Vessel-to-Infrastructure (V2I) communication have many benefits [10]. Moreover, to make vessel control from the quayside possible, the communication methods and data stream capacity have to be improved [39].

The Internet is one of the options for V2V and V2I communication. To solve the problems of limited coverage and low access speed, SpaceX [6] tries to achieve worldwide internet connectivity with the launch of 800 low-orbit satellites in 2019 (D2.20). Google [64] plans to launch 180 satellites to provide the earth with a worldwide Internet connectivity (D2.19). Also, Facebook [49] is doing test-trails with solar-based drones to provide future worldwide internet (D2.17). Another way to achieve a better Internet connection is by making smarter use of the current availability's. The ESA [29] is now working on a two-way communication device between vessel and shore, which uses WiFi, 3G/4G, Very Small Aperture Terminal (VSAT) and INMARSAT (D2.18). Mu and Zhou [62] and Harada [31] proposed ad hoc networks between vessels to provide an Internet connection further of the coast. Ejaz et al. [14] proposed a meshing network between neighbouring vessels (D2.14), but in a more conceptual way. A switching device that chooses the cheapest and fastest option from the available sources at that moment can also achieve smarter use of the Internet resources. Mu et al. [32] proposed a device that switches between the different types of the Internet (D2.15): increases the data stream in case of WiFi connection and decreases it in the case of the satellite connection. Sumić et al. [47] thought to make a better use of terrestrial Internet sources would be more efficient (D2.13) (Table 4).

According to these developments, we can find many possible developments that can achieve better maritime communication, and they are likely to happen on a very short notice.

3.3   Motion Controller

For vessels, the motion control part is challenging, as the sailing is highly influenced by environmental disturbances, such as wind and water conditions.

GE marine solutions [16] designed a Vessel Control System (VSC) which is able to provide full remote supervisory control and monitoring of all ship systems (D3.5). In the Netherlands there is currently an inland vessel, called the MSC Saluté, sailing semi-autonomous (D3.4) [37]. It can follow an earlier recorded track by changing rudder angles. The MSC Saluté was able to follow the recorded track, but still needed some interventions of the captain when there are disturbances. Sørensen and Breivik [45] did a comparative research on four different control methods through simulations. Although all results seemed promising, the so-called $\mathcal{L}_1$ adaptive backstepping method with command governor achieved the best results (D3.3). Alfi et al. [2] make use of the well known

Table 4. Developments in communication

Label	Description	New on-board?	Technology Existing	Cooperation between vessels	New technology	Concept Principle	Simulation	Laboratory prototype	Prototype	TRL	Ref.
D2.13	Use of terrestrial Internet	✓	✓			✓				2	[47]
D2.14	Multi-hop network	✓		✓		✓				3	[14]
D2.15	Switching between Internet sources	✓	✓				✓			6	[32]
D2.16	Mesh network among neighbouring vessels	✓		✓			✓			6	[31, 62]
D2.17	Worldwide Internet Facebook				✓				✓	7	[49]
D2.18	ESA two way communication		✓						✓	8	[29]
D2.19	Worldwide Internet Google				✓				✓	9	[64]
D2.20	Worldwide Internet SpaceX				✓				✓	9	[6]

$H\infty$ performance formula that is able to follow waypoints (D3.2). Zhu et al. [63] researched the possibility of capturing the highly nonlinear dynamics of a vessel in a simplified model (D3.1).

To conclude, the maneuverability of vessels is poor and the reaction time is extremely long. Moreover, the steering commands react differently in the different wind or water conditions. Therefore, accurate control of large cargo vessels could be possible, but there are still some big steps to make.

3.4   Hardware

Engine. One of the advantages of autonomous sailing is reducing emissions. The expectation is that cargo vessel will also become more environmentally friendly. The development of engines from cargo vessels is mainly based on the environmental friendliness of it. On short-term, this means the upcoming of hybrid or LNG powered vessel. Eventually, the engines could change into hydrogen or fully electric engines.

The Royal Academy of Engineering [40] carried out a multi-discipline study on different types of engines. On the short-term, they expect that diesel will still remain important propulsion method and LNG powered vessels will follow shortly (D4.4). As alternative propulsion methods, they introduced gas turbines and hybrid propulsion. The AMS [4] designed the newest LNG-electric driven inland tanker, called the Ecotanker III (D4.3). It already achieved a CO_2 reduction of 20% to 25%. AMS designed eight environmental friendly inland vessels in total, which are all currently in use. Guangzhou Vesselyard International Company Ltd. designed a full-electric cargo vessel and is now able to travel a distance of 80 km without charging (D4.2) [26]. Geertsma et al. [17] investigated the possibilities for future propulsion and concluded that hybrid propulsion would be a good alternative for the future (D4.1).

Sensors. SA relies on the information provided by the sensors. As the above section about SA already described the combination of sensors, this section focuses on individual sensors and supporting sensors as wind measuring sensors. Yalcin et al. [58] did a comparative research between radar, lidar, and ultrasound (D4.8). The results showed that lidar was the best option for obstacle recognition. Radar and ultrasound can also measure distance, but the obstacle identification is more difficult. The CoVadem-project [8] created a network where inland vessels can share their depth measurements (D4.7). When there are accurate depth measurements, vessel companies can optimize their routing and loading. Sakib [41] created a low-cost digital Inertial Measurement Unit (IMU) that provides the captain with a detailed hydrographic survey to make traveling safer for container vessels (D4.6). With the findings from Yalcin et al. [58], we can find that although the results of lidar imaging are really accurate, the technology is extremely expensive (D4.5).

3.5   Overall Developments

This section presents the developments of some technologies that needs the cooperation of the above-mentioned subsystems, in particular, the technologies related to maneuvering and maintenance (Table 5).

Table 5. Developments in vessel maneuvering

Label	Description	Asisting devices?	Method		Concept				TRL	Ref.
			Multi-shooting	ANN	Principle	Simulation	Laboratory prototype	Prototype		
D5.1	Platooning	✓			✓				1	[35]
D5.2	ANN for berthing			✓		✓			5	[21]
D5.3	ANN for berthing	✓		✓			✓		6	[50]
D5.4	Quasi real-time control		✓				✓		7	[30]
D5.5	Laser ranging and docking	✓						✓	8	[36]

Vessel Maneuvering. When a vessel is able to berth (semi-) autonomously, one or two sailors could be removed from the vessel, which would reduce the operational costs. Mizuno et al. [30] proposed a quasi real-time optimal control scheme for automatic berthing. By using a multiple shooting method, they were able to even berth a small vessel at sea (D5.4). In [50], berthing vessels autonomously is discussed. An Artificial Neural Network (ANN) is applied with the help of auxiliary devices such as a thruster and a tugboat (D5.3). Another technology of autonomous berthing is using an ANN to learn to berth from the actual captain. A problem with this technology is that it only works on one specific port. Im and Nguyen [21] carried out a research on the ability of this system to also be able to berth in other ports. However, it was only able to berth from one approaching direction. Easier berthing can also be achieved from the quayside. Perkovic et al. [36] developed a laser ranging and laser docking system on the quayside that gives a hydrographic survey on the Port of Koper to the vessel (D5.5). Recently, NMT introduced a new Maritime European innovative project called NOVIMAR, where platooning of cargo vessels is investigated [35] (D5.1). A manned vessel acts as a leading vessel and the rest of the vessels that follow the vessel autonomously.

Maintenance. One of the main problems of autonomous sailing, especially for deep-sea vessels, is that no crew on board for maintenance purposes. An ASV needs a maintenance strategy that enables it to sail for weeks or months without breaking down. Lazakis et al. [24] introduced the Inspection Capabilities

Table 6. Table with developments.

	TRL 1	TRL 2	TRL 3	TRL 4	TRL 5	TRL 6	TRL 7	TRL 8	TRL 9
Navigation	D1.6 [40] ('12) Situation awareness		D1.1 [18] ('15) State estimation; D1.7 [27] ('16) Multiple sensors for SA	D1.2 [5] ('16) Obstacle detection; D1.8 [59] ('13) Lidar for SA	D1.9 [18] ('15) Radar for SA		D1.3 [60] ('16) Obstacle detection; D1.4 [56] ('14) Wind measuring; D1.5 [49] ('17) Sensor fusion; D1.10 [29] ('15) APHA for obstacle detection		
Guidance		D2.13 [48] ('15) Use of terrestrial internet	D2.5 [52] ('16) Plot PAD to create CA; D2.6 [25] ('16) TBA for CA; D2.14 [14] ('19) Multi-hop network		D2.7 [61] ('15) Distributed CA over vessels; D2.8 [55] ('17) Real-time DCPA and TCPA	D2.9 [57] ('14) Evolutionary algorithm; D2.10 [22] ('16) Collision avoidance by MPC; D2.12 [19] ('17) Path re-planning; D2.15 [33] ('12) Switching between Internet sources; D2.16 [32,63] ('15) Mesh network among neighbouring vessels	D2.17 [50] ('14) Worldwide internet Facebook	D2.1 [15] ('17) Collision Avoidance; D2.2 [43,54,9] ('10) Heuristic search algorithms; D2.3 [23,58] ('09) Potential field methods; D2.11 [20] ('14) Collision Avoidance; D2.18 [30] ('14) ESA two way communication	D2.4 [34] ('03) Line-of-Sight; D2.19 [65] ('14) Worldwide internet Google; D2.20 [6] ('16) Worldwide internet SpaceX
Control		D3.1 [64] ('17) Simplified control model			D3.2 [2] ('15) H_∞ controller	D3.3 [46] ('15) L_∞ adaptive back-stepping with command governor	D3.4 [38] ('17) Semi-Autonomous inland vessel		D3.5 [16] ('16) Vessel automation system
Hardware			D4.1 [17] ('17) Hybrid propulsion	D4.5 [59] ('13) Disadvantages lidar		D4.6 [42] ('17) Low-cost Hydrographic Survey		D4.2 [26] ('17) Full-electric propulsion '17; D4.3 [4] ('17) LNG-Electric propulsion '17; D4.7 [8] ('17) Network for depth-measurements	D4.4 [41] ('13) LNG and Hybrid propulsion; D4.8 [59] ('13) Lidar best sensor
Overall	D5.1 [36] ('17) Platooning; D5.6 [40] ('12) Maintenance strategy			D5.7 [39] ('16) Maintenance Framework	D5.2 [21] ('16) ANN for berthing	D5.3 [51] ('17) ANN for berthing	D5.4 [31] ('12) Quasi real-time control for berthing; D5.8 [24] ('16) INCASS innovative maintenance system	D5.5 [37] ('15) Laser ranging and docking for berthing	

a) In each cell, from left to right: label, sources, year of publication. Next line gives the short description.
b) The grey marked area highlights the developments that will become commercial on a short term.

for Enhanced Vessel Safety (INCASS) project about an innovative maintenance system. This is a combination of software and hardware to make maintenance smarter (D5.8). Rødseth [38] proposed a framework for an unmanned engine room in the MUNIN project (D5.7). At last Rødseth and Burmeister [39] stated the importance of better maintenance strategies for ASVs (D5.6). Therefore, some research has been done for the maintenance of autonomous vessels, but that this particular aspect of autonomous sailing still needs more efforts.

3.6   Short-Term Development

Table 6 presents an overview of technology developments towards autonomous navigation. Developments appreciated with TRL 7, 8 or 9 are expected to become commercial on a shorter term (in 5 to 10 years).

The quality of SA increases as the software to detect obstacles becomes better. In the coming years, it will have a good supporting role for the captain. The ability to communicate at sea will make a big increase as multiple companies are setting up worldwide Internet coverage. Vessels will also increase the communication between each other and the quayside. The critical factor in removing crew from the board is at the moment the berthing of the vessel. Therefore a big increase of research towards autonomous berthing is expected, which could lead to a step towards autonomous sailing. The number of sustainable vessels will increase. The engines will be driven by LNG, Hybrid or Electrical propulsion.

4   Conclusions

Existing research provides a track from existing vessels to a remote-controlled vessel with reduced crews, an unmanned remote-controlled vessel, and at the end, a fully autonomous vessel. The first step is to equip existing vessels to realize autonomous sailing. In this paper, we focus on the technologies that make existing vessels "smarter". A categorization of technologies is provided based on the basic architecture of ASV: Navigation, Guidance, Control, and Hardware. An overview of the development of the technologies in each category is presented. The Technology Readiness Level (TRL) is applied to indicate whether these technologies are likely to become commercial in the shorter term.

Based on the analysis, the developments of technologies will bring about a lot of changes on board to make existing vessels smarter in the next 5 to 10 years. Firstly, from the perspective of hardware, the accuracy of the sensors is expected to be improved while reducing the costs. This is the basis of a better situation awareness. Secondly, software-based systems could achieve a breakthrough. Data fusion and situation awareness come to mature and could provide the OOW with more accurate information. With the development of Computational Logistic, ship routing and scheduling can be accomplished without human. More Navigation assistance devices will be equipped on board to facilitate safe navigation. Autonomous trajectory tracking and autonomous collision avoidance system under the supervision of the OOW is expected to be implemented in the near future. Thirdly, autonomous motion control is also promising. Unmanned engine room in large merchant vessels will be realized in the foreseeable future. As the increasing concerns on emission, hybrid propulsion may be in the majority. Last but not the least, the considerable investment on the worldwide Internet will greatly promote the communication and cooperation between vessels and also infrastructures, which also provides the basis for remote control.

Acknowledgment. This research is partially supported by SmartPort project 'TET-SP: Autonomous shipping in the Port of Rotterdam' 2017 and the China Scholarship Council under Grant 201406950041.

References

1. Situation awareness (2018). https://en.wikipedia.org/wiki/Situation_awareness
2. Alfi, A., Shokrzadeh, A., Asadi, M.: Reliability analysis of h-infinity control for a container ship in way-point tracking. Appl. Ocean Res. **52**, 309–316 (2015)
3. All about AIS: History of AIS (2012). http://www.allaboutais.com/index.php/en/aisbasics1/ais-history
4. AMS Transport Crewing Solutions: Factsheet AMS (2017). https://ams-bv.com/images/pdf/Factsheet-NL-AMS-DEF.pdf
5. Asvadi, A., Premebida, C., Peixoto, P., Nunes, U.: 3D Lidar-based static and moving obstacle detection in driving environments: an approach based on voxels and multi-region ground planes. Robot. Auton. Syst. **83**, 299–311 (2016)
6. Brodkin, J.: SpaceX plans worldwide satellite internet with low latency, gigabit speed (2016). https://arstechnica.com/information-technology/2016/11/spacex-plans-worldwide-satellite-internet-with-low-latency-gigabit-speed/
7. Campbell, S., Naeem, W., Irwin, G.: A review on improving the autonomy of unmanned surface vehicles through intelligent collision avoidance manoeuvres. Annu. Rev. Control **36**(2), 267–283 (2012)
8. Challenge, S.S.: Covadem: coöperatieve vaardieptemetingen (2017). https://smartshippingchallenge.nl/initiatieven-en-innovaties/covadem-cooperatieve-vaardieptemetingen
9. Chen, L., Negenborn, R.R., Lodewijks, G.: Path planning for autonomous inland vessels using A*BG. In: Paias, A., Ruthmair, M., Voß, S. (eds.) ICCL 2016. LNCS, vol. 9855, pp. 65–79. Springer, Cham (2016). https://doi.org/10.1007/978-3-319-44896-1_5
10. Chen, L., Hopman, H., Negenborn, R.R.: Distributed model predictive control for vessel train formations of cooperative multi-vessel systems. Transp. Res. Part C Emerg. Technol. **92**, 101–118 (2018)
11. Christiansen, M., Fagerholt, K., Nygreen, B., Ronen, D.: Ship routing and scheduling in the new millennium. Eur. J. Oper. Res. **228**(3), 467–483 (2013)
12. Christiansen, M., Fagerholt, K., Ronen, D.: Ship routing and scheduling: status and perspectives. Transp. Sci. **38**(1), 1–18 (2004)
13. Das, B., Subudhi, B., Pati, B.B.: Cooperative formation control of autonomous underwater vehicles: an overview. Int. J. Autom. Comput. **13**(3), 199–225 (2016)
14. Ejaz, W., Manzoor, K., Kim, H.J., Jang, B.T., Jin, G.J., Kim, H.S.: Two-state routing protocol for maritime multi-hop wireless networks. Comput. Electr. Eng. **39**(6), 1854–1866 (2013)

15. Escario, J., Jimenez, J., Giron-Sierra, J.: Optimization of autonomous ship maneuvers applying swarm intelligence. In: Proceedings of 2010 IEEE International Conference on Systems Man and Cybernetics, pp. 2603–2610 (2010)

16. GE Corperation: GE's marine solutions (2016). http://www.gepowerconversion.com/industries/marine/vessel-automation

17. Geertsma, R.D., Negenborn, R.R., Visser, K., Hopman, J.J.: Design and control of hybrid power and propulsion systems for smart ships: a review of developments. Appl. Energy **194**, 30–54 (2017)

18. Hermann, D., Galeazzi, R., Andersen, J.C., Blanke, M.: Smart sensor based obstacle detection for high-speed unmanned surface vehicle. In: Proceedings of 10th IFAC Conference on Manoeuvring and Control of Marine Craft, MCMC 2015, Copenhagen, 24–26 August 2015, vol. 48, pp. 190–197 (2015)

19. Hu, L., Naeem, W., Rajabally, E., Watson, G., Mills, T., Bhuiyan, Z., Salter, I.: COLREGs-compliant path planning for autonomous surface vehicles: a multiobjective optimization. In: Proceedings of 20th IFAC World Congress, vol. 50, pp. 13662–13667 (2017)

20. Hyundai Heavy Industries Co.: Hyundai heavy develops new collision avoidance system for ships (2014). http://www.ship-technology.com/news/newshyundai-heavy-develops-new-collision-avoidance-system-for-ships-4326091/

21. Im, N.K., Nguyen, V.S.: Artificial neural network controller for automatic ship berthing using head-up coordinate system. Int. J. Naval Architect. Ocean Eng. **10**(3), 235–249 (2018)

22. Johansen, T.A., Cristofaro, A., Perez, T.: Ship collision avoidance using scenario-based model predictive control. In: Proceedings of 10th IFAC Conference on Control Applications in Marine Systems, CAMS 2016, Trondheim, Norway, 13–16 September, vol. 49, pp. 14–21 (2016)

23. Khatib, O.: Real-time obstacle avoidance for manipulators and mobile robots. Int. J. Robot. Res. **5**(1), 90–98 (1986)

24. Lazakis, I., Dikis, K., Michala, A.L., Theotokatos, G.: Advanced ship systems condition monitoring for enhanced inspection, maintenance and decision making in ship operations. Transp. Res. Procedia **14**(Suppl. C), 1679–1688 (2016)

25. Lazarowska, A.: A trajectory base method for ship's safe path planning. Procedia Comput. Sci. **96**(2016), 1022–1031 (2016)

26. Leary, K.: China has launched the world's first all-electric cargo ship (2017). https://futurism.com/china-launched-worlds-first-all-electric-cargo-ship/

27. Liu, Z., Zhang, Y., Yu, X., Yuan, C.: Unmanned surface vehicles: an overview of developments and challenges. Annu. Rev. Control **41**, 71–93 (2016)

28. Ma, F., Wu, Q., Yan, X., Chu, X., Zhang, D.: Classification of automatic radar plotting aid targets based on improved fuzzy c-means. Transp. Res. Part C Emerg. Technol. **51**(2015), 180–195 (2015)

29. Mediamobil Communication GmbH and ESA: SASS@Sea - satellite based system and services for broadband applications at sea (2014). https://business.esa.int/projects/sasssea

30. Mizuno, N., Uchida, Y., Okazaki, T.: Quasi real-time optimal control scheme for automatic berthing. In: Proceedings of 10th IFAC Conference on Manoeuvring and Control of Marine Craft, MCMC 2015, Copenhagen, 24–26 August 2015, vol. 48, pp. 305–312 (2015)

31. Mu, L.: A hybrid network for maritime on-board communications. In: 2012 IEEE 8th International Conference on Wireless and Mobile Computing, Networking and Communications (WiMob), pp. 761–768 (2012)

32. Mu, L., Prinz, A.: Delay-oriented data traffic migration in maritime mobile communication environments. In: Proceedings of 4th International Conference on Ubiquitous and Future Networks (ICUFN 2012), Phuket, Thailand, pp. 417–422 (2012)
33. Naeem, W., Sutton, R., Ahmad, M.: LQG/LTR control of an autonomous underwater vehicle using a hybrid guidance law. In: Proceedings of 1st IFAC Workshop on Guidance and Control of Underwater Vehicles, Newport, UK, pp. 31–36 (2003)
34. NASA: Technology readiness level (2012). https://www.nasa.gov/directorates/heo/scan/engineering/technology/txt_accordion1.html
35. Netherlands Maritime Technology: NMT coördineert nieuw maritiem europees innovatieproject 'NOVIMAR' (2017). https://maritimetechnology.nl/nmt-coordineert-nieuw-maritiem-europees-innovatieproject-novimar/
36. Perkovic, M., Gucma, M., Luin, B., Gucma, L., Brcko, T.: Accommodating larger container vessels using an integrated laser system for approach and berthing. Microprocess. Microsyst. **52**, 106–116 (2017)
37. Rijnmond, R.: Zelfvarend binnenschip komt eraan (2017). https://www.rijnmond.nl/nieuws/161984/Zelfvarend-binnenschip-komt-eraan
38. RØdseth, H.: Predictive maintenance for autonomous ship - emerging trends in Industrie 4.0 (2016). https://www.cml.fraunhofer.de/content/dam/cml/de/documents/Sonstiges/Autonomous_Ship_Industrie_4.pdf
39. Rødseth, Ø.J., Burmeister, H.C.: Developments toward the unmanned ship. In: Proceedings of International Symposium Information on Ships-ISIS, vol. 201, pp. 30–31 (2012)
40. Royal Academy of Engineering: Future ship powering options. Technical report, Royal Academy of Engineering (2013)
41. Sakib, S.: Implementation of digital imu for increasing the accuracy of hydrographic survey. Procedia Eng. **194**, 386–393 (2017)
42. Sariff, N., Buniyamin, N.: An overview of autonomous mobile robot path planning algorithms. In: Proceedings of the 4th Student Conference on Research and Development, Selangor, Malaysia, pp. 183–188 (2006)
43. Schiaretti, M., Chen, L., Negenborn, R.R.: Survey on autonomous surface vessels: Part I - a new detailed definition of autonomy levels. In: Bektaş, T., Coniglio, S., Martinez-Sykora, A., Voß, S. (eds.) Computational Logistics. LNCS, vol. 10572, pp. 219–233. Springer, Cham (2017). https://doi.org/10.1007/978-3-319-68496-3_15
44. Schiaretti, M., Chen, L., Negenborn, R.R.: Survey on autonomous surface vessels: Part II - categorization of 60 prototypes and future applications. In: Bektaş, T., Coniglio, S., Martinez-Sykora, A., Voß, S. (eds.) Computational Logistics. LNCS, vol. 10572, pp. 234–252. Springer, Cham (2017). https://doi.org/10.1007/978-3-319-68496-3_16
45. Sørensen, M.E.N., Breivik, M.: Comparing nonlinear adaptive motion controllers for marine surface vessels. In: Proceedings of 10th IFAC Conference on Manoeuvring and Control of Marine Craft, MCMC 2015, Copenhagen, 24–26 August 2015, vol. 48, pp. 291–298 (2015)
46. Statheros, T., Howells, G., Maier, K.M.: Autonomous ship collision avoidance navigation concepts, technologies and techniques. J'. Navig. **61**(1), 129–142 (2008). https://doi.org/10.1017/S037346330700447X
47. Sumić, D., Peraković, D., Jurcević, M.: Optimizing data traffic route for maritime vessels communications. Procedia Eng. **100**, 1286–1293 (2015)
48. Tesla: Autopilot (2017). https://www.tesla.com/en_GB/autopilot?redirect=no
49. The Guardian: Facebook drone that could bring global internet access completes test flight (2017). https://www.theguardian.com/technology/2017/jul/02/facebook-drone-aquila-internet-test-flight-arizona

50. Tran, V.L., Im, N.: A study on ship automatic berthing with assistance of auxiliary devices. Int. J. Naval Archit. Ocean Eng. 4(3), 199–210 (2012)
51. Tsou, M.C.: Multi-target collision avoidance route planning under an ECDIS framework. Ocean Eng. **121**, 268–278 (2016)
52. United Nations Conference on Trade and Development: Merchant fleet by flag of registration and by type of ship, annual, 1980–2018 (2018). http://unctadstat. unctad.org/wds/TableViewer/tableView.aspx?ReportId=93
53. Uras, T., Koenig, S.: An empirical comparison of any-angle path-planning algorithms. In: Proceedings of the 8th Annual Symposium on Combinatorial Search, Ein Gedi, Israel, pp. 206–210 (2015)
54. Wang, X., Liu, Z., Cai, Y.: The ship maneuverability based collision avoidance dynamic support system in close-quarters situation. Ocean Eng. **146**, 486–497 (2017)
55. Wolken-Möhlmann, G., Gottschall, J., Lange, B.: First verification test and wake measurement results using a Ship-LIDAR system. Energy Procedia **53**, 146–155 (2014)
56. Xu, Q.: Collision avoidance strategy optimization based on danger immune algorithm. Comput. Ind. Eng. **76**, 268–279 (2014)
57. Xue, Y., Clelland, D., Lee, B., Han, D.: Automatic simulation of ship navigation. Ocean Eng. **38**(17–18), 2290–2305 (2011)
58. Yalcin, O., Sayar, A., Arar, O., Akpinar, S., Kosunalp, S.: Approaches of road boundary and obstacle detection using LIDAR. In: Proceedings of 1st IFAC Workshop on Advances in Control and Automation Theory for Transportation Applications, vol. 46, pp. 211–215 (2013)
59. Young, J., Simic, M.: Lidar and monocular based overhanging obstacle detection. Procedia Comput. Sci. **60**, 1423–1432 (2015)
60. Zhang, J., Zhang, D., Yan, X., Haugen, S., Soares, C.G.: A distributed anti-collision decision support formulation in multi-ship encounter situations under COLREGs. Ocean Eng. **105**, 336–348 (2015)
61. Zheng, H., Negenborn, R.R., Lodewijks, G.: Robust distributed predictive control of waterborne AGVs–a cooperative and cost-effective approach. IEEE Trans. Cybern. **48**(8), 2449–2461 (2018)
62. Zhou, M.T., Harada, H.: Cognitive maritime wireless mesh/ad hoc networks. J. Netw. Comput. Appl. **35**(2), 518–526 (2012)
63. Zhu, M., Hahn, A., Wen, Y.Q., Bolles, A.: Identification-based simplified model of large container ships using support vector machines and artificial bee colony algorithm. Appl. Ocean Res. **68**, 249–261 (2017)
64. Zolfagharifard, E.: Now Google is going to dominate space: search giant to launch 180 satellites to provide internet access for the entire planet, sources claim (2014). http://www.dailymail.co.uk/sciencetech/article-2646039/Googles-plans-world-domination-Search-giant-launch-180-satallites-bring-internet-access-ENTIRE-planet.html

A UAV-Driven Surveillance System to Support Rescue Intervention

Danilo Cavaliere[1] , Vincenzo Loia[2] , and Sabrina Senatore[1]([⊠])

[1] Dipartimento di Ingegneria dell'Informazione ed Elettrica e Matematica Applicata,
Universitá degli Studi di Salerno, Fisciano, Italy
{dcavaliere,ssenatore}@unisa.it
[2] Dipartimento di Scienze Aziendali, Management e Innovation Systems,
Universitá degli Studi di Salerno, Fisciano, Italy
loia@unisa.it

Abstract. In recent years, the intelligent surveillance systems have attracted many application domains, due to the increasing demand on security and safety. Unmanned Areal Vehicles (AUVs) represent the reliable, low-cost solution for mobile sensor node deployment, localization, and collection of measurements.

This paper presents a surveillance UAV-based system, aimed at understanding the scene situation by collecting raw data from the environment (by exploiting some possible sensor modalities: CCTV camera, infrared camera, thermal camera, radar, etc.), processing their fusion and yielding a semantic, high-level scenario description. UAV is able to recognize objects and the spatio-temporal relations with other objects and the environment. Moreover, UAV is able to individuate alerting situations and suggest a recommended intervention to humans. A Fuzzy cognitive map model is indeed, injected in the UAV: from the semantic description of the scenario, the UAV is able to deduct casual effect of occurring situations, that enhances the scenario understanding, especially when alarming situations are discovered.

Keywords: Situation understanding · Situation awareness
Fuzzy cognitive maps · Semantic Web

1 Introduction

The increasing growth of commercially available unmanned aerial systems (UAVs) provides a reliable, low-cost platform for mobile sensor deployment, easy to use in many application domains: from military to surveillance and reconnaissance tasks; to journalism to commercial delivery; to disaster monitoring to rescue operation and healthcare. UAVs are also used for some Earth observations missions: they provide very interesting synergies in monitoring applications (agriculture/vegetation, volcanos, etc.) and very risky situations (disasters, earthquakes, forest-fires, rescue tasks, etc.).

© Springer Nature Switzerland AG 2018
R. Cerulli et al. (Eds.): ICCL 2018, LNCS 11184, pp. 124–135, 2018.
https://doi.org/10.1007/978-3-030-00898-7_8

UAV-based systems represent a new paradigm in high-level landscape description data collection. They can operate either under remote control by a human operator or autonomously by on-board computers. Supported by computer vision, and machine learning techniques, UAV is able to automatically analyze image, video, audio or other type of surveillance data. It is the remote flying "eye", even though its cognitive capabilities are limited to reporting mobile and fixed objects appearing in the scene: generally it is not able to "understand" what it sees by its camera.

In complex and dynamic environments, unmanned vehicles should be autonomous in strict sense, that means they should have a human-like mental model to carefully perceive the environment, locate environmental elements, understand the evolving situation (i.e., fire), make a decision, and eventually reporting solutions to human operators (i.e., call for fire department).

The rationale behind this approach is to make UAV "aware" of what it sees by camera: raw data collected by remote sensing technologies are translated in higher level information. Semantic Web technology provides indeed, a scenario description at semantic level: bounding boxes from video analysis becomes object categories (i.e., cars, river, person, etc.); detected flames in the tree become fire in the wood. The semantic support allows UAV to recognize evolving scenario involving object-actors and the occurring events. UAV recognizes not just individual objects in the scene, but also the interaction among (fixed or mobile) objects and with the environment. It can understand the scenario and get a synthetic view, described as many succeeding events such as "person walking in the park"; "car running on the road", and alerting situations, such as "broken car on the road", "fire in the wood". This knowledge level is further enriched by an upper new level, targeted at a decision making task: what action to take when an alarming situation is detected. The UAV is equipped with a cognitive model, i.e., the fuzzy cognitive map (FCM), used for causal knowledge acquisition and reasoning processes. The FCM model aims at solving decision-making problems, and then, allows the UAV to deduct the appropriate actions to make when an alarming situation occurs. Our UAV indeed, is able to recognize alarming situation, by giving a reliability degree, i.e., a value that describes how likely the situation can be considered alarming and suggest the rescue intervention to cope with the situation.

The paper is organized as follows. Section 2 provides an overview of the principal works and application domains where the UAV assumes a crucial role; Sect. 3 gives an overall description of the system, with an outlook to each component compounding the system. A case study, presented in Sect. 4, evidences the potential of a UAV-based rescue-recommender. A simulation of the FCM model shows the applicability of the approach and the effectiveness of the whole system to detect alerts and suggest proper interventions. Conclusion closes the paper.

2   Related Work

The description of scenario from UAV video is a very challenging topic in literature. Some approaches employ pre-fixed scene regions for scenario classification
[21], but they are not sufficient to interpret dynamic and evolutionary scenarios
from UAV videos. Bridging robust object recognition with high-level event detection could be a promising solution for scenario understanding. The single tasks
of detecting objects and events are not easy; in Video Surveillance domain many
works exclusively aim at studying single-task among object tracking [19], object
recognition [18], object activity and event detection [16]. This mainly depends
on several issues related to the drone-embedded camera and its drone-related
movements. In fact, the mobile camera provides further issues to object tracking, preventing the use of standard good-performing methods used with fixed
camera [17]. Additionally, the movements of the camera also affect the high-
level interpretation of the situation. The main issue for a high-level modeling of
the scenario is the lack of reference points, which undermines the understanding
of object activities, as well as motion blur, complicates object recognition [13].
Video scene showing different kinds of outside environments evolves quickly. The
multitude and variety of distinct environments filmed by the UAV make the use
of Machine Learning methods not enough to cope with scenario detection due
to the great amounts of samples required for model training [14]. Furthermore,
these methods do not exhibit cognitive capabilities to recognize events and scene
object action. A solution is to employ Cognitive Science methodologies to model
higher-level information on the environment and fuse it with heterogeneous data
detected by Computer Vision. High-level contextual knowledge could support
UAV in acquiring situation awareness.

Semantics empower Intelligent Systems to accomplish cognitively complex
tasks. Semantics is based on ontologies modeling high-level knowledge on the
video scene. Ontology reasoning allows collection and deduction of new facts
on the scenario [3]. Main trends use semantics to fuse data on the video scene
retrieved from different sources [6]. The building of a high-level context to support scenario interpretation is crucial to enhance the knowledge on the scenario
and support decision makers in dealing with sensor imprecision [20]. Some solutions exploit a priori knowledge on the scene to propose novel ad-hoc scene
ontologies [9] aimed at inferring events and object activities [15]. These systems
are designed for specific applications targeted at certain domains and environments. There is the need to build generic high-level models suitable to cope with
the modeling of various contexts in distinct kinds of outside environment. Some
trends [1,2] model knowledge on locations at different levels of granularities (i.e.
cities, regions, states), which improves context-based reasoning. Our approach
addresses model adaptability by querying external sources, such as Google Maps
service to acquire data on the environment and model different kinds of environment with an upper ontology.

Semantics could be used to infer knowledge on the current scenario, but it
can not be used to make predictions and assess scenario evolution [5]. In order to
provide assessment of the scenario evolution, Fuzzy Cognitive Map models are

interesting solutions. Fuzzy Cognitive Maps (FCMs) add Fuzzy Logic to neural network in order to reason at very high level, over causal relations among concepts, which depict a mental model [10]. FCMs are generally employed to support applications on soft reasoning domains (i.e. organization theory, political sciences, business management) [7], these models are particularly used to deal with uncertainty and support decision-makers [8]. FCM builds a mental landscape based on causal-reasoning, representing human-like reasoning, aimed at supporting the activation of multiple concepts. These features support the detection of real scenarios, as composed by multitudes of co-occurring events. Generally, domain experts concern with the design of FCMs for specific applications. Fusing various distinct FCMs improves the knowledge base and the cognitive power of the overall system [10]. Our model, indeed aims at exploiting data from different sensors, as well as detected events and activities to build a unique FCM modeling different aspects on the scenario. General trends on FCM focus on searching new solutions to improve views on the scene domain, as well as generate easy-readable maps of the scenario [11,12]. In order to provide easy-to-read maps, make the process quicker and focus on potential alerting situations, our approach instead, builds and feeds the FCM on the basis of the events detected by the ontology reasoning. Many approaches combine ontologies with FCMs to design hierarchies of concepts, and build ontologies [4,12] or to build FCMs from ontologies, focussing on ontology matching to solve semantic ambiguity [11].

3   Overview

Figure 1 shows the logical overview of the UAV cognitive model; it is split in three main layers of knowledge. The remote sensing systems embedded in the drone perceive the raw data from the different source sensors: video camera, air sensor, GPS, infrared camera, etc. This layer, the *Data acquisition* yields a fusion of all the collected data, that are then re-processed at higher level, in order to extract high level conceptualization of the original data; the generated knowledge level enables the UAV to recognize objects, their relations with the environments and with the other objects appearing in the scenes. This level, called *Situation understanding* is also in charge of producing an enhanced scene perception: thanks to a cognitive model, UAV can recognize possible alerting situations appearing or near to appear in the scene evolution. Discovered critical situations produce an alert with a reliability degree that can generate a distress signal, i.e., a call for a rescue intervention, that is one specific for the alert type, from the feasible *Alert rescuers*.

3.1   Data Acquisition

The UAV is provided with several sensors, in order to percept the data about the scene objects, that could be people, vehicle, animals or something else pushed or carried by humans. Among the sensors employed by the UAV, the fundamental

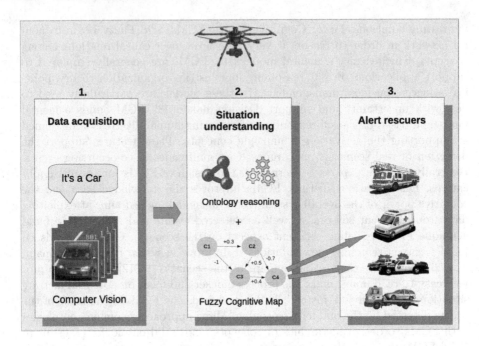

Fig. 1. The layer-based mental model of the UAV

ones are cameras of different types, such as HD action cameras, HD cameras with wide-angle lens and infra-red cameras. The main goal of this module is to percept the mobile objects from the video; to this purpose, video tracking algorithms are employed to detect and track objects present in the video sequence. Each object (or track) is marked with a bounding box and a unique ID identifying the same object through video frames. The tracking output is a configuration file with information about bounding boxes, such as their dimensions, position, as well as computed data on its speed and direction.

The configuration file also contains data about the scene object label, which indicates the identity of the object. The label is provided by classification algorithms used to distinguish the tracked objects in people, vehicles or other.

3.2 Situation Understanding

The raw data about the tracked objects are passed to the *Situation Understanding* component. This component achieves two main steps, respectively, targeted at detecting the current scenario and asses its possible evolution in the near future. As the first step, the tracking and classification data are processed to build high-level knowledge about the tracked scene objects. In order to accomplish this task, a scene ontology is employed: the TrackPOI ontology, introduced in [3]. This ontology model defines the *Track* class to semantically represent each tracked object moving in the scene. Special properties relate the objects to some of their own features (i.e. speed, direction and dimensions). The TrackPOI ontology also models knowledge about the fixed places of the environment, where the scene objects move. Knowledge about the places is retrieved by Google Maps service so that each place in the area which the drone flies over, such as road, park, square, etc. is detected and described as an individual of the *POI* class. Spatio-temporal relations among the objects, and with the environment further feed the asserted knowledge. Ontology reasoning over the collected knowledge allows the detection of higher-level events, such as "people crossing the road", "people meeting", "vehicle accelerating" or "car stopping", etc. In a nutshell, the reasoning provides a high-level description of the scenario as a collection of happened events.

The next step of *Situation Understanding* is to provide an evaluation of the eventual scenario evolution. To this purpose, Fuzzy Cognitive Maps are employed. Fuzzy Cognitive Maps (FCMs) are mental model representations close to the human way of reasoning. FCM represents high-level knowledge as composed by high-level concepts related by causal-effect relationships. So that, knowledge modeling about scenario can also include the effect that one concept has on another. Our FCM models people and vehicle features and relations, as well as the alerting events that could happen, as FCM concepts. The FCM is represented as a weighted graph, where each node represents a concept and the relations among them are described by directed weighted edges. The edge weight is a fuzzy value representing the impact that a concept has on the connected other concept. Just for example, Let c_1 and c_2 be two concepts and given a directed edge from c_1 to c_2, the edge weight expresses how c_1 impacts on c_2. The weight sign represents if c_1 causes an increase or a decrease on the value of c_2.

The knowledge retrieved by the ontology reasoning, along with the sensor-retrieved data, feed input values of the FCM concepts, so that they represent the initial scenario state. Starting from these concept values, the FCM is updated to infer new concept values, representing an eventual scenario evolution, by running the model according to the following formula:

$$C_i^t = f \left(\sum_{j=1}^{n} C_j^{t-1} W_{ji} + C_i^{t-1} \right) \tag{1}$$

with $j \neq i$, C_i^t is the updated value for concept i at iteration t, C_j^{t-1} represents all the other concept values related to concept i at iteration t. W_{ji} indicates the edge weight between concept i and j. C_i^{t-1} refers to the value for concept i at iteration $t - 1$ and f is a threshold function to squash the result in $[0, 1]$ interval. The final case study, presented in Sect. 4, shows how the combination of ontology reasoning and FCM simulation support the detection of the current scenario and the assessment of possible scenario evolutions.

3.3   Alert Rescuers

The FCM execution is a process that iteratively converges towards a stable situation. At the end of this process, new values are assigned to the concepts, according to the initial activation concept configuration. Our model aims at assessing alerting situations, so new concept values express occurred events, or, in other words, which alerting events are most likely to happen or have already happened. The FCM also infers which kind of relief effort is more suited to call, according to the detected event type. Analysing the relations between the detected events and other concepts, related to the events, the FCM detects the causes behind each detected event. The *Situation Understanding* component forwards these information to the *Alert rescuers* component, which depending on the results, alerts the most suitable rescuers accordingly. This component builds a report in a high-level language, illustrating the event occurred and the main causes detected by the system.

Fig. 2. The case study scenario: crossroad accident

4   Case Study

The main functionalities of our smart UAV model are further demonstrated in a case study. The case study shows how the model works on a real scenario taken from a drone video. The scenario is shown in Fig. 2, which shows a video frame presenting an accident scene set in the middle of a crossroad. The white car in the middle is the main vehicle involved, as it is stopped in the opposite direction of travel. Many other cars are stopped or move slowly because of the stopped car and its pieces present on the road. Some people moving around or crossing the road are also present in the scene. The scene is highly risky, because the visibility of many oncoming vehicles is compromised by the stopped vehicles and traffic. People safeness is put at risks as well for the same reason, because they walk on the road where cars and trucks are running at high speed with little visibility.

Fig. 3. Data acquisition: people and vehicles are detected and identified by using video tracking and classification

Furthermore, the stopped vehicle on the road constraints some vehicles to come into the oncoming traffic, generating other risky situations.

The UAV flies over roads where this kind of situation could happen. As first step, the UAV detects the people and vehicles as mobile pixels present in the scene employing video tracking algorithms. Then, it employs classification algorithms to distinguish the objects between people and vehicles (see Fig. 3). The second step preliminarily retrieves information on the fixed environmental features. In details, position and identity of places are acquired from Google Maps service; in this scenario, the crossroad, as well as the kind and names of the roads in the area, are returned. Some other places are also returned, such as the service station, visible in the upper right of the figure, and other places in crossroad surrounding. All discovered places and tracks data are translated into the semantic assertions by the TrackPOI ontology, feeding a knowledge base enclosing also the spatio/temporal relations. Reasoning over the built knowledge base generates higher-level events occurring in the current scenario. The output of the ontology-reasoning for the scenario is shown in Fig. 4.

Main inferred knowledge about the occurring events is partly reported in the white box on the left side of the figure and displayed as labels on people or vehicle. The annotation shows for each scene object, the events they triggered. Let us notice that many *running on road* events are detected for many vehicles. These events state which vehicle is currently running on which road (i.e. POI_1). Other numerous events, among the detected, include "decelerating cars". In fact, many cars coming from the right, decelerate because of the stopped car in the middle of the crossroad.

Spatio/temporal relations among the scene objects are also detected and represented by special properties, such as *near*. Vehicle Track_12, for example, is near a person (Track_15).

Exploiting these simple spatio/temporal relations, more complex spatio/temporal relations and events, such as *goingTowards* or *crossing*, are also found. These events are detected by employing rules including complex relations over time, space, and context. As an example, let us consider the crossing person (i.e. Track_15) and the car (Track_16) which goes towards him. The car moving towards the crossing people could be a potential alerting situation, even though, generally the semantics does not make predictions, but supports an overall scenario understanding. The combination of these events indeed provides a meaningful high-level full description of what is currently happening in the current scenario. This high-level picture composed of events will be used to support scenario assessment through FCMs. In fact, once high-level events have been detected, our system triggers the automatic building of an FCM on the whole scenario. The FCM is composed of several sub-FCMs modeling different aspects of the scenario, which strictly depend on the detected environment and inferred alerting events.

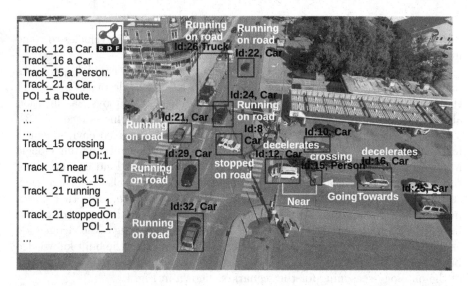

Fig. 4. Situation understanding: spatio/temporal relations among people and vehicles, together with high-level events are detected through ontology reasoning

In this case, the FCM models the people and vehicle activities, and the events occurring in the scenario. The simple events detected by the ontology reasoning, feed FCM concepts and edges by providing the initial values, representing the current scenario. These input values trigger the execution of the FCM simulation. The process runs several iterations until convergence is reached. After the process execution, the final FCM reports new values for some output concepts representing alerting events. These values provide an evaluation about the concepts that are likely to happen. In particular, concepts assuming high values will represent the events that have most chances to occur in the near future. Some other output FCM concepts represent the most suitable rescuers to call.

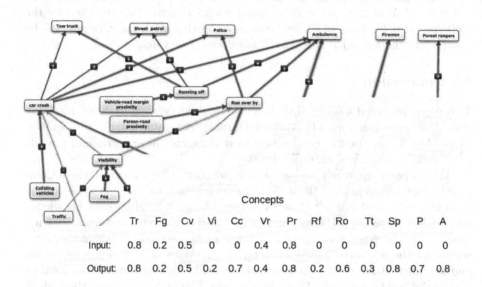

Fig. 5. Situation understanding: FCM simulation evaluates possible scenario evolutions and the most suited rescuers to alert. The FCM concepts: Traffic (Tr), Fog (Fg), Colliding vehicles (Cv), Visibility (Vi), Car crash (Cc), Vehicle-road margin proximity (Vr), Person-road proximity (Pr), Running off (Rf), Run over by (Ro), Tow truck (Tt), Street patrol (Sp), Police (P), Ambulance (A)

The FCM simulation result is shown in Fig. 5: the activated concepts assume values from sensors (i.e. *fog*) as well as from spatio/temporal relations and events detected by the ontology reasoning (i.e. *Person-road proximity*). These events are shown as input in figure. The main alerting events that could occur to people and vehicles are also represented, among them there are *Car crash*, *Running off* and *Run over by*. After FCM execution on the scenario, high values are revealed for the concepts *Run over by* (0.6) and *Car crash* (0.7). The concepts shown on top of the figure represent the different kinds of rescuers that could be alerted: *Tow Truck, Street patrol, Police, Ambulance, Firemen, Forest rangers*. The causal-effect relations among the occurred alerting events and rescuers determine the most suitable rescue intervention. In fact, the high values reported by *Run over by* and *Car crash* concepts, representing the detected events, have a direct impact on *Street patrol* (0.8), *Ambulance* (0.8) and *Police* (0.7) concepts, for which high values are inferred. According to these results, the three relief efforts will be alerted and provided with reports on the occurred event (i.e. *Run over by* and *Car crash*) along with the main causes behind it. The reported causes are represented by the name and value of the FCM concepts that causally influence the event detection. For instance, the value of *Run over by* depends by high values for *Traffic* that causes a decrease in *visibility*, so that these information will be provided to the *Police* and *Ambulance*, alerted for the eventual *Run over by* event. The same applies to the *Car crash* event, whose values depend on the reduced *Visibility* and vehicles going towards each other, modeled by the *Colliding vehicles* concept. These concepts together with their values will be provided to the alerted *Police*, *Street patrol* and *Tow truck*.

5   Conclusion

This paper proposed a smart UAV-based surveillance system to detect alarming real-life scenario taken from UAV video. Ontology reasoning and Fuzzy Cognitive Maps (FCMs) are used in conjunction to understand the scenario and provide assessment of possible scenario evolutions.

The proposed approach presents a multi-level architecture model where Computer Vision techniques are the underlying module to detect mobile scene objects from the video and classify them in people and vehicle. High-level knowledge elicited by the raw track data, and place information retrieved from external sources, are described semantically by ontological assertions on the scene. Reasoning on the acquired knowledge supports the detection of high-level events and the interaction among the scene objects. The detected events are used to trigger the building of an FCM on the scenario. The FCM represents the highest knowledge level that, from the data from the lower level, deduct alerting events, that are likely to happen. FCM aims at improving the general understanding of the events by suggesting the most suited rescue intervention to call, according to the revealed event.

References

1. Bernad, J., Bobed, C., Mena, E., Ilarri, S.: A formalization for semantic location granules. Int. J. Geogr. Inf. Sci. **27**(6), 1090–1108 (2013). https://doi.org/10.1080/13658816.2012.739691
2. Bobed, C., Ilarri, S., Mena, E.: Exploiting the semantics of location granules in location-dependent queries. In: Catania, B., Ivanović, M., Thalheim, B. (eds.) ADBIS 2010. LNCS, vol. 6295, pp. 73–87. Springer, Heidelberg (2010). https://doi.org/10.1007/978-3-642-15576-5_8
3. Cavaliere, D., Loia, V., Saggese, A., Senatore, S., Vento, M.: Semantically enhanced UAVs to increase the aerial scene understanding. IEEE Trans. Syst. Man Cybern. Syst. **PP**(99), 1–13 (2017). https://doi.org/10.1109/TSMC.2017.2757462
4. Chauvin, L., Genest, D., Loiseau, S.: Ontological cognitive map. In: 2008 20th IEEE International Conference on Tools with Artificial Intelligence, vol. 2, pp. 225–232, November 2008. https://doi.org/10.1109/ICTAI.2008.42
5. Crispim-Junior, C.F., et al.: Semantic event fusion of different visual modality concepts for activity recognition. IEEE Trans. Pattern Anal. Mach. Intell. **38**(8), 1598–1611 (2016). https://doi.org/10.1109/TPAMI.2016.2537323

6. D'Aniello, G., Gaeta, M., Hong, T.P.: Effective quality-aware sensor data management. IEEE Trans. Emerg. Topics Comput. Intell. **2**(1), 65–77 (2018). https://doi.org/10.1109/TETCI.2017.2782800

7. Glykas, M.: Fuzzy cognitive strategic maps in business process performance measurement. Expert Syst. Appl. **40**(1), 1–14 (2013). https://doi.org/10.1016/j.eswa.2012.01.078

8. Glykas, M.: Fuzzy Cognitive Maps: Advances in Theory, Methodologies, Tools and Applications, 1st edn. Springer, Heidelberg (2010). https://doi.org/10.1007/978-3-642-03220-2

9. Gómez-Romero, J., Patricio, M.A., García, J., Molina, J.M.: Ontology-based context representation and reasoning for object tracking and scene interpretation in video. Expert Syst. Appl. **38**(6), 7494–7510 (2011). https://doi.org/10.1016/j.eswa.2010.12.118

10. Kosko, B.: Fuzzy cognitive maps. Int. J. Man-Mach. Stud. **24**(1), 65–75 (1986). https://doi.org/10.1016/S0020-7373(86)80040-2

11. Lee, D.H., Lee, H.: Construction of holistic fuzzy cognitive maps using ontology matching method. Expert Syst. Appl. **42**(14), 5954–5962 (2015). https://doi.org/10.1016/j.eswa.2015.03.020

12. Lee, H., Kwon, S.J.: Ontological semantic inference based on cognitive map. Expert Syst. Appl. **41**(6), 2981–2988 (2014). https://doi.org/10.1016/j.eswa.2013.10.029

13. Li, X., Lu, H.: Object tracking based on local learning. In: 2012 19th IEEE International Conference on Image Processing, pp. 413–416, September 2012. https://doi.org/10.1109/ICIP.2012.6466883

14. Li, Y., Guo, Y., Kao, Y., He, R.: Image piece learning for weakly supervised semantic segmentation. IEEE Trans. Syst. Man Cybern. Syst. **47**(4), 648–659 (2017). https://doi.org/10.1109/TSMC.2016.2623683

15. Meditskos, G., Kompatsiaris, I.: iknow: ontology-driven situational awareness for the recognition of activities of daily living. Pervasive Mob. Comput. **40**, 17–41 (2017). https://doi.org/10.1016/j.pmcj.2017.05.003. http://www.sciencedirect.com/science/article/pii/S157411921630195X

16. Min, W., Zhang, Y., Li, J., Xu, S.: Recognition of pedestrian activity based on dropped-object detection. Sig. Process. **144**, 238–252 (2018). https://doi.org/10.1016/j.sigpro.2017.09.024

17. Piccardi, M.: Background subtraction techniques: a review. In: 2004 IEEE International Conference on Systems, Man and Cybernetics (IEEE Cat. No.04CH37583), vol. 4, pp. 3099–3104 vol. 4, October 2004. https://doi.org/10.1109/ICSMC.2004.1400815

18. Rangel, J.C., Martínez-Gómez, J., Romero-González, C., García-Varea, I., Cazorla, M.: Semi-supervised 3D object recognition through CNN labeling. Appl. Soft Comput. **65**, 603–613 (2018). https://doi.org/10.1016/j.asoc.2018.02.005

19. Smedt, F.D., Hulens, D., Goedemé, T.: On-board real-time tracking of pedestrians on a UAV. In: 2015 IEEE Conference on Computer Vision and Pattern Recognition Workshops (CVPRW), pp. 1–8, June 2015. https://doi.org/10.1109/CVPRW.2015.7301359

20. Snidaro, L., García, J., Llinas, J.: Context-based information fusion: a survey and discussion. Inf. Fusion **25**, 16–31 (2015). https://doi.org/10.1016/j.inffus.2015.01.002

21. Yuan, Y., Mou, L., Lu, X.: Scene recognition by manifold regularized deep learning architecture. IEEE Trans. Neural Netw. Learn. Syst. **26**(10), 2222–2233 (2015). https://doi.org/10.1109/TNNLS.2014.2359471

Container Handling and Container Terminals

A Literature Review on Container Handling in Yard Blocks

Filip Covic[✉]

Institute for Operations Research, HBS Hamburg Business School,
University of Hamburg, Hamburg, Germany
`filip.covic@uni-hamburg.de`

Abstract. The yard block of a container terminal is the central point of synchronisation for asynchronous container flows between the transport by deep sea vessels and the hinterland. The structure of the block stipulates that containers are stacked onto each other while only the topmost container can be accessed directly by a yard crane. This stacking restriction shapes the general framework of container handling in yard blocks. The optimisation of stacking configurations within a block has been a continuously growing stream of research in container terminal planning leading to a diverse set of problem definitions, modelling techniques and solution approaches. In this regard, an extensive literature review is conducted in this work surveying container handling problems in the time-span 1997–2018. For this purpose, the major handling problems and their versions are conceptually defined. Next, the surveyed studies are systematically classified and compared based on key properties for practical yard block planning as well as for a theoretical analysis of container stacking. Afterwards, the work is aligned with other literature reviews in this field and a conclusion is given in order to provide a comprehensive overview of container handling problems. Eventually, this should support the identification of open questions in existing problems and the initialisation of new research streams in future research.

Keywords: Container terminal · Container handling
Yard operations · Literature review

1 Introduction

The containerisation of multi-modal transport and its standardisation have spurred international trade and worldwide interconnectivity of production and supply chains as major drivers of globalisation (WTO 2013). According to estimates by the *United Nations Conference on Trade and Development* (UNCTA 2017), approximately 80% of global trade volume and more than two thirds of the respective value are moved by deep-sea vessels and seaports. This makes

---

This paper is based on work also presented in (Covic 2018).

© Springer Nature Switzerland AG 2018
R. Cerulli et al. (Eds.): ICCL 2018, LNCS 11184, pp. 139–167, 2018.
https://doi.org/10.1007/978-3-030-00898-7_9

the container terminal the central point for the handling of containerised and non-containerised cargo worldwide.

As a result of a continuously growing demand for container transport, terminal operators face problems of managing an increased throughput, scarce land availability at seaports and an intense competition between terminals to provide efficient service for shipping companies. The container terminal can be described as an aggregate of organisational units where their point of interaction is stipulated by the purpose of container handling. Specifically, the largest share of container handling occurs within the yard block. It is the principle storage system for containers, while they dwell in the terminal. In this regard, it is imperative to examine the characteristics of container handling in yard blocks and how they relate to other activities that must be performed in order to guarantee intact block operations (Covic 2018).

Following from these developments, container terminal planning has experienced an increasing interest in the scientific literature for the past 20 years. Solutions to planning problems have been devised to support practical decision-making of terminal operators as well as to understand the fundamental principles governing the decision problems from a theoretical perspective. This can be also seen from the extensive list of works providing (literature) overviews of terminal operations as a whole on the one hand, or dealing with a specialised focus on specific operational areas of the terminal on the other hand. For the former, overviews include the works by Vis and Koster (2003); Günther and Kim (2006); Steenken et al. (2004); Stahlbock and Voß (2008a); Brinkmann (2011) and Schwarze et al. (2012) who provide comprehensive classification schemes, (scientometric) literature reviews and design and operational principles of container terminal activities. Examples for the latter category of overviews are Stahlbock and Voß (2008b) and Carlo et al. (2014b) for transport operations in terminals, Bierwirth and Meisel (2010) and Bierwirth and Meisel (2015) for an overview of the Berth Allocation Problem and Quay Crane Assignment and Scheduling Problem, Boysen and Stephan (2016) and Boysen et al. (2017) for the Yard Crane Scheduling Problem, Carlo et al. (2015) for general waterside operations and Lange et al. (2017) for hinterland operations. Apart from these, simulation of terminal operations is a popular tool to cope with the stochastic, dynamic and complex underlying environment (cf. Kemme, 2012). For this reason, Angeloudis and Bell (2011) and Dragović et al. (2017) provide overviews of simulation approaches in container terminal planning with a focus on modelling techniques in the former and a meta-analysis of the relevant literature in the latter.

Similarly, overviews of container handling in yard blocks can be found in (Luo et al. 2011; Caserta et al. 2011a; Carlo et al. 2014a). While the first two studies address the surveyed problems either on a conceptual or on an analytical scale, the third overview is concerned with a literature review of handling operations in the time-span 2004–2012. However, the scope of this review encompasses the entire terminal and the provided classification scheme is more strongly oriented towards a broader and abstract view on general aspects of container storage within the terminal. In contrast, the literature review in this work specifically defines five major container handing problems on yard block level by which a classification scheme is proposed that covers all handling operations in the block. The concentrated focus on distinct and characteristic handling problems and the crisp identification and formal differentiation of central aspects of these problems enable a thorough and detailed examination of the literature based on eleven key properties. By this approach, it is possible to identify and relate the key research streams and developments about container handling in yard blocks over the time-span 1997–2018. The remainder of this work is structured as follows. First, a short description of terminal operations is given which makes it possible to give a formal conceptual definition of container handling problems covering yard block operations. Afterwards, a systematic literature overview is given for 61 studies from the literature. Based on this, the properties are individually evaluated and central findings are illustrated by selecting important papers. Finally, a conclusion and outlook with open questions for further research are given.

2   Terminal and Container Handling Operations

In general, container terminal optimisation can be approached on different levels while the classification and terminology of the planning hierarchies is subject to variation in the literature (cf. Steenken et al. 2004). A possible distinction of terminal operations is proposed by Günther and Kim (2006). They distinguish the *terminal design level* which is related to the setup of new terminals by making decisions on the desired automation level, the yard block dimensions and layout and the crane and internal vehicle systems among others. *Operative planning* is conducted in the framework of routing policies for internal vehicles, storage policies for containers, berth allocation and crane assignments among others. Lastly, *real-time control* represents decision-making within an *online* environment in which, for instance, a slot assignment of a container must be performed and the corresponding job has to be scheduled for a yard crane immediately. The operational synchronisation of the decisions and processes for the three hierarchies is performed by a central electronic interface, called Terminal Operating System (TOS), which manages all strategies, controls the associated terminal operations and stores all equipment and container data (Choi et al. 2003).

Problems within these planning levels occur in three major areas into which the container terminal can be divided (Steenken et al. 2004):

- *Waterside*: it encompasses the berthing zones for vessels, the quay cranes loading and unloading vessels and the horizontal vehicles connecting the berthing area with the yard blocks. In this area, export containers may be retrieved from the block and transferred to vessels by internal vehicles (e.g., *Automated-Guided-Vehicle* (AGV)). Import containers may be delivered from vessels to the block by internal vehicles.
- *Yard area*: it is the central interface between waterside and landside operations. All terminals are composed of several yard blocks managed by yard

cranes for temporary storage of containers as their delivery and retrieval are not fully synchronised.

- *Landside*: it connects the terminal with the hinterland operations of *external trucks* (XTs) and trains. In this area, export containers may be delivered from the hinterland to the block by XTs. Import containers may be retrieved from the block and transported to the hinterland by XTs and trains. Arrival information on external trucks may be obtained through a Terminal Appointment System (TAS).

The working mechanism of a terminal with vertically aligned blocks (front-end) is illustrated in Fig. 1 showing the respective operational areas described above. The optimisation of the individual processes in each area follows lower level objectives of the entire terminal that contribute to an increase of the *gross crane rate* (GCR) of the quay cranes being the overall measurement for productivity of the terminal (Petering et al. 2009). Shipping carriers operating deep-sea vessels are considered as the most important customers by container terminals. The priority to minimise their berthing times is highest due to the financial incentives in terms of revenue generation for the terminal (Petering et al. 2009; Davis 2009). Thus, the planning within the local problems occurring in the

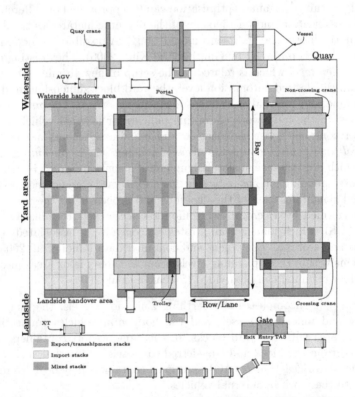

Fig. 1. Container terminal environment (front-end layout)

container terminal areas must be evaluated with reference to the global terminal productivity measured by the GCR.

Taking a closer look at yard block operations, a general distinction can be made between three problem types. While all problem types are in some way focused on the optimisation of yard block productivity by minimising the number of re-handling moves at retrieval, the means of how this is achieved by the solution to the problems are different for each type. Hence, a formal distinction can be made as proposed in the following.

1. *Storage problems*: this type directly deals with the initial storage of containers. By solving this kind of problems, the aim is to find the best possible slot for stacking containers at the beginning of their dwelling period in the terminal and yard block. The *Storage Space Allocation Problem (SSAP)* (Zhang et al. 2003) and the *Container Stacking Problem (CSP)* (Dekker et al. 2007) are problems of this type.

2. *Re-handling problems*: in problems of this type, there is a continuous optimisation of a single yard block over the planning periods in terms of making suitable slots available for incoming containers while mis-overlaid containers are unblocked for retrieval. Characteristically, incoming and outgoing containers are both considered and the solution to these problems must account for the interaction effects of container storage and retrieval. The dynamic versions of the *Blocks Relocation Problem (BRP)* (Tang et al. 2015), the *Pre-marshalling Problem (PMP)* and the *Re-marshalling Problem (RMP)* (Covic 2017) belong to this group of problems.

3. *Retrieval problems*: within these problems, two objectives can be stated. Firstly, the optimal sequence for retrieving containers from the yard block is to be found. Alternatively, it is searched for the optimal sequence of preparatory handling moves within the block, such that the new configuration matches the retrieval sequence as closely as possible. Incoming containers are not allowed in this case. The static versions of the *Blocks Relocation Problem (BRP)* (Caserta et al. 2012), the *Pre-marshalling Problem (PMP)* (Lee and Hsu 2007) and the *Re-marshalling Problem (RMP)* (Kang et al. 2006a) belong to this group of problems.

The *Storage Space Allocation Problem (SSAP)* deals with the container assignment on a macro-level. It is the problem that must be primarily solved when a container reaches the yard area. The SSAP is defined in Definition 1.

Definition 1. *The Storage Space Allocation Problem is to assign an incoming container to a yard (sub-)block such that yard block productivity is maximised (the number of future re-handling moves is minimised).*

Within this scope, the SSAP is complementary to the *Container Stacking Problem (CSP)* as stated in Definition 2.

Definition 2. *The Container Stacking Problem is to assign an incoming container to a slot in a yard block such that the number of future re-handling moves is minimised.*

The correction of mis-overlays resulting from initial stacking is performed by re-handling operations that can either occur before or during the time of retrieval. In this context, the resulting *Blocks Relocation Problem (BRP)* is stated in Definition 3.

Definition 3. *The Blocks Relocation Problem is to find a minimal sequence of container handling moves such that a given container configuration is emptied. The static version refers to the case of not considering incoming containers during container re-handling. The dynamic version refers to the case of considering incoming containers during container re-handling while the set of containers dwelling in the block before is retrieved.*

In the two remaining problems and their respective versions, the goal is to completely prevent shuffle moves at the time of retrieval. The *Pre-marshalling Problem (PMP)* as defined in Definition 4 is solved on the bay-level where no movement of the crane portal is needed.

Definition 4. *The Pre-marshalling Problem is to find a minimal sequence of container marshalling moves within a bay of a yard block, in order to transform an initial bay configuration into a target configuration, such that the number of shuffle moves during future retrieval operations is minimised. The static version refers to the case of not considering incoming containers during container re-handling. The dynamic version refers to the case of considering incoming containers during container re-handling.*

Finally, the *Re-marshalling Problem (RMP)* with reference to Definition 5 extends to the entire yard block to identify pre-emptive shuffle moves. Note that container movement distances during retrieval may substantially contribute to vehicle waiting times at handover areas.

Definition 5. *The Re-marshalling Problem is to find a minimal sequence of container marshalling moves within a yard block, in order to transform an initial block configuration into a target configuration, such that the number of shuffle moves (and yard crane movement) during future retrieval operations is minimised. The static version refers to the case of not considering incoming containers during container re-handling. The dynamic version refers to the case of considering incoming containers during container re-handling.*

3   Literature Review

The multi-problem structure of container handling is reflected in Table 1 where an overview of the current state of the literature with reference to main properties of problem scope, parameters, data, modelling and solution approaches is given. Here, the studies deemed to be of importance for the examination in this work are listed chronologically and categorised individually according to the defined research properties. Based on the extensive overview in Table 1, the main contributions and principle research directions are reviewed with respect to the selected classification properties. In this way, historic developments, active research streams and open questions in the scientific literature are identified. A short synopsis of the core findings from the tabular literature overview is provided for each classification property.

The first column of Table 1 states the authors and year the research was published. Secondly, the examined container handling problem is stated in line with the problem definitions in Sect. 2. In the third column, the aggregation level of the analysed problems is categorised in terms of the degree of detail of terminal operations the planning is concerned with. Depending on the problem scope, it is common in the literature to aggregate decisions about container stacking to block or sub-block level and to pass on the results to a separate model for disaggregating the hierarchical decisions for exact storage slots. With regard to the fourth classification property, the main rational of container terminal optimisation is the increase of yard block productivity as defined in Sect. 2. In this context, it is stated how different studies operationalise this superordinate goal into objectives with reference to the respective container handling problem. Next, one of the main issues are uncertain or inaccurate data about containers at initial storage. Particularly, time data of container delivery and retrieval are of interest for efficient container handling. Hence, the fifth column analyses if the uncertainty of container time data is accounted for. The sixth column deals with the time horizon for planning. Predominantly, the container handling problems, which do not allow new container deliveries during handling of a set of containers, are not (multi-)periodic. On the other hand, by employing a TAS, for instance, a rolling horizon approach could be applied due to continuously updated container data which warrants a repeated solution process. Column seven specifies the container types for which the studied model and solution methods are applicable. It is possible to define the container handling problems for one or more container types depending on the storage and stacking strategy employed in a yard block. Next, container handling problems, which are solely focused on re-handling, usually do not allow new containers to be stored in the relevant bay or block. In the literature, these problems are termed as static while problems allowing new containers to enter are referred to as dynamic. Column eight makes the distinction between these problems. Naturally, container allocation and re-handling encompasses container movement which is performed by a yard crane. In column nine, the hierarchy level is declared on which scheduling decisions of yard crane movement are considered while the type of yard crane (capacity) modelling is provided in column ten. Finally, columns eleven and twelve describe the modelling techniques and the corresponding solution approaches employed in the literature for the respective container handling problem[1].

---

[1] For reasons of compactness, abbreviations within the classification properties are used in Table 1.

Problem Scope

In general, the nature of research in the container handling literature is split into isolated problems. Kim (1997) initiates the discussion on container re-handling by proposing a framework for estimating the expected number of shuffle moves. From this starting point, the discussion on container handling branches out into four major problems which are mainly targeted separately in the literature. Originally, the focus is put on appropriate initial allocation strategies on different aggregation levels. As it is rarely possible to provide an initial allocation configuration without precipitating future re-handling operations, the Blocks Relocation Problem (BRP) is defined and solved by dynamic programming by Kim and Bae (1998). Algebraic model formulations are provided by Wan et al. (2009) and Caserta et al. (2012) for the static BRP where no incoming containers are allowed during shuffle operations. The former model is extended by Tang et al. (2015) to include incoming containers with reference to the dynamic BRP. Exact approaches to the BRP can be identified as the current trend in the literature on container handling as demonstrated by Galle et al. (2018); Tanaka and Mizuno (2018) and Tricoire et al. (2018). To avoid shuffling during retrieval operations for the purpose of reducing vehicle waiting times at handover areas, Kim et al. (2000) introduce the Re-marshalling Problem (RMP) which they solve by decomposition. To reduce computational effort and accommodate different yard block layouts, Lee and Hsu (2007) algebraically model a special version of the RMP, the Pre-marshalling Problem (PMP), which is defined for one bay only. The PMP is investigated extensively in subsequent works, e.g., by Bortfeldt and Forster (2012), who devise lower bounds, by Tanaka and Tierney (2018), who improve the bounds of the latter and Tierney et al. (2017), who solve the PMP to optimality by applying A* and IDA* algorithms. In contrast, the RMP is scarcely covered in the literature[2]. Encountering modelling difficulties due to encompassing multiple bays and non-uniform formal definitions of the RMP, there does not exist a comprehensive algebraic model for the RMP in the literature. It can be observed that the current state of the literature has targeted

---

Objective: N_{shu} - number of shuffle moves, N_{PM} - number of pre-marshalling moves, N_{RM} - number of re-marshalling moves, ω - vehicle waiting time, cap_{YC} - yard crane capacity, Dis_{YC} - yard crane movement distance, Dis_{IT} - internal vehicle movement distance

Time data: D - deterministic, S - stochastic (terms in brackets denote that either time data consideration is not explicitly stated or that time data is pre-processed deterministically or that subsets of time data are considered to be stochastic.)

Time horizon: 1 - one period, M - multiple periods, R - rolling horizon

Container type: I - import containers, E - export containers, T - transshipment containers

Modelling: LP - linear program, IP - (linear) integer program, NLIP - non-linear integer program, DP - dynamic programming

Solution method: SA - Simulated Annealing, GA - Genetic Algorithm, B&B - Branch and Bound.

[2] See also the literature review by Carlo et al. (2014a) that mentions only two studies on the RMP.

container handling as isolated problems extensively while the RMP has not been formalised yet. However, these problems do not appear isolated in nature and are highly dependent on each other in daily yard block operations. First efforts to integrate storage and shuffle optimisation by formulation of the dynamic BRP can be observed (cf. Akyüz and Lee 2014; Tang et al. 2015) but there does not exist a holistic model incorporating the implications of decisions resulting from the solution of the different handling problems. This can also be regarded as a reason for the reluctance to formulate the RMP as an algebraic model.

Aggregation Level

In Table 1, it is shown that the isolation of problems does not only concern the scope of container handling problems but also the aggregation level of assigning containers to a position in the terminal. The majority of works is focused either on the block level or the stack level but does not account for the relevancy of available or suitable slots in the block when assigning a container to a block and vice versa. While Zhang et al. (2003) develop a solution method for the storage space allocation on block level without specifying a specific slot in the block, Duinkerken et al. (2001) and Dekker et al. (2007) describe container stacking strategies on the stack level and evaluate single block performance. Although these two approaches are highly dependent on each other, only Ries et al. (2014) target both problems holistically albeit not providing any algebraic modelling formulation. In general, the SSAP is defined either on block level or on sub-block level by grouping containers according to their characteristics to make them interchangeable and, thus, not requiring to designate a specific slot. The microscopic version of the SSAP, which is often referred to as Container Stacking Problem (CSP) in the literature, presupposes the allocation to a block in order to find a suitable individual target slot for a container. Moreover, it can be observed that certain assumptions are introduced for container grouping in the scope of the RMP to enable an assignment of containers to specific bays only (cf. Yu et al. 2009) or where empty stacks are required to which containers are re-marshalled (cf. Kang et al. 2006a).

Objective

The majority of studies focuses on the minimisation of the respective type of re-handling in accordance with the handling problem that is targeted. If at all, minimisation of yard crane movement distance and the balance of yard crane capacity are mainly addressed in the scope of the SSAP (e.g., Han et al. (2008) for distance and Sharif and Huynh (2013) for capacity) or the CSP (e.g., Chen and Lu (2012) for distance and Borgman et al. (2010) for capacity). Apart from this, yard block occupancy is also considered as objective in terms of balancing the crane workload (e.g., Jin et al. (2016) for the SSAP and Park et al. (2011) for the CSP). There are fewer studies that consider multiple performance indicators and the relationship between re-handling moves, vehicle waiting times and capacity utilisation. These include Duinkerken et al. (2001), Dekker et al. (2007), Borgman et al. (2010) and Asperen et al. (2013) building a joint research stream

Table 1. Literature overview of container handling problems in yard blocks

Author	Problem scope	Aggregation level	Objective	Time data	Time horizon	Container type	Dynamic	Scheduling	Yard crane modelling	Modelling	Solution method
Kim (1997)	BRP	stack	$E(N_{shu})$	(D)	1	I	✓	-	-	-	algebraic formula, regression
Kim and Bae (1998)	RMP	sub	N_{RM}, Dis_{YC}	D	1	E		block	two YC with interference	DP (bay matching), Transport LP (movement), TSP with precedence constraints (sequencing)	decomposition
Kim et al. (2000)	BRP	stack	$E(N_{shu})$	-	1	E		block	-	-	DP^a, classification for decision trees
Duinkerken et al. (2001)	CSP	stack	N_{shu}, Dis_{YC}, equipment utilisation	-	M	I/E	✓		velocity	simulation	heuristic stacking rules
Preston and Kozan (2001)	BRP	stack	berthing time	D	M	E		block	travel time	IP	GA
Kim and Park (2003)	BRP	sub	Dis_{YC}	D	R	E	✓	term	movement cost	IP with optional extensions	Least-Duration-of-Stay Heuristic, Sub-gradient Heuristic
Zhang et al. (2003)	SSAP	block	block occupancy, Dis_{IT}	(D)	R	I/E/T	✓	term	parallel servers	NLIP (linearised), Transport LP	decomposition

(continued)

Table 1. (*continued*)

Author	Problem scope	Aggregation level	Objective	Time data	Time horizon	Container type	Dynamic	Scheduling	Yard crane modelling	Modelling	Solution method
Murty et al. (2005b)	CSP	stack	$E(N_{shu})$	(D)	M	I/E/T	✓	block	-	decision-support-system (real-world application Hongkong International Terminal)	best-fit algorithm based on the smallest expected shuffle index
Kang et al. (2006a)	RMP	sub	N_{RM}	D	1	E	-	block	-	-	Partial-Order-Graphs, SA
Kang et al. (2006b)	CSP	stack	N_{shu}	-	M	E	✓	-	-	simulation	SA for choosing best stacking strategy
Kim and Hong (2006)	BRP	stack	N_{shu}	D	1	I/E/T	-	bay	-	-	B& B, Expected Reshuffle Heuristic
Kozan and Preston (2006)	CSP	stack	berthing time	D	M	(I)/E	✓	block	travel time	IP for container transfer and slot allocation	iterative solution of one model with input data from other model, GA, Tabu Search, GA-Tabu-hybrid
Lim and Xu (2006)	CSP	stack	block occupancy	D	M	I/E/T	✓	-	-	LP	Critical-Shaking-Neighbourhood Heuristic
Dekker et al. (2007)	CSP	stack	N_{shu}, crane workload, block occupancy	(D)	M	I/E/T	✓	-	velocity	simulation	heuristic online stacking rules

(*continued*)

Table 1. (*continued*)

Author	Problem scope	Aggregation level	Objective	Time data	Time horizon	Container type	Dynamic	Scheduling	Yard crane modelling	Modelling	Solution method
Lee and Hsu (2007)	PMP	stack	N_{PM}	D	M^b	I/E/T	-	bay	-	Multi-Commodity-Flow	constraint relaxation and variation
Han et al. (2008)	SSAP	sub	$capy_C$	D	M	E/T	✓	-	movement capacity per shift	IP	sequential Tabu Search with pre-calculated lower bounds
Caserta et al. (2009)	BRP	stack	N_{shu}	D	1	I/E/T	-	bay	-	binary description of stacking area in matrix form	rule-based look-ahead heuristic
Caserta and Voß (2009b)	PMP	stack	N_{PM}	D	1	I/E/T	-	bay	-	-	local search embedded in Corridor Method
Lee and Chao (2009)	PMP	stack	N_{PM}	D	1	E	-	bay	-	IP for movement sequence	neighbourhood search, solving IP in neighbourhood
Park et al. (2009)	RMP	stack	N_{RM}	D	1	I/E/T	-	block	-	-	cooperative co-evolutionary algorithm

(*continued*)

Table 1. (*continued*)

Author	Problem scope	Aggregation level	Objective	Time data	Time horizon	Container type	Dynamic	Scheduling	Yard crane modelling	Modelling	Solution method
Wan et al. (2009)	BRP	stack	N_{shu}	D	M[c]	E	(✓)	block	-	IP for BRP without incoming containers	IP-based heuristic for BRP with incoming containers, comparison with Murty et al. (2005a)[d] Kim and Hong (2006)
Yu et al. (2009)	RMP	sub	$Disy_C$	-	1	E	-	block	moving distance	Assignment IP, Multi-Commodity-Flow	Assignment IP exact solution (CPLEX), GA
Borgman et al. (2010)	CSP	stack	$w_{XT}, cap_{YC}, N_{shu}$, block occupancy	(D)	M[e]	I/E/T	✓	-	-	simulation	heuristic online stacking rules
Zhao and Goodchild (2010)	BRP	stack	N_{shu}	(D)	M	I	-	bay	-	simulation	rule-based heuristic
Caserta et al. (2011b)	BRP	stack	N_{shu}	D	1	I/E/T	-	bay	-	-	dynamic programming embedded in Corridor Method
Choe et al. (2011)	RMP	sub	N_{RM}, YC interference	D	1	I/E/T	-	block	-	-	Partial-Order-Graphs, SA, depth-limited B& B search
Park et al. (2011)	CSP	stack	N_{shu}, block occupancy	D	M	I/E/T	✓	-	moving distance and interference	simulation	dynamically adjusting stacking rules

(continued)

Table 1. (*continued*)

Author	Problem scope	Aggregation level	Objective	Time data	Time horizon	Container type	Dynamic	Scheduling	Yard crane modelling	Modelling	Solution method
Saurí and Martín (2011)	SSAP	sub	N_{shu}	S	M	I	✓	-	-	stochastic algebraic model	heuristic stacking rules
Bortfeldt and Forster (2012)	PMP	stack	N_{PM}	D	1	E	-	bay	-	bay layout description in matrix form	tree search heuristic with pre-calculated lower bounds
Caserta et al. (2012)	BRP	stack	N_{shu}	D	M^f	I/E/T	-	bay	-	two versions of IP (BRP I, BRP II^g)	rule-based heuristic
Chen and Lu (2012)	CSP	stack	Dis_{YC}, N_{shu}	-	M	E	✓	-	-	NLIP (linearised) for container allocation to sub-block	exact linearised NLIP solution for sub-block allocation, hybrid sequence algorithm for container stacking
Expósito-Izquierdo et al. (2012)	PMP	stack	N_{PM}	D	1	I/E/T	-	bay	-	-	Lowest-Priority-First Heuristic
Huang and Lin (2012)	PMP	stack	N_{PM}	D	1	I/E/T/	-	bay	-	-	labelling heuristics

(*continued*)

Table 1. (*continued*)

Author	Problem scope	Aggregation level	Objective	Time data	Time horizon	Container type	Dynamic	Scheduling	Yard crane modelling	Modelling	Solution method
Jiang et al. (2012)	SSAP	sub	block occupancy, $capy_C$	D	1	E/T	✓	-	-	IP for yard template generation, IP for space allocation, IP for workload assignment	balancing protocol for integrating decisions obtained by IPs
Lee et al. (2012)	SSAP	block	$Disy_C$	D	M	T	✓	term	aggregated within terminal processing capacity	IP	2-level Heuristic: LP-relaxation, neighbourhood search, Tabu Search
Ünlüyurt and Aydin (2012)	BRP	stack	$N_{shu}, Disy_C$	D	1	I/E/T	-	bay	travel time	-	tailored B&B, B&B-based greedy heuristic, Difference Heuristic
Asperen et al. (2013)	RMP	stack	$\omega_{XT}, capy_C, N_{shu}$	(D)	M[h]	I	-	-	-	simulation	heuristic online stacking rules
Kemme (2013)	CSP	stack	N_{shu}, ω	S	M	I/E/T/	-	block	multi-RMG, scheduling, routing, capacity	simulation	Combined Cost Function
Petering and Hussein (2013)	BRP	stack	N_{shu}	D	M	I/E/T	-	bay	-	IP (BRP III)[i]	look-ahead heuristic

(*continued*)

Table 1. (*continued*)

Author	Problem scope	Aggregation level	Objective	Time data	Time horizon	Container type	Dynamic	Scheduling	Yard crane modelling	Modelling	Solution method
Rendl and Prandtstetter (2013)	PMP	stack	N_{PM}	D	M	I/E/T	-	bay	-	constraint programming	rule-based search heuristic
Sharif and Huynh (2013)	SSAP	block	balance cap_{YC}, Dis_{IT}	S	M	I/E	-	term	-	terminal as undirected graph	Ant Colony Optimisation
Yu and Qi (2013)	CSP	stack	$E(\omega)$	(D)	R	I	✓	block	travel time	three IPs for SSAP based on different segregation strategies	decomposition for SSAP: Convex Cost Network Flow algorithm, DP, rule-based heuristic for RMP
Akyüz and Lee (2014)	BRP	stack	N_{shu}	D	M	I/E	✓	bay	-	IP	heuristic index-based rules, IP-based heuristic, Beam Search
Jovanovic and Voß (2014)	BRP	stack	N_{shu}	D	1	I/E/T	-	bay	-	-	improvement of heuristic in Caserta et al. (2011b) by look-ahead mechanism
Ku (2014)	BRP	stack	N_{shu}	(S)^j	1	I/E/T	-	bay	-	stochastic dynamic programming	Expected Reshuffle Heuristic
Ries et al. (2014)	SSAP	stack	N_{shu}, Dis_{YC}	S	M	I	✓	block	travel time	Fuzzy Logic framework	2-phase Fuzzy Logic rule-based approach

(continued)

Table 1. (*continued*)

Author	Problem scope	Aggregation level	Objective	Time data	Time horizon	Container type	Dynamic	Scheduling	Yard crane modelling	Modelling	Solution method
Zehendner and Feillet (2014)	BRP	stack	N_{shu}	D	M^k	I/E/T	-	bay	-	IP^l	Column Generation
Zhen (2014)	SSAP	sub	$capY_C$	(S)	1	E/T	✓	-	yard crane movement capacity per shift	deterministic IP for real-time storage	decision support system for coping with uncertainties
Tang et al. (2015)	BRP	stack	N_{shu}	D	M^m	I/E/T	✓	block	-	IP^n	rule-based heuristics
Hottung and Tierney (2016)	PMP	stack	N_{PM}	D	1	I/E/T	-	bay	-	-	(learning) Biased Random-Key GA
Jin et al. (2016)	SSAP	sub	$Dis y_C$	D	M^o	I/E/T	✓	term	handling capacity	IP	Divide-and-Conquer, Harmony Search, Constraint Satisfaction Search
Ku and Arthanari (2016a)	BRP	stack	N_{shu}	D	1	I/E/T	-	bay	-	-	Abstraction Method, Bi-directional Search

(*continued*)

Table 1. (*continued*)

Author	Problem scope	Aggregation level	Objective	Time data	Time horizon	Container type	Dynamic	Scheduling	Yard crane modelling	Modelling	Solution method
Ku and Arthanari (2016b)	BRP	stack	$E(N_{shu})$	S	1	I	-	bay	-	stochastic dynamic programming	Tree-Search, Abstraction Heuristic, Expected Reshuffle Index Heuristic
Tierney and Voß (2016)	PMP	stack	N_{PM}	S	1	I/E/T	-	bay	-	(robust) constraint programming	relaxation, IDA*
Covic (2017)	RMP	stack	$N_{shu}, (N_{RM}), \omega$	S	M	I/E/T	✓	block	multi-RMG, scheduling, routing, capacity	simulation	(online) Fuzzy Complex Queries
Jovanovic et al. (2017)	PMP	stack	N_{PM}	D	1	I/E/T	-	bay	-	-	combination of heuristics by Murty et al. (2005b); Caserta et al. (2011b); Expósito-Izquierdo et al. (2012); Ünliyurt and Aydin (2012); Jovanovic nad Voß (2014)
Tierney et al. (2017)	PMP	stack	N_{PM}	D	1	E	-	bay	-	-	A*, IDA*, branching and symmetry breaking rules
Ting and Wu (2017)	BRP	stack	N_{shu}	D	1	I/E/T	-	bay	-	-	Beam Search

(*continued*)

Table 1. (*continued*)

Author	Problem scope	Aggregation level	Objective	Time data	Time horizon	Container type	Dynamic	Scheduling	Yard crane modelling	Modelling	Solution method
Wang et al. (2017)	PMP	stack	N_{PM}	D	1	I/E/T	-	bay	-	-	feasibility-based heuristic
Zehendner et al. (2017)	BRP	stack	N_{shu}	(S)	1	I/E/T	-	bay	-	-	online levelling heuristic with performance guarantee, competitive analysis (with offline)
Galle et al. (2018)	BRP	stack	N_{shu}	D	1	I/E/T	-	bay	-	IP building on binary encoding of Caserta et al. (2009)	exact solution of IP by exploiting properties of binary encoding and optimality
Tanaka and Mizuno (2018)	BRP	stack	N_{shu}	D	1	I/E/T	-	bay	-	-	B& B with new lower bounds and dominance properties
Tanaka and Tierney (2018)	PMP	stack	N_{PM}	D	1	I/E/T	-	bay	-	-	iterative deepening B& B, greedy heuristic completing partial solutions, improved lower bounds of Bortfeldt and Forster (2012)

(*continued*)

Table 1. (*continued*)

Author	Problem scope	Aggregation level	Objective	Time data	Time horizon	Container type	Dynamic	Scheduling	Yard crane modelling	Modelling	Solution method
Tricoire et al. (2018)	BRP	stack	N_{shu}	D	1	I/E/T	-	bay	-	-	look-ahead heuristic based on Petering and Hussein (2013), Rake Search (breadth-first tree search with construction heuristics), Pilot Method, B& B with new lower bounds

[a]See Zhang et al. (2010) for corrections of dynamic programming formulation.
[b]Must be sufficiently large to solve the model.
[c]Equals the number of containers to be retrieved.
[d]Analogous to Murty et al. (2005b).
[e]No global optimisation or explicit planning.
[f]Must be sufficiently large to solve the model, alternative formulation with one period provided.
[g]A correction of the BRP II model and an alternative formulation BRP II-A are provided by Zehendner et al. (2015).
[h]No global optimisation or explicit planning.
[i]Reformulation of Caserta et al. (2012).
[j]Deterministic for vessels, stochastic for XTs.
[k]Must be sufficiently large to solve the model.
[l]Based on Caserta et al. (2012).
[m]Equals number of container to be retrieved.
[n]Reformulation of Wan et al. (2009).
[o]One day, subdivided into 12-hour periods.

on (online) stacking rules and the effects on different yard block performance indicators.

Time Data

One of the central aspects in container handling is the availability of data to classify individual containers. As described above, this is particularly important for avoiding mis-overlays of containers if container delivery and retrieval data are available. Most of the research either approximates the retrieval sequence of containers by their destination vessel or weight group (e.g., Chen and Lu (2012); Kang et al. (2006b)) or assumes deterministic given retrieval orders for the containers. Only few reviewed works cope with uncertainty of container time data like Rendl and Prandtstetter (2013) or Tierney and Voß (2016), who develop robust constraint programming models, Borjian et al. (2013), who develop a stochastic programming model and solve it with dynamic programming, or Ku (2014), who use deterministic time data for vessels but stochastic data for XTs in combination with a TAS. In addition, Zehendner et al. (2017) develop an online heuristic where the retrieval time of only a certain amount of containers is known in advance.

Dynamic Problems and Time Horizon

Due to fact that the static BRP, PMP and RMP do not allow incoming containers while performing the necessary re-handling moves, models and solution approaches are mostly formulated dependent on configuration stages instead of time-dependent periods. In many cases, where these problems are formulated as multi-periodic models, a sufficiently large artificial time horizon must be stipulated to ensure a feasible solution (cf. Caserta et al. (2012) for discussion of BRP; Park et al. (2009) for the RMP; Lee and Hsu (2007) for the PMP). As for most of these problems only the deterministic retrieval order is given, no specific time point must be considered when stacking or rearranging containers. In contrast, when allowing incoming containers, the difference between the retrieval time of the newly delivered container and the retrieval time of containers already dwelling in the block must be accounted for.

Container Types

Export and import containers have different characteristics regarding the accuracy and availability of their data. Therefore, it is common in the literature to assume an allocation strategy of containers to blocks where different types of containers are segregated in designated sub-blocks or stacks. For instance, Wan et al. (2009) target only export containers stating that delivery and retrieval data of these containers are more accurate due to available stowage plans. Lee et al. (2012) only consider transshipment containers in order to focus on water-side operations and inter- and intra-terminal movement. On the other hand, Asperen et al. Asperen et al. (2013) use import containers only in order to analyse the impact of TAS information on general stacking strategies.

Scheduling and Yard Crane Modelling

The explicit modelling and solution of container movement scheduling and sequencing can be found infrequently in the literature dealing with container handling problems reviewed in this work. More often, only the translated movement time or cost is considered (if at all) without an explicit consideration of movement restrictions (cf. Kim and Park 2003; Kozan and Preston 2006; Lee et al. 2012; Jin et al. 2016). Instead, complete scheduling problems are studied separately from container handling problems within their own domain Bierwirth and Meisel (2015). In this context, it is assumed that the container handling problems have been solved beforehand and are able to provide the necessary input. Still, the BRP, PMP and RMP can be regarded as scheduling problems because a sequence must be determined to move containers with respect to their stacking order before and after the movement. However, a common version of the BRP is the restricted version which allows shuffling only for containers that are in the same stack as the container to be retrieved in order to reduce the solution space of the problem (see Caserta et al. (2012) and Petering and Hussein (2013) for the discussion of the restricted and unrestricted BRP). In the restricted version, the scheduling decisions to be made are limited as the movement sequence of containers corresponds to the stacking sequence of the containers stacked on top of the container to be retrieved. In contrast, a decision about the optimal schedule in the scope of the PMP and RMP needs to be included as the movement sequence directly influences the number of needed moves to transform a given configuration into a desired one. Hence, the container movement sequence within a bay (PMP) or block (RMP) and for the unrestricted BRP usually results from the solution of these problems (see Borjian et al. (2013) for the BRP, Lee and Hsu (2007) for the PMP, Kang et al. (2006a) for the RMP). Particularly, according to the definition of the RMP above, the optimisation of the yard crane movement distance is inherent to this problem. Thus, movement times and yard crane capacity must be included as long yard crane distances may be induced otherwise (Kim et al. 2000). Except for a couple of papers like (Yu et al. 2009) and (Yu and Qi 2013), capacity considerations are generally not part of the optimisation for the BRP, PMP and RMP as can be observed from Table 1. This may be another reason why the RMP has not been formally modelled in the literature until now as the yard crane movement has considerably less influence on rearranging containers within a bay instead of a block. In the scope of the SSAP on block and sub-block level, it is more common to include yard crane movement times and capacity because one of the main objectives of these problems is to minimise the movement time and balance yard crane workload between the blocks (cf. Kim and Park 2003; Jiang et al. 2012; Sharif and Huynh 2013; Zhen 2014).

Modelling and Solution Method

In general, two modelling approaches are most common in the literature on container handling. Principally for the SSAP, container handling problems are often formulated as a set of sub-problems formulated as algebraic models.

In this context, prior to any model formulation, the parts of the entire problem are decomposed into multiple formulations with characteristic properties to facilitate the solution process. For instance, Kim et al. (2000) provide three different models including dynamic programming for bay matching, a Transport LP for container movement and a TSP with precedence constraints for the subsequent job sequencing. Similar approaches can be observed in Zhang et al. (2003) who provide a non-linear integer programming model for determining the number of containers assigned to each block and afterwards a Transport LP to match specific containers with designated blocks. This same idea is followed by Kozan and Preston (2006) for the SSAP and Yu et al. (2009) for the RMP on sub-block level for container transfer and location allocation. Jiang et al. (2012) formulate three integer programming models for yard template generation, slot allocation and yard crane workload assignment. It is also common to formulate some sub-problems as integer program where special properties are exploited for the subsequent solution process while other sub-problems are directly targeted by heuristics without providing an algebraic model (cf. Lee and Chao 2009).

The BRP has seen considerable development in algebraic formulation approaches as single complete integer programming model starting with Wan et al. (2009), followed by two formulations by Caserta et al. (2012), reformulations of the latter by Petering and Hussein (2013); Zehendner et al. (2015) and a reformulation of the first by Tang et al. (2015). The BRP is proved to be NP-hard by Caserta et al. (2011a) while the feasibility problem is also NP-complete for the related general container loading problem as shown by Bruns et al. (2016). For this reason, rule-based heuristics are the major solution approaches for solving container stacking problems.

Otherwise, simulation is another prevalent modelling approach to cope with uncertainty and computational complexity of container handling problems. This modelling technique is primarily applied for the CSP in order to test a multitude of heuristic stacking rules. First studies are performed by Duinkerken et al. (2001) followed by Dekker et al. (2007). Borgman et al. Borgman et al. (2010) build on the simulation model and the general stacking ideas of the latter by developing online stacking rules that are computationally efficient for real-time stacking decisions. Asperen et al. Asperen et al. (2013) use the same simulation model to apply and test the performance of these stacking rules in the framework of a TAS for import containers. Other modelling techniques include a stack layout description as matrix form (Caserta et al. 2009; Bortfeldt and Forster 2012; Galle et al. 2018), constraint programming (Rendl and Prandtstetter 2013; Tierney and Voß 2016) and fuzzy logic framework (Ries et al. 2014; Covic 2017).

4   Conclusion

In conclusion, it can be seen that container handling in yard blocks has developed into numerous research streams over the past 20 years. This observation highlights the acknowledged importance of optimising the processes of yard block operations in order to contribute to overall terminal productivity. The results of this review can be encapsulated as follows. While the BRP seems to be the prime focus within container handling problems, the RMP has experienced the least attention, probably due to the increased planning complexity of reconciling stacking and movement decisions. Still, the optimisation of general re-handling in all its forms can be stated as the main objective of the literature where decisions are made on stack level. Moreover, the majority of studies deals with static problems (no incoming containers) which are based on deterministic container and vehicle arrival data. Due to a strong focus on optimisation within a bay of a yard block, yard crane movement does not seem to be a major aspect of consideration. If crane movement is considered at all, it is rather done conceptually without sophisticated or detailed modelling. With regard to the modelling techniques and solution approaches, algebraic modelling formulations seem to be the most common approach for modelling container handling while simulation is also a popular modelling tool for larger systems. When solving the problems, heuristics dominate with a strong orientation towards a particular category of heuristics that are based on container stacking rules.

From this conclusion it can be inferred that the full scope of the RMP (in particular, by analytical or algebraic techniques) is scarcely addressed in the literature. An integrated model, where interaction effects between all types of container handling problems with the RMP are regarded, may support an optimisation approach contributing to global terminal productivity. In this context, it is imperative to take data uncertainty of containers into account in order to reflect the true nature of container terminal planning. Thus, future research may be concerned with the intricate task of aligning the integrative model (solution approach) with the stochastic input data. By accomplishing this, it may help container terminal research to increase its potential for direct practical application and benefit. In this sense, the literature on yard block optimisation must be also integrated into the (digital) administrative systems of terminal operations in practice. Namely, the interfaces and practical implementation of the TOS, being the backbone of terminal management and the data warehouse for containers and equipment, must be taken into account. In this context, the nature of the studied decision problems and appropriate solution approaches must be suited to the applicability within and the reality of the TOS (cf. Choi et al. 2003; Vis and Koster 2003; Petering 2011).

In essence, it seems worthwhile to extract key principles of theoretical research for application in real-life terminal operations. Specifically, practical solution approaches must be apt for an implementation into an online environment where proper decisions about container handling operations can be made in real-time. Thus, it is advisable to find a balance between exact methods providing opportunities for additional theoretical insights and (meta-)heuristic approaches giving the necessary computational flexibility for applications. One possible direction for achieving this balance may be the combination of exact and meta-heuristic techniques by means of *math-heuristics* (Caserta and Voß 2009a) with some studies already embracing this approach (e.g., Caserta and Voß (2009b)).

References

Akyüz, M.H., Lee, C.Y.: A mathematical formulation and efficient heuristics for the dynamic container relocation problem. Naval Res. Logistics **61**(2), 101–118 (2014)

Angeloudis, P., Bell, M.G.H.: A review of container terminal simulation models. Marit. Policy Manage. **38**(5), 523–540 (2011)

Asperen, E., van Borgman, B., Dekker, R.: Evaluating impact of truck announcements on container stacking efficiency. Flex. Serv. Manuf. J. **25**(4), 543–556 (2013)

Bierwirth, C., Meisel, F.: A survey of berth allocation and quay crane scheduling problems in container terminals. Eur. J. Oper. Res. **202**(3), 615–627 (2010)

Bierwirth, C., Meisel, F.: A follow-up survey of berth allocation and quay crane scheduling problems in container terminals. Eur. J. Oper. Res. **244**(3), 675–689 (2015)

Borgman, B., van Asperen, E., Dekker, R.: Online rules for container stacking. OR Spectrum **32**(3), 687–716 (2010)

Borjian, S., Manshadi, V.H., Barnhart, C., Jaillet, P.: Dynamic Stochastic Optimization of Relocations in Container Terminals. Working paper, MIT (2013)

Bortfeldt, A., Forster, F.: A tree search procedure for the container pre-marshalling problem. Eur. J. Oper. Res. **217**(3), 531–540 (2012)

Boysen, N., Stephan, K.: A survey on single crane scheduling in automated storage/retrieval systems. Eur. J. Oper. Res. **254**(3), 691–704 (2016)

Boysen, N., Briskorn, D., Meisel, F.: A generalized classification scheme for crane scheduling with interference. Eur. J. Oper. Res. **258**(1), 343–357 (2017)

Brinkmann, B.: Operations systems of container terminals: a compendious overview. In: Böse, J. (ed.) Handbook of Terminal Planning. Operations Research/Computer Science Interfaces Series, vol. 49, pp. 25–39. Springer, New York (2011). https://doi.org/10.1007/978-1-4419-8408-1_2

Bruns, F., Knust, S., Shakhlevich, N.V.: Complexity results for storage loading problems with stacking constraints. Eur. J. Oper. Res. **249**(3), 1074–1081 (2016)

Carlo, H.J., Vis, I.F.A., Roodbergen, K.J.: Storage yard operations in container terminals: literature overview, trends, and research directions. Eur. J. Oper. Res. **235**(2), 412–430 (2014a)

Carlo, H.J., Vis, I.F.A., Roodbergen, K.J.: Transport operations in container terminals: literature overview, trends, research directions and classification scheme. Eur. J. Oper. Res. **236**(1), 1–13 (2014b)

Carlo, H.J., Vis, I.F.A., Roodbergen, K.J.: Seaside operations in container terminals: literature overview, trends, and research directions. Flex. Serv. Manuf. J. **27**(2–3), 224–262 (2015)

Caserta, M., Voß, S.: Metaheuristics: intelligent problem solving. In: Maniezzo, V., Stützle, T., Voß, S. (eds.) Matheuristics. Annals of Information Systems, vol. 10, pp. 1–38. Springer, Boston (2009a). https://doi.org/10.1007/978-1-4419-1306-7_1

Caserta, M., Voß, S.: A corridor method-based algorithm for the pre-marshalling problem. In: Giacobini, M. (ed.) EvoWorkshops 2009. LNCS, vol. 5484, pp. 788–797. Springer, Heidelberg (2009b). https://doi.org/10.1007/978-3-642-01129-0_89

Caserta, M., Schwarze, S., Voß, S.: A new binary description of the blocks relocation problem and benefits in a look ahead heuristic. In: Cotta, C., Cowling, P. (eds.) EvoCOP 2009. LNCS, vol. 5482, pp. 37–48. Springer, Heidelberg (2009). https://doi.org/10.1007/978-3-642-01009-5_4

Caserta, M., Schwarze, S., Voß, S.: Container rehandling at maritime container terminals. In: Bóse, J.W. (ed.) Handbook of Terminal Planning. Operations Research/Computer Science Interfaces Series, vol. 49, pp. 247–269. Springer, New York (2011a). https://doi.org/10.1007/978-1-4419-8408-1_13

Caserta, M., Voß, S., Sniedovich, M.: Applying the corridor method to a blocks relocation problem. OR Spectrum, **33**(4), 915–929 (2011b)

Caserta, M., Schwarze, S., Voß, S.: A mathematical formulation and complexity considerations for the blocks relocation problem. Eur. J. Oper. Res. **219**(1), 96–104 (2012)

Chen, L., Lu, Z.: The storage location assignment problem for outbound containers in a maritime terminal. Int. J. Prod. Econ. **135**(1), 73–80 (2012)

Choe, R., Park, T., Oh, M.S., Kang, J., Ryu, K.R.: Generating a rehandling-free intra-block remarshaling plan for an automated container yard. J. Intell. Manuf. **22**(2), 201–217 (2011)

Choi, H.R., Kim, H.S., Park, B.J., Park, N.K., Lee, S.W.: An ERP approach for container terminal operating systems. Marit. Policy Manage. **30**(3), 197–210 (2003)

Covic, F.: Re-marshalling in automated container yards with terminal appointment systems. Flex. Serv. Manuf. J. **29**(3–4), 433–503 (2017)

Covic, F.: Container Handling in Automated Yard Blocks Based on Time Information. Ph.D thesis, University of Hamburg (2018)

Davis, P.: Container terminal reservation systems. In: 3rd Annual METRANS National Urban Freight Conference, Long Beach, CA (2009)

Dekker, R., Voogd, P., van Asperen, E.: Advanced methods for container stacking. In: Kim, K.H., Günther, H.-O. (eds.) Container Terminals and Cargo Systems, pp. 131–154. Springer, Heidelberg (2007). https://doi.org/10.1007/978-3-540-49550-5_7

Dragović, B., Tzannatos, E., Park, N.K.: Simulation modelling in ports and container terminals: literature overview and analysis by research field, application area and tool. Flex. Serv. Manuf. J. **29**(1), 4–34 (2017)

Duinkerken, M.B., Evers, J.J.M., Ottjes, J.A.: A simulation model for integrating quay transport and stacking policies on automated container terminals. In: Proceedings of the 15th European Simulation Multiconference, pp. 909–916, Prague (2001)

Expósito-Izquierdo, C., Melián-Batista, B., Moreno-Vega, M.: Pre-marshalling problem: heuristic solution method and instances generator. Expert Syst. Appl. **39**(9), 8337–8349 (2012)

Galle, V., Barnhart, C., Jaillet, P.: A new binary formulation of the restricted container relocation problem based on a binary encoding of configurations. Eur. J. Oper. Res. **267**(2), 467–477 (2018)

Günther, H.O., Kim, K.H.: Container terminals and terminal operations. OR Spectrum **28**(4), 437–445 (2006)

Han, Y., Lee, L.H., Chew, E.P., Tan, K.C.: A yard storage strategy for minimizing traffic congestion in a marine container transshipment hub. OR Spectrum **30**(4), 697–720 (2008)

Hottung, A., Tierney, K.: A biased random-key genetic algorithm for the container pre-marshalling problem. Comput. Oper. Res. **75**, 83–102 (2016)

Huang, S.H., Lin, T.H.: Heuristic algorithms for container pre-marshalling problems. Comput. Ind. Eng. **62**(1), 13–20 (2012)

Jiang, X., Lee, L.H., Chew, E.P., Han, Y., Tan, K.C.: A container yard storage strategy for improving land utilization and operation efficiency in a transshipment hub port. Eur. J. Oper. Res. **221**(1), 64–73 (2012)

Jin, J.G., Lee, D.H., Cao, J.X.: Storage yard management in maritime container terminals. Transp. Sci. **50**(4), 1300–1313 (2016)

Jovanovic, R., Voß, S.: A chain heuristic for the blocks relocation problem. Comput. Ind. Eng. **75**, 79–86 (2014)

Jovanovic, R., Tuba, M., Voß, S.: A multi-heuristic approach for solving the pre-marshalling problem. Cent. Eur. J. Oper. Res. **25**(1), 1–28 (2017)

Kang, J., Oh, M.-S., Ahn, E.Y., Ryu, K.R., Kim, K.H.: Planning for intra-block remarshalling in a container terminal. In: Ali, M., Dapoigny, R. (eds.) IEA/AIE 2006. LNCS (LNAI), vol. 4031, pp. 1211–1220. Springer, Heidelberg (2006a). https://doi.org/10.1007/11779568_128

Kang, J., Ryu, K.R., Kim, K.H.: Deriving stacking strategies for export containers with uncertain weight information. J. Intell. Manuf. **17**(4), 399–410 (2006b)

Kemme, N.: Effects of storage block layout and automated yard crane systems on the performance of seaport container terminals. OR Spectrum **34**(3), 563–591 (2012)

Kemme, N.: Design and Operation of Automated Container Storage Systems. Contributions to Management Science, 1st edn. Physica, Heidelberg (2013). https://doi.org/10.1007/978-3-7908-2885-6

Kim, K.H.: Evaluation of the number of rehandles in container yards. Comput. Ind. Eng. **32**(4), 701–711 (1997)

Kim, K.H., Bae, J.W.: Re-marshalling export containers in port container terminals. Comput. Ind. Eng. **35**(3–4), 655–658 (1998)

Kim, K.H., Hong, G.P.: A heuristic rule for relocating blocks. Comput. Oper. Res. **33**(4), 940–954 (2006)

Kim, K.H., Park, K.T.: A note on a dynamic space-allocation method for outbound containers. Eur. J. Oper. Res. **148**(1), 92–101 (2003)

Kim, K.H., Park, Y.M., Ryu, K.R.: Deriving decision rules to locate export containers in container yards. Eur. J. Oper. Res. **124**(1), 89–101 (2000)

Kozan, E., Preston, P.: Mathematical modelling of container transfers and storage locations at seaport terminals. OR Spectrum **28**(4), 519–537 (2006)

Ku, D.: Rehandling problem of pickup containers under truck appointment system. In: Hwang, D., Jung, J.J., Nguyen, N.-T. (eds.) ICCCI 2014. LNCS (LNAI), vol. 8733, pp. 272–281. Springer, Cham (2014). https://doi.org/10.1007/978-3-319-11289-3_28

Ku, D., Arthanari, T.S.: On the abstraction method for the container relocation problem. Comput. Oper. Res. **68**, 110–122 (2016a)

Ku, D., Arthanari, T.S.: Container relocation problem with time windows for container departure. Eur. J. Oper. Res. **252**(3), 1031–1039 (2016b)

Lange, A.K., Schwientek, A., Jahn, C.: Reducing truck congestion at ports - classification and trends. In: Jahn, C., Kersten, W., Ringle, C.M. (eds.) Digitalization in Maritime and Sustainable Logistics, pp. 37–58. epubli, Berlin (2017)

Lee, D.H., Jin, J.G., Chen, J.H.: Terminal and yard allocation problem for a container transshipment hub with multiple terminals. Transp. Res. Part E Logistics Transp. Rev. **48**(2), 516–528 (2012)

Lee, Y., Chao, S.L.: A neighborhood search heuristic for pre-marshalling export containers. Eur. J. Oper. Res. **196**(2), 468–475 (2009)

Lee, Y., Hsu, N.Y.: An optimization model for the container pre-marshalling problem. Comput. Oper. Res. **34**(11), 3295–3313 (2007)

Lim, A., Xu, Z.: A critical-shaking neighborhood search for the yard allocation problem. Eur. J. Oper. Res. **174**(2), 1247–1259 (2006)

Luo, J., Wu, Y., Halldorsson, A., Song, X.: Storage and stacking logistics problems in container terminals. OR Insight **24**(4), 256–275 (2011)

Murty, K.G., Liu, J., Wan, Y., Linn, R.: A decision support system for operations in a container terminal. Decis. Support Syst. **39**(3), 309–332 (2005a)

Murty, K.G., et al.: Hongkong international terminals gains elastic capacity using a data-intensive decision-support system. Interfaces, **35**(1), 61–75 (2005b)

Park, K., Park, T., Ryu, K.R.: Planning for remarshaling in an automated container terminal using cooperative coevolutionary algorithms. In: Proceedings of the 2009

ACM symposium on Applied Computing, SAC 2009, , Honolulu, HI, pp. 1098–1105 (2009)

Park, T., Choe, R., Kim, Y.H., Ryu, K.R.: Dynamic adjustment of container stacking policy in an automated container terminal. Int. J. Prod. Econ. **133**(1), 385–392 (2011)

Petering, M.E.H.: Decision support for yard capacity, fleet composition, truck substitutability, and scalability issues at seaport container terminals. Transp. Res. Part E Logistics Transp. Rev. **47**(1), 85–103 (2011)

Petering, M.E.H., Hussein, M.I.: A new mixed integer program and extended look-ahead heuristic algorithm for the block relocation problem. Eur. J. Oper. Res. **231**(1), 120–130 (2013)

Petering, M.E.H., Wu, Y., Li, W., Goh, M., de Souza, R.: Development and simulation analysis of real-time yard crane control systems for seaport container transshipment terminals. OR Spectrum **31**(4), 801–835 (2009)

Preston, P., Kozan, E.: An approach to determine storage locations of containers at seaport terminals. Comput. Oper. Res. **28**(10), 983–995 (2001)

Rendl, A., Prandtstetter, M.: Constraint models for the container pre-marshaling problem. In: Katsirelos, G., Quimper, C.G. (eds.) 12th International Workshop on Constraint Modelling and Reformulationm ModRef, Uppsala, vol. 13, pp. 44–56 (2013)

Ries, J., González-Ramírez, R.G., Miranda, P.: A fuzzy logic model for the container stacking problem at container terminals. In: González-Ramírez, R.G., Schulte, F., Voß, S., Ceroni Díaz, J.A. (eds.) ICCL 2014. LNCS, vol. 8760, pp. 93–111. Springer, Cham (2014). https://doi.org/10.1007/978-3-319-11421-7_7

Saurí, S., Martín, E.: Space allocating strategies for improving import yard performance at marine terminals. Transp. Res. Part E Logistics Transp. Rev. **47**(6), 1038–1057 (2011)

Schwarze, S., Voß, S., Zhou, G., Zhou, G.: Scientometric analysis of container terminals and ports literature and interaction with publications on distribution networks. In: Hu, H., Shi, X., Stahlbock, R., Voß, S. (eds.) ICCL 2012. LNCS, vol. 7555, pp. 33–52. Springer, Heidelberg (2012). https://doi.org/10.1007/978-3-642-33587-7_3

Sharif, O., Huynh, N.: Storage space allocation at marine container terminals using ant-based control. Expert Syst. Appl. **40**(6), 2323–2330 (2013)

Stahlbock, R., Voß, S.: Operations research at container terminals: a literature update. OR Spectrum **30**(1), 1–52 (2008a)

Stahlbock, R., Voß, S.: Vehicle routing problems and container terminal operations - an update of research. In: Golden, B., Raghavan, S., Wasil, E. (eds.) The Vehicle Routing Problem: Latest Advances and New Challenges. Operations Research/Computer Science Interfaces, vol. 43, pp. 551–589. Springer, Boston (2008b). https://doi.org/10.1007/978-0-387-77778-8_25

Steenken, D., Voß, S., Stahlbock, R.: Container terminal operation and operations research - a classification and literature review. OR Spectrum **26**(1), 3–49 (2004)

Tanaka, S., Mizuno, F.: An exact algorithm for the unrestricted block relocation problem. Comput. Oper. Res. **95**, 12–31 (2018)

Tanaka, S., Tierney, K.: Solving real-world sized container pre-marshalling problems with an iterative deepening branch-and-bound algorithm. Eur. J. Oper. Res. **264**(1), 165–180 (2018)

Tang, L., Jiang, W., Liu, J., Dong, Y.: Research into container reshuffling and stacking problems in container terminal yards. IIE Trans. **47**(7), 751–766 (2015)

Tierney, K., Voß, S.: Solving the robust container pre-marshalling problem. In: Paias, A., Ruthmair, M., Voß, S. (eds.) ICCL 2016. LNCS, vol. 9855, pp. 131–145. Springer, Cham (2016). https://doi.org/10.1007/978-3-319-44896-1_9

Tierney, K., Pacino, D., Voß, S.: Solving the pre-marshalling problem to optimality with A* and IDA*. Flex. Serv. Manuf. J. **29**(2), 223–259 (2017)

Ting, C.J., Wu, K.C.: Optimizing container relocation operations at container yards with beam search. Transp. Res. Part E Logistics Transp. Rev. **103**, 17–31 (2017)

Tricoire, F., Scagnetti, J., Beham, A.: New insights on the block relocation problem. Comput. Oper. Res. **89**, 127–139 (2018)

UNCTAD, United Nations Conference on Trade and Development. Review of Maritime Transport 2017. UNCTAD/RMT/2017, United Nations, Geneva (2017)

Ünlüyurt, T., Aydin, C.: Improved rehandling strategies for the container retrieval process. J. Adv. Transp. **46**(4), 378–393 (2012)

Vis, I.F.A., de Koster, R.: Transshipment of containers at a container terminal: an overview. Eur. J. Oper. Res. **147**(1), 1–16 (2003)

Wan, Y., Liu, J., Tsai, P.C.: The assignment of storage locations to containers for a container stack. Naval Res. Logistics **56**(8), 699–713 (2009)

Wang, N., Jin, B., Zhang, Z., Lim, A.: A feasibility-based heuristic for the container pre-marshalling problem. Eur. J. Oper. Res. **256**(1), 90–101 (2017)

WTO, World Trade Organization. Factors shaping the future of world trade. World trade report, WTO Publications, Geneva (2013)

Yu, M., Qi, X.: Storage space allocation models for inbound containers in an automatic container terminal. Eur. J. Oper. Res. **226**(1), 32–45 (2013)

Yu, V.F., Cheng, H.Y., Ting, H.I.: Optimizing re-marshalling operation in export container terminals. In: Proceedings of the Asia Pacific Industrial Engineering & Management Systems Conference, APIEMS 2009, Kitakyushu, pp. 2934–2938 (2009)

Zehendner, E., Feillet, D.: Branch and price approach for the container relocation problem. Int. J. Prod. Res. **52**(24), 7159–7176 (2014)

Zehendner, E., Caserta, M., Feillet, D., Schwarze, S., Voß, S.: An improved mathematical formulation for the blocks relocation problem. Eur. J. Oper. Res. **245**(2), 415–422 (2015)

Zehendner, E., Feillet, D., Jaillet, P.: An algorithm with performance guarantee for the online container relocation problem. Eur. J. Oper. Res. **259**(1), 48–62 (2017)

Zhang, C., Liu, J., Wan, Y., Murty, K.G., Linn, R.L.: Storage space allocation in container terminals. Transp. Res. Part B Methodological **37**(10), 883–903 (2003)

Zhang, C., Chen, W., Shi, L., Zheng, L.: A note on deriving decision rules to locate export containers in container yards. Eur. J. Oper. Res. **205**(2), 483–485 (2010)

Zhao, W., Goodchild, A.V.: The impact of truck arrival information on container terminal rehandling. Transp. Res. Part E Logistics Transp. Rev. **46**(3), 327–343 (2010)

Zhen, L.: Storage allocation in transshipment hubs under uncertainties. Int. J. Prod. Res. **52**(1), 72–88 (2014)

A New Lower Bound for the Block Relocation Problem

Tiziano Bacci$^{(\boxtimes)}$, Sara Mattia, and Paolo Ventura

IASI - CNR, Via dei Taurini 19, 00185 Rome, Italy
{tiziano.bacci,sara.mattia,paolo.ventura}@iasi.cnr.it

Abstract. In this paper we deal with the restricted Block (or Container) Relocation Problem. We present a polynomial time algorithm to calculate a new lower bound for the problem.

Keywords: Block relocation · Container relocation · Lower bound

1 Introduction

A container terminal is a facility where containers are transferred between different transportation vehicles. Containers that wait for a truck, or a train, or a cargo ship, are stored in an area called *yard*. Since the yard is limited in space, the containers are piled into stacks. The height of each stack, i.e., the maximum number of containers that can be piled one above the other, is constrained by the height of the yard cranes used to move the containers. Typically, a container yard stores at the same time thousands of containers grouped into hundreds of stacks with a storage capacity which may be up to 10 containers [6]. Since a stack is accessible only from the top, when a container has to be retrieved from the storage area, any container located above it has to be moved into another stack with a *reshuffle operation*. Reshuffle operations are costly and time-consuming. Then, given a retrieval order of the containers, it is crucial to find a way to reallocate all the reshuffled containers so as to minimize the total number of reshuffle operations. This problem is known as the Block Relocation Problem (BRP). Here, we consider the restricted BRP, where only containers above the next one that has to be retrieved can be reshuffled.

Figure 1 gives an example of the BRP with an initial yard consisting of 3 stacks, 4 available slots for each stack, and 7 stored containers. Starting from the initial yard, the sequence of movements of an optimal solution is reported. At each step, the next container to be moved with a reshuffle or a retrieval operation is highlighted in gray. The minimum number of reshuffles required is six: containers 7, 3, 4 are reshuffled in order to retrieve container 1; container 4 is reshuffled to retrieve container 2; then, containers 7 and 6 are reshuffled to retrieve container 4.

Recent surveys on optimization problems arising in the management of container terminals can be found in [3,12]. In this context, the BRP problem is

© Springer Nature Switzerland AG 2018
R. Cerulli et al. (Eds.): ICCL 2018, LNCS 11184, pp. 168–174, 2018.
https://doi.org/10.1007/978-3-030-00898-7_10

known to be NP-hard [4], as it generalizes the Mutual Exclusion Scheduling [2,7]. Being the BRP problem both theoretically and computationally hard to solve, many heuristic approaches have also been introduced (see [5,8,10,11,14,17] among the others). On the other hand, few contributions in the literature deal with lower bounds for the problem, although they can be used to certify the quality of heuristic approaches or to reduce the search space in exact algorithms [13,17]. In this paper we present a new lower bound for the BRP problem. Experimental results show the effectiveness of our procedure. Throughout the paper, the words *container* and *block* will be used interchangeably.

In Sect. 2, we give a survey on the different lower bounds presented in the literature. In Sect. 3, we introduce the new lower bound. In Sect. 4, we compare our method with the lower bounds existing in the literature. In Sect. 5, we give the conclusions.

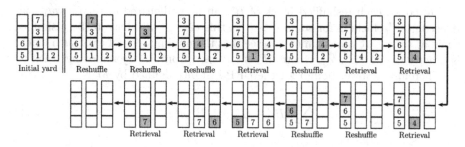

Fig. 1. A representation of an optimal solution for the Block Relocation Problem.

2   Lower Bounds for the Restricted BRP

All the lower bounds for the BRP problem existing in the literature can be calculated by iteratively solving relaxations of the *Generalized Minimum Blocking Items Problem* (GMBIP), whose definition is given in the following.

Let $M \in \mathbb{Z}^{w \times h}$ be a yard with w stacks of height h, and let $M(j,k)$ denote the block allocated in the k-th position of stack j ($M(j,k) = 0$ if the slot is empty). Now, let B be a set of n blocks that have to be located in the available slots of the stacks of M, according to a given order ϕ (denote by $\phi(i)$ the i-th block to be located). Finally, let $\overline{M}$ be the yard obtained from M once all the blocks of B have been allocated. $\overline{M}$ is said to be *feasible* if it is compatible with ϕ, i.e. for each couple $b, b' \in B$ with $\phi^{-1}(b) < \phi^{-1}(b')$, b is not located above b' in $\overline{M}$. Furthermore, we say that block $r = \overline{M}(j,k)$ is *r'-blocking* if it is located in some slot above $r' < r$ and that r is *blocking* if it is r'-blocking for some r' in $\overline{M}$. Given an input instance defined by (M, B, ϕ), the GMBIP problem is to find the feasible configuration $\overline{M}$ that minimizes the total number of blocking blocks of B. We denote by $G^*(M, B, \phi)$ such a minimum value.

The GMBIP problem is NP-hard in general, as it slightly generalizes the Minimum Blocking Items Problem (MBIP). In fact, in MBIP the initial yard is always empty. The computational complexity of MBIP has been analyzed in [1]. We now describe how to calculate a lower bound for the BRP by iteratively solving instances of GMBIP. Consider a BRP instance, with a set $\{1,\ldots,n\}$ of blocks located in a yard M. Recall that, at each time i, block i (located in stack say t^i) is retrieved from M and all the i-blocking blocks are reshuffled from t^i. Then, let $M^0 = M$ and, for each $i = 1,\ldots,n$, let B^i be the set of i-blocking blocks of M^{i-1} taken in the order ϕ^i from the top to the bottom, and let M^i be obtained from M^{i-1} by removing the block i and all the blocks in B^i. Now observe that:

- $M^n(j,k) = 0$, for all $j \in \{1,\ldots,w\}$ and $k \in \{1,\ldots,h\}$;
- block i could not be present in M^{i-1};
- all the blocks in B^i have to be reshuffled at time i in any solution of the input BRP instance defined by M.

Moreover, each reshuffled block of B^i can be reallocated in such a way that it becomes i'-blocking for some $i' > i$. In this case, it will have to be reshuffled again at time i'. It is not difficult to see that the minimum number of such blocks is exactly $G^*(\tilde{M}^i, B^i, \phi^i)$, where $\tilde{M}^i$ is obtained from M^i by removing stack t^i.

Therefore, a lower bound for the BRP instance defined by M is

$$\sum_{i=1}^{n}(|B^i| + G^*(\tilde{M}^i, B^i, \phi^i)). \tag{1}$$

As already mentioned, all the lower bounds for the BRP presented in the literature are derived from (1) substituting, at each iteration i, $G^*(\tilde{M}^i, B^i, \phi^i)$ with some lower bound.

In particular, the lower bound LB_K, introduced by Kim and Hong [9], uses 0 as a lower bound for $G^*(\tilde{M}^i, B^i, \phi^i)$. In other cases, such a lower bound on $G^*(\tilde{M}^i, B^i, \phi^i))$ is defined as the optimal value of some relaxed variant of GMBIP. Zhu et al. [17] defined LB_Z by solving a variant of GMBIP where both restrictions on the order ϕ of the incoming blocks as well as on the capacity h of the stacks are relaxed. The optimal value of such a relaxed problem, denoted here by $G^Z(M, B, \phi)$, is calculated in $O(n)$. Tanaka and Takii [13] proposed a $O(2^n)$ algorithm to solve the GMBIP variant (here denoted by $GMBIP^T$) obtained by relaxing the capacity restriction on the stacks of the current yard. We call $G^T(M, B, \phi)$ the optimal value of this problem and LB_T the corresponding lower bound for the BRP problem.

In the next section, we present a new lower bound for the BRP problem, obtained by solving to optimality (in $O(n\,log(n)+w\,log(w))$ time) the relaxation of GMBIP obtained by removing the restriction on the order ϕ.

3   A New Lower Bound for the Restricted BRP

In this section, we introduce a new lower bound for the BRP problem. Let $GMBIP^B$ be the GMBIP variant where the optimal configuration $\overline{M}$ does not need to be feasible with respect to the input order ϕ. We denote by $G^B(M, B, \phi)$ its optimal value and by LB_B the corresponding lower bound for BRP, obtained according to (1). As $0 \leq G^Z(M, B, \phi) \leq G^T(M, B, \phi), G^R(M, B, \phi)$, the following holds

$$LB_K \leq LB_Z \leq LB_T, LB_B \tag{2}$$

Note that, as the GMBIP^T and GMBIP^B are defined on two different relaxations of GMBIP, there does not exists a theoretical dominance relation between $G^T(M, B, \phi)$ and $G^B(M, B, \phi)$ (and therefore between LB_T and LB_B).

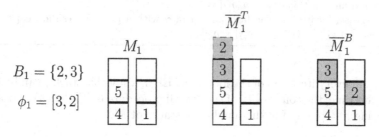

Fig. 2. An instance (M_1, B_1, ϕ_1) with $G^T(M_1, B_1, \phi_1) = 0 < G^B(M_1, B_1, \phi_1) = 1$.

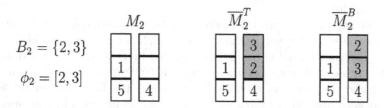

Fig. 3. An instance (M_2, B_2, ϕ_2) with $G^B(M_2, B_2, \phi_2) = 0 < G^T(M_2, B_2, \phi_2) = 1$.

In Figs. 2 and 3, we present two input instances (M_1, B_1, ϕ_1) and (M_2, B_2, ϕ_2) with $G^T(M_1, B_1, \phi_1) < G^B(M_1, B_1, \phi_1)$ (Fig. 2) and $G^T(M_2, B_2, \phi_2) > G^B(M_2, B_2, \phi_2)$ (Fig. 3). In the figures, $\overline{M}_i^T$ and $\overline{M}_i^B$ represent the optimal solutions of GMBIP^T and GMBIP^B, respectively, obtained from the input instance (M_i, B_i, ϕ_i), for each $i = 1, 2$. Observe that, in Fig. 2, the optimal configuration $\overline{M}_1^T$ (of value 0) does not satisfy the restriction on the capacity of the first stack. On the other hand, in Fig. 3, in the optimal solution $\overline{M}_2^B$ (of value 0), block 2 is located above block 3, although 2 precedes 3 in ϕ_2.

Now, let δ_j be the residual capacity of each stack j of M and let σ_j be the smallest index of a block located in j. The GMBIP^B problem is then the minimum cost assignment problem where one wants to assign every block $b \in B$ in some available stack j of capacity δ_j. Here, each assignment (b, j) has cost 1, if $b > \sigma_j$, and 0, otherwise. In the following, we present an exact algorithm to solve GMBIP^B.

Algorithm 1. Algorithm for solving the GMBIP^B

STEP 0: Set G^B = 0.

STEP 1: Construct the vector ω, in which the blocks of B are ordered non increasingly with respect to their index.

STEP 2: Construct the vector γ, in which the stacks of M are ordered non increasingly with respect to the value of σ.

STEP 3:

for $i = 1, \ldots, n$ do

    let $b = \omega[i]$;

    locate b in the first stack j of γ with residual capacity $\delta[j] > 0$ and such that $\sigma[j] > b$, if any; in this case, set $\delta[j] = \delta[j] - 1$;

    otherwise, locate b in the last stack j of γ with a positive residual capacity; in this case, set G^B = G^B + 1, $\delta[j] = \delta[j] - 1$.

Since, at every iteration of the loop in STEP 3, a suitable stack j can be found in constant time, the complexity of the overall procedure is $O(n \log(n) + w \log(w))$ time, being $n = |B|$ and w the number of stacks of M.

4   Computational Results

In this section, we present some computational results that show the effectiveness of our algorithm.

For the experiments, we used two datasets. The first one includes six sets of instances already known in the literature [5,11,14–17] and provided by the authors. The corresponding results are reported in Table 1. The second dataset contains instances that we generated according to two parameters, the number of stacks and the height of each stack. Following [16], for each pair (w, h), we randomly generated 50 instances with $w \times h - (h - 1)$ blocks. The corresponding results are reported in Table 2.

Each row of the tables is related to a group of instances. In Table 1, each group corresponds to one of the datasets in the literature (indicated in the first column), with the exception of the dataset presented in [5], where we distinguish among small-medium and large size instances. In Table 2 the instances are grouped according to their size.

For each row, n, w and h indicate the number (or range) of blocks, stacks, and stack height, respectively, while I is the number of instances in the group. Then, columns LB_K, LB_Z, LB_T, and LB_B correspond to the lower bounds presented in the previous sections. For each of these columns, *Value* is the average value and *Time* is the average computing time (in seconds) on the instances of the group. The symbol $*$ is used to indicated that some instance of the corresponding group could not be solved within the time limit of one hour.

The computational results show that, as expected, $LB_B \geq LB_K$ and $LB_B \geq LB_Z$ (see (2)). Moreover, they also show that, in practice, $LB_T > LB_B$. However, the computational times required to solve GMBIP^T are, in particular on the large instances, much higher than the ones needed for solving GMBIP^B. Indeed, recall that the algorithm proposed by Tanaka and Takii is exponential in the number of reshuffled blocks. Therefore, even if it produces better bounds, LB_T cannot be used in practice within an algorithm that solves BRP real-size instances. Hence, LB_B seems to present a good compromise between quality of the solution and computational time, as it always outperforms the values provided by LB_K and LB_Z in, essentially, the same computing time.

Table 1. Comparative analysis on different lower bounds obtained on six datasets taken from the literature

SET	n	w	h	I	LB_K		LB_Z		LB_T		LB_B	
					Value	Time	Value	Time	Value	Time	Value	Time
[5]	[9, 100]	[3, 10]	[5, 12]	840	18.37	0.00	22.04	0.00	22.87	0.00	22.05	0.00
	10000	100	102	40	9485.92	0.00	11177.08	0.00	*	*	11177.52	0.00
[11]	[70, 720]	[16, 160]	[6, 8]	14	213.11	0.00	215.29	0.00	215.61	0.00	215.33	0.00
[14]	[3, 7]	[6, 36]	[4, 7]	8000	8.00	0.00	9.53	0.00	9.81	0.00	9.54	0.00
[15]	[3, 21]	6	[2, 5]	600	2.93	0.00	3.16	0.00	3.20	0.00	3.16	0.00
[16]	[7, 133]	[3, 12]	[3, 12]	4000	32.71	0.00	39.01	0.00	41.13	0.00	39.11	0.00
[17]	[15, 69]	[6, 10]	[3, 7]	12500	21.46	0.00	25.87	0.00	26.59	0.00	25.93	0.00

Table 2. Comparative analysis on different lower bounds on randomly generated instances

n	w	h	I	LB_K		LB_Z		LB_T		LB_B	
				Value	Time	Value	Time	Value	Time	Value	Time
46	10	5	50	24.20	0.00	27.18	0.00	27.72	0.00	27.32	0.00
91	10	10	50	63.36	0.00	79.16	0.00	82.46	0.00	79.30	0.00
451	10	50	50	406.54	0.00	592.18	0.00	*	*	592.18	0.00
496	100	5	50	268.84	0.00	272.80	0.00	273.46	0.00	272.92	0.00
991	100	10	50	697.40	0.00	715.70	0.00	723.32	0.01	716.28	0.00
4951	100	50	50	4502.16	0.00	5014.68	0.00	*	*	5014.76	0.00
4996	1000	5	50	2713.62	0.00	2718.08	0.00	2718.86	0.00	2718.08	0.00
9991	1000	10	50	7059.04	0.00	7081.30	0.00	7089.14	0.14	7081.44	0.00
49951	1000	50	50	4552.20	0.01	46134.12	0.01	*	*	46134.30	0.01

5   Conclusions

We introduced a new lower bound (LB_B) for the restricted Block Relocation Problem, that is widely studied in the context of logistics of containers in container terminals. We also presented an algorithm to calculate LB_B in polynomial time. Computational results showed that our lower bound is very effective and it is able to produce good values even on large instances. This suggests that it could be successfully integrated into heuristic algorithms for solving real-size instances, both to certify the quality of the solutions as well as to limit the search space.

Acknowledgements. The authors have been partially supported by Ministry of Instruction University and Research (MIUR) with the program PRIN 2015, project "SPORT - Smart PORt Terminals", code 2015XAPRKF; project "Nonlinear and Combinatorial Aspects of Complex Networks", code 2015B5F27W; project "Scheduling cuts: new optimization models and algorithms for cutting, packing and nesting in manufacturing processes", code 20153TXRX9.

References

1. Bacci, T., Mattia, S., Ventura, P.: Some complexity results for the minimum blocking items problem. In: Sforza, A., Sterle, C. (eds.) ODS 2017. PROMS, vol. 217, pp. 475–483. Springer, Cham (2017). https://doi.org/10.1007/978-3-319-67308-0_48

2. Bonomo, F., Mattia, S., Oriolo, G.: Bounded coloring of co-comparability graphs and the pickup and delivery tour combination problem. Theor. Comput. Sci. **412**(45), 6261–6268 (2011)

3. Carlo, H., Vis, I., Roodbergen, K.: Storage yard operations in container terminals: literature overview, trends, and research directions. Eur. J. Oper. Res. **235**(2), 412–430 (2014)

4. Caserta, M., Schwarze, S., Voß, S.: A mathematical formulation and complexity considerations for the blocks relocation problem. Eur. J. Oper. Res. **219**, 96–104 (2012)

5. Caserta, M., Voß, S., Sniedovich, M.: Applying the corridor method to a blocks relocation problem. OR Spectr. **33**(4), 915–929 (2011)

6. Günter, H.O., Kim, K.H.: Container Terminals and Automated Transport Systems. Springer, Heidelberg (2005). https://doi.org/10.1007/b137951

7. Jansen, K.: The mutual exclusion scheduling problem for permutation and comparability graphs. Inf. Comput. **180**, 71–81 (2003)

8. Jovanovic, R., Voß, S.: A chain heuristic for the blocks relocation problem. Comput. Ind. Eng. **75**, 79–86 (2014)

9. Kim, K.H., Hong, G.P.: A heuristic rule for relocating blocks. Comput. Oper. Res. **33**, 940–954 (2006)

10. Ku, D., Arthanari, T.: On the abstraction method for the container relocation problem. Comput. Oper. Res. **68**(Suppl. C), 110–122 (2016)

11. Lee, Y., Lee, Y.L.: A heuristic for retrieving containers from a yard. Comput. Oper. Res. **47**, 1139–1147 (2010)

12. Lehnfeld, J., Knust, S.: Loading, unloading and premarshalling of stacks in storage areas: survey and classification. Eur. J. Oper. Res. **239**, 297–312 (2014)

13. Tanaka, S., Takii, K.: A faster branch-and-bound algorithm for the block relocation problem. IEEE Trans. Autom. Sci. Eng. **13**(1), 181–190 (2016)

14. Unluyurt, T., Aydin, C.: Improved rehandling strategies for the container retrieval process. J. Adv. Transp. **46**(4), 378–393 (2012)

15. Wan, Y., Liu, J., Tsai, P.C.: The assignment of storage locations to containers for a container stack. Nav. Res. Logist. **56**(8), 699–713 (2009)

16. Wu, K.C., Ting, C.J.: A beam search algorithm for minimizing reshuffle operations at container yards. In: International Conference on Logistics and Maritime Systems, Busan, Korea, 15–17 September 2010 (2010)

17. Zhu, W., Qin, H., Lim, A., Zhang, H.: Iterative deepening A* algorithms for the container relocation problem. IEEE Trans. Autom. Sci. Eng. **9**, 710–722 (2012)

The Standard Capacity Model: Towards a Polyhedron Representation of Container Vessel Capacity

Rune Møller Jensen[✉] and Mai Lise Ajspur

IT University, Rued Langaards Vej 7, 2300 Copenhagen S, Denmark
rmj@itu.dk

Abstract. Container liner shipping is about matching spare capacity to cargo in need of transport. This can be realized using cargo flow networks, where edges are associated with vessel capacity. It is hard, though, to calculate free capacity of container vessels unless full-blown non-linear stowage optimization models are applied. This may cause such flow network optimization to be intractable. To address this challenge, we introduce the Standard Capacity Model (SCM). SCMs are succinct linear capacity models derived from vessel data that can be integrated in higher order optimization models as mentioned above. In this paper, we introduce the hydrostatic core of the SCM. Our results show that it can predict key parameters like draft, trim, and stress forces accurately and thus can model capacity reductions due to these factors.

Keywords: Container vessel capacity · Stowage planning
Linear modelling

1 Introduction

Container liner shipping is a major driver of the world economy [4]. Today, there are more than 5000 container vessels in the world [14], mostly sailing on cyclic services with published fixed weekly schedules and freight rates. Liner shipping companies adjust these service networks and their fleet over the year to fit seasonal trends and long-term developments in the world economy, but they seldom make fleet and network changes due to current cargo on the network and known bookings. For that reason, it is a central objective to maximize the utilization of the service network, as any free capacity in the network is a business opportunity.

Previous work has studied how to apply revenue management methods in the liner shipping industry similar to the ones successfully applied in the airline industry (e.g., [16]). This has turned out to be challenging in practice. A major obstacle is to compute the free capacity of a container vessel. Although surprising

This research is supported by the Danish Maritime Fund, Grant No. 2016-064.

R. Cerulli et al. (Eds.): ICCL 2018, LNCS 11184, pp. 175–190, 2018.
https://doi.org/10.1007/978-3-030-00898-7_11

at first, it is not simply the number of vacant slots on the vessel, since a large number of local and global constraints may cause slots to be impossible to use. These constraints include: stacking limitations due to different length, height, power need (reefer containers), and dangerous content of containers; limited volume, weight, and securing capacity of container stacks; vessel hydrostatics like stability requirements and stress force limitations; containers blocking each other due to different port of discharge; capacity preserving stowage patterns; and work balancing of quay cranes. It is recognized by leading economists that this problem blocks a paradigm change in liner shipping. According to Stopford, the ability to match spare capacity to cargo in need of transportation on the fly would allow the "Uberisation" of the freight business [5]. Today, the higher sales and cargo flow functions in liner shipping companies are unable to make these matchings. The spare capacity of a container vessel is often simply calculated as its maximum volume, weight, and reefer container capacity subtracted the capacity taken up by on board cargo without consideration of losses due stowage restrictions and rules. This can cause great over-estimates of the free capacity of the vessels [3].

In the last two decades, a number of automated stowage planning methods have been published (e.g., [1,7,8,10,12,15]). The input to these methods is the arrival condition of the vessel and a list of containers to load, and the output is a stowage plan. As such, these methods are unable to compute the spare capacity of the vessel, since the containers to load are assumed to be known. Several of the contributions, though, apply optimization models, where the containers to load can act as decision variables rather than constants (e.g., [1,3,10]). These models can be used to compute the spare capacity of a vessel. In practice, though, they can be challenging to apply in higher functions such as sales and cargo flow. The stowage planning problem is NP-hard [2], even in its various abstract versions [13]. This means that the stowage optimization models can take long time to solve, which also happens in practice (e.g., [10]). Since it can take more than five hours to generate a stowage plan manually, this is an acceptable evil in stowage planning. In higher functions, on the other hand, capacity models can be parts of larger optimization models which require that they are scalable. For instance, in capacity and uptake management, a cargo flow network could be used to match cargo demand with spare capacity. In such a network modelling several weeks of a major trade line, there are thousands of edges representing voyage legs, and each of these needs to be associated with a capacity model.

To address this challenge, this paper introduces the *Standard Capacity Model* (SCM). The SCM is based on several insights from previous work on stowage planning optimization. First, a significant source of the complexity and inaccessibility of these models is the spatial misalignment of data describing container vessels. To clear this, the SCM interpolates vessel data to align with the endpoints of each bay. Second, stowage optimization models have many details that can be abstracted away in capacity calculations. To this end, the granularity of the SCM can be adjusted. At the finest level, each bay forms a *section*. At coarser levels, adjacent sections are merged. Third, a previous study of vessel hydrostatics show that these can be accurately approximated by linear functions for a fixed displacement [11]. Container vessels at normal drafts, however, are near box-formed. This opens for a linear formulation of the hydrostatic equilibrium equations at any displacement that until now has not been exploited.

The intractable elements of stowage planning include separation rules of containers with dangerous goods and the fact that quay cranes only can discharge containers from the top of stacks [2]. In more abstract capacity models, though, it may be possible to express some of these combinatorial aspects as linear trade-offs. In particular, a significant body of industrial work shows that surprisingly many highly complex aspects of stowage planning can be linearly expressed [9]. Our objective is in time to mature the SCM with these advanced linear models. In this paper, we focus on the hydrostatic core of the SCM that to our knowledge is the first linear approximation of the hydrostatic equilibrium of a container vessel for variable displacement. Our results show that the hydrostatic model is able to predict key parameters like draft, trim, and stress forces with a sufficient accuracy for practical application even for coarse standard capacity models.

The remainder of this paper is organized as follows. We define the problem in Sect. 2 and introduce the SCM in Sect. 3. In Sect. 4, we evaluate the prediction accuracy of the SCM, and finally in Sect. 5 we conclude and discuss directions of future work.

2 Problem Formulation

Container vessels mainly transport ISO containers with the dominating lengths 20′, 40′, and 45′, while the containers usually are 8′ wide. There are two common heights: standard 8′6″ (DC) and high-cube 9′6″ (HC). Containers have corner fittings that allow them to be stacked about 10 high. 45DC and 20HC are rare. *Reefer* containers are refrigerated and need external power. *Out-of-gauge* (OOG) containers have irregular dimensions (e.g., open top containers with cargo sticking up). Flatracks are flat containers to carry non-containerized cargo (break-bulk). *DG* containers contain dangerous goods such as fireworks and chemicals. They must be placed according to complex separation rules and may not be allowed near reefers since these are spark generators.

As shown in Fig. 1, the cargo space of a container vessel is divided into *bays*, which each consists of stacks (rows) of cells. A cell is divided into a fore and aft *slot* and can accordingly hold one 40′ (or 45′) container or two 20′ containers. Some cells have power plugs allowing reefers to be stowed. Each bay is divided into stowage areas above and under *hatch covers* that separate on deck and below deck cells. A vessel has a cargo securing manual that details how the vessel can be stowed securely. The precise position of bays, fuel tanks and ballast water tanks are provided by the shipyard that build the vessel. The yard also provides details about the *lightship*, which is the vessel without cargo, fuel or ballast water. This information can be given as a set of blocks with known mass and center of gravity as shown in Fig. 1. From this data, the resulting center of gravity of the vessel can be calculated.

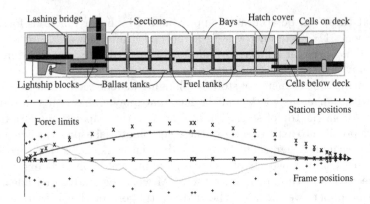

Fig. 1. Top: the structure of a cellular container vessel and an example of a section partitioning that can be used by the SCM. Bottom: an example of the shear forces (light blue curve) and bending moments (dark blue curve) along a vessel. Light blue plus signs and dark blue crosses are the associated force limits given by the classification society for a set of frame positions. (Color figure online)

The Bonjean table of the vessel can be used to compute its center of buoyancy. For a set of cross-sections called *stations* along the vessel, the table gives the submerged area as a function of the distance from the keel to the water line (*draft*) of the station. A vessel is in hydrostatic equilibrium when the center of buoyancy and gravity are vertically aligned. In this condition, the vessel floats at rest in the water at a stable draft and *trim*. Trim is the difference between aft and fore draft of the vessel (i.e., nose up is positive trim). The total weight of a vessel is referred to as its *displacement* and has different summer and winter limits depending on sea location. Many ports such as Hamburg have significant tide dependent draft limits. Fuel efficient trims are typically around -2 m (i.e., nose down).

While the sum of buoyancy and gravity forces are vertically aligned at hydrostatic equilibrium, the forces acting on the vessel are usually distributed unevenly over the hull. Figure 2 shows an example of the resulting gravity and buoyancy forces and how they would cause sections of the vessel to change draft if they could move freely. The counteracting forces in the hull that prevents such movement are referred to as *stress forces.* The critical stress forces acting on a vessel are *shear forces* (SF), *bending moments* (BM), and *torsion moments* (TM). These forces are defined relative to a cross-section of the vessel. Consider the cross-section indicated by the white diamond in Fig. 2. SF at the cross-section is the sum of forces fore of the cross-section.[1] BM is the sum of forces each multiplied with the longitudinal distance to them. TM is caused by the distribution of forces over the center line. It is defined like BM using the transversal distance

---

[1] SF can just as well be defined as the sum of forces aft of the cross-section. The reason is that since the vessel is at hydrostatic equilibrium, the two forces must be equal, but with opposite sign.

to the force. SF and BM measure how much the forces try to shear and bend the cross-section, while TM measures how much they try to twist it. The classification society of the vessel defines minimum and maximum limits of these forces for a number of frame positions along the vessel. Figure 1 shows an example of SF and BM forces. Notice that frame and station positions are misaligned. A ship typically has higher gravity forces than buoyancy forces in the bow and stern. Consequently, SF forces are positive aft and negative fore, while BM is high midship.

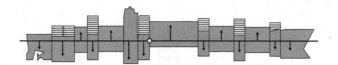

Fig. 2. An example of the resulting forces (black arrows) acting in the longitudinal direction at hydrostatic equilibrium.

3   The Standard Capacity Model

The purpose of the Standard Capacity Model (SCM) is to: (1) simplify the data representation of vessels by aligning all data points to a reference system defined by sections; (2) simplify the capacity constraints of vessels by a linear polyhedron approximation; and (3) provide a model with an adjustable level of detail. As mentioned in the introduction, the key idea of the SCM is to partition the vessel into sections that are aligned with bays. At the finest level of detail, sections hold at most one bay. At coarser levels, some sections are merged. As an example, Fig. 1 shows a partitioning of a vessel into six sections, where the largest sections aggregate three bays each. The choice of sections depends on the application. For large cargo flow models, it may only be computationally tractable with a few sections per vessel. The choice also depends on the cellular structure of the vessel. A section partitioning also should be made with stowage trade-offs in mind (e.g., cluster bays with same reefer plug and lashing bridge arrangement).

This paper focuses on the main building block of the SCM which, to our knowledge, is the first linear approximation to a hydrostatic model of a container vessel that allows variable displacement. We model the hydrostatic equilibrium of forces acting on the vessel in the longitudinal direction. This enables the SCM to model core parameters such as draft, trim, BM, and SF, and the approach is possible to extend in the transversal and vertical direction to model list, TM, and metacentric height. For this purpose, we need to approximate the relations between the variables of the SCM (see Table 1) as linear equations.

The mass of a section is the sum of masses of lightship blocks, ballast water, fuel, and cargo within the boundaries of the section. If a block (e.g., a ballast water tank) extends beyond the section, only the mass of the fraction within the section is included in the sum. If we assume that all gravity forces act from the longitudinal mid-point of the section, the resulting gravity force clearly can be expressed as a linear function of the cargo and ballast water in the section.[2]

The buoyancy of a section depends on the draft of the section rather than its weight. It can be estimated from the Bonjean table of the vessel. Recall that the Bonjean table for each station gives the submerged area of a cross-section at the station as a function of the mid-ship draft at even keel. Figure 3 shows the Bonjean data of a 15000 TEU container vessel with a representative fine form hull. Notice that the curves are shown over the complete operational draft range of the vessel. The lightship draft is about four meters and the maximum summer draft is about 16 m.

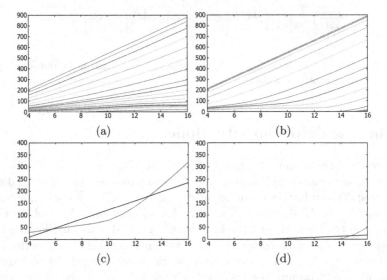

Fig. 3. Subermged area of fore (a) and aft (b) stations as a function of mid-ship draft at even keel. Two examples of linearisations of aft curves (c, d).

The mid-ship stations (top curves in graph (a) and (b)) have the largest submerged areas. Since the vessel has vertical sides at drafts above four meters in this part, the submerged areas grow linearly with the draft from this level. The curves for fore sections are slightly non-linear, while they are significantly non-linear for the aft sections (lower curves graph (b)). The reason is that the full stern only touches the water at maximum draft.

Despite these non-linearities, the dominating shape of the vessel is box-formed with vertical sides within the operational draft area. This means that the horizontal surface formed by the water line approximately has fixed shape such that the longitudinal moment needed to achieve a particular trim adjustment (say plus one meter) is constant for different displacements. This conclusion

---

[2] In future versions of the SCM, this point may be divided in the transversal and vertical direction to estimate TM and metacentric height.

seems in conflict with previously published data shown in Fig. 4(a). The graph shows trim as a function of displacement (i.e., draft) and longitudinal center of gravity (lcg) for the same vessel. For a fixed lcg (i.e., fixed longitudinal moment) this graph shows a highly non-linear relation between trim and displacement. A closer inspection of the graph, however, reveals that the displacement range is far out of operational levels which are above a fuelled ship of about 75 K tons and below maximum summer displacement of 218 K tons. Between 100K and 218K, we do see a rather linear relation between trim and displacement as expected from the analysis above. Due to this, we approximate the vessel as box-formed sections. To this end, the SCM uses a linear approximation to all Bonjean curves. Two examples of the linearisations of aft curves are shown in Fig. 3(c, d). Assume that the two boundaries of section k lie between station i and $i + 1$ and $j - 1$ and j, respectively, as shown in Fig. 4(b). Further assume that the submerged area of station v is A_v according to the linearisation above. We then have that the buoyancy of section k in tons is

$$\frac{d}{2} \sum_{v=i}^{j-1} f_v l_v (A_{v+1} + A_v), \tag{1}$$

where d is the density of salt water, f_v is the longitudinal fraction of station v to $v + 1$ within the boundary of section k, and l_v is the distance between the two stations. At maximum summer displacement of the example vessel, this approximation underestimates the buoyancy with about 3.06%. This is probably due to the slightly convex shape of the hull. Consequently we adjust the linearisation such that it predicts the correct summer displacement.

Below we present the SCM as an LP feasible set (i.e., a polyhedron). Table 1 contains the explanation of symbols for sets, constants, and variables used in the model. Sections are numbered from the bow (i.e., $\mathcal{S} = \{1, 2, \ldots\}$). They form a complete physical partitioning of the vessel such that all of its parts belong to a section and no part belongs to two sections. Section borders are aligned with

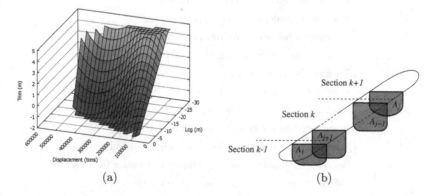

(a)   (b)

Fig. 4. (a) Trim as a function of displacement and lcg [11]. (b) Bouyancy approximation of a section.

bays, and a section cannot divide a bay. An example of a section partitioning with six sections is shown with green boxes in Fig. 1. Let $L_s^G = \frac{L_s}{2} + \sum_{s'>s} L_{s'}$ denote the distance from the stern to the center of gravity (mid-point) of section s. Further, let $L_s^F = \sum_{s'>s} L_{s'}$ denote the distance to the fore boundary of section s (reference point for stress forces). A container type $\tau \in T$ is a triple (l, r, w), where $l \in \{20, 40, 45\}$ is the length of the container, $r \in \{RF, NR\}$ is the reefer property of the container (reefer or non-reefer), $w \in \{9, 14, 29\}$ is the weight class of the container expressed as the average weight of containers in the class in metric tons. The buoyancy linearisation constants Φ_s, Ψ_s, and Θ_s of section s are approximated using (1). The domain of all variables is $\mathbb{R}_0^+$. In particular, this means that we relax the integrality of c_s^τ. Previous work shows negligible impact of this relaxation in practice [10], due to the large number containers in each bay (near 1000 on average on modern vessels). Also notice that none of the variables are identified as decision variables. This is on purpose since any subset of the variables can act as decision variables depending on the application of the SCM.

Table 1. Sets, constants, and variables used in the SCM.

S	Set of sections
T	Set of container types
L_s	Length of section s in meters
W_s^0	Lightship weight of section s in tons
$SF_s^{+/-}$	Positive and negative shear force limit in tons
$BM_s^{+/-}$	Positive and negative bending moment limit in tons meters
W_s	Container weight capacity of section s in tons
V_s	Container volume capacity of section s in TEU
R_s	Number of reefer plugs in section s
Φ_s, Ψ_s, Θ_s	Buoyancy linearisation constants of section s
W_τ	Weight of container type τ in tons
V_τ	Volume of container type τ in TEU
R_τ	Indicates whether container type τ is refer
d	Draft aft in meters at the stern of the vessel
tr	Trim of the vessel in meters
w_s	Weight of section s in tons
b_s	Buoyancy of section s in tons
r_s	Resulting force acting on section s in tons
sf_s	Shear force between section $s - 1$ and s in tons
bm_s	Bending moment between section $s - 1$ and s in tons meters
t_s	Weight of tank content of section s in tons
c_s^τ	Number of containers of type τ in section s

The SCM is a polyhedron over the variables defined by the following linear equations and inequalities.

$$b_s = \Phi_s d + \Psi_s tr + \Theta_s \qquad \forall s \in \mathcal{S} \qquad (2)$$

$$w_s = W_s^0 + t_s + \sum_{\tau \in \mathcal{T}} W_\tau c_s^\tau \qquad \forall s \in \mathcal{S} \qquad (3)$$

$$r_s = b_s - w_s \qquad \forall s \in \mathcal{S} \qquad (4)$$

$$\sum_{s \in \mathcal{S}} w_s = \sum_{s \in \mathcal{S}} b_s \qquad \sum_{s \in \mathcal{S}} L_s^G r_s = 0 \qquad (5)$$

$$sf_s = \sum_{s' < s} r_{s'} \qquad \forall s \in \mathcal{S} \setminus \{1\} \qquad (6)$$

$$bm_s = \sum_{s' < s} (L_{s'}^G - L_s^F) r_{s'} \qquad \forall s \in \mathcal{S} \setminus \{1\} \qquad (7)$$

$$\text{SF}_s^- \le sf_s \le \text{SF}_s^+ \qquad \forall s \in \mathcal{S} \setminus \{1\} \qquad (8)$$

$$\text{BM}_s^- \le bm_s \le \text{BM}_s^+ \qquad \forall s \in \mathcal{S} \setminus \{1\} \qquad (9)$$

$$\sum_{\tau \in \mathcal{T}} W_\tau c_s^\tau \le W_s \qquad \forall s \in \mathcal{S} \qquad (10)$$

$$\sum_{\tau \in \mathcal{T}} V_\tau c_s^\tau \le V_s \qquad \forall s \in \mathcal{S} \qquad (11)$$

$$\sum_{\tau \in \mathcal{T}} R_\tau c_s^\tau \le R_s \qquad \forall s \in \mathcal{S} \qquad (12)$$

Equation (2) defines the buoyancy of section s as a linear expression over aft draft d and trim tr. The linearisation coefficients Φ_s and Ψ_s and constant Θ_s are estimated from the linearisation of the Bonjean curves using Eq. (1) to calculate buoyancy of a section. Equation (3) defines the weight of section s as the sum of the lightship fraction within the section, the weight of fluids in tanks of the section, and the weight of cargo stowed in the section. Equation (4) defines the resulting vertical force r_s acting on section s. Since the positive direction is upward, buoyancy counts positive, while gravity counts negative. The two equalities of Eq. (5) ensure that the vessel is in hydrostatic equilibrium. The first equation says that at hydrostatic equilibrium, the total buoyancy of the hull must equal the total weight of the vessel. Otherwise, it must go to a higher or lower draft to be in equilibrium. Also at hydrostatic equilibrium, the sum of longitudinal moments of any cross-section must be zero. Otherwise, the vessel must go to a higher or lower trim to be in equilibrium. The second equation expresses the constraint for the cross-section at origo (the stern). Equation (6) defines the shear force at the fore boundary of section s. Since the shear force is the sum of resulting forces acting fore of this cross-section, we add the resulting forces of all sections in front of the point. Notice that we do not compute shear force at the fore boundary of the first section. This boundary is at the very tip of the vessel, where the shear force by definition is zero. Equation (7) defines the bending moment at the fore boundary of section s. We now have to multiply the resulting force with the distance $L_{s'}^G - L_s^F$ to it. Again, we do not compute bending moment for the fore boundary of the first section, since it is zero. The limits of shear force and bending moment are ensured by Eq. (8) and (9). The last three inequalities are stowage capacity constraints. Equation (10)–(12) ensure that the weight, volume, and reefer requirements of containers stowed in a section are within the capacity of the section. As shown in industrial projects [9], these constraints can be extended with advanced linear trade-offs between container types and weight classes. We plan to integrate these constraints into the SCM in future work.

4   Experimental Results

The purpose of the experiments is to evaluate the hydrostatic core of the SCM introduced in this paper. Specifically, we investigate the accuracy of the model's hydrostatic parameters as a function of given weight distributions, as well as the accuracy of the model in terms of the number of sections in the section partitioning.

The experiments are based on the 15000 TEU vessel introduced in the last section. For this vessel, we have access to the hydrostatic table approved by its classification society. For a given lcg and displacement, we can use this table to find the associated trim and draft at hydrostatic equilibrium and compare with the values predicted by the SCM. The table, however, does not include the stress forces over the vessel. To find these, we construct a vessel condition corresponding to the equilibrium and use an approved loading computer of the vessel [6] to calculate the forces that we then compare to the ones predicted by the SCM.

We have chosen three different weight levels at 100%, 80%, and 60% of maximum summer displacement. Notice that since about 35% of the weight of the vessel is steel and fuel, the vessel is usually less than half full by volume of cargo at 60% of maximum displacement. For each of the three displacement levels, we use 10 different cargo weight distributions over its bays corresponding to an operational lcg range. Water ballast tanks are assumed to be empty, while all other tanks are assumed to be 70% full by volume.[3]

The real hydrostatic equilibrium of the vessel has been approximated as follows. We first compute the displacement and lcg of the vessel using the longitudinal positions of lightship blocks, tanks, and bays stowed with one of the cargo weight distributions. Since these parameters decide the hydrostatic equilibrium of the vessel in the longitudinal dimension, we can lookup the associated draft and trim in the hydrostatic table.[4]

The equations of the SCM model have been implemented in Java and solved with the JAMA matrix package for given weight distributions. The CPU time

---

[3] These constant weight blocks of tanks are added to the lightship blocks in these experiments.

[4] Due to the sparsity of the hydrostatic table, in practice we interpolate the trim and draft from nearby entries.

required for these computations is negligible (less than one second in all cases). From these computations, we get the trim, draft (adjusted from aft to mid-ship draft), and stress forces predicted by the SCM. Lcg is non-linear in the SCM variables and therefore not included in the model. For a given cargo weight distribution, however, we can compute the underlying lcg of the SCM, since it assumes that all weights of a section s act from their approximated center of gravity, L_s^G.

4.1   Variable Displacement, Fixed Number of Sections

In the first set of experiments, we use the most detailed version of the SCM, where each section at most holds a single bay. This model has 26 sections. Table 2 shows the trim, draft, and lcg predicted by the SCM for 100%, 80%, and 60% of maximum displacement. The draft predictions are quite accurate. The highest deviation is about 40 cm and only seen at 100% of maximum displacement.

The correlations between the real trim and lcg and the predicted trim and lcg are shown in Fig. 5. As depicted in Fig. 5(c), the lcg prediction of the SCM is highly accurate for all weight distributions. This is not a trivial result. We have that the longitudinal position of cargo weight is at the center of sections independently of the number of sections. This, however, is not the case for lightship and tank blocks that usually are misaligned with section boundaries. What the results show is that impact on lcg at this level of detail is negligible.

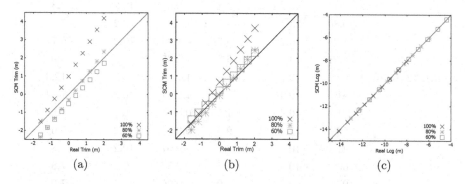

(a)   (b)   (c)

Fig. 5. (a–b) Correlation between real and predicted trim for two linearisation choices of the Bonjean curves. (c) Correlation between real and predicted lcg. In each case, a fixed partitioning with 26 sections and 60%, 80%, and 100% of maximum displacement were used.

An accurate lcg prediction is needed for an accurate trim prediction of the SCM. The trim prediction, however, also includes error caused by the buoyancy approximation. The SCM trim predictions shown in Fig. 5(a) uses the linearisation of Bonjean curves partly shown in Fig. 3(c, d). The trim predictions are very accurate for 80% and 60% of maximum displacement. Keep in mind that the vessel is almost 400 meters long, so the differences of about 30 cm is an angular error

Table 2. Trim, draft, and lcg predicted by the SCM for 100%, 80%, and 60% of maximum displacement using a partitioning with 26 sections.

Disp. (ton)	Real Values			SCM Values		
	Trim (m)	Draft (m)	Lcg (m)	Trim (m)	Draft (m)	Lcg (m)
218788 (100%)	2.01	15.6	−14.9	4.19	16.0	−14.9
	1.61	15.6	−14.1	3.55	16.0	−14.2
	1.20	15.6	−13.3	2.90	16.0	−13.4
	0.80	15.7	−12.6	2.26	16.0	−12.6
	0.40	15.7	−11.8	1.63	16.0	−11.9
	0.00	15.7	−11.0	0.99	16.1	−11.1
	−0.41	15.7	−10.3	0.36	16.1	−10.3
	−0.80	15.8	−9.5	−0.26	16.1	−9.6
	−1.21	15.8	−8.8	−0.88	16.1	−8.8
	−1.61	15.8	−8.0	−1.50	16.1	−8.1
175030 (80%)	2.01	13.1	−13.0	2.35	13.1	−13.1
	1.61	13.1	−12.2	1.82	13.1	−12.3
	1.20	13.1	−11.4	1.27	13.1	−11.4
	0.80	13.2	−10.6	0.73	13.1	−10.6
	0.39	13.2	−9.8	0.20	13.1	−9.8
	0.00	13.2	−9.0	−0.32	13.2	−9.0
	−0.40	13.2	−8.3	−0.82	13.2	−8.3
	−0.80	13.2	−7.5	−1.34	13.2	−7.5
	−1.21	13.2	−6.7	−1.85	13.2	−6.7
	−1.61	13.3	−6.0	−2.35	13.2	−6.0
131272 (60%)	2.02	10.4	−12.3	1.71	10.2	−12.3
	1.61	10.4	−11.4	1.25	10.2	−11.4
	1.21	10.4	−10.5	0.80	10.2	−10.5
	0.80	10.4	−9.6	0.36	10.2	−9.6
	0.41	10.5	−8.7	−0.07	10.2	−8.8
	0.00	10.5	−7.8	−0.53	10.2	−7.8
	−0.39	10.5	−7.0	−0.96	10.2	−7.0
	−0.80	10.5	−6.1	−1.39	10.2	−6.1
	−1.21	10.5	−5.2	−1.84	10.3	−5.2
	−1.60	10.5	−4.4	−2.25	10.3	−4.4

of less than 0.1%. We attribute the higher error at 100% of maximum displacement to the underestimate of the buoyancy of the stern. To test this hypothesis, we changed the linearisation to best fit within a displacement range corresponding to between 60% and 100% maximum displacement. Since this range starts at about 10 m draft, the linearisation of the stern curves become more accurate (e.g., see Fig. 3(c)). The resulting trim prediction is shown in Fig. 5(b) and shows significant accuracy improvement as expected.

4.2   Variable Number of Sections, Fixed Displacement

In the second set of experiments, we fix the displacement to 80% of maximum, while the numbers of sections vary from 26 to 4. The trim and lcg predictions are shown in Tables 3 and 4, respectively. The correlations between the real trim and lcg and the predicted trim and lcg for these experiments are shown in Fig. 6(a)

Table 3. Trim predicted by the SCM for 80% of maximum displacement using six different partitionings.

Real Values			SCM Trim for 26 to 4 Sections					
Trim (m)	Draft (m)	Lcg (m)	26	13	10	8	6	4
2.01	13.11	−13.02	2.35	2.40	2.54	1.65	0.78	0.39
1.61	13.13	−12.21	1.82	1.86	2.02	1.12	0.24	−0.16
1.20	13.15	−11.38	1.27	1.31	1.47	0.57	−0.30	−0.68
0.80	13.16	−10.58	0.73	0.79	0.95	0.00	−0.88	−1.20
0.39	13.18	−9.79	0.20	0.26	0.42	−0.50	−1.37	−1.71
0.00	**13.20**	**−9.01**	**−0.32**	**−0.26**	**−0.11**	**−1.01**	**−1.87**	**−2.22**
−0.40	13.21	−8.25	−0.82	−0.76	−0.61	−1.54	−2.40	−2.72
−0.80	13.23	−7.47	−1.34	−1.28	−1.12	−2.03	−2.93	−3.20
−1.21	13.25	−6.71	−1.85	−1.79	−1.66	−2.57	−3.44	−3.79
−1.61	13.26	−5.96	−2.35	−2.29	−2.14	−3.08	−3.94	−4.29

Table 4. Lcg predicted by the SCM for 80% of maximum displacement using six different partitionings.

Real Values			SCM Lcg for 26 to 4 Sections					
Trim (m)	Draft (m)	Lcg (m)	26	13	10	8	6	4
2.01	13.11	−13.02	−13.06	−13.12	−13.21	−11.88	−11.14	−11.09
1.61	13.13	−12.21	−12.25	−12.33	−12.42	−11.09	−10.31	−10.23
1.20	13.15	−11.38	−11.42	−11.49	−11.61	−10.27	−9.48	−9.43
0.80	13.16	−10.58	−10.62	−10.72	−10.84	−9.43	−8.59	−8.63
0.39	13.18	−9.79	−9.82	−9.91	−10.05	−8.68	−7.84	−7.85
0.00	**13.20**	**−9.01**	**−9.04**	**−9.13**	**−9.26**	**−7.92**	**−7.08**	**−7.06**
−0.40	13.21	−8.25	−8.29	−8.38	−8.51	−7.13	−6.28	−6.28
−0.80	13.23	−7.47	−7.51	−7.60	−7.76	−6.41	−5.46	−5.54
−1.21	13.25	−6.71	−6.74	−6.83	−6.95	−5.60	−4.69	−4.64
−1.61	13.26	−5.96	−6.00	−6.09	−6.23	−4.84	−3.92	−3.86

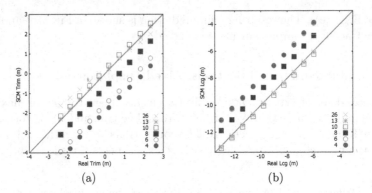

Fig. 6. Correlations between real and predicted trim (a) and lcg (b) for a fixed 80% of maximum displacement and six different section partitionings.

and (b), respectively. As depicted Fig. 6(b), the lcg positions predicted by the SCM are systematically off for the coarser partitionings with 8, 6, and 4 sections with a fixed amount. This error may be due to the misalignment of lightship blocks and tanks which will be more significant at courser levels of the model. The trim results shown in Fig. 6(a) are off correspondingly. Since each section partitioning forms an independent SCM model, it should be possible to reduce its trim error (and related stress force error) by adjusting the fixed position of the longitudinal center of gravity of its sections.

4.3   Stress Forces

In order to evaluate the stress forces predicted by the SCM, we construct a condition of the vessel at 80% of maximum summer displacement in an approved loading computer [6] corresponding to row six (bold) of Tables 3 and 4. The real trim of this condition is zero and the SCM predicts it to be −0.32 meters.

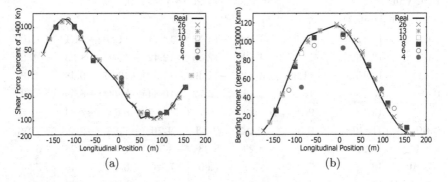

Fig. 7. Shear force (a) and bending moment (b) predictions for a fixed 80% of maximum displacement and six different section partitionings.

Despite this buoyancy inaccuracy of the SCM, the stress force predictions are remarkably accurate even for coarse partitionings with 8 and 6 sections. The results are shown in Fig. 7. The solid curves are the real forces calculated by the loading computer. Notice that the shear force curve is uneven. This is expected given the lightship weight distribution of the vessel. A bending curve is usually smooth even for an uneven weight distribution. The impression of the curve at -50 m lcg is due to a missing measure point over the accommodation of the vessel.

5 Conclusion and Future Work

In this paper, we have introduced the Standard Capacity Model (SCM). The objective of our work on the SCM is to provide a polyhedron representation of container vessel capacity that can be integrated in higher order optimization models like cargo flow networks for uptake and revenue management. Further, the aim is to enable the modelling of vessel capacity and key parameters substantially more accurate than is done today without sacrificing tractability. In this paper, we have introduced the hydrostatic core of the SCM, which to our knowledge is the first polyhedron approximation to hydrostatic equilibria of container vessels that allow variable displacement.

Our results show that the box-shaped hull approximation of sections applied by the SCM is realistic in typical sailing conditions and leads to accurate draft, trim, and stress force predictions also for coarse section partitionings. The results are well within the precision needed for practical application in the liner shipping industry. In future work, we plan to extend the model with advanced linear trade-offs between container types and weight classes shown in industrial projects [9]. We also consider applying regression analysis to find the longitudinal center of gravity of each section with minimum trim error such that the systematic errors seen in Fig. 6 can be reduced. A similar approach can be used to make a linear approximation to the metacentric height (i.e., transversal stability) of the vessel.

References

1. Ambrosino, D., Paolucci, M., Sciomachen, A.: A MIP heuristic for multi poty stowage planning. Transp. Res. Procedia **10**, 725–734 (2015)
2. Avriel, M., Penn, M., Shpirer, N.: Container ship stowage problem: complexity and connection to the coloring of circle graphs. Discrete Appl. Math. **103**, 271–279 (2000)
3. Delgado, A.: Models and algorithms for container vessel stowage optimization. Ph.D. thesis, IT University of Copenhagen (2013)
4. Economist: The humble hero, May 2013
5. Economist: Thinking outside the box, April 2018
6. Interschalt: MACS3 loading computer. http://navis.com
7. Kang, J., Kim, Y.: Stowage planning in maritime container transportation. J. Oper. Res. Soc. **53**(4), 415–426 (2002)

8. Li, F., Tian, C., Cao, R., Ding, W.: An integer linear programming for container stowage problem. In: Bubak, M., van Albada, G.D., Dongarra, J., Sloot, P.M.A. (eds.) ICCS 2008. LNCS, vol. 5101, pp. 853–862. Springer, Heidelberg (2008). https://doi.org/10.1007/978-3-540-69384-0_90

9. Optivation: Mathematical cargomix optimization model for the K-class (2013)

10. Pacino, D., Delgado, A., Jensen, R.M., Bebbington, T.: Fast generation of near-optimal plans for eco-efficient stowage of large container vessels. In: Böse, J.W., Hu, H., Jahn, C., Shi, X., Stahlbock, R., Voß, S. (eds.) ICCL 2011. LNCS, vol. 6971, pp. 286–301. Springer, Heidelberg (2011). https://doi.org/10.1007/978-3-642-24264-9_22

11. Pacino, D., Delgado, A., Jensen, R., Bebbington, T.: An accurate model for seaworthy container vessel stowage planning with ballast tanks, pp. 17–32 (2012)

12. Parreño, F., Pacino, D., Alvarez-Valdes, R.: A GRASP algorithm for the container stowage slot planning problem. Transp. Res. Part E Logist. Transp. Rev. **94**, 141–157 (2016)

13. Tierney, K., Pacino, D., Jensen, R.: On the complexity of container stowage planning problems. Discret. Appl. Math. **169**, 225–230 (2014)

14. UNCTAD: Review of maritime transport 2016. United nations conference on trade and development UNCTAD (2016)

15. Wilson, I., Roach, P.: Principles of combinatorial optimization applied to containership stowage planning. J. Heuristics **5**, 403–418 (1999)

16. Zurheide, S., Fischer, K.: A revenue management slot allocation model for liner shipping networks. Marit. Econ. Logist. **14**(3), 334–361 (2012)

Crane Intensity and Block Stowage Strategies in Stowage Planning

Dario Pacino$^{(\boxtimes)}$ (iD)

DTU Management Engineering, Technical University of Denmark,
Kongens Lyngby, Denmark
darpa@dtu.dk

Abstract. The increasing size of container vessels is raising the complexity of daily operations of both the carrier and the terminal. This paper focuses on stowage planning, the problem of assigning container to positions in a vessel. In particular, it studies the implementation of known planning strategies within an optimisation framework. Block stowage and crane intensity are presented and mathematically modelled on a simplified version of the problem. An experimental evaluation, on a large set of novel benchmark instances, shows that even in this simplified version the problem is not trivially solved. A matheuristic based on large neighbourhood search is presented, which is able to find a solution to all instances in short computational times.

1 Introduction

The container shipping industry has continuously grown in the past many years, and though it now experiencing a period with little growth [13], the complexity of the daily planning operations is still very high. The use of mega-vessels (now able of carrying more than 20,000 containers), is not only having an impact on port operations, but it is also making the cargo planning of the vessel a very complex and time-consuming task. This task is known as stowage planning, and it is often performed by the carrier a few hours before calling each port. In the past decade, the number of academic works on stowage planning has increased showing a continuous interest from the community. Solution approaches are divided between theoretical works, where a deeper understanding of specific optimisation challenges is sought (e.g. [3,5,10,12]), and applied approached where heuristic and decomposition methods aim at solving rich stowage planning problems that can be implemented in practice (i.e. [1,2,6–8]). Theoretical works are characterised by simple definitions of the problem where e.g. only one container size is assumed and where stability constraints are ignored. Those works, however, focus on some particular combinatorial challenges such as allowing containers to be shuffled along the route or understanding computational complexity. The work presented in this paper belongs to this category, though it is motivated by more applied issues. While academic works tend to look at optimal conditions, practical stowage planners are used to work with uncertain data and

© Springer Nature Switzerland AG 2018
R. Cerulli et al. (Eds.): ICCL 2018, LNCS 11184, pp. 191–206, 2018.
https://doi.org/10.1007/978-3-030-00898-7_12

rules-of-thumb. It would then be reasonable to assume that a first professional implementation of a decision support system for stowage planning would follow current planning practices. It turns out that current planning practices face hard combinatorial problems that have not yet been studied and for which, current state-of-the-art methods are not applicable. One of such problems comes from the concept of *crane intensity*. Crane intensity is an estimation of the number of cranes that a container terminal needs to use to handle a vessel. Crane intensity is calculated by dividing the total number of container moves by the number of moves the longest crane will perform. The longest crane is the crane that has the most moves assigned. To better explain this, consider the example in Fig. 1. The figure shows a vessel which is divided into bays. The number within each bay represents the number of container moves (load or discharge) that will have to be performed. Since two handling cranes cannot work on adjacent bays, the intervals below the vessel represent all the possible combination of adjacent bays and their respective workload. In this example, the longest crane has a total of 150 moves resulting in a crane intensity of 3. The longest crane can be seen as the handling operations makespan as described in [6]. In the remainder of the paper, we will refer to the longest crane as the makespan.

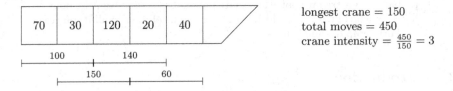

longest crane = 150
total moves = 450
crane intensity = $\frac{450}{150} = 3$

Fig. 1. Example of crane intensity calculation.

Crane intensity is not used by stowage planner as a KPI, it instead used as a target performance. The planners know by experience that at a given port, the vessel is usually serviced by a specific number of cranes, f.ex. 3. By forcing a stowage plan to have a crane intensity of 3, the container assignment will be forced to distribute containers along the bays such that the 3 cranes are fully utilised[1].

Another important concept is *block stowage*, which referrers to the practice of dividing a bay into logical sub-section to which only containers with the same discharge port can be assigned. This rule is used to avoid overstowage (which occurs when a container with a later discharge port is stowed over one with an earlier discharge port), and to improve container handling at the port. When containers with the same discharge port are clustered together, it is easier for the terminal to implement more advanced handling operations such as dual-cycling (interchanging load and discharge operations) and tandem lifts (moving more than one container at the time). The modelling of block stowage is already

---

[1] Note that this is a rule-of-thumb used by the industry.

present in the scientific work of e.g. [1,4]. To the best of the author's knowledge, no research has been published on the modelling of crane intensity and its combination to a block stowage policy.

This paper presents the Block Stowage Problem with Crane Intensity (BSPCI), a simplified stowage planning problem that only focuses on the combinatorial interplay between the block stowage strategy and the targeting of a specific crane intensity. With this new problem, we aim at finding efficient solution methods which can build the foundation for more rich problem definitions. We propose a mathematical formulation and a matheuristic based on the Large Neighborhood Search (LNS) framework. We test the mathematical formulation and the heuristic approach on a benchmark of 600 instances. The results show that the LNS is able to find solutions for all the instances where the mathematical model fails. For the remaining instances, the LNS can reach solutions that are either better or within 10% from the best-known solution in 75% of the cases.

The remainder of the paper is organised as follows. Section 2 formally introduces the problem and the compact formulation, followed by Sect. 3 where the design of the matheuristic is presented. Section 4 discussed the computational results before conclusions are drawn in Sect. 5.

2   Background and Problem Definition

A container vessel is a commercial vehicle designed to sail with standardised cargo containers. The most common containers (ISO containers) are 8' wide, 8.6' high, and 20', 40' or 45' long. There exists also a number of containers with such as refrigerated containers, tanks for special cargo. Each container is stowed on a cargo hold called a bay. A bay can hold multiple containers arranged in stacks. A cell indicates the position of a 40' long container and it is identified by a stack number and a tier number (the vertical index in the stack). The stowage of containers in a bay is subject to a number of physical constraints, i.e. maximum weight limits, availability of power plugs, and capacity limits. Aside from these stacking constraints, a loaded vessel must also be seaworthy, meaning that it must be stable while sailing, that it does not run aground when calling a port etc. We refer the reader to [6,7] for a detailed description of all the constraints governing stowage planning in practice. In this paper, we focus on a simplified version of the problem where only the main combinatorial aspects regarding crane intensity are taken into account.

We assume the bays of the vessel to be composed of a number of blocks. Each block can be seen as a logical grouping of stacks (not necessarily adjacent). The vessel travels on a predefined route, where containers can be loaded or discharged. The number of containers to transport from each port to any other port is known in advance. A block is only allowed to stow container destined to the same port. The port assignment of the block is not predetermined and is thus a part of the decision. All containers are of the same size and are coupled with an origin/destination port. The ship is assumed empty in the first port and after it arrives at the last port. At each port, all the containers destined to that port are discharged, and container destined to the following ports are loaded. At each port, the total number of operations in a bay is given by the number of load and discharge operations to be performed on that bay. Container moves are performed by the cranes of the container terminal, and no two cranes can work on adjacent bays. We disregard stacking and stability constraints aside from limiting the capacities of the blocks.

The BSPCI aims at finding an assignment of containers to blocks throughout the entire route. The assignment has to minimise the sum of the absolute difference between the found and the given crane intensity at each port.

Let us describe the problem more formally by first introducing the mathematical notation.

Sets

$P = \{1, 2, \dots, n\}$	The set of visited ports, where n is the last port.
$P_i^j \subseteq P$	The set of ports between port $i \in P$ and $j \in P$.
C	The set of blocks.
B	The set of adjacent bay pairs.
$C_b \subset C$	The set of blocks belonging to the same pair of adjacent bays $b \in B$.
T	The set of all pair of origin/destination ports $\{(i, j) \mid i, j \in P, i < j\}$.

Coefficients

q_c	The capacity of block $c \in C$.
tms_i	The target makespan at port $i \in P$.
$\bar{c}_i$	The cost for exceeding the target makespan at port $i \in P$.
$\underline{c}_i$	The cost for not reaching the target makespan at port $i \in P$.
t_{ij}	The number of containers to transport from port $i \in P_1^{n-1}$ to port $j \in P_{i+1}^n$.
$\hat{t}_{ij} = \sum_{k \in P_1^i} t_{kj}$	The total number of j-containers on board upon leaving port $i \in P$. A j-container is a container destined to port $j \in P$.

Decision variables

$x_{ij}^c \in \mathbb{Z}_+$	The number of j-containers ($j \in P_{i+1}^n$) loaded in block $c \in C$ at port $i \in P_1^{n-1}$
$y_{ij}^c \in \mathbb{B}$	A binary variable equal to 1 if at least one j-container is stowed in block $c \in C$ upon leaving port $i \in P_1^{n-1}$
$z_i \in \mathbb{R}_+$	The makespan at port $i \in P$
$\delta_i^b \in \mathbb{R}_+$	Auxiliary variable equal to the difference between z_i and the number of operations in the adjacent bays $b \in B$ performed at port $i \in P$
$\beta_i^b \in \mathbb{B}$	Auxiliary variable equal to 1 if and only if $\delta_i^b > 0$ for $b \in B$
$u_i \in \mathbb{R}_+$	A variable equal to the deviation from the target makespan if, at port i, z_i is strictly less than tms_i and 0 otherwise
$o_i \in \mathbb{R}_+$	A variable equal to the deviation from the target makespan if, at port i, z_i is strictly greater than tms_i and 0 otherwise

With the presented notation, the BSPCI can be formulated as the following mixed-integer program:

$$z^* = \min \sum_{i \in P_1^n} (\bar{c}_i o_i + \underline{c}_i u_i) \tag{1}$$

$$\text{s.t.} \sum_{k \in P_1^i} \sum_{c \in C} x_{kj}^c = \hat{t}_{ij} \qquad\qquad (i,j) \in \mathcal{T} \tag{2}$$

$$y_{ij}^c \leq \sum_{k \in P_1^i} x_{kj}^c \leq q_c y_{ij}^c \qquad\qquad (i,j) \in \mathcal{T} \quad c \in C \tag{3}$$

$$\sum_{j \in P_{i+1}^n} y_{ij}^c \leq 1 \qquad\qquad i \in P_1^{n-1} \quad c \in C \tag{4}$$

$$\delta_1^b + \sum_{c \in C_b} \sum_{j \in P_2^n} x_{1j}^c = z_1 \qquad\qquad b \in B \tag{5}$$

$$\delta_n^b + \sum_{c \in C_b} \sum_{i \in P_1^{n-1}} x_{in}^c = z_n \qquad\qquad b \in B \tag{6}$$

$$\delta_i^b + \sum_{c \in C_b} \left(\sum_{r \in P_1^{i-1}} x_{ri}^c + \sum_{j \in P_{i+1}^n} x_{ij}^c \right) = z_i \qquad i \in P_2^{n-1} \quad b \in B \tag{7}$$

$$\delta_i^b \leq M \beta_i^b \qquad\qquad i \in P \quad b \in B \tag{8}$$

$$\sum_{c \in \hat{C}} \beta_i^c \leq |\hat{C}| - 1 \qquad\qquad i \in P \tag{9}$$

$$z_i + u_i - o_i = tms_i \qquad\qquad i \in P \tag{10}$$

$$x_{ij}^c \in \mathbb{Z}_+ \qquad\qquad (i,j) \in \mathcal{T} \quad c \in C \tag{11}$$

$$y_{ij}^c \in \mathbb{B} \qquad\qquad (i,j) \in \mathcal{T} \quad c \in C \tag{12}$$

$$z_i \in \mathbb{R}_+ \qquad\qquad i \in P \tag{13}$$

$$\delta_i^b \in \mathbb{R}_+ \qquad\qquad i \in P \quad b \in B \tag{14}$$

$$\beta_i^b \in \mathbb{B} \qquad\qquad i \in P \quad b \in B \tag{15}$$

$$u_i, o_i \in \mathbb{R}_+ \qquad\qquad i \in P \tag{16}$$

The objective function (1) aims to minimize the weighted sum of the number of ports where the makespan is not equal to the target makespan. Since all containers must be stowed, the total number of moves is constant and the crane intensity measure can be translated to a target makespan. Constraints (2) guarantee that all container transports are satisfied. Constraints (3) are block capacity constraints and act as on-off constraints for the y-variables based on the values of the x-variables. Constraints (4) are block stowage constraints ensuring that each block can contain only containers with the same discharge port. Constraints (5)–(9) set all z-variables equal to the makespan of the number of operations for each port $i \in P_1^n$; in particular, constraints (5) concern port 1, constraints (6) port n, and constraints (7) ports 2 to $n - 1$. Constraints (8)–(9) ensure that at least one

β-variable is equal to 0 for each port i, thus setting the corresponding δ_i^c equal to zero and the makespan (i.e., variable z_i) equals the number of operations in the subset of four consecutive blocks. Constraints (10) set u_i and o_i equal to the difference between z_i and tms_i for each port $i \in P$. Constraints (11)–(16) define the range of the decision variables.

3 LNS Based Matheuristic

In Sect. 4, we show that the proposed formulation is not applicable to efficiently solve the BSPCI, thus heuristic methods are sought. It is important to note, however, that a number of instances can indeed be solved by the mathematical formulation. This insight has inspired us to use a mathematical-based heuristic to solve the BSPCI. We adopted the LNS framework where, given an initial solution, at each iteration, a part of the current solution is destroyed using a destroy operator. A repair heuristic is then used to rebuild the solution. This process is iterated until a termination criterion is met. Since the LNS framework is well-known, we refer the reader to [9] for a more in-depth description of the framework and its extension. The remainder of the section will, instead, present how each of the main LNS components has been adapted to solve the BSPCI.

3.1 Initial Solution

Finding an initial solution to the BSPCI is not trivial. Given that blocks have different capacities, it is not simple to analytically identify the number of blocks needed, moreover, this decision is made more difficult by the fact that a discharge port will also need to be assigned. We propose a 2-phase approach where first a mathematical model identifies the number of blocks to be used and, subsequently, a heuristic procedure assigns containers to the blocks.

Since it is reasonable to assume that many blocks in a container vessel will have the same capacity, let Q be the set of available block capacities, and C_q be the number of blocks of capacity $q \in Q$. We define a j-block to be a block assigned to only hold j-containers (where $j \in P$ and a j-container is a container destined to port j). The decision variable of the model, $x_{ij}^q \in \mathbb{Z}_+$, identifies the number of j-blocks with capacity q to be added to the j-blocks used in ports previous to i. The model is then formulated as follows:

$$\max \sum_{i \in P_1^{n-1}} \sum_{j \in P_{i+1}^n} \sum_{q \in Q} x_{ij}^q \tag{17}$$

s.t.

$$\sum_{h \in P_1^i} \sum_{q \in Q} q x_{hj}^q \geq \hat{t}_{ij} \quad \forall (i,j) \in \mathcal{T} \tag{18}$$

$$\sum_{h \in P_1^i} \sum_{j \in P_{i+1}^n} x_{hj}^q \leq C_q \quad \forall i \in P_1^{n-1}, q \in Q \tag{19}$$

$$x_{ij}^q \in \mathbb{Z} \quad \forall q \in Q, (i,j) \in \mathcal{T} \tag{20}$$

The objective function (17) maximise the number of used blocks. It is not strictly necessary to solve the model as an optimisation problem, but we believe that this will give more flexibility during the subsequent heuristic search since containers can be distributed to more blocks thus making it easier to stay within the target makespan. Constraints (18) ensures that at each port we assign enough j-blocks to fulfil the container demand $\hat{t}_{ij}$. The number of used blocks is then restricted by Constraints (19). Finally, the domain of the variables is defined in Constraint (20).

The mathematical model of the first phase effectively identifies how many block to use at each port for a specific discharge destination. In the second phase, Algorithm 1 uses this information to assign containers to each block, at every port.

The algorithm assigns containers to blocks starting from the first port and continuing in order (line 2). At each port, the set of discharge ports are sorted according to the number of moves to be performed such that the discharge port with the most containers is assigned first (line 3). The actual container assignment starts in line 4. We start by assigning the total number of container move (at the current port) to an auxiliary variable L. So long as L is positive, it means that we still have containers moves to perform (line 5). We keep count of the containers destined to discharge port d to be loaded/unloaded at port p, and we keep a sorted list of blocks ($\hat{C}$). The list sorts the blocks first by descending objective cost and then by ascending capacity. A block has a positive objective cost if the block is part of the adjacent bays defining the makespan. The cost of the block is then equal to the part of the objective cost for port p. The list is composed of the available blocks for port $p \in P$ and discharge port $d \in D$ ($\hat{C}_{pd}$). The set of available blocks is computed with a simple procedure. Blocks are evaluated sequentially starting from the first one. If a block has any remaining capacity and has been assigned to discharge port d (or it has not yet been assigned) it is included in the set. We keep adding blocks to the set until we reach the amount identified in the solution of the first phase. The first block in the list is then selected for container assignment (line 7). If the block is empty, it first needs to be assigned to the selected discharge port (lines 8–9). Lines 10–13 assign containers to the block. Since a solution with minimum cost is one that reaches the target makespan, we first check if the block is affecting the makespan (line 10). If this is not the case we assign as many containers as we have available, though at most the amount needed to reach the makespan or the capacity of the block (line 11). Otherwise, we load as many containers as the capacity of the block allows. The actual assignment is performed in line 14, which the updates all the necessary variables. The main idea behind this procedure is of trying to first load containers in blocks until the makespan is reached. The remaining containers are then greedily assigned where capacity is available. Once an initial solution is found, the mathematical model is run for 10 s to warmstart the LNS.

Algorithm 1. 2nd phase of the initial solution procedure

```
1  D_bp = 0;// Discharge port assignment for block b at port p
2  for each port p ∈ P do
3  │   for each discharge port d ∈ D sorted by moves do
4  │   │   L = t̂_pd; // Containers to load
5  │   │   while L > 0 do
6  │   │   │   Ĉ = sort(Ĉ_pd) by cost and capacity;
7  │   │   │   b = POP(Ĉ);
8  │   │   │   if block b is empty then
9  │   │   │   │   D_bi = d   ∀i ∈ P_p^d;
10 │   │   │   if makespan is not reached then
11 │   │   │   │   L = min(ΔMk, q_b, L);
12 │   │   │   else
13 │   │   │   │   L = min(q_b, L);
14 │   │   │   ASSIGN_ LOAD(L, b, p, d);
```

3.2 Repair Operator

In our adaptation of the LNS framework, we use a single repair operator based on the mathematical model described in Sect. 2. As we will see in the next section, a number of destroy operators are used to select which parts of the solution have to be removed. Let $\hat{X}$ be the set of variable assignments which has to be re-evaluated, where $(c, i, j) \in \hat{X}$ represents the indexes relative to the variable assignment for block $c \in C$, at port $i \in P_{n-1}^1$ for discharge port $j \in P_{i+1}^n$. Assume that $\bar{X}$ is the set of all variable assignment indexes in the current solution, the repair operator solves the model from Sect. 2 with the following additional variable fixings:

$$x_{ij}^c = v_{ij}^c \quad \forall (c, i, j) \in \bar{X} \setminus \hat{X} \tag{21}$$

where v_{ij}^c is value assigned to variable x_{ij}^c in the current solution.

3.3 Destroy Operators

Six destroy operators have been designed for the BSPCI, each targeting special parts of the problem. The operators are selected at random at each iteration and they can be roughly classified as random and cost-based, and are described in the following.

Random destroy. This is the simplest of the destroy operators, where variable assignments are simply selected at random, thus

$$\hat{X} = \{(c, i, j) | (c, i, j) \in \bar{X}, r_{ij}^c \leq \rho_1\},$$

where $r_{ij}^c \in [0, 1]$ is a random value for each variable assignment indexed by $x \in \bar{X}$, and $\rho_1 \in [0, 1]$ is a parameter of the algorithm.

Random bin destroy. This operator selects bins (sets of adjacent bays) at random and relaxes all the variable assignments related to the selected bins, thus

$$\hat{X} = \left\{(c,i,j)|(c,i,j) \in \bar{X}, c \in C_b, b \in B, r_b \le \rho_2\right\},$$

where $r_b \in [0,1]$ is a random value for each bin, and $\rho_2 \in [0,1]$ is a parameter of the algorithm.

Random discharge port assignment destroy. This operator targets blocks with a specific discharge port assignment. Let $d \in P$ be a discharge port selected uniformly at random. The set of relaxed assignments is then

$$\hat{X} = \left\{(c,i,j)|(c,i,j) \in \bar{X}, j = d, r_x \le \rho_3\right\}$$

where $r_x \in [0,1]$ is a random value for each variable assignment index $x \in \bar{X}$, and $\rho_3 \in [0,1]$ is a parameter of the algorithm. We use r_x in order to limit the size of the relaxed solution.

Random Block destroy. The two previous destroy operators can be seen as a version of the Shawn-removal technique [11] used in vehicle routing, where block assignments (either by discharge port or by bin association) are relaxed together. This version of the operator is more basic and only selects blocks at random, thus

$$\hat{X} = \left\{(c,i,j)|(c,i,j) \in \bar{X}, c \in C, r_c \le \rho_4\right\},$$

where $r_c \in [0,1]$ is a random value for each block, and $\rho_4 \in [0,1]$ is a parameter of the algorithm.

Random Port destroy. This operator aims at re-optimising the portion of the solution related to a specific port. Given a port p selected uniformly at random,

$$\hat{X} \subseteq \left\{(c,i,j)|(c,i,j) \in \bar{X}, i = p, r_x \le \rho_5\right\},$$

where $r_x \in [0,1]$ is a random value for variable assignment index, and $\rho_5 \in [0,1]$ is a parameter of the algorithm.

Cost based bin destroy. This and the next three destroy operators are cost based versions of the operators we have already seen. This particular case is an extension of the *Random bin destroy* operator. The aim is to make the random selection biased toward bins that, if changed, might have an impact on the objective function. For each bin $b \in B$ at every port $i \in P$, we calculate an impact factor $f_{bi} = \frac{z_{bi}}{z_i}$ where z_{bi} is the total number of moves to be performed in bin b at port i. In order to mitigate the impact of the bias, we also draw a random number $r_b \in [0,1]$ to be combined with the impact factor. The set of relaxed variable assignments is then

$$\hat{X} = \left\{(c,i,j)|(c,i,j) \in \bar{X}, c \in C_b, b \in B, \frac{f_{bi} + r_b}{2} \le \rho_6\right\}$$

Cost based discharge port assignment destroy. Using the same cost impact factor as in the previous operator (f_{bi}), this destroy operator implements a cost based version of the *Random discharge port assignment destroy* by defining

$$\hat{X} = \left\{ (c,i,j) | (c,i,j) \in \bar{X}, j = d, \frac{\max_{b \in B_c} (f_{bi}) + r_x}{2} \leq \rho_7 \right\}$$

where B_c is the set of bins including block c.

Cost based block destroy. This destroy operator is very similar to the previous one, with the difference that we do not restrict the variable assignment selection to specific discharge ports. More formally the set of variable assignments to relax is

$$\hat{X} = \left\{ (c,i,j) | (c,i,j) \in \bar{X}, \frac{\max_{b \in B_c} (f_{bi}) + r_x}{2} \leq \rho_8 \right\}.$$

Cost based move destroy. This is another relational destroy operator. The main idea is to first relax variables assignments of the bin that determines the makespan at a port. Let $p^M \in P$ be a randomly selected port with a positive impact on the objective function, and $b^M \in B$ be the bin that defined the makespan at port p^M. The first set of variable assignment to relax is then indexed by

$$\hat{X}^B = \left\{ (c,i,j) | (c,i,j) \in \bar{X}, i = p^M, c \in C_b^M \right\}.$$

In order to improve the solution we now need to relax variable assignments that will allow us to either add or remove containers for the b^M bin. This can be achieved by relaxing blocks that have the same discharge port as those in the bin b^M, and which have available capacity. Let $D(x)$ be the discharge port assigned to block c of the triplet $x = (c,i,j)$. The variable assignment we want to relax are the indexed by

$$\hat{X}^R = \bigcup_{d \in \{D(x) | x \in \hat{X}'\}} \left\{ (c,i,j) | (c,i,j) \in \bar{X}, i = p^M, j = d, \sum_{j \in P_i} x_{ij}^c < q_c \right\}.$$

Finally we also include variable assignments for blocks that are empty

$$\hat{X}^E = \left\{ (c,i,j) | (c,i,j) \in \bar{X}, i = p^M, \sum_{j \in P_i} x_{ij}^c = 0 \right\}$$

and the union of all these sets defines the set of variable assignments to relax

$$\hat{X} = \hat{X}^B \cup \hat{X}^R \cup \hat{X}^E.$$

3.4 Acceptance and Termination Criteria

A new solution s' is accepted if its objective value ($f(s')$) is better than that of the current solution s. Non-improving solutions that have the same objective value as the current solution are accepted only if they have not been visited before. An hash-key is generated for each solution and used to individually identify already visited configurations. A time limit of 300 s has been selected as termination criteria for the heuristics.

4 Computational Results

The LNS matheuristic and the mathematical formulations have been tested on a 2.30 GHz Intel Xeon E5 Processor with 128 GB of RAM. The heuristic has been implemented using C++ and all models have been solved using CPLEX version 12.8.

The experiments are based on a randomly generated set of 600 benchmark instances composed of 8 vessels with a capacity to carry from 1,200 to 18,000 containers (see Table 1 for details). For each of these vessels, an instance group is generated assuming a route visiting 5, 10, 15, 20 and 25 ports. For each of these groups, sub-groups are created to target specific origin/destination patterns: long distance, short distance and mixed distances as defined in [3]. Five instances are then generated for each sub-group[2]. According to our industrial collaborator, exceeding the target makepasan should be double as expensive as not reaching it, we have thus assigned $\bar{c}_i = 2$ and $\underline{c}_i = 1$ for all ports $i \in P$.

Table 1. Vessels' capacity. Each bay is assumed to be composed of two blocks. A larger one representing the outer stacks, and a smaller one representing the central stacks.

	Vessels							
	A	B	C	D	E	F	G	H
Bays	10	10	18	18	20	20	24	24
Bay capacity	120	210	180	300	420	600	624	750
Larger block capacity	80	140	120	200	280	400	416	500
Smaller block capacity	40	70	60	100	140	200	208	250
Vessel capacity	1,200	2,100	3,240	5,400	8,400	12,000	14,976	18,000

4.1 Evaluation of the Compact Model

The large set of benchmark instances has been designed to evaluate the impact each instance feature has on the solution of the compact formulation. With a time limit of one hour, the model is able to find optimal solutions for only 20 instances. Feasible solutions are found for 431, while 149 instances are not solved. Though the formulation is able to find a large number of feasible solutions, the average gap to the lower bound is ca. 65%. Figure 2 shows a histogram of the gap distribution among the solved instances, where it is possible to see that most solutions have more than 50% gap.

---

[2] The instances can be obtained upon request to the author.

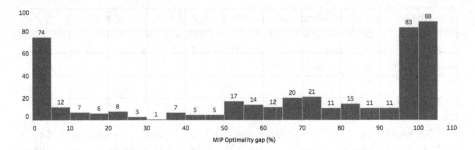

Fig. 2. Histogram representing the distribution of the gap between feasible solutions and CPLEX lower bound.

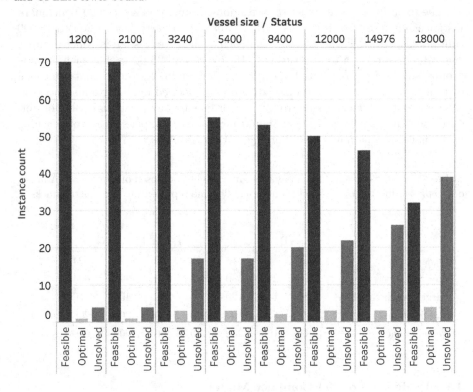

Fig. 3. Solved solutions by vessel size.

Figures 3 and 4 show the distribution of feasible, optimal and unsolved instances with respect to vessel size and number of visited ports, respectively. As expected, the larger the vessels and the number of visited ports the harder the problem it is to solve with the compact formulation. Tests have also been run to identify patterns between the size of the problem and the optimality gap, however, they did not produce any results worth mentioning.

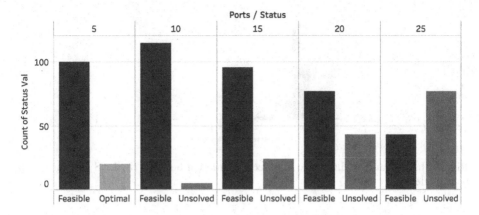

Fig. 4. Solved solutions by number of visited ports.

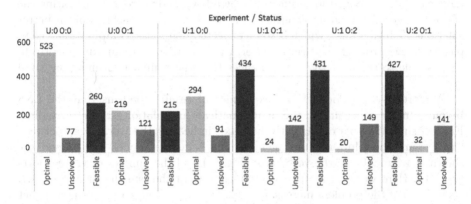

Fig. 5. Sensitivity analysis of the objective cost coefficients.

More interesting are the results obtained when the objective weights are changed. Figure 5 presents an histogram for different combinations of cost coefficients. E.g. column "U:1 O:2" represents a cost coefficient assignment of $\underline{c}_i = 1$ and $\overline{c}_i = 2$ for all ports $i \in P$. As expected, simple feasibility problems (U:0 O:0) are easier to solve and the more components we add to the objective function the harder it is to find optimal solutions. For all of the presented combinations there exists a number of instances for which a solution is never found.

4.2   Evaluation of the LNS

Compared to the compact formulation, the LNS algorithm is able to find feasible solutions for all problem instances i.e. 149 instances more than the mathematical model. For the instances where the compact formulation finds optimal solutions, the matheuristic is able to match those with an average optimality gap of 0.2%. Figure 6 depicts an analysis of the quality of the solutions found by the LNS and

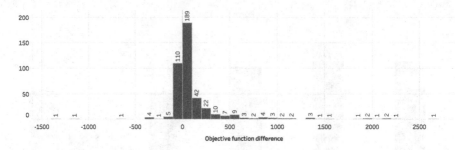

Fig. 6. Histogram over difference in objective function between the LNS and the compact model.

the feasible solutions found by the compact formulation. As is can be seen, in most cases, the two solutions differ only by a few hundred containers during the entire voyage. There do exist outliers where either approach finds much better solutions than the other, which is to be expected due to the computational complexity of the problem. Notice, however, that the LNS is an anytime algorithm that is able to solve all the 149 instances that the compact formulation could not.

A more in-depth look at the experimental results confirms that the LNS becomes more efficient as the mathematical formulation starts degrading. Figure 7 shows two bar charts with the y-axis indicating the average gap between the LNS and the solution found the by the mathematical model. The changes are shown as a function of the vessel size (Fig. 7a) and the number of visited ports (Fig. 7b). Here we see the opposite tendency than the one shown in Sect. 4.1. As the size of the problem increases, the LNS reduces its gap since the compact formulation has a harder time finding solutions. Note that due to the outliers

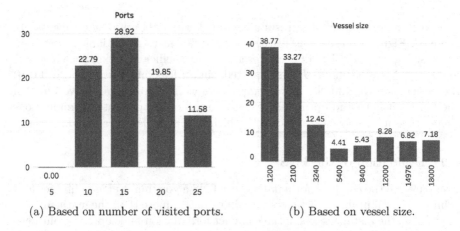

(a) Based on number of visited ports.   (b) Based on vessel size.

Fig. 7. LNS average gap to the solution found by the compact model.

the average gap is not a good indication of the quality of the LNS. Figure 7 can be reproduced using the median as a measure, which will result in the same conclusion but where the highest median gap is only 10%.

5 Conclusions

In this paper, we have presented a new variation of the stowage planning problem which includes block stowage and crane intensity strategies. Though the problem was greatly simplified, due to its combinatorial nature, it has been proven hard to solve. A compact formulation was presented and a matheuristic based on the LNS framework was implemented to solve the problem. Experiments on a randomly generated set of benchmark instances have shown that the LNS can be used to quickly find solution comparable to those of the mathematical formulation. Further research is, however, needed to both improve the LNS approach and to extend it to a more rich problem definition.

Acknowledgments. The author would like to thank the Danish Maritime Foundation for supporting this research under the project 2015-119 DTU Transport, Dynastow. Thanks are also due to Roberto Roberti for the fruitful discussions about the mathematical formulations.

References

1. Ambrosino, D., Paolucci, M., Sciomachen, A.: A MIP heuristic for multi port stowage planning. Transp. Res. Procedia **10**, 725–734 (2015)
2. Ambrosino, D., Paolucci, M., Sciomachen, A.: Computational evaluation of a MIP model for multi-port stowage planning problems. Soft Comput. **21**(7), 1753–1763 (2017)
3. Avriel, M., Penn, M., Shpirer, N., Witteboon, S.: Stowage planning for container ships to reduce the number of shifts. Ann. Oper. Res. **76**, 55–71 (1998)
4. Christensen, J., Pacino, D.: A matheuristic for the Cargo Mix Problem with Block Stowage. Transp. Res. Part E Logist. Transp. Rev. **97**, 151–171 (2017)
5. Ding, D., Chou, M.C.: Stowage planning for container ships: a heuristic algorithm to reduce the number of shifts. Eur. J. Oper. Res. **246**(1), 242–249 (2015)
6. Pacino, D., Delgado, A., Jensen, R.M., Bebbington, T.: Fast generation of near-optimal plans for eco-efficient stowage of large container vessels. In: Böse, J.W., Hu, H., Jahn, C., Shi, X., Stahlbock, R., Voß, S. (eds.) ICCL 2011. LNCS, vol. 6971, pp. 286–301. Springer, Heidelberg (2011). https://doi.org/10.1007/978-3-642-24264-9_22
7. Pacino, D., Delgado, A., Jensen, R.M., Bebbington, T.: An accurate model for seaworthy container vessel stowage planning with ballast tanks. In: Hu, H., Shi, X., Stahlbock, R., Voß, S. (eds.) ICCL 2012. LNCS, vol. 7555, pp. 17–32. Springer, Heidelberg (2012). https://doi.org/10.1007/978-3-642-33587-7_2
8. Parreño, F., Pacino, D., Alvarez-Valdes, R.: A GRASP algorithm for the container stowage slot planning problem. Transp. Res. Part E Logist. Transp. Rev. **94**, 141–157 (2016)

9. Pisinger, D., Ropke, S.: Large neighborhood search. In: Gendreau, M., Potvin, J.Y. (eds.) Handbook of Metaheuristics, pp. 399–419. Springer, Boston (2010). https://doi.org/10.1007/978-1-4419-1665-5_13

10. Roberti, R., Pacino, D.: A decomposition method for finding optimal container stowage plans (2018, accepted manuscript to appear in Transportation Science)

11. Shaw, P.: Using constraint programming and local search methods to solve vehicle routing problems. In: Maher, M., Puget, J.-F. (eds.) CP 1998. LNCS, vol. 1520, pp. 417–431. Springer, Heidelberg (1998). https://doi.org/10.1007/3-540-49481-2_30

12. Tierney, K., Pacino, D., Jensen, R.M.: On the complexity of container stowage planning problems. Discrete Appl. Math. **169**, 225–230 (2014)

13. UNCTAD: Review of maritime transport 2017 (2017)

Skipping the Storage Phase in Container Transshipment Operations

M. Flavia Monaco[1]([⊠])⑩ and Marcello Sammarra[2]⑩

[1] Dipartimento di Ingegneria Informatica, Modellistica, Elettronica e Sistemistica,
Università della Calabria, Rende, Italy
monaco@dimes.unical.it
[2] Istituto di Calcolo e Reti ad Alte Prestazioni,
Consiglio Nazionale delle Ricerche, Rende, Italy
sammarra@icar.cnr.it

Abstract. The paper deals with the problem of scheduling the loading/discharging operations of two simultaneously berthed vessels, assuming that some of the containers discharged from a vessel must be directly loaded on the other one. For these containers also the stowage position must be decided. The aim is to minimize the time needed to complete all the operations required by the involved vessels. For this problem we present a mathematical model, a heuristic algorithm and discuss the computational results on a set of randomly generated instances.

Keywords: Container terminal · Direct transshipment · Stowage plan

1 Introduction

At a maritime terminal the conventional transshipment flow of containers follows the quay-yard-(yard)-quay cycle, where the yard-to-yard movements concern possible housekeeping operations aimed at reconfiguring the yard and recovering storage spaces. From the operative point of view, the storage of the containers in the yard is essential to decouple in time the ingoing and outgoing container flows. Therefore the discharging and loading operations (of the same containers) are independent and can be planned and scheduled separately and efficiently. On the other hand, the yard is a critical resource, due to its limited capacity, and terminal planners are concerned with the reduction of the sojourn time of the containers in the yard (dwell-time) because that could increase the terminal throughput. The dwell-times have also a relevant economic impact for the shipping operators, since they contribute to determine the port fees. Therefore, reducing the dwell-times is a common target for the two main operators of the transshipment market. In view of that, we investigate the feasibility of a new operative transshipment modality, called *Direct Transshipment*.

The research has been supported by Ministero dell'Istruzione, Università e Ricerca under the PRIN 2015 research program - Grant 2015XAPRKF - Smart PORt Terminals.

R. Cerulli et al. (Eds.): ICCL 2018, LNCS 11184, pp. 207–221, 2018.
https://doi.org/10.1007/978-3-030-00898-7_13

We consider two vessels, simultaneously berthed at not necessarily adjacent berths, and we assume that some of the containers discharged from each of them must be directly loaded into the other, while the rest of the cargo follows the conventional transshipment flow (quay-yard-quay). Clearly, in the direct transshipment modality the unloading and loading operations are no longer independent and the related scheduling processes are concurrent: each container to be directly transshipped represents two dependent tasks (unloading/loading), to be executed by different machines (quay cranes) operating on different vessels, linked by a strict precedence relationship. In order to fully take advantage from this operative modality, the stowage decisions for the directly transshipped containers must become a degree of freedom for the planners. This is to say that the stowage positions of such containers and the cranes scheduling will be determined concurrently. The Direct Container Transshipment Problem (DCTP) is then the problem of scheduling all the vessel operations while deciding the stowage positions for the containers to be directly transshipped, so as to minimize the overall service time of the vessels.

The paper is organized as follows. The scientific literature related to the DCTP is discussed in Sect. 2. In Sects. 3 and 4 we state and formulate the problem. The solution algorithm is presented in Sect. 5. Section 6 discusses the computational experience. Finally conclusions are drawn in Sect. 7.

2   Related Works

The direct transshipment of containers between vessels seems to be a relatively new problem in the literature concerning the management of container terminals. In a more wide research stream, it has some similarities with the Cross Docking policy at a distribution terminal in a logistic network [4]. The main analogy between them is the common need of synchronizing the arrival and departure sequences of the carriers, while having low or possibly zero inventory levels.

The effectiveness of the direct transshipment of containers between different means of transport has been investigated in [1] for the case of rail hubs, in [7] for the case of trains and trucks, and in [3] for the case of mother ships and barges. Only few papers refer to the direct ship-to-ship transshipment. In [8] the authors consider a one-way direct transshipment between a mother vessel and a feeder, which "moor at the same berth and utilize the same handling equipment". Some researchers have recently studied an analogous problem in the case of mobile (or offshore) harbours, which are floating platforms equipped with portal cranes and are used to perform unloading and loading operations in the open sea (see [13]).

The only work that recognizes the effectiveness of the direct transshipment of containers between vessels is described in [15]. The authors present a model that integrates the berth allocation of the vessels, the yard space assignment, and the direct transshipment plan; the aim is to minimize the operative costs of trucks and yard cranes, as well as the delay cost of the vessels.

The DCTP we address in this paper, formerly introduced in [11], completely differs from that described in [15]; we deal with the operative management of the direct transshipment of containers between vessels, so integrating the scheduling of the quay cranes allocated to the vessels and the stowage of the directly transshipped containers.

3 Problem Statement

In order to derive the mathematical model for the DCTP, we first need to introduce the adopted notation and to detail the complex interactions between the main decision components of the problem, from which the constraints originate. As noted before, the management of the direct transshipment operative modality calls for the integration of two decision processes: the scheduling of the cranes operating on the two vessels and the stowage of the containers to be directly transshipped. Therefore, we will model it as a Quay Crane Scheduling Problem QCSP (see e.g. [2,14]) on a virtual vessel, with stowage constraints.

3.1 Notation

The mathematical model for the DCTP relies on the following main entities, that are vessels ($V = \{A, B\}$), tasks ($\Omega^v, v \in V$), and quay cranes ($Q^v, v \in V$). For each vessel v, the set of tasks consists of three disjoint subsets $\Omega^v = G^v \cup D^v \cup L^v$. G^v is the set of groups of containers, to be loaded or discharged, following the conventional transshipment flow, also called conventional tasks; D^v is the set of single containers discharged from v and directly transshipped to the other vessel; L^v is the set of containers directly transshipped from the other vessel and to be loaded into v. In the following we will refer to a container to be directly transshipped as a DT container. The vessels V can be seen as a single virtual vessel, with task set $\Omega = \Omega^A \cup \Omega^B$ and crane set $Q = Q^A \cup Q^B$.

In Table 1 we detail the characteristics and the attributes of vessels, tasks, and cranes.

Some of the entries in Table 1 need to be further explained. Each element θ of the sets Θ^v actually represents a single stowage slot in terms of $(bay, row, tier)$ coordinates; moreover each $\theta \in \Theta^v$ is able to accommodate only a subset of the containers in L^v, as indicated by the class-based stowage plan (see [12]). Therefore we define by $\Theta_i^v \subseteq \Theta^v, i \in L^v$ the set of slots where container i can be stowed, and by $H_i^v \subset H^v$ the set of bays where a stowage slot compatible with the container $i \in L^v$ is located, that is $H_i^v = \{b \in H^v \mid b = b_\theta, \theta \in \Theta_i^v\}$, where b_θ is the bay coordinate corresponding to θ. In passing, we observe that H_i^v can also be defined for tasks $i \in G^v \cup D^v$, being in this case $H_i^v = \{b(i)\}$ a singleton. Finally, we define $\Theta^v(b, i)$ the set of slots $\theta \in \Theta_i^v$ compatible with i and located in the bay b.

Due to physical restrictions, there are tasks whose processing can not overlap in time. For example, in the same bay unloading tasks always precede the loading ones, while in adjacent bays simultaneous processing of tasks by different cranes is forbidden, due to safety issues. This is to say that, for each vessel, some temporal restrictions on task pairs are defined. They impose either precedence

Table 1. Notation

Vessels	
a^v	Arrival time of vessel $v \in V$
Θ^v	Set of slots available for stowing containers directly transshipped to the vessel $v \in V$
H^v	Set of bays of vessel $v \in V$
Tasks	
p_i	Processing time of task $i \in \Omega^v$
$b(i)$	Bay coordinate of task $i \in \Omega^v$
Cranes	
r_k	Release time of crane $k \in Q^v$
$[s_k, f_k]$	Set of adjacent bays where crane k is allowed to operate
l_k^0	Initial bay-position of crane k ($s_k \leq l_k^0 \leq f_k$)
$\hat{t}$	Time a crane takes to travel from a bay to an adjacent one
$t_{b_1 b_2}$	Time a crane takes to travel between two generic bays: $t_{b_1 b_2} = \hat{t}|b_1 - b_2|$
t_{kb}^0	Time needed to the crane k to reach the bay b from its starting bay-position: $t_{kb}^0 = \hat{t}|l_k^0 - b|$
$Q(b)$	The set of cranes k such that $b \in [s_k, f_k]$

or non-simultaneity constraints on the processing of the tasks. More clearly, if there is a precedence between tasks i and j, then the processing of j can not start before the processing of i has been completed. Conversely, a non-simultaneity relationship between i and j imposes that either i must precede j, or j must precede i. To model these relations we need to extend our notation.

Precedence Relationships: A first set of precedence relationships is given by

$$\Phi = \bigcup_{v \in V} \Phi^v \cup \bar{\Phi}$$

where

$$\Phi^v = \{(i,j) \mid i \to j, \ i,j \in G^v \cup D^v\} \quad v \in V$$

relates to pairs of tasks belonging to the same vessel. Φ^v basically expresses precedences due to the operations the tasks require and can be populated using the stowage plan. Conversely

$$\bar{\Phi} = \{(i,j) \mid i \to j, (i \in D^A, j \in L^B) \vee (i \in D^B, j \in L^A)\}$$

are the precedence relationships between discharging and loading operations on different vessels for the DT containers.

Further precedence relationships for a given vessel are needed to guarantee that the directly transshipped containers will be loaded according to the stowage plan. To this aim we define

$$\Phi_1^v = \{(i,\theta) \mid i \rightarrow \theta, i \in G^v \cup D^v, \theta \in \Theta^v\} \qquad v \in V$$
$$\Phi_2^v = \{(\theta,i) \mid \theta \rightarrow i, \theta \in \Theta^v, i \in G^v\} \qquad v \in V$$
$$\Phi_3^v = \{(\theta_1,\theta_2) \mid \theta_1 \rightarrow \theta_2, \theta_1,\theta_2 \in \Theta^v\} \qquad v \in V$$

The sets Φ_1^v and Φ_2^v induce precedence relationships between a task whose stowage position is known and the container that will be stowed into the ship-slot θ. The sets Φ_3^v define precedence relationships between pairs of ship-slots due to their relative positions within the same bay. As a consequence they will induce, at runtime, also a set of precedence relations on the containers that will be stowed there.

Non Simultaneity Relationships: Assumed δ to be the safety distance between two adjacent cranes, expressed in number of bays, we define the set of bays that cannot be operated simultaneously by different cranes as follows

$$\Psi^v = \{(b_1,b_2) \mid b_1,b_2 \in H^v, b_1 < b_2, b_2 - b_1 \leq \delta\} \qquad v \in V$$

The above sets allow to impose the non-simultaneity constraints between each pair of tasks in close bays, even for the loading ones whose stowage bay must be decided by the model. However, as disclosed in [2], the sets Ψ^v are not sufficient to model the interferences between non adjacent cranes working on the vessel v. To this aim we define

$$\Delta_{b_1 b_2}^{hk}(v) = \max\left\{\hat{t}\left((\delta+1)(k-h) - (b_2 - b_1)\right), 0\right\} \quad b_1,b_2 \in H^v, h,k \in Q^v, v \in V$$

as the minimum time to elapse between the processing of any task located in the bay b_1 and any task in bay b_2 by cranes h and k, respectively. Therefore

$$\hat{\Delta}(v) = \left\{(b_1,b_2,h,k) \mid \Delta_{b_1 b_2}^{hk}(v) > 0\right\}$$

is the set of all combinations of bays and cranes that cause interferences on the vessel v. Observe that if $(b_1,b_2) \in \Psi^v$, then $(b_1,b_2,h,k) \in \hat{\Delta}(v)$ for all cranes $h,k \in Q^v$ such that $h \in Q(b_1), k \in Q(b_2)$. Thus, $\hat{\Delta}(v)$ is a generalization of Ψ^v and extends the corresponding definition in [2].

3.2   The Mathematical Model

The DCTP model involves both binary and continuous variables, as detailed in Table 2. Note that, as $H_i^v = \{b(i)\}$, then $\alpha_{ib(i)} = 1$, $\alpha_{ib} = 0 \; \forall b \neq b(i)$ are input data for all tasks $i \in G^v \cup D^v$, $v \in V$.

The constraints to be imposed for each vessel $v \in V$ can be stated as follows:

Table 2. Variables

Continuous variables	
$c_i \geq 0$	Completion time of task $i \in \Omega^v, v \in V$
$\sigma_{ij} \geq 0$	Transfer time of a DT container from the discharging to the loading bay, $(i,j) \in \bar{\Phi}$
$w^v \geq 0$	Makespan of vessel $v \in V$
$w = \max\limits_{v \in V} w^v$	The makespan of the virtual vessel
Binary variables	
$x_{ijk} = 1$	If tasks $i, j \in \Omega^v$ are performed consecutively by crane k, $v \in V$
$z_{ij} = 1$	If task $i \in \Omega^v$ is completed before the processing of task $j \in \Omega^v$ starts, $v \in V$
$y_{i\theta} = 1$	If container $i \in L^v$ is stowed in the slot $\theta \in \Theta_i^v$, $v \in V$
$\alpha_{ib} = 1$	If task $i \in \Omega^v$ must be handled in a bay $b \in H_i^v$, $v \in V$

Crane Routing Constraints: Constraints (1) to (4) define the sequence of tasks performed by each crane. Note that 0 and T are dummy tasks with $p_0 = p_T = 0$, $\Omega_0^v = \Omega^v \cup \{0\}$, $\Omega_T^v = \Omega^v \cup \{T\}$ and $x_{0Tk} = 1$ corresponds to an empty sequence for crane k.

$$\sum_{j \in \Omega_T^v} x_{0jk} = 1 \qquad\qquad k \in Q^v \qquad (1)$$

$$\sum_{i \in \Omega_0^v} x_{iTk} = 1 \qquad\qquad k \in Q^v \qquad (2)$$

$$\sum_{k \in Q^v} \sum_{j \in \Omega_T^v} x_{ijk} = 1 \qquad\qquad i \in \Omega^v \qquad (3)$$

$$\sum_{j \in \Omega_T^v} x_{ijk} - \sum_{j \in \Omega_0^v} x_{jik} = 0 \qquad i \in \Omega^v, k \in Q^v \qquad (4)$$

Stowage Constraints: Constraints (5) and (6) assign a stowage position, in terms of slot and bay, to the DT containers, while constraints (7) and (8) state the relationships between y's and α's, x's and α's variables, respectively. Note that, for a fixed $i \in L^v$, summing up constraints (7) on the compatible bays $b \in H_i^v$ and taking into account constraints (6), one gets $\sum_{\theta \in \Theta_i^v} y_{i\theta} = 1$. Constraints (8) impose that if a task must be performed in a bay b, it has to be assigned to a crane able to operate that bay.

$$\sum_{i \in L^v} y_{i\theta} = 1 \qquad\qquad \theta \in \Theta^v \qquad (5)$$

$$\sum_{b \in H_i^v} \alpha_{ib} = 1 \qquad\qquad i \in L^v \qquad (6)$$

$$\sum_{\theta \in \Theta(i,b)} y_{i\theta} = \alpha_{ib} \qquad\qquad i \in \mathrm{L}^v, b \in H_i^v \qquad (7)$$

$$\sum_{k \in Q(b)} \sum_{j \in \Omega^v} x_{ijk} \geq \alpha_{ib} \qquad\qquad i \in \Omega^v, b \in H_i^v \qquad (8)$$

Completion Time Constraints: The completion times of the tasks are computed through constraints (9)–(12). Here and in what follows M is a big constant.

$$c_i - p_i \geq a^v \qquad\qquad\qquad\qquad i \in \Omega^v \quad (9)$$

$$r_k - c_j + \sum_{b \in H_j^v} \alpha_{jb} t_{l_k^0 b} + p_j \leq M(1 - x_{0jk}) \qquad j \in \Omega^v, k \in Q^v \quad (10)$$

$$c_i - c_T \leq M(1 - x_{iTk}) \qquad\qquad\qquad i \in \Omega_0^v, k \in Q^v \quad (11)$$

$$c_i + \hat{t} \left| \sum_{b \in H_i^v} b\alpha_{ib} - \sum_{b \in H_j^v} b\alpha_{jb} \right| + p_j - c_j \leq M(1 - x_{ijk}) \quad i,j \in \Omega^v, k \in Q^v \quad (12)$$

The non linearity in constraints (12) can be easily handled replacing each of them with the following set of constraints

$$c_i + (\alpha_{ib_1} + \alpha_{jb_2} - 1)t_{b_1 b_2} + p_j - c_j \leq M(1 - x_{ijk}) \quad b_1 \in H_i^v, b_2 \in H_j^v \quad (13)$$

Precedence Constraints: Constraints (14)–(18) impose the precedence relationships between pairs of tasks.

$$c_i + p_j - c_j \leq 0 \qquad\qquad (i,j) \in \Phi^v \qquad\qquad\qquad (14)$$

$$c_i + p_j - c_j \leq M(1 - y_{j\theta}) \qquad j \in \mathrm{L}^v, \theta \in \Theta_j^v, (i,\theta) \in \Phi_1^v \qquad (15)$$

$$c_j + p_i - c_i \leq M(1 - y_{j\theta}) \qquad j \in \mathrm{L}^v, \theta \in \Theta_j^v, (\theta,i) \in \Phi_2^v \qquad (16)$$

$$c_i + p_j - c_j \leq M(2 - y_{i\theta_1} - y_{j\theta_2}) \quad i,j \in \mathrm{L}^v, \theta_1 \in \Theta_i^v, \theta_2 \in \Theta_j^v, (\theta_1, \theta_2) \in \Phi_3^v$$
$$(17)$$

$$c_i + \sigma_{ij} + p_j - c_j \leq 0 \qquad\qquad (i,j) \in \bar{\Phi} \qquad\qquad\qquad (18)$$

In particular, (14) define the precedence between tasks whose stowage position is known, while (15) to (18) take into account precedence relationships involving DT containers to be loaded. The variables σ_{ij} in (18) are defined through equations (19), where l_A^F is the last bay of vessel A, d_{AB} is the inter-vessel distance expressed in number of bays, and τ is the time a straddle carrier takes to cover a ship-bay (see [5]).

$$\sigma_{ij} = \tau \left(\sum_{v \in V} \sum_{b \in H_j^v} b\alpha_{jb} - b(i) + l_A^F + d_{AB} \right) \qquad (i,j) \in \bar{\Phi} \qquad (19)$$

Non Simultaneity Constraints: The relations between the completion times of the tasks and the $z's$ variables are stated by constraints (20), (21), and (22). They impose a partial time-ordering on the tasks of the same vessel. Actually, for each pair of tasks, either i precedes j ($z_{ij} = 1, z_{ji} = 0$), or j precedes i ($z_{ji} = 1, z_{ij} = 0$), or, finally, the processing of i and j overlap ($z_{ij} = z_{ji} = 0$). Note that if i and j are tasks located in too close bays, constraints (22) avoid that they are processed simultaneously.

$$c_i + p_j - c_j \leq M(1 - z_{ij}) \qquad\qquad i,j \in \Omega^v \quad (20)$$

$$c_j - p_j - c_i \leq M z_{ij} \qquad\qquad i,j \in \Omega^v \quad (21)$$

$$z_{ij} + z_{ji} \geq \alpha_{ib_1} + \alpha_{jb_2} - 1 \quad i,j \in \Omega^v, b_1 \in H_i^v, b_2 \in H_j^v, (b_1, b_2) \in \Psi^v \quad (22)$$

Non Interference Constraints: For each pair of tasks $i, j \in \Omega^v$ and for each pair of compatible bays $b_1 \in H_i^v, b_2 \in H_j^v$, the following constraints must hold for each pair of cranes h and k that would cause interference working simultaneously on bays b_1 and b_2, that is $(b_1, b_2, h, k) \in \hat{\Delta}(v)$:

$$\sum_{u \in \Omega_0^v} x_{uih} + \sum_{u \in \Omega_0^v} x_{ujk} + \alpha_{ib_1} + \alpha_{jb_2} \leq 3 + z_{ij} + z_{ji} \qquad (23)$$

$$c_i + \Delta_{b_1 b_2}^{hk}(v) + p_j - c_j \leq M \left(5 - \alpha_{ib_1} - \alpha_{jb_2} - z_{ij} - \sum_{u \in \Omega_0^v} x_{uih} - \sum_{u \in \Omega_0^v} x_{ujk} \right)$$
$$(24)$$

$$c_j + \Delta_{b_1 b_2}^{hk}(v) + p_i - c_i \leq M \left(5 - \alpha_{ib_1} - \alpha_{jb_2} - z_{ji} - \sum_{u \in \Omega_0^v} x_{uih} - \sum_{u \in \Omega_0^v} x_{ujk} \right)$$
$$(25)$$

Objective Function Definition. The objective function to be minimized is a linear combination of two conflicting functions: the makespan of the virtual vessel, defined by (26)–(27), and the average waiting time for the DT containers (28).

$$c_i \leq w^v \quad i \in \Omega^v, v \in V \qquad (26)$$

$$w^v \leq w \quad v \in V \qquad (27)$$

$$\min \lambda w + \mu \frac{1}{|\bar{\Phi}|} \sum_{(i,j) \in \bar{\Phi}} (c_j - p_j - \sigma_{ij} - c_i) \qquad (28)$$

4   Refinement of the DCTP Model

As motivated in [2,10], the search of feasible solutions of the QCSP can be limited to the unidirectional schedules, where all the cranes move from the bow to the stern of the vessel, or in the opposite direction. Therefore, also in the DCTP model (1)–(11), (13)–(28) it is possible to impose the one-way movement of all the cranes allocated to the single vessels. To this aim we introduce two new binary variables: $\gamma^v = 1$ if the cranes in Q^v move from the bow to the stern, $v \in V$, and a set of additional constraints (see [10]):

$$\sum_{b \in H_i^v} b\alpha_{ib} - \sum_{b \in H_j^v} b\alpha_{jb} \leq M(1 - x_{ijk}) + M(1 - \gamma^v) \quad i,j \in \Omega^v, \ v \in V \quad (29)$$

$$\sum_{b \in H_j^v} b\alpha_{jb} - \sum_{b \in H_i^v} b\alpha_{ib} \leq M(1 - x_{ijk}) + M\gamma^v \qquad i,j \in \Omega^v, \ v \in V \quad (30)$$

Note that (29), (30) extend the corresponding constraints (16), (17) in [10], also to the DT containers to be loaded, whose stowage bay is unknown.

5 Solution Algorithm

The formulation of the DCTP (1)–(11), (13)–(30) as a QCSP on a virtual vessel with side constraints naturally drives to design a solution algorithm by suitably modifying the Tabu Search Algorithm for the QCSP described in [10].

Given a feasible stowage plan for the DT containers, our algorithm iterates over feasible solutions constructed by a two-phases approach:

1. *Routing phase*: for each crane, a feasible sequence of tasks is determined taking into account precedence, one-way and cranes' operative range constraints.
2. *Scheduling phase*: the completion time of the tasks is computed imposing the non simultaneity and non interference constraints, and the precedence constraints related to the DT containers.

A feasible schedule for the cranes can be represented by a disjunctive graph with node set $\Omega \cup \{0, T\}$ (see [10]), where disjunctive edges model the non simultaneity and the non interference constraints. To perform the scheduling phase and evaluate the makespan for the virtual vessel, we have to find the critical path from 0 to T on such a disjunctive graph. This problem is, in general, $\mathcal{N}P$-hard, while in our case it becomes easier to solve. The one-way assumption, in fact, uniquely identify the orientation of the disjunctive edges giving rise to an acyclic graph.

To describe the Tabu Search algorithm for the DCTP, it is sufficient to specify the memory mechanism and the neighbourhood structure. We adopt the attributive memory mechanism, meaning that a solution is declared tabu if at least one of the attributes describing that solution is tabu [6]. In order to introduce the neighbourhood structure, let us denote by $(\bar{x}, \bar{y})$ a given feasible solution in terms of the main scheduling variables (x) and stowage variables (y). We define *swap move* the swapping of the stowage positions of two containers in $L^v, v \in V$; $N_1(\bar{y})$ is the set of all feasible stowage configurations obtained from $\bar{y}$ by performing a swap move. Furthermore, we define *shift move* the shifting of a task currently assigned to the crane k to an adjacent crane ($k - 1$ or $k + 1$), and $N_2(\bar{x})$ as the set of all feasible schedules obtained from $\bar{x}$ by performing a shift move. The neighborhood of $(\bar{x}, \bar{y})$ can now be defined as follows:

$$\mathcal{N}(\bar{x}, \bar{y}) = \{(\bar{x}, y) \mid y \in N_1(\bar{y})\} \cup \{(x, \bar{y}) \mid x \in N_2(\bar{x})\} \quad (31)$$

6   Computational Experience

The Tabu Search Algorithm (TSA) has been implemented in C++. The stopping criterion is based on a maximum number of iterations equal to 2000. The tabu tenure has been set to 15 iterations; the diversification penalty has been set equal to 0.05. The tests have been carried out on a machine equipped with a 3.1 GHz Intel Core i5 CPU and 16 GB of RAM. TSA has been tested on a set of instances randomly generated as described in the next subsection.

6.1   Instance Generator Algorithm

Let I an instance of the standard QCSP defined by: number of conventional tasks (T), number of cranes (Q), and number of bays (B). In such an instance each conventional task is characterized by a processing time p and a bay location b. Given two QCSP instances, say I_A and I_B, called *seed* instances, an instance I_{AB} of the DCTP can be constructed as follows. First, we assume that a task i of I_A or I_B with processing time p_i is a group of p_i containers. Let c be the number of container classes and n_c the number of DT containers of class c in the instance I_{AB}. For each instance I_A and I_B, apply the following algorithm (DCTP-G):

1. Randomly select a class c and a bay b.
2. Randomly select in the bay b a conventional task i of class c; let p_i its processing time.
3. If $n_c \leq p_i$, replace the task i with $n_c + 1$ tasks, where the first tasks represent n_c DT containers, while the last one, if any, represents a residual conventional task i' with processing time $p_{i'} = p_i - n_c$. *GO TO* 1.
4. If $n_c > p_i$, replace the selected conventional task i with p_i DT containers; set $n_c = n_c - p_i$, $b = b + 1$ and *GO TO* 2.

To generate the seed instances, we have adopted the instance generator *QCSPgen* developed by Meisel and Bierwirth in [9]. The interested reader is referred to [9] for more details on the *QCSPgen* algorithm. Here we just mention that, among the input parameters of *QCSPgen*, we have set the distribution of the tasks within the vessel to be uniform; the density of precedence relationships among tasks of the same bay to be one, meaning that within the same bay all the tasks are sorted to reflect the stowage constraints. Finally we have set the crane safety distance to be one bay. We have generated two sets of instances to represent two kind of vessels: mother vessels and feeder vessels. The dimensions of the seed instances are reported in Table 3, where *NoI* indicates the number of instances of each type generated.

The seed-instances of Table 3 are combined each other and become the input for the DCTP-G algorithm, either as I_A or as I_B, together with the number of container classes c and the number of DT containers per class, n_c, giving rise to 33 DCTP instances as detailed in Table 4.

Table 3. Dimensions of the seed instances.

Type	Bays	Tasks	Cranes	NoI
Mother	20	10	3	1
Mother	20	15	3	1
Mother	20	20	4	1
Feeder	10	10	2	5

Table 4. Description of the DCTP instances.

Code	I_A	I_B	c	$\sum_c n_c$	NoI	Direct transshipment flow
MF	Mother	Feeder	2	240	15	$A \to B$
FM	Feeder	Mother	2	150	15	$A \to B$
MM	Mother	Mother	2	250	3	$A \to B$ (115), $A \leftarrow B$ (125)

6.2   Lower Bounds for the DCTP

In order to evaluate the effectiveness of TSA, we need to compute lower bounds for the DCTP. The most natural way to achieve this aim is to consider a relaxed DCTP model obtained by removing constraints (5) to (8) related to the stowage decision. The resulting DCTP-SR relaxed problem, consisting of two standard QCSPs linked by constraints (18), is hard to solve to optimality. Actually standard ILP solvers easily run out of memory, due to the high number of constraints and variables. For these reasons we also relax constraints (18), getting the DCTP-WR problem that decomposes in two QCSPs no longer dependent on each other, and relatively easy to be solved by a standard ILP solver.

6.3   Analysis of the Results

In the following Tables 5, 6, and 7, we summarize the computational results obtained by solving the instances described in Table 4 by our TSA, and the corresponding DCTP-WR problems by ILOG Cplex 12.6.2., setting $\lambda = \mu = 1$ in the objective function. For each instance, in the leftmost columns of the result Tables we report: the data related to the number of bays and cranes (B-Q) of each vessel; the number of tasks T in the form (DT containers - conventional tasks). Observe that the instance code $I_A I_B - \alpha - \beta - \gamma - \delta$ is useful to recognize the seed instances generating it, being $\alpha - \beta$ the number of bays and tasks of I_A and $\gamma - \delta$ the number of bays and tasks of I_B. Then we report the makespan of the two vessels, both for the DCTP-WR and DCTP models, marking in bold the maximum between them. As for the DT containers, in the second last column (AvgD) we report the average waiting time between the discharging and the loading operation. Finally, in the last column, we report the gap computed

Table 5. Results on Mother-Feeder instances.

Code	Vessel A		Vessel B		LB		UB			Gap%
	$B-Q$	T	$B-Q$	T	w^A	w^B	w^A	w^B	$AvgD$	
MF-20-10-10-10-1	20-3	240-10	10-2	240-6	**1482**	259	**1530**	1328	0.00	3.24
MF-20-10-10-10-2	20-3	240-10	10-2	240-6	**1482**	266	**1482**	1385	26.29	1.77
MF-20-10-10-10-3	20-3	240-10	10-2	240-7	**1482**	267	**1484**	1392	101.40	6.98
MF-20-10-10-10-4	20-3	240-10	10-2	240-5	**1482**	267	**1482**	1376	4.17	0.28
MF-20-10-10-10-5	20-3	240-10	10-2	240-6	**1482**	300	**1482**	1425	13.83	0.93
MF-20-15-10-10-1	20-3	240-13	10-2	240-7	**1482**	259	**1485**	555	77.54	5.43
MF-20-15-10-10-2	20-3	240-13	10-2	240-3	**1482**	266	**1485**	555	31.67	2.34
MF-20-15-10-10-3	20-3	240-13	10-2	240-5	**1482**	277	**1482**	543	59.23	4.00
MF-20-15-10-10-4	20-3	240-13	10-2	240-6	**1482**	284	**1482**	523	24.43	1.65
MF-20-15-10-10-5	20-3	240-13	10-2	240-7	**1482**	300	**1485**	580	60.14	4.26
MF-20-20-10-10-1	20-4	240-20	10-2	240-6	**1926**	259	**1926**	817	21.50	1.12
MF-20-20-10-10-2	20-4	240-20	10-2	240-6	**1926**	266	**1926**	759	26.29	1.37
MF-20-20-10-10-3	20-4	240-20	10-2	240-7	**1926**	277	**1926**	768	55.93	2.90
MF-20-20-10-10-4	20-4	240-20	10-2	240-5	**1926**	267	**1926**	750	4.17	0.22
MF-20-20-10-10-5	20-4	240-20	10-2	240-6	**1926**	300	**1926**	799	13.83	0.72

Table 6. Results on Feeder-Mother instances.

Code	Vessel A		Vessel B		LB		UB			Gap%
	$B-Q$	T	$B-Q$	T	w^A	w^B	w^A	w^B	$AvgD$	
FM-10-10-20-10-1	10-2	150-8	20-3	150-10	259	**1457**	373	**1607**	358.99	34.93
FM-10-10-20-10-2	10-2	150-8	20-3	150-10	266	**1457**	277	**1607**	325.00	32.60
FM-10-10-20-10-3	10-2	150-8	20-3	150-10	277	**1457**	295	**1607**	319.41	32.22
FM-10-10-20-10-4	10-2	150-6	20-3	150-10	273	**1606**	273	**1606**	534.82	33.30
FM-10-10-20-10-5	10-2	150-8	20-3	150-10	300	**1457**	335	**1607**	338.03	33.50
FM-10-10-20-15-1	10-2	150-8	20-3	150-15	259	**1471**	425	**1616**	413.81	37.99
FM-10-10-20-15-2	10-2	150-8	20-3	150-15	266	**1471**	277	**1640**	367.69	36.48
FM-10-10-20-15-3	10-2	150-8	20-3	150-15	277	**1471**	295	**1640**	362.10	36.10
FM-10-10-20-15-4	10-2	150-6	20-3	150-15	273	**1414**	273	**1532**	144.85	18.59
FM-10-10-20-15-5	10-2	150-8	20-3	150-15	300	**1471**	335	**1640**	380.72	37.37
FM-10-10-20-20-1	10-2	150-8	20-4	150-18	259	**1926**	425	**1926**	87.93	4.57
FM-10-10-20-20-2	10-2	150-8	20-4	150-18	266	**1926**	266	**1926**	57.54	2.99
FM-10-10-20-20-3	10-2	150-8	20-4	150-18	277	**1926**	295	**1926**	63.68	3.31
FM-10-10-20-20-4	10-2	150-6	20-4	150-19	273	**1926**	284	**1926**	35.54	1.85
FM-10-10-20-20-5	10-2	150-8	20-4	150-18	300	**1926**	335	**1926**	76.65	3.98

Table 7. Results on Mother-Mother instances.

Code	Vessel A		Vessel B		LB		UB			Gap%
	$B-Q$	T	$B-Q$	T	w^A	w^B	w^A	w^B	$AvgD$	
MM-20-10-20-15-1	20-3	240-10	20-3	240-14	**1482**	**1482**	1482	**1600**	16.10	9.05
MM-20-10-20-20-2	20-3	240-10	20-4	240-20	1482	**1926**	1607	**1927**	158.04	8.20
MM-20-15-20-20-3	20-3	240-13	20-4	240-18	1482	**1926**	**1972**	1926	0.67	2.42

as $Gap\% = 100 \times (UB - LB)/LB$ where $LB = \max\{w^A, w^B\}$ and $UB = \max\{w^A, w^B\} + AvgD$. We do not report the computation times, since they are about 300 s almost uniformly on all the instances.

A first oversight to the Tables shows that TSA gives satisfactory results on all but the first two groups of instances in Table 6. Actually, very often the overall makespan coincides with its lower bound, and the average delay of the DT containers is small enough, resulting in a gap that does not exceed 7% for the M-F instances, 5% for the last group of F-M instances, and 10% for the M-M instances. In particular, the results obtained on the M-F instances (Table 5) can be motivated as follows: the makespan of the virtual vessel is always attained at the mother vessel, therefore the delay of the DT containers can be reduced by, eventually, letting the cranes working on the feeder vessel wait. This is because the makespan of the feeder can increase without affecting the objective function. This phenomenon is especially evident in the first and fourth instances of the first group, and in the fourth instance of the last group in Table 5.

But the delay of the DT containers could be unavoidable. In fact, it is strictly related to their class and to the class-based stowage plan of both vessels, that is to the data set of the seed instances and to the random procedure implemented by DCTP-G, whose output could likely be an instance not enough suitable for the Direct Transshipment modality. Looking, for example, at the third instance of the first group in Table 5, we observe that the average delay of DT containers is relatively high, in spite of the remarkable growth of the makespan of the feeder vessel. As for the F-M instances, most of them seem to belong to the class of instances for which the Direct Transshipment is not a convenient approach. Actually, even for the last (end best) group of instances in Table 6, we can observe that the delay of the DT containers is unavoidable. This is very clear for the instance FM-10-10-20-20-2, where the makespan of both vessels is computed in an optimal way.

The results in Table 7, related to the bi-directional flow of DT containers between mother vessels, exhibit both the characteristics discussed before. The minimum value of AvgD is attained at the third instance, for which the makespan of vessel A, that was the minimum in the relaxed problem, grows so much to become the makespan of the virtual vessel in DCTP. In the second instance, instead, it is evident that the waiting time for the DT containers can not be reduced any further.

7   Conclusions

In this paper we have addressed the Direct Container Transshipment Problem (DCTP), that is the problem of scheduling the loading/discharging operations of two vessels sharing the berthing time windows, and assuming that some containers discharged from a vessel must be directly loaded on the other one, completely skipping the storage phase in the yard. The aim is to minimize a linear combination of the time needed to complete all the operations and the average waiting time for the directly transshipped containers. The DCTP integrates two operative decision processes: the scheduling of the quay cranes and the stowage of the container directly transshipped. For this problem we have described a mixed integer linear model and we have derived a Tabu Search heuristic algorithm. We have tested the heuristic algorithm on a set of randomly generated instances. The algorithm is able to find feasible solutions of good quality in almost all the considered instances within a short amount of computation time, despite the intrinsic hardness of the problem.

The DCTP generalizes the Quay Crane Scheduling Problem (QCSP). Actually, when no direct transshipment operation has to be performed, the DCTP separates into two non-standard QCSP, where for some export containers also the stowage position must be decided. Vice-versa, if the stowage plans of the two involved vessels are completely known, the resulting DCTP reduces to two independent QCSP, one for each vessel.

The direct transshipment of containers allows to reduce yard congestions and, at the same time, the storage costs. From this point of view, it seems certainly a profitable modality for the terminal management. As for the shipping line companies, the saving in the storage costs must be evaluated in connection with possible increasing of the berthing times. This is the focus of future developments.

References

1. Alicke, K.: Modeling and optimization of the intermodal terminal Mega Hub. OR Spectr. **24**(2), 1–18 (2002)
2. Bierwirth, C., Meisel, F.: A fast heuristic for quay crane scheduling with interference constraints. J. Sched. **12**(4), 345–360 (2009)
3. Blumenhagen, D.: Containerization and Hinterland traffic. Marit. Policy Manag. **8**(3), 197–206 (1981)
4. Chiarello, A., Gaudioso, M., Sammarra, M.: Truck synchronization at single door cross-docking terminals. OR Spectr. **40**(2), 395–447 (2018)
5. Garro, A., Monaco, M.F., Russo, W., Sammarra, M., Sorrentino, G.: Agent-based simulation for the evaluation of a new dispatching model for the straddle carrier pooling problem. Simulation **91**(2), 181–202 (2015)
6. Gendreau, M., Potvin, J.Y.: Tabu search. In: Gendreau, M., Potvin, J.Y. (eds.) Handbook of Metaheuristics. ISOR, vol. 146, pp. 41–59. Springer, Boston (2010). https://doi.org/10.1007/978-1-4419-1665-5_2
7. Lee, B.K., Jung, B.J., Kim, K.H., Park, S.O.: A simulation study for designing a rail terminal in a container port. In: Perrone, L.F., Wieland, F.P. et al. (eds.) Proceedings of the 38th Conference on Winter Simulation, pp. 1388–1397 (2006)
8. Lian, C., Hwang, H., Gen, M.: A berth allocation planning problem with direct transhipment consideration. J. Intell. Manuf. **23**(6), 2207–2214 (2012)
9. Meisel, F., Bierwirth, C.: A unified approach for the evaluation of quay crane scheduling models and algorithms. Comput. Oper. Res. **38**(3), 683–693 (2011)
10. Monaco, M.F., Sammarra, M.: Quay crane scheduling with time windows, one-way and spatial constraints. Int. J. Shipp. Transp. Logist. **3**(4), 454–474 (2011)
11. Monaco, M.F., Sammarra, M.: The direct ship-to-ship container transshipment problem at a maritime terminal. In: Gunther, H.-O., Kim, K.-H., Kopfer, H. (eds.) International Conference on Logistics and Maritime Systems, LOGMS2012, pp. 79–89 (2012)

12. Monaco, M.F., Sammarra, M., Sorrentino, G.: The terminal-oriented ship stowage planning problem. Eur. J. Oper. Res. **239**(1), 256–265 (2014)

13. Nam, H., Lee, T.: A scheduling problem for a novel container transport system: a case of mobile harbor operation schedule. Flex. Serv. Manuf. J. **25**, 576–608 (2013)

14. Sammarra, M., Cordeau, J.-F., Laporte, G., Monaco, M.F.: A tabu search algorithm for the quay crane scheduling problem. J. Schedul. **10**(4–5), 327–336 (2007)

15. Zeng, Q., Feng, Y., Chen, Z.: Optimizing berth allocation and storage space in direct transshipment operations at container terminals. Marit. Econ. Logist. **19**(3), 474–503 (2017)

Alternative Performance Indicators for Optimizing Container Assignment in a Synchromodal Transportation Network

M. R. Ortega del Vecchyo[1,2], F. Phillipson[1(✉)], and A. Sangers[1]

[1] TNO, PO Box 96800, 2509 JE The Hague, The Netherlands
frank.phillipson@tno.nl
[2] Delft University of Technology, Delft, The Netherlands

Abstract. Several different attributes are deemed important in the container-to-mode assignment on a synchromodal transportation network. This paper proposes a way to quantify several of this different attributes: Robustness, Flexibility and Customer Satisfaction. These attributes are used as alternative objectives when optimizing the container assignment in a Synchromodal Transportation Network, modelling it as a Minimum Cost Multi-Commodity Flow on a Space-Time Network.

Keywords: Synchromodal logistics
Minimum cost multi commodity flow · Space-time graphs · Objectives
Robustness · Flexibility

1 Introduction

The focus in this paper is on the problem of allocating containers or container flows to different modalities as part of a synchromodal logistic approach. Flow of goods on a synchromodal, intermodal or multimodal network can be modeled via a multi-commodity flow problem on a special kind of graph called *space-time network (STN)* or space-time graph, as shown in [7]. In the case where only cost is considered, it can be modeled as a *Minimum Cost Multi-Commodity Flow (MCMCF)* problem. This problem is a generalization of the minimum cost flow problem to the case where there are multiple flows with multiple origin-destination (OD) pairs. The container flow problem we are considering is synchromodal because as new information becomes available, new solutions can be derived.

The question "what to optimize?" in such networks has been scarcely addressed in the literature, despite it being a recognized problem in [10,13]. In [10] it is proposed that cost, service, frequency, service time, delivery reliability, flexibility and safety are all performance indicators. In [12], customer responsiveness and quality as objectives are also objectives. Next to this, in supply chain logistics in general, there is a growing attention for environmental risks

© Springer Nature Switzerland AG 2018
R. Cerulli et al. (Eds.): ICCL 2018, LNCS 11184, pp. 222–235, 2018.
https://doi.org/10.1007/978-3-030-00898-7_14

and sustainability [1,3,6,8,15,16]. Note that we do not require the alternative objectives or performance indicators to be the objective value. They also can be used as constraint to guarantee a certain (minimum or maximum) value.

In most papers however, cost of the operation and service time are still the only used objectives, and other attributes are neglected [14]. As it is stated in [5], many transportation planning problems are solved via a deterministic optimization-based tool where the lowest-cost solution is chosen. However, the used forecasts can be very inaccurate and realizations may lead to new plans that has to be changed drastically, or might be unfeasible. In the literature these problems are sometimes addressed using the terms reliability, flexibility, robustness and resilience, where different terms can be used for similar things. In [9] definitions are proposed as an attempt to encompass consistently the meaning intended in other papers for each concept:

- Robustness is the ability to endure foreseen and unforeseen changes in the environment without adapting.
- Flexibility is the ability to react to foreseen and unforeseen changes in the environment in a pre-planned manner.
- Agility is the ability to react to unforeseen changes in the environment in an unforeseen and unplanned manner.
- Resilience is the ability to survive foreseen and unforeseen changes in the environment that have a severe and enduring impact.

In this work, in the context of transportation planning, we will use the following definitions based on meanings explained both implicitly and explicitly on several sources such as [4,5,9,11,14]:

- Robustness is the capacity of a plan to overcome uncertain events or disturbances in the future and still be carried over as planned.
- Flexibility is the capacity of a plan to adapt to uncertain events or disturbances, when these force the plan not to be able to be carried on anymore.

We propose, define and compare different performance indicators that are in play on a synchromodal transportation chain. The remainder of this paper is organised as follows. In the next section the theoretical basis of this work is presented. Next, in Sect. 3 the attributes that are considered are built from concept to implementation. Finally, in Sect. 4 we will present the conclusions and give directions for further research.

2   Minimum Cost Multicommodity Flow on Space-Time Graphs

In this section we introduce a modelling framework that we need in the remainder of the paper: minimum cost multicommodity flow on space time graphs [2].

On a graph (G, A) with n nodes and m arcs, where each arc (i, j) has capacity $u_{ij} > 0$, the multicommodity flow problem is a network flow problem with K

commodities with demand of flow d_k, $k = 1, ..., K$, between different source nodes s_k and sink nodes t_k. This problem is concerned with finding a feasible flow. The MCMCF problem is the problem of finding a minimum cost feasible flow.

A formulation of the MCMCF problem is as follows. Let $P(k)$ be the set of all directed simple paths on G from s_k to t_k, $C(P)$ the cost of the path $P \in \cup_k P(k)$, that is, the sum of all the costs of arcs $(i,j) \in P$. Then the MCMCF problem can be formulated as

$$\min \sum_k \sum_{P \in P(k)} C(P)x_P,$$

$$\sum_k \sum_{P \in P(k)} x_P \delta_{ij}(P) \leq u_{ij} \quad \text{for all } (i,j) \in A, \tag{1}$$

$$\sum_{P \in P(k)} x_P = d_k \quad \text{for all } k,$$

$$x_P \geq 0 \quad \text{for } P \in \cup_k P(k),$$

where

$$\delta_{ij}(P) = \begin{cases} 1 & \text{if } (i,j) \in P \\ 0 & \text{if } (i,j) \notin P. \end{cases}$$

In the previous formulation, there is one decision variable x_P for each path between an Origin-Destination (OD) pair, for each OD pair.

The MCMCF can be applied to a space-time graph. The idea behind a space-time graph, as its name suggests, is that every node represents a location at a specific time, and arcs represent a change of state. They are meant to show the characteristics of an underlying graph G with node set S as time changes discretely from 1 to T where each of these discrete times is referred to as a time-stamp.

Formally, we say that a graph G is a STN (or space-time graph) if its node set is of the form $S \times \{1, 2, ..., T\}$ for some $T \in \mathbb{Z}^+$ and some set S and every arc $((a,p),(b,q)) \in A(G)$ satisfies $p < q$. We refer to the node (a,p) as location a at time p, and to T as the time horizon of G.

Additionally, we consider the following assumptions for our problem:

1. At every time-stamp, there is an unlimited number of trucks going from any location to any other location. These trucks are more expensive and quicker than any other means of transportation.
2. Truck price is fixed and is the same for every OD pair.
3. Every number of containers has the possibility of remaining idle in a given location with no additional cost.
4. Only one arc from (A,t) to (B,s) is allowed.

3   Attributes

Earlier we gave the new definitions of Robustness and Flexibility. These are important to cope with uncertain events. In practice, an "uncertain event" on a transportation network can come in many different forms: disturbances in handling times upon arrival on a terminal, arrival of new orders, assignment of time slots for arrival of certain modes, and so on. Note that the relevance of these uncertainties usually varies depending on the different time-scales. Planners often deal with these uncertainties via strategic behavior, that is, by acknowledging these uncertainties and taking them into consideration when making their decisions. In this study, we restrict ourselves to consider the uncertain events of travel times and handling times on terminals. Therefore, the definitions of the concepts can be read as follows:

- Robustness is the capacity of a plan to overcome delays in travel times and handling times on terminals and still be carried on as planned.
- Flexibility is the capacity of a plan to adapt to delays in travel times and handling times on terminals when these force the plan not to be able to be carried on anymore.

Next to these two, we take into account the lateness of a whole plan via the attribute customer satisfaction. These three attributes will be defined in the following sections and for each a numerical example will be used to show the impact on the MCMCF problem.

3.1   Robustness

To illustrate the meaning of robustness in our model, we first show how we would like to quantify robustness for a simple case. In a STN with a number of orders, we want to give a numerical value to each solution of the problem, that is, a value per transportation plan. In the case of a single order, we may assign a value to a path. Consider a path P such as the one in Fig. 1 with an OD pair $((A,0), (C,5))$, that is, with a source on location A at time 0 and sink on location C at time 5. The robustness value is meant to represent how likely this plan can endure despite delays in travel times and handling times, that is, we need to see how likely the transportation mode from location B at time 3 to location C at time 4 (arc from $(B,3)$ to $(C,4)$) will be able to take place for path P despite delay in travel time from A to B and handling times at B. If the resource doing the trip $((A,0), (B,1))$ is also the one on the trip $((B,3), (C,4))$ then there will be no handling at location B, thus the flow of containers through this path will certainly make this connection. Otherwise, the handling at location B will depend on these factors:

- The number of containers going through arc $((A,0), (B,1))$, since all of these will be handled at location B, which is, in this case, the flow going through path P.
- The number of timestamps available from the estimated time of arrival to arc B, which is 1, and the time of departure of the trip to C, which is 3.

This kind of link is what we refer to as an event, and we define the robustness of a plan with respect to the robustness of these events.

Definition 1. *For a given path P on a space-time graph, we say that $e = ((A, t_0), (B, t_1), (B, t_2))$ is an **event** of the path P if the path $((A, t_0), (B, t_1), (B, t_1 + 1), ..., (B, t_2), (C, t_3))$ for some $C \neq B$ and $B \neq A$ is a sub-path of P, and the resource of the trip $((A, t_0), (B, t_1))$ is a different resource than the one of trip $((B, t_2), (C, t_3))$. Also, $e = ((A, t_0), (B, t_1), (B, t_2))$ is an event of P if the path $((A, t_0), (B, t_1), (B, t_1 + 1), ..., (B, t_2))$ is a sub-path of P and (B, t_2) is the last node on P. If the event is of the latter form we refer to it as the **last event** of P. We use the short notation $e \in P$ to denote that the event e is an event of the path P. For a path-based multi-commodity flow problem Pr on a space-time graph, we say that e is an event of the problem Pr if it is an event of a path P of an OD pair in Pr. We use the short notation $e \in Pr$ to denote that the event e is an event of the problem Pr. If x_P is the flow variable of a path P, and F is a solution to Pr, the flow on an event is defined as $F_e = \sum_{P \in P(e)} x_P$ where $P(e) = \{P \in \cup_k P(k) | ((A, t_0), (B, t_1)) \in P\}$ (see Eq. 1 for notation).*

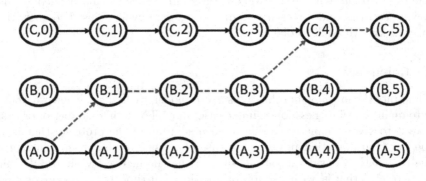

Fig. 1. Robustness of a path (in dotted blue). (Color figure online)

In the path of Fig. 1, if the resource of edge $((A, 0), (B, 1))$ is a different resource from the one in edge $((B, 3)(C, 4))$, the path has the event: $e_1 = ((A, 0), (B, 1), (B, 3))$. In any case, the last event $e_2 = ((B, 3), (C, 4), (C, 5))$ in on the path. Our main assumption when determining robustness of an event is that the information in the three elements that constitute the event and the flow of the event are necessary and sufficient to determine the robustness of the event. More specifically, we determine the robustness of an event via a measure of robustness, which depends on the amount of flow f and the number of timestamps t.

Definition 2. *We say that a function r' is a **measure of robustness** if r' : $\mathbb{R}^+ \times \mathbb{Z}^+ \to [0, 1]$ and the following holds:*

– $r'(0, t) = 1$ for all t, $\lim_{f \to \infty} r'(f, t) = 0$ for any fixed t, $r'(f, t)$ is a decreasing function of f for any fixed t.

- $r'(f, 0) = \epsilon$ for $\epsilon > 0$ a number close to zero, $\lim_{t \to \infty} r'(f, t) = 1$ for any fixed f, $r'(f, t)$ is an increasing function of t for any fixed f.

We define the robustness $r(e, f)$ of an event $e = ((A^e, t_0^e), (B^e, t_1^e), (B^e, t_2^e))$ with flow f as $r(e, f) = r'(f, t_2^e - t_1^e)$. If the first argument f is omitted then $r(e) = r'(F_e, t_2^e - t_1^e)$.

Thus, the two variables of the function $r'(f, t)$ are thought of as denoting the amount of flow of the first arc of the event and the timestamps available. The properties of the measure of robustness attempt to be the minimum requirement we would expect for a way to measure the robustness of such an event: if the amount of flow is small with respect to the number of timestamps, then quite likely (probability close to 1) the event will be a success in terms of arrival before departure time of the next transportation mode, whereas if the amount of flow is large with respect to the number of timestamps, then it might be difficult to make this connection. We should note that whenever a specific flow is considered large or small depends of course on the units considered for each timestamp, but in any case, the equalities and the limits must hold. The reason for the small value $r'(f, 0) = \epsilon$ is that just-in-time connections might not be very robust, but they can still be made.

Definition 3. *Let F be a solution flow for a path-based multi-commodity flow problem Pr on a space-time graph. We define the* **robustness of the solution** *$R(F)$ as the product of the robustness of the events of the plan. that is*

$$R(F) = \prod_{e \in Pr} r(e) = \prod_{e \in Pr} r'(F_e, t_2^e - t_1^e)$$

Thus, in order to quantify the robustness, a robustness measure r' must be specified. We propose the function

$$r'(f, t) = \begin{cases} e^{-\lambda \frac{f}{t}} & \text{if } t > 0 \\ e^{-\lambda \frac{f}{0.5}} & \text{if } t = 0 \end{cases}$$

with $\lambda > 0$ a parameter to be specified depending on the units that represent each timestamp. For simplicity, if an event e is such that $t_2^e - t_1^e = 0$ we write $\frac{F_e}{t_2^e - t_1^e}$ when we actually mean $\frac{F_e}{0.5}$. Then robustness is defined as

$$R(F) = \prod_{e \in Pr} r'(F_e, t_2^e - t_1^e) = \prod_{e \in Pr} e^{-\lambda \frac{F_e}{t_2^e - t_1^e}}.$$

Maximizing the robustness function is the same as maximizing the logarithm of the robustness function, such that

$$\log R(F) = \log \prod_{e \in Pr} e^{-\lambda \frac{F_e}{t_2^e - t_1^e}} = \sum_{e \in Pr} \log e^{-\lambda \frac{F_e}{t_2^e - t_1^e}}$$

$$= \sum_{e \in Pr} -\lambda \frac{F_e}{t_2^e - t_1^e} = -\lambda \sum_{e \in Pr} \frac{F_e}{t_2^e - t_1^e}.$$

Since $\lambda > 0$, maximizing the robustness function is equivalent to minimizing

$$\sum_{e \in Pr} \frac{F_e}{t_2^e - t_1^e}. \tag{2}$$

This expression is linear with respect to the flow path-based variables x_P, and the sum depends only on the events of the problem, which are independent of the solution proposed. Therefore this expression can be constructed as a linear objective on a linear program (LP). We refer to the expression 2 as the **robustness expression.**

Notice that $R(F)$ tends to decrease if the number of events in the problem $|\{e \in Pr\}|$ increases. For this reason, in order to treat instances of different sizes in a similar way, it is practical to introduce the **geometric mean robustness of the solution** $MR(F)$ defined as

$$MR(F) = \left(\prod_{e \in Pr} r(e) \right)^{\frac{1}{|\{e \in Pr\}|}}.$$

Then we obtain, using the same robustness measure as before

$$\log MR(F) = \frac{\log \prod_{e \in Pr} e^{-\lambda \frac{F_e}{t_2^e - t_1^e}}}{|\{e \in Pr\}|} = \frac{-\lambda}{|\{e \in Pr\}|} \sum_{e \in Pr} \frac{F_e}{t_2^e - t_1^e}.$$

The objective $\sum_{e \in Pr} \frac{F_e}{t_2^e - t_1^e}$ is the linear expression that represents robustness when the measure of robustness is chosen to be $r'(f, t) = e^{-\lambda \frac{f}{t}}$. Robustness and mean robustness is maximized when this expression is minimized, regardless of the $\lambda > 0$. The equalities

$$\sum_{e \in Pr} \frac{F_e}{t_2^e - t_1^e} = -\frac{\log R(F)}{\lambda}$$

and

$$\sum_{e \in Pr} \frac{F_e}{t_2^e - t_1^e} = -\frac{\log MR(F)|\{e \in Pr\}|}{\lambda}$$

allow us to calculate the robustness and the mean robustness of a solution, and give a value for the robustness expression when we are aiming for a robustness or mean robustness value. Also, a good estimate of λ should be chosen when the unit of the timestamps is determined, from the interpretation given to $r'(f, t) = e^{-\lambda \frac{f}{t}}$.

In order to gauge the influence of robustness, we present some results about the impact of the robustness expression as a constraint on the MCMCF problem on a space-time graph. Suppose we want to solve an instance with time horizon

$T = 80$, 10 terminals, 200 orders, and that each time-stamp represents one hour. We assume that quite certainly (with probability .90) 10 containers can be handled in one hour, that is to say

$$r'(f,t) = e^{-\lambda(10)} > 0.90 \implies -10\lambda > \log(0.90)$$

$$\implies \lambda < 0.0105.$$

We fix $\lambda = 0.01$. By solving the MCMCF problem without involving robustness, we obtain an optimal solution F_1 with a mean robustness of

$$MR(F_1) = \exp\left(\frac{-\lambda}{|\{e \in Pr\}|} \sum_{e \in Pr} \frac{F_e}{t_2^e - t_1^e}\right) = .7761$$

and a cost $C(F_1) = 204,399$. If we wish to improve the mean robustness slightly to 0.78, we input the constraint $\sum_{e \in Pr} \frac{F_e}{t_2^e - t_1^e} = -\frac{\log MR(F)|\{e \in Pr\}|}{\lambda} \leq -\frac{\log(.78)|\{e \in Pr\}|}{\lambda} = 7602.9175$ and rerun the solver. This constraint will guarantee the solution will have a mean robustness of at least 0.78.

This change, as little as it may seem, already brings some significant differences on the new solution F_2. F_1 and F_2 differ in the transportation planning of 21 out of the 200 orders, despite the fact that $C(F_2) = 204,400.96 \approx C(F_1) + 1$ and both have same number of trucks used, that is, the plan is altered without compromising cost.

Comparison between increasingly robust solutions reveals the following tendencies about the more "average robust" solutions:

- They tend to prefer paths that have less connections between different resources.
- They tend to prefer earlier arrival.
- They tend to prefer paths connected by the same resource.

The first characteristic may be helpful to prevent from possible handling costs incurred in terminals. The second characteristic can be beneficial so that future resources are allocated for future uncertain happenings, but it can affect if there are costs for long idle times at destination terminals.

3.2   Flexibility

The second attribute proposed is flexibility, which was defined as the capacity of a plan to adapt to delays in travel times and handling times on terminals, when these delays force the plan not to be able to be carried out anymore. To calculate the flexibility of a single path (simple case with a single commodity, and one path carries all the flow) such as the one described in Fig. 2, we first identify those links in the path that could be problematic in terms of flexibility as we have defined it. As in the case of robustness, we refer to these problematic links as events. In this path, with OD pair $((A,0),(C,5))$, if there was a delay on the transportation arc from $(A,1)$ to $(B,2)$ or on the handling time at B

such that the connection with the arc $(B, 3)$ to $(C, 4)$ is lost (in this case, the delay made the trip arrive at time 4), there is still the possibility to take the arc from $(B, 4)$ to $(C, 5)$. The flexibility of this path is defined in terms of the cost of this alternative route with respect to the cost of the original route. In the case where there is more than one event on the path, the flexibility of the path is done with respect to the cost of the alternative routes corresponding to each of these events. In order to state unambiguously the flexibility of a flow, a series of definitions are necessary. The definition of event is still as in Definition 1, except that in this context, last events are not considered events.

Definitions

- For a path P on an STN and an event $e = ((A, t_1), (B, t_2), (B, t_3))$ on the path, we define the subpath P_e with respect to e as the subpath of P that contains all the nodes from (B, t_3) onward. In the case of the example in Fig. 2, for the event $e = ((A, 1), (B, 2), (B, 3))$ the subpath defined is $P_e = ((B, 3), (C, 4), (C, 5))$.
- For a solution F of a multi-commodity flow problem on a STN G, we denote by $G \backslash F$ the STN G whose arcs' capacity have been lowered according to the flow of F, that is, the capacity of an arc in $G \backslash F$ is the capacity of the arc on G minus the flow passing through that arc on F.
- For a pair of nodes (A, t_1) and (B, t_2) on a space-time graph G and a positive real number r, we denote by $\mathrm{mincost}((A, t_1), (B, t_2), r)_G$ the cost of the optimal solution of the minimum cost flow problem with source node (A, t_1), sink node (B, t_2) and flow r in G.
- For a path P with flow x_P of a solution F of a multi-commodity flow problem on a STN G and an event $e = ((A, t_1), (B, t_2), (B, t_3))$ on the path, we define the anti-flexibility $\varphi_{G \backslash F}(e, x_P)$ of the event as the least cost that would be incurred if the trip scheduled from A at time t_1 to B at time t_2 would arrive one timestamp after time t_3 to B. That is,

$$\varphi_{G \backslash F}(e, x_P)$$

$$= \mathrm{mincost}((B, t_3 + 1), (S_P, t_P), x_P)_{G \backslash F} - C(P_e) x_P.$$

Here, $C(P_e)$ is the cost of the subpath P_e and (S_P, t_P) is the last node on P. Notice the dependency of the min-cost algorithm on the solution flow F as well as on G, that is, the capacity of the arcs on G are lowered corresponding to the flow F. We call the above anti-flexibility because $\varphi_{G \backslash F}(e, x_P)$ decreases as the flexibility of the event increases, according to our definition of flexibility.

- For a solution flow F of a path-based multi-commodity flow problem on a space-time graph G and a robustness function r, we define its anti-flexibility $\phi_G(F)$ as

$$\phi_G(F) = \sum_{P \in F, x_P > 0} \sum_{e \in P} \varphi_{G \backslash F}(e, x_P)(1 - r(e))$$

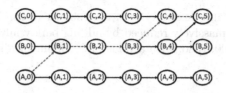

Fig. 2. Flexibility of a path (in dotted blue) (Color figure online)

In our case, the robustness function r used is the one implied by the exponential robustness measure $r'(f,t) = e^{-\lambda \frac{f}{t}}$, as shown in the previous section. The last expression is a sum of all the incurred costs that could happen on the plan from delays, this is the expression we seek to minimize, however, it is far from linear in terms of the flow variables of the paths. Notice also that in order to calculate the anti-flexibility of an event of a path P, the value of the flow variable x_P must be known in advance, which is of course not the case. In addition, a constraint whose coefficients involve solving several min-cost problems can be very computationally heavy. Thus a linear expression that overcomes these challenges is sought for in order to include it in a LP formulation. For this purpose, the following linearization is constructed.

Definition 4. *For a path P on a STN G and an event $e = ((A,t_1),(B,t_2),$ $(B,t_3))$ on the path, we define the* linear anti-flexibility of the event $\iota_G(e)$ *as*

$$\iota_G(e) = C(P_e^d) - C(P_e),$$

where P_e^d is the shortest (least costly) path on G from $(B,t_3 + 1)$ to (S_P,t_P). For a path P on a STN G, and a robustness function r, we define the linear anti-flexibility *of the path $\iota_G(P)$ as*

$$\iota_G(P) = \sum_{e \in P} \iota_G(e)(1 - r(e,c)),$$

where c is an arbitrary fixed number, preferably close to the average flow of a path. We fix c to be the lowest possible order size.

Now, the linear anti-flexibility of a path is a coefficient relatively easy to calculate. With it, we define the *linear anti-flexibility expression*

$$\sum_P \iota_G(P)x_P. \tag{3}$$

To explore the effect of the linear anti-flexibility expression obtained, we consider an MCMCF problem on an instance with the same characteristics as the one in the robustness example. The problem without any linear anti-flexibility constraint yields a solution F_1 with an anti-flexibility value $\phi_G(F_1) = 6078$, a linear anti-flexibility of $\sum_P \iota_G(P)x_P = 5075$ and a cost of $C(F_1) = 153,655$. By adding the linear flexibility as a constraint, and changing the constraint value, we obtain costs and anti-flexibility values as shown in the Table 1. Note that the anti-flexibility is massively reduced by adding barely any cost with a linear flexibility value of 3000. Also, with a value of 2000, anti-flexibility is reduced by more than half.

Table 1. Trade-off between anti-flexibility and cost

	Linear anti-flexibility	Cost	Anti-flexibility
F_1	5075	153,655	6079
F_2	3000	153,843	3502
F_3	2000	154,285	2665
F_4	1000	155,724	1567
F_5	650	156,471	1341

In terms of how different the solution are, F_2 has a different plan for 34 orders when compared to F_1, whereas F_4 has 39 orders with a different plan with respect to plan F_1. This suggests that the linear anti-flexibility constraint affects the plan considerably.

Comparison between the solutions reveals the following tendencies about the more "flexible" solutions:

- They tend to prefer trips with a cheap backup alternative in the future (notice that flexibility of a path can be negative, meaning that the backup route is cheaper).
- They tend to prefer single link trips, or trips from the same voyage.

As it was defined, anti-flexibility represents the expected extra costs that will be incurred on the plan, assuming there is a full refund for the arcs that were planned to be used but were not reached on time. Of course this refund does not necessarily take place, making the anti-flexibility more of a lower bound on the expected extra costs. If the expected costs are to be minimized, then it may be appropriate to minimize the lower bound on this expected cost, that is, the sum of costs and anti-flexibility. Unfortunately, anti-flexibility cannot be put on the LP, so one must rely on the use of the linear anti-flexibility to reach a low value for the sum of costs and anti-flexibility.

For the case of one commodity that can be served with a single path, given a collection of possible solution paths, observe that if the anti-flexibility of a path is minimized, then the path obtained might have negative values, meaning that the backup paths are cheaper than the solution path. If this paths are much cheaper, then the anti-flexibility is much lower. Of course this has to be avoided, and it shows that the anti-flexibility is an expression that comes with trade-offs. Anti-flexibility measures how cost-effective the backup plans are with respect to the chosen plan, and although it is good to have these alternatives cost-effective, it is probably not good to have them much cheaper than the chosen plan.

3.3   Customer Satisfaction

From the point of view of a customer, perhaps the most important thing of an order is its timely delivery. However, as our meetings with different stakeholders in practice revealed, this attribute is dependent on the client. That is to say, some clients might have no problem if their order arrives later than the agreed arrival time, whereas for others it might be crucial to have it on time. In our model we consider this customer dependent lateness as an independent attribute. We fix a maximum amount of lateness that can be allowed for each order, and in order to measure customer satisfaction of a solution plan, we observe how late the arrival of each order is, taking into account the priority of each client. Since the order cannot be considered delivered until every container has arrived to the destination, we make the following definitions:

Definition 5. *For a solution flow F of a multi-commodity flow problem Pr on a space-time graph, and an order $o \in Pr$ of the problem, we define the delivery time of the order $d(o)$ as the maximum of the arrival times of the containers on that order.*

Definition 6. *For each order o with an OD pair, and $t \in \{0, 1, 2, ..., r\}$ a number of timestamps, we refer to the satisfaction $s(o, t) \in [0, 1]$ as the number that reflects how satisfied the customer of order o will be if the order arrives t timestamps after the due time. For a fixed o, we assume $s(o, t)$ to be decreasing on t. The maximum number of lateness $r < T$ is fixed for computational ease.*

Notice that this can be extended to a case where there is also penalty for early arrival, or even further, for arrival at any specific timestamp per order. However, this is not done in this paper.

Definition 7. *For a solution flow F of a multi-commodity flow problem Pr on a space-time graph and a family of numbers $w(o) \in [0, 1]$ such that $\sum_{o \in Pr} w(o) = 1$, we define the customer satisfaction as*

$$(\sum_{o \in Pr} s(o, t_o)w(o))^2, \tag{4}$$

where t_o is the delay in number of timestamps of order o.

Customer satisfaction can be implemented via several indicator variables, one per order, per number of timestamps delayed. However, given the addition of several binary variables, considering customer satisfaction comes with a computational burden.

As with the previous attributes, we use an instance with 200 orders to see the effect of the customer satisfaction expression as a constraint, with randomly generated weights. The maximum number of delayed timestamps r is set to 10. Without any constraint related to customer satisfaction, we obtain a solution F_1 with a customer satisfaction value of .8190. By increasingly constraining the value of customer satisfaction, we obtain the costs summarized in the following table.

	Customer satisfaction	Cost
F_1	.8190	60082
F_2	.8525	60094
F_3	.9	60548
F_4	.95	62474
F_5	.98	64488

The table shows that substantial cost reduction can be obtained at the expense of customer satisfaction, i.e. timely delivery. Overall, by adding the attribute of customer satisfaction to the model, and stretching the possibilities of delayed containers, we obtain a new range of solutions and a way to compare their effectiveness.

4   Conclusions

We constructed linear formulations of three different attributes: robustness, flexibility and customer satisfaction. These formulations can be used as objectives in optimizing a synchromodal transportation problem. As optimization environment we used the Minimum Cost Multi-Commodity Flow formulation on a space-time network. As expected the alternative objectives have a trade-off with other objectives, such as cost. To handle this, further research is done on multi-objective optimization, where various objectives are considered together and a pareto-front of alternative optimal solutions are presented. Another direction that is being investigated is the use of robust optimization methods within the synchromodal transportation approach.

Acknowledgement. This work has been carried out within the project 'Complexity Methods for Predictive Synchromodality' (Comet-PS), supported by NWO (the Netherlands Organisation for Scientific Research), TKI-Dinalog (Top Consortium Knowledge and Innovation) and the Early Research Program 'Grip on Complexity' of TNO (The Netherlands Organisation for Applied Scientific Research).

References

1. Ahluwalia, P.K., Nema, A.K.: Multi-objective reverse logistics model for integrated computer waste management. Waste Manage. Res. **24**(6), 514–527 (2006)
2. Andersen, J., Crainic, T., Christiansen, M.: Service network design with asset management: Formulations and comparative analyses. Transp. Res. Part C Emerg. Technol. **17**(2), 197–207 (2009)
3. Baykasoğlu, A., Subulan, K.: A multi-objective sustainable load planning model for intermodal transportation networks with a real-life application. Transp. Res. Part E Logistics Transp. Rev. **95**, 207–247 (2016)
4. Beuthe, M., Bouffioux, C.: Analysing qualitative attributes of freight transport from stated orders of preference experiment. J. Transp. Econ. Policy (JTEP) **42**(1), 105–128 (2008)

5. Caplice, C., Jauffred, F.: Balancing robustness and flexibility in transportation networks (2014). http://ctl.mit.edu/sites/ctl.mit.edu/files/caplice-SCB-MarApr2014.pdf

6. Caramia, M., Dell'Olmo, P.: Multi-objective Management in Freight Logistics. Increasing Capacity, Service Level and Safety with Optimization Algorithms. Springer, London (2008). https://doi.org/10.1007/978-1-84800-382-8

7. Crainic, T.: Service network design in freight transportation. Eur. J. Oper. Res. **122**(2), 272–288 (2000)

8. Govindan, K., Paam, P., Abtahi, A.R.: A fuzzy multi-objective optimization model for sustainable reverse logistics network design. Ecol. Ind. **67**, 753–768 (2016)

9. Husdal, J.: A conceptual framework for risk and vulnerability in virtual enterprise networks. In: Managing Risk in Virtual Enterprise Networks: Implementing Supply Chain Principles, p. 1 (2010)

10. Ishfaq, R., Sox, C.R.: Intermodal logistics: the interplay of financial, operational and service issues. Transp. Res. Part E Logistics Transp. Rev. **46**(6), 926–949 (2010)

11. Miller-Hooks, E., Zhang, X., Faturechi, R.: Measuring and maximizing resilience of freight transportation networks. Comput. Oper. Res. **39**(7), 1633–1643 (2012)

12. Ramezani, M., Bashiri, M., Tavakkoli-Moghaddam, R.: A new multi-objective stochastic model for a forward/reverse logistic network design with responsiveness and quality level. Appl. Math. Model. **37**(1), 328–344 (2013)

13. Riessen, B.v., Negenborn, R., Dekker, R.: Synchromodal container transportation: an overview of current topics and research opportunities. In: Computational Logistics (2015)

14. SteadieSeifi, M., Dellaert, N.P., Nuijten, W., Van Woensel, T., Raoufi, R.: Multimodal freight transportation planning: a literature review. Eur. J. Oper. Res. **233**(1), 1–15 (2014)

15. Tuzkaya, G., Kilic, H.S., Aglan, C.: A multi-objective supplier selection and order allocation model for green supply chains. J. Mil. Inf. Sci. **4**(3), 87–96 (2016)

16. Xifeng, T., Ji, Z., Peng, X.: A multi-objective optimization model for sustainable logistics facility location. Transp. Res. Part D Transp. Environ. **22**, 45–48 (2013)

Vehicle Routing and Multi-modal Transportation

The Cost of Continuity
in the Collaborative Pickup
and Delivery Problem

Margaretha Gansterer[✉], Richard F. Hartl, and Sarah Wieser

Department for Business Administration, University of Vienna,
Oskar-Morgenstern-Platz 1, 1090 Vienna, Austria
{margaretha.gansterer,richard.hartl,sarah.wieser}@univie.ac.at

Abstract. We assess the potential total profit in collaborative pickup
and delivery problems, where carriers are willing to exchange transportation requests. For this, we design an adaptive large neighborhood search
method that is used to generate solutions of publicly available but yet
unsolved test instances. Our computational study reveals that collaboration profits might go up to 40% of the initial total profit, but typically
come with unevenly distributed workloads. Such solution are of course
not acceptable in practice. Thus, the aim of this study is to elaborate
on the cost of continuity, i.e. the possibility for carriers to not deviate
too much from their initial situations. Carriers might, for instance, not
be willing to give up on some of their customers or want to stay with
minimum profits. The rational behind keeping customers is that carriers might have long-term or particular valuable customer relationships,
which they do not want to abandon. We discuss different types of continuity constraints and assess their impact on the total collaboration profit.
Our computational study shows that even in the presence of continuity
constraints remarkable total collaboration profits can be achieved.

Keywords: Logistics · Collaboration · Centralized planning
Adaptive large neighborhood search

1 Introduction

The transportation industry is to a high degree inefficient, costly and not sustainable. It puts an enormous burden on both society and the environment,
e.g. [5,26]. Besides that, it is extremely competitive and companies are facing
very low profit margins, making it hard to stay in business. To increase efficiency, carriers establish collaborations, where parts of their logistics operations
are planned jointly. By increasing efficiency, collaborations also serve ecological
goals. It is well known that transportation is one of the main contributors of

Supported by FWF the Austrian Science Fund (Projectnumbers P27858-G27 and
P26973-N15).

R. Cerulli et al. (Eds.): ICCL 2018, LNCS 11184, pp. 239–252, 2018.
https://doi.org/10.1007/978-3-030-00898-7_15

CO_2 emissions [4]. Thus, public authorities are encouraging companies to join horizontal collaborations. They not only aim at reduced emissions of harmful substances, but also on reduced road congestion and noise pollution. Hence, it is not surprising that collaborative vehicle routing is an active research area of high practical importance. It has recently been identified by [38] to be one of the big trends in transportation. Nevertheless, some carriers are still reluctant to join collaborative partnerships, since they do not want to reveal sensitive information or are afraid to lose some of their customers. However, the benefits of joint logistics planning are convincing: improvements of up to 30% have been reported [19,39].

If collaborative decisions are made by a central authority having full information, this is referred to as *centralized collaborative planning*. An example for such a central authority might be an online platform providing services for collaborative decision making [12]. In our study, we focus on centralized decision making problems occurring in the less than truckload pickup and delivery market. Thus, we assume customer requests to have specified origins and destinations, and several requests can be served within the same tour. These problems arise, for instance, in the small parcel industry. In this branch of the transportation industry, collaborative planning is of particular importance since shipments from different customers can be moved on the same vehicle. This gives carriers much flexibility to share customer requests among each other [2]. In Fig. 1 we illustrate the investigated setting with three carriers.

We assume a central authority having full information, aiming at an efficient distribution of customer requests to carriers. The problem has been introduced by [8]. In [20] the problem is extend by workload constraints, taking an even distribution of workload among the participants into account. The authors compare different exact solution approaches for the problem with and without workload constraints. However, only small test instances (15 transportation requests) can be solved with exact methods. No efficient solution approaches for larger instances have been presented so far. We develop an adaptive large neighborhood search (ALNS) algorithm to solve the problem. The aim of this study is to elaborate on the cost of continuity, i.e. the possibility for carriers to not deviate too much from their initial situations. Carriers might, for instance, not be willing to give up on some of their customers. The rational behind keeping customers is that carriers might have long-term or particular valuable customer relationships, which they do not want to abandon. Also carriers might insist on a minimum individual profit.

Thus, the contribution of our study is threefold:

- we are the first to develop a solution method for the centrally planned collaborative pickup and delivery problem with continuity constraints,
- based on an extensive set of test instances, we quantify the huge potential of carrier collaborations,
- we elaborate on the cost of continuity. This is of particular relevance for carriers who do not want to share all their customers with coalition partners or insist on a minimum individual profit.

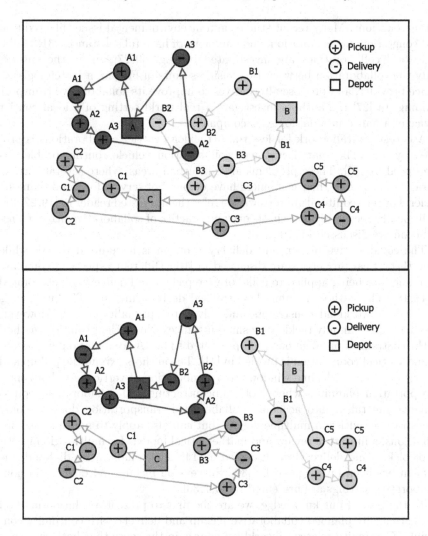

Fig. 1. The collaborative pickup and delivery problem of 3 carries (A, B, C). The upper part shows the pre-collaborative setting. An efficient redistribution of customer requests to carriers is shown in the lower part [19].

The remainder of the paper is organized as follows. We provide a literature review in Sect. 2. The problem description is given in Sect. 3 and the solution approach in Sect. 4. We describe the computational study in Sect. 5. Conclusions are summed up in Sect. 6.

2   Literature Review

Collaborative vehicle routing is intensively discussed in recent literature. Some early works were presented by [11,25], who provide some general insights into carrier coalitions. Many recent studies aim at environmental issues like reduced road congestion, noise pollution, and emissions of harmful substances [31,34,37]. Real-world collaborations are investigated by, e.g., [9,27]. In [9] the authors study the collaboration between two business units of Fritom, a Dutch logistics service provider, and propose alternatives to improve its collaborative transport planning. In [27] the authors investigate a real-world setting of a local courier service of a multi-national logistics company.

A two stage framework for less than truckload carrier collaborations is proposed by [32]. The shared customer collaboration vehicle routing problem is introduced by [15]. These problems arise in urban areas, where several carriers operate and some of their customers have demand of service for more than one carrier. For related literature reviews, we refer the interested reader to [19,22,40]. Additionally, recent advances in theory and practice of collaborative urban transportation are discussed in [10].

The collaborative pickup and delivery problem is introduced by [8], while game theoretical properties are discussed in [16]. Different aspects of combinatorial auctions being applied to collaborative pickup and delivery are elaborated in [17,18]. The authors assume decentralized decision making. The problem is extended by workload constraints and solved to optimality in [20]. However, these methods can only tackle very small instances. No efficient solution method for the centralized problem has been presented so far. A combination of horizontal and vertical cooperation is tackled in [41]. The authors extend the pickup and delivery problem with time windows to a combination of vertical and horizontal transportation planning, where both subcontracting and collaborative request exchange are taken into account. Collaborative transportation does not only orchestrate activities in multiple depots, but can also apply to multiple regions. Definitions, a literature review and mathematical models for multi-region multi-depot pickup and delivery are discussed in [14]. The collaborative dial-and-ride problem has been investigated by [30]. Surveys on pickup and delivery in non-collaborative settings are presented by [6,7,33].

To the best of our knowledge, we are the first to present a solution method for the centrally planned collaborative pickup and delivery with continuity constraints. Continuity relates to workload equity, in the sense that both measures target on a "fair" distribution of workload. A survey and analysis on workload equity in routing problems is given in [29]. The authors state that most works consider tour length as the equity metric. Fewer papers consider workload in terms of the demand served per tour. The authors reveal that literature almost exclusively focus on equity in terms of tour length, but the papers reporting on applications also include cases in which other factors are equally or more relevant. They claim that further research should place a greater emphasis on broader definitions of workload, like the number of stops as this seems to be the primary determinant of workload in the small parcel delivery sector. We agree with this and claim that this is of particular relevance in collaborative vehicle routing, where carriers are typically very critical in accepting a reallocation of customers. They might, for instance, have some customers which they do not want to exchange with their competitors. We contribute to close this gap by measuring continuity, which can be defined as fairness, in the number of customers kept by the initial owner.

3   Problem Description

We investigate the collaborative pickup and delivery problem, meaning that each request is associated with a prespecified origin and destination. The problem belongs to the class of traveling salesman problems with precedence constraints. Each carrier has a depot and at each depot the carrier has to solve the single vehicle case of the vehicle routing problem with pickups and deliveries [33]. This is also known as one-to-one pickup and delivery problem [6]. Related mathematical models, where the problem is based on a Hamiltonian tour formulation can be found in [20,28].

In collaborative vehicle routing, solutions typically need to fulfill special characteristics in order to be accepted by the participants. These characteristics might be minimum or maximum workload constraints or a certain number of customers visited by each carrier [20]. We introduce new characteristics by assuming that carriers want to keep (i) a given number of their "own" customers, or (ii) a minimum fraction of the number of customers they initially held, or (iii) a minimum revenue (or profit). We extend the model presented in [20] by the following constraints in order to ensure that each carrier yields a minimum revenue:

n number of customers
m number of depots
P set of pickup vertices, $P = \{1, .., n\}$
D set of delivery vertices, $D = \{n + 1, .., 2n\}$
W set of depot vertices, $W = \{2n + 1, .., 2n + m + 1\}$
N set of all vertices, $N = P \cup D \cup W \{1, .., 2n + m + 1\}$
x_{ij} decision variable indicating whether arc ij is used or not
φ_i revenue when serving customer i ($\varphi_i = 0$ at pickup nodes, and $\epsilon_i > 0$ at delivery nodes)
Ω_i minimum revenue for tours at depot i
r_i total revenue when arriving at customer i.

$$r_j \leq r_i + \varphi_i + M(1 - x_{ij}) \quad \forall i \in N\backslash\{2n + m + 1\}, j \in N \tag{1}$$
$$r_j \geq r_i + \varphi_i - M(1 - x_{ij}) \quad \forall i \in N\backslash\{2n + m + 1\}, j \in N \tag{2}$$
$$r_i - r_{i-1} \geq \Omega_i \quad \forall i \in W \mid i > 2n + 1 \tag{3}$$

Constraints (1) and (2) are required to sum up the revenue along the route. In the following constraint (3) we ensure that a minimum revenue is yielded. If such constraints are used to ensure a minimum profit per carrier, the travel cost for reaching a delivery node have to be included in constraints (1) and (2). If carriers consider a solution acceptable, if a minimum or maximum number of customers is included in a tour, φ_i is set to 1 for all pickup nodes $i, i \in P$.

4   Solution Method

In order to generate solutions for the collaborative pickup an delivery problem, we develop an ALNS algorithm. ALNS is a renowned metaheuristic which goes back to [36]. Neighborhood search-based methods are well known to be powerful and fast tools for various classes of vehicle routing problems, e.g. [1,13,23,35].

ALNS algorithms explore the solution space by iteratively destroying and repairing solutions. We use three operators to remove nodes from a tour (destroy phase) and refer to these operators as set RO:

1. random removal,
2. worst removal randomized as suggested in [36],
3. related removal randomized [36].

For repairing solutions we apply the following operators (referred to as set IO):

1. Greedy Insertion, where requests (both pickup and delivery nodes) are iteratively inserted in the best possible position of the routes [21],
2. Greedy Insertion with noise (i.e. a certain flexibility, determined by a cost function, in selecting the best insertion of a request) according to [21],
3. Greedy Insertion where the probability of inserting a request in a tour increases with its current constraint violation,
4. 2-Regret Insertion 1: best position in best tour versus best position in second best tour,
5. 2-Regret Insertion 2: best position versus second best position over all tours,
6. 2-Regret Insertion 2, where the probability of inserting a request in a tour increases with its current constraint violation.

An initial feasible solution is constructed according to Algorithm 1, which uses the initial distribution ($S_{InitialDistribution}$) of requests to carriers as an input. All continuity measures relate to the initial distribution of customers, since the latter reflects the initial situation of each carrier. If the continuity constraint implies that each carrier has to yield a minimum profit, we cannot guarantee that the construction heuristic finds a feasible solution. In this case, we use the initial distribution as a starting solution. The initial distribution originates in the data instances. Each carrier has a set of customers spread around the depot, where the degree of customer area overlap influences the distance to the depot.

As it is shown in Algorithm 1, for generating initial solutions, we first assume that none of the requests, where R defines the set of all requests, is assigned to a carrier. These requests are denoted as *unfulfilled requests*. We iteratively assign them to tours using insert operator (5) from above. Obviously, if customer areas have a low degree of overlap, requests are with a high probability assigned to their original owners, since the insertion cost to the tours of other players are relatively high. If there exists tours, where continuity constraints are violated (i.e. constraint violation V_t of tour t is not zero), we try to repair these tours. This is done by identifying the tours facing the maximum violations (these tours are denoted as *eligbile tours* and are assigned to set T), as well as all unfulfilled requests that - being inserted in the tour - would neutralize the constraint violations. These are denoted as *eligible requests*. We again use insertion operator (5) to select and insert these requests.

Algorithm 1. Construction Heuristic

Input: $S_{InitialDistribution}$

    Initial solution $S = \{S_1, ..., S_m\}$

2: list of unfulfilled requests $U \leftarrow R$

    while $U \neq \varnothing$ **do**

4:    **if** $\nexists\, S_t$ that violates a given continuity constraint, $\forall t = \{1, ..., m\}$, **then**

        select and insert request $i \in U$ to S_t by applying insertion operator (5)

6:    **else**

        for all tours $t = \{1, .., m\}$ determine constraint violation V_t and highest constraint violation $V_{max} = \max\limits_{t=\{1,..,m\}} \{V_t\}$

8:        include all tours t with violation V_{max} in set of eligible tours T

        include all requests $i \in U$ that would reduce constraint violation in set of eligible requests ER_t, $\forall t \in T$,

10:        select and insert request $i \in ER_t$ to S_t by applying insertion operator (5) for all $t \in T$

        end if

12:    delete request i from U

    end while

14: **if** S is *feasible* **then**

    return S

16: **else**

    return $S_{InitialDistribution}$

18: **end if**

The ALNS procedure is given in Algorithm 2. Starting with initial solution S (see Algorithm 1), we iteratively try to improve the incumbent solution S_{best} by working on current solution S'. As suggested by [36], we select removal and insertion operators using weight-based probabilities (W^+ and W^-), which are updated after each iteration. For this we use the following three scenarios: (i) the operator used in this iteration found a new global best solution, (ii) the operator used in this iteration found a solution that was not accepted before and is better than the current solution, and (iii) the operator used in this iteration found a solution that that was not accepted before and is worse than the current solution but was accepted. For details on the update process we refer the reader to [36].

A linear threshold acceptance criterion is applied such that we accept solutions if they are less than Z percent ($Z = Z_{start}, ..., Z_{end}$) worse than the incumbent solution. During the search, solutions with violations of continuity constraints may be accepted with a certain probability but are penalized according to [24]. The probability of accepting an infeasible solution decreases with the number of iterations performed and $SBest$ is only updated when a better feasible solution is found. The algorithm stops after a given number of iterations.

Algorithm 2. Adaptive Large Neighborhood Search

Input: Initial feasible solution S
1: $S_{best} \leftarrow S$
2: **while** *stopping criterion* is not met **do**
3:      $S' \leftarrow S$
4:      select number of nodes $q \in \{q_{Min}, q_{Max}\}$ to be removed
5:      select removal operator $ro \in RO$ based on weights W^-
6:      select insertion operator $io \in IO$ based on weights W^+
7:      remove q requests by applying $ro(S')$
8:      reinsert removed requests by applying $io(S')$
9:      **if** S' is *accepted* **then**
10:         $S \leftarrow S'$
11:     **end if**
12:     **if** S' is *feasible* and $f(S') < f(S_{best})$ **then**
13:         $S_{best} \leftarrow S'$
14:     **end if**
15:     update W^- and W^+
16: **end while**
17: **return** S_{best}

---

5   Computational Study

For our computational study, we use self-created instances based on data proposed in [8]. For these, we generate equidistant carrier depots with a distance of 200. Requests are randomly generated within a radius of 150, 200, and 300 from their carrier's depot. This is in line with [8], where 3 different degrees of customer area overlaps are proposed, i.e. if the radius is large, customer areas are strongly overlapping. Each carrier initially holds either 10 or 15 pickup and delivery requests. We refer to these scenarios as GH_O1, GH_O2, and GH_O3, respectively. We generate 20 instances for each scenario. All instances are publicly available (http://prolog.univie.ac.at/research/FairInst/Fairness_instances. zip).

In our computational study, we set q_{Min}, q_{Max} to $0.05 * n$ and $0.35 * n$, respectively [3]. T_{start} is 0.1, while T_{end} is set to 0.05. We stop the search procedure after 100 iterations.

5.1   Total Collaboration Gain

Using the proposed ALNS approach, the potential total collaboration gain can be quantified. We do this by comparing the initial total profit (prior to collaboration) against the total profit after the reallocation of customers. The total profit is calculated as the total revenue minus the travel cost, which is based on distances. Computational results for instances GH_O1-GH_O3 are summarized in Table 1.

The results correspond to real-world case studies available in the literature, where collaboration profits of around 30% are reported [19]. However, we show

Table 1. Total percentage collaboration profit. We report the average gap between the initial and the collaborative solution.

Instances	30 customers	45 customers
GH_O1	11.55%	15.81%
GH_O2	27.55%	23.28%
GH_O3	42.94%	39.11%
Average	27.35%	26.07%

that depending on the degree of customer region overlaps, potential collaboration gains can even go beyond that: almost 40% for instances with 45 customers being distributed in strongly overlapping customer regions. At this point we want to refer the interested reader to [22], where methods for sharing these collaboration profits or costs are elaborated.

We observe that the potential profit increases with the number of customers but decreases with the degree of competition (i.e. overlap of regions). This is based on the fact that a carrier with many customers has a lot of flexibility to build profitable tours. New requests can easily be integrated. Thus, these carriers have less desire to collaborate with other players. Interestingly, this observation does not hold for instances GH_O1, where carriers have a low degree of competition, i.e. their customer areas do not have much overlap. This effect might be based on the fact, that a higher number of customers comes with an increased probability to have customers which are close to the depot of the competitor. This seems to have a strong influence, which results in the observation that (in contrast to instances GH_O2 and GH_O3) a higher number of customers leads to a higher collaboration gain.

While potential collaboration profits are huge, it is reported that centralized planned collaborations are rarely accepted in practice [19]. Carriers do not want to reveal sensitive information, and they are not willing to share the full set of their customers. In the following, we quantify the cost of continuity constraints. For this, we assume that carriers might consider solutions acceptable, where a certain continuity with respect to carriers' initial situations is kept.

5.2 The Cost of Continuity

Let us first assume that carriers want to keep some of their initial customers. This is a natural assumption, since carriers typically do not want to share all their customers with coalition partners [19]. In Table 2 we display the distribution of customers if no continuity constraints are considered.

The results reveal that customers are unevenly distributed. In particular in instances with high customer region overlap (GH_O3), one carrier serves almost all carriers. Obviously, in instances with low overlap (GH_O1) this effect is reduced. However, we see that even in these instances the uneven distribution is significantly increased if more customers are available. This observation

Table 2. Distribution of customers in ALNS solution without continuity constraints. We report the average maximum and minimum share of customers.

Instances	n = 30		n = 45		Average	
	Max	Min	Max	Min	Max	Min
GH_O1	56.67%	18.67%	74.33%	10.78%	65.50%	14.72%
GH_O2	95.00%	1.33%	94.11%	1.22%	94.56%	1.28%
GH_O3	100.00%	0.00%	97.56%	0.78%	98.78%	0.39%
Average	83.89%	6.67%	88.67%	4.26%	86.28%	5.46%

reinforces the assumption from above, that in case of low overlap, more customers have a positive influence on potential collaboration profits.

However, if carriers have strongly overlapping customer regions, one carrier serves almost all customers. This is of course not acceptable in real-world collaborations. We thus emphasize that continuity constraints are needed in order to generate appropriate solutions. In the following we quantify the loss in collaboration profit if such constraints are considered.

In Table 3 we give an overview on the observed decrease in collaboration profit depending on the number of customers kept.

Table 3. The cost of keeping $\frac{1}{3}$ or $\frac{2}{3}$ of the initial customers. In the first column (no), we report the total percentage collaboration profit without continuity constraints The two following columns show the percentage decrease in total collaboration profit if $\frac{1}{3}$ and $\frac{2}{3}$ of customers are kept, respectively. Results are given for 30 and 45 customers.

Instances	n = 30			n = 45			Average		
	No	$\frac{1}{3}$ kept	$\frac{2}{3}$ kept	No	$\frac{1}{3}$ kept	$\frac{2}{3}$ kept	No	$\frac{1}{3}$ kept	$\frac{2}{3}$ kept
GH_O1	11.55%	−3.27%	−6.75%	15.81%	−8.32%	−11.76%	13.68%	−5.80%	−9.26%
GH_O2	27.55%	−12.82%	−19.64%	23.28%	−13.75%	−20.00%	25.42%	−13.29%	−19.82%
GH_O3	42.94%	−26.72%	−36.22%	39.11%	−21.37%	−32.17%	41.03%	−24.04%	−34.19%
Average	27.35%	−14.27%	−20.87%	26.07%	−14.48%	−21.31%	26.71%	−14.38%	−21.09%

The results show that the desire to keep customers is particularly expensive if customer regions are strongly overlapping (GH_O3). On average, the total collaboration profit decreases by about 24%. This is not the case if customer regions are more isolated (GH_O1), where most of the customers are not attractive to coalition partners, anyway. In this case, keeping one third of the customers decreases total collaboration profit by less than 6%. An interesting observation is that keeping a second third of customer comes with relatively low additional cost increase (e.g. around 6% points in instance set GH_O2). This can be explained by the fact that the second set of customers can be selected such that it is close to the first customer. Thus, the additional travel cost for holding the second set of customers can be kept on a low level.

In Table 4 we compare the results assuming that carriers want to keep a given number of customers, irrespective whether these specific customers where in their initial tours.

Table 4. The cost of keeping $\frac{1}{3}$ or $\frac{2}{3}$ of the initial customers. In the first column (no), we report the total percentage collaboration profit without continuity constraints. The two following columns show the percentage decrease in total collaboration profit if $\frac{1}{3}$ and $\frac{2}{3}$ of customers are kept, respectively. Results are given for 30 and 45 customers.

Instances	n = 30			n = 45			Average		
	No	$\frac{1}{3}$ kept	$\frac{2}{3}$ kept	No	$\frac{1}{3}$ kept	$\frac{2}{3}$ kept	No	$\frac{1}{3}$ kept	$\frac{2}{3}$ kept
GH_O1	11.55%	−1.85%	−5.55%	15.81%	−6.15%	−9.16%	13.68%	−4.00%	−7.35%
GH_O2	27.55%	−10.97%	−17.78%	23.28%	−8.77%	−14.93%	25.42%	−9.87%	−16.36%
GH_O3	42.94%	−18.44%	−28.34%	39.11%	−16.37%	−24.39%	41.03%	−17.40%	−26.37%
Average	27.35%	−10.42%	−17.22%	26.07%	−10.43%	−16.16%	26.71%	−10.42%	−16.69%

We observe that, not surprisingly, keeping a *number* of customers is less costly than keeping specific customers. On average, if carriers want to have not less than a third of the number of customer they had before collaborating, decreases the average total percentage collaboration profit by around 10%. If they want to have two thirds, additional 6% points are lost.

In Table 5 we compare results where carriers want to yield a minimum profit.

Table 5. The cost of having 80% or 90% of the initial profit. In the first column (no), we report the total percentage collaboration profit without this constraint. The two following columns show the percentage decrease in total collaboration profit if 80% and 90% of the initial profit has to be ensured, respectively. Results are given for 30 and 45 customers.

Instances	n = 30			n = 45			Average		
	No	80%	90%	No	80%	90%	No	80%	90%
GH_O1	11.55%	−4.47%	−6.68%	15.81%	−5.61%	−10.55%	13.68%	−5.04%	−8.62%
GH_O2	27.55%	−18.34%	−22.67%	23.28%	−13.80%	−17.64%	25.42%	−16.07%	−20.16%
GH_O3	42.94%	−27.23%	−33.05%	39.11%	−22.68%	−27.71%	41.03%	−24.95%	−30.38%
Average	27.35%	−16.68%	−20.80%	26.07%	−14.03%	−18.63%	26.71%	−15.35%	−19.72%

The results show that also profit-based continuity constraints decrease average total percentage collaboration profit significantly. However, even if carriers want to have at least 90% of their initial profits, the potential total gain is remarkable. In the presence of effective and fair profit sharing methods [22], individual profits of carriers are of course additionally increased. Thus, it can be concluded that, even if carriers are not willing to share all their customers or insist on high profits (prior to profit sharing), huge collaboration profits can be achieved.

6   Conclusion

In this study we assessed the cost of continuity in collaborative pickup and delivery problems. For doing so, we developed an ALNS approach to solve the centralized decision problem introduced in [20]. To the best of our knowledge, no efficient solution method for this problem has been presented so far.

Using the proposed algorithm we were able to quantify the potential total profit in centrally planned collaborations. For the publicly available test instances, these potential collaboration profit goes beyond up to 40%, which is significantly higher than it is reported in literature. Another interesting observation was that for some sets of instances the collaboration profit increases with the number of customers, which is in a sense non-intuitive.

However, we show that these solutions lead to very unevenly distributed workload among the participating carriers. This might explain why carriers, even if collaboration profits are known to be very high, are cautious to join centrally planned coalitions. Another reason is that they do not want to offer all their customers for possible exchanges with coalition partners. In our computational study we quantified the impact of letting carriers keep initial customers or profits. We observed that the cost of such constraints depends on the degree of customer region overlap and the number of customers. We show that even if carriers insist on relatively high numbers of customers or individual profits, the total collaboration profits are remarkable.

Our study emphasized the strength of collaborative vehicle routing. It hopefully stimulates to do more research in this relevant and interesting topic. The proposed solution method can be used to conduct further experiments and by this gain more interesting insights into the field of centralized collaborative planning.

References

1. Amous, M., Toumi, S., Jarboui, B., Eddaly, M.: A variable neighborhood search algorithm for the capacitated vehicle routing problem. Electron. Notes Discret. Math. **58**, 231–238 (2017)
2. Archetti, C., Speranza, M., Vigo, D.: Vehicle routing problems with profits. In: Toth, P., Vigo, D. (eds.) Vehicle Routing: Problems, Methods, and Applications, pp. 273–297. MOS-SIAM Series on Optimization (2014)
3. Azi, N., Gendreau, M., Potvin, J.Y.: An adaptive large neighborhood search for a vehicle routing problem with multiple routes. Comput. Oper. Res. **41**, 167–173 (2014)
4. Ballot, E., Fontane, F.: Reducing transportation CO_2 emissions through pooling of supply networks: perspectives from a case study in french retail chains. Prod. Plan. Control **21**(6), 640–650 (2010)
5. Bektaş, T., Demir, E., Laporte, G.: Green vehicle routing. In: Psaraftis, H. (ed.) Green Transportation Logistics. ISOR, pp. 243–265. Springer, Cham (2016). https://doi.org/10.1007/978-3-319-17175-3_7
6. Berbeglia, G., Cordeau, J.F., Gribkovskaia, I., Laporte, G.: Static pickup and delivery problems: a classification scheme and survey. TOP **15**(1), 1–31 (2007)

7. Berbeglia, G., Cordeau, J.F., Laporte, G.: Dynamic pickup and delivery problems. Eur. J. Oper. Res. **202**(1), 8–15 (2010)
8. Berger, S., Bierwirth, C.: Solutions to the request reassignment problem in collaborative carrier networks. Transp. Res. Part E: Logist. Transp. Rev. **46**, 627–638 (2010)
9. Buijs, P., Alvarez, J.A.L., Veenstra, M., Roodbergen, K.J.: Improved collaborative transport planning at dutch logistics service provider fritom. Interfaces **46**(2), 119–132 (2016)
10. Cleophas, C., Cottrill, C., Ehmke, J.F., Tierney, K.: Collaborative urban transportation: recent advances in theory and practice. Eur. J. Oper. Res. (2018). https://doi.org/10.1016/j.ejor.2018.04.037
11. Cruijssen, F., Bräysy, O., Dullaert, W., Fleuren, H., Salomon, M.: Joint route planning under varying market conditions. Int. J. Phys. Distrib. Logist. Manag. **37**(4), 287–304 (2007)
12. Dai, B., Chen, H.: Mathematical model and solution approach for carriers collaborative transportation planning in less than truckload transportation. Int. J. Adv. Oper. Manag. **4**, 62–84 (2012)
13. Defryn, C., Sörensen, K.: A fast two-level variable neighborhood search for the clustered vehicle routing problem. Comput. Oper. Res. **83**, 78–94 (2017)
14. Dragomir, A.G., Nicola, D., Soriano, A., Gansterer, M.: Multidepot pickup and delivery problems in multiple regions: a typology and integrated model. Int. Trans. Oper. Res. **25**(2), 569–597 (2018)
15. Fernandez, E., Roca-Riu, M., Speranza, M.G.: The shared customer collaboration vehicle routing problem. Eur. J. Oper. Res. **265**(3), 1078–1093 (2018). https://doi.org/10.1016/j.ejor.2017.08.051
16. Gansterer, M., Hartl, R., Vetschera, R.: The cost of incentive compatibility in auction-based mechanisms for carrier collaboration. Networks (2018, forthcoming)
17. Gansterer, M., Hartl, R.F.: Request evaluation strategies for carriers in auction-based collaborations. OR Spectr. **38**(1), 3–23 (2016)
18. Gansterer, M., Hartl, R.F.: Bundle generation in combinatorial transportation auctions. Working Paper (2017)
19. Gansterer, M., Hartl, R.F.: Collaborative vehicle routing: a survey. Eur. J. Oper. Res. **268**(1), 1–12 (2018)
20. Gansterer, M., Hartl, R.F., Salzmann, P.E.H.: Exact solutions for the collaborative pickup and delivery problem. Central Eur. J. Oper. Res. **26**(2), 357–371 (2018)
21. Ghilas, V., Demir, E., Woensel, T.V.: An adaptive large neighborhood search heuristic for the pickup and delivery problem with time windows and scheduled lines. Comput. Oper. Res. **72**, 12–30 (2016)
22. Guajardo, M., Rönnqvist, M.: A review on cost allocation methods in collaborative transportation. Int. Trans. Oper. Res. **23**(3), 371–392 (2016). https://doi.org/10.1111/itor.12205
23. Haddad, M.N., et al.: Large neighborhood-based metaheuristic and branch-and-price for the pickup and delivery problem with split loads. Eur. J. Oper. Res. (2018). https://doi.org/10.1016/j.ejor.2018.04.017
24. Kovacs, A., Parragh, S., Hartl, R.: A template-based adaptive large neighborhood search for the consistent vehicle routing problem. Networks **63**(1), 60–81 (2014)
25. Krajewska, M., Kopfer, H.: Collaborating freight forwarding enterprises. OR Spectr. **28**(3), 301–317 (2006)
26. Lin, C., Choy, K., Ho, G., Chung, S., Lam, H.: Survey of green vehicle routing problem: past and future trends. Expert Syst. Appl. **41**(4, Part 1), 1118–1138 (2014)

27. Lin, C.: A cooperative strategy for a vehicle routing problem with pickup and delivery time windows. Comput. Ind. Eng. **55**(4), 766–782 (2008)

28. Lu, Q., Dessouky, M.: An exact algorithm for the multiple vehicle pickup and delivery problem. Transp. Sci. **38**(4), 503–514 (2004)

29. Matl, P., Hartl, R.F., Vidal, T.: Workload equity in vehicle routing problems: a survey and analysis. Transp. Sci. **52**(2), 239–260 (2018). https://doi.org/10.1287/trsc.2017.0744

30. Molenbruch, Y., Braekers, K., Caris, A.: Benefits of horizontal cooperation in dial-a-ride services. Transp. Res. Part E: Logist. Transp. Rev. **107**, 97–119 (2017)

31. Montoya-Torres, J.R., Muñoz-Villamizar, A., Vega-Mejia, C.A.: On the impact of collaborative strategies for goods delivery in city logistics. Prod. Plan. Control **27**(6), 443–455 (2016)

32. Nadarajah, S., Bookbinder, J.: Less-than-truckload carrier collaboration problem: Modeling framework and solution approach. J. Heuristics **19**, 917–942 (2013)

33. Parragh, S., Dörner, K., Hartl, R.: A survey on pickup and delivery problems. Part II: transportation between pickup and delivery locations. J. für Betriebswirtschaft **58**, 21–51 (2008)

34. Pérez-Bernabeu, E., Juan, A.A., Faulin, J., Barrios, B.B.: Horizontal cooperation in road transportation: a case illustrating savings in distances and greenhouse gas emissions. Int. Trans. Oper. Res. **22**(3), 585–606 (2015)

35. Polacek, M., Benkner, S., Doerner, K.F., Hartl, R.F.: A cooperative and adaptive variable neighborhood search for the multi depot vehicle routing problem with time windows. Bus. Res. **1**(2), 207–218 (2008)

36. Ropke, S., Pisinger, D.: An adaptive large neighborhood search heuristic for the pickup and delivery problem with time windows. Transp. Sci. **40**(4), 455–472 (2006)

37. Sanchez, M., Pradenas, L., Deschamps, J.C., Parada, V.: Reducing the carbon footprint in a vehicle routing problem by pooling resources from different companies. NETNOMICS: Econ. Res. Electron. Netw. **17**(1), 29–45 (2016)

38. Speranza, M.G.: Trends in transportation and logistics. Eur. J. Oper. Res. **264**(3), 830–836 (2018)

39. Vanovermeire, C., Sörensen, K., Breedam, A.V., Vannieuwenhuyse, B., Verstrepen, S.: Horizontal logistics collaboration: decreasing costs through flexibility and an adequate cost allocation strategy. Int. J. Logist. Res. Appl. **17**(4), 339–355 (2014)

40. Verdonck, L., Caris, A., Ramaekers, K., Janssens, G.K.: Collaborative logistics from the perspective of road transportation companies. Transp. Rev. **33**(6), 700–719 (2013)

41. Wang, X., Kopfer, H., Gendreau, M.: Operational transportation planning of freight forwarding companies in horizontal coalitions. Eur. J. Oper. Res. **237**(3), 1133–1141 (2014)

A Matheuristic Approach to the Pickup and Delivery Problem with Time Windows

Carlo S. Sartori[✉] and Luciana S. Buriol

Instituto de Informática,
Universidade Federal do Rio Grande do Sul, Porto Alegre, Brazil
{cssartori,buriol}@inf.ufrgs.br

Abstract. In this work, the Pickup and Delivery Problem with Time Windows is studied. It is a combinatorial optimization problem, in which the objective is to construct the best set of vehicle routes while respecting side constraints, such as precedence between locations to be visited, and the time to service them. To tackle this problem, a matheuristic based on Iterated Local Search method is proposed, with an embedded Set Partitioning Problem that is iteratively solved to recombine routes of previously found solutions. Results indicate the approach works well for a standard benchmark set of instances from the literature. A number of new best-known solutions has been found.

Keywords: Matheuristic · Pickup and delivey problem
Time windows · Iterated local search

1 Introduction

The *Vehicle Routing Problem* (VRP) is a well-known combinatorial optimization problem used to model scenarios on transportation and logistics, with significant economic importance [10]. Its objective is to build the best set of vehicle routes to service a set of requests, such that all side constraints are satisfied. Given the number of real-world applications, several models and variations have been proposed to best describe each particular scenario.

One such variation is the *Pickup and Delivery Problem* [20], in which requests are pairs of pickup and delivery locations. Goods have to be transported from the pickup location to the corresponding delivery location, by the same vehicle route. Constraints include the pickup to be serviced before its delivery, and the maximum capacity of vehicles, which should not be exceeded.

When locations have to be serviced within a given period, the problem is known as *Pickup and Delivery Problem with Time Windows* (PDPTW) [8]. In this case, vehicles cannot start service before the beginning, or after the end of the time window of a given location.

In this work, the PDPTW is considered. It is a $\mathcal{NP}$-Hard problem, which generalizes both the *Classical Vehicle Routing Problem* and the *Vehicle Routing Problem with Time Windows*. Additionally, it can be applied to various real-world

© Springer Nature Switzerland AG 2018
R. Cerulli et al. (Eds.): ICCL 2018, LNCS 11184, pp. 253–267, 2018.
https://doi.org/10.1007/978-3-030-00898-7_16

scenarios, such as product delivery, bulk product transportation, dial-a-ride problems, courier services, airline scheduling, bus routing, and logistics and maintenance support [15]. Thus, it is reasonable to consider that efficient methods to solve the PDPTW can be used for related problems in transportation.

This work proposes the study of a hybrid method, using metaheuristic and exact solutions of a *Set Partitioning Problem* (SPP) to solve the PDPTW. Such combination is referred in the literature as *matheuristic* [5]. The metaheuristic components are based on the work of [7,19], and include the *Adaptive Guided Ejection Search* and the *Large Neighborhood Search*. They are embedded into an *Iterated Local Search* framework [13] with a SPP solver. Moreover, a tuning procedure is performed to decide the best parameter values and components to be used. Experimental results show this approach performs well when compared to a *state-of-the-art* method.

The remainder of the article goes as follows. In Sect. 2 the PDPTW is formally defined, and basic notations are introduced. A literature review is presented in Sect. 3. Our proposed method is detailed in Sect. 4, and experiments and results are discussed in Sect. 5. The work is concluded in Sect. 6.

2   Problem Definition

An instance of the PDPTW is defined on a graph $G = (V, E)$, where V is the set of $n = |V|$ vertices and E the set of edges. Let $P \subset V$ be the set of pickup locations, and set $D \subset V$ be the delivery locations, $P \cap D = \emptyset$, and the set of requests is $R = P \cup D$. There is a single depot denoted by 0, and $V = \{0\} \cup R$. A request is a pair (p, d), where each $p \in P$ has only one paired $d \in D$. Edges (i, j) have an associated cost c_{ij}, and time t_{ij} to travel between locations i and j.

Every location $i \in V$ has a time window of the form $[e_i, l_i]$, where e_i is the earliest time service can start at location i, and l_i is the latest time. There is also an associated waiting time w_i, or how long the vehicle takes to complete the service at that location. By definition, the time window of the depot has values $e_0 = 0$, and l_0 as the maximum time a route can take.

A homogeneous fleet of vehicles is available at the depot, from where they start and end their routes. Every vehicle has an associated maximum capacity Q, which is the maximum amount of load it can carry at once. Each location has a demand, or the total amount of goods to be picked up or delivered. Given any pickup and delivery pair (p, d), their demands q are such that $q_d = -q_p$, and pickups are non-negative, $q_p \geq 0$. In other words, everything that has been collected has to be dropped at the corresponding location.

A solution to the PDPTW is a set of routes $s = \{r_1, \ldots, r_m\}$. Each route is a sequence of locations to be visited, starting and ending at the depot, denoted by $r = \{v_0, v_1, \ldots, v_h, v_{h+1}\}$, where $v_k \in R, k = 1, \ldots, h$, $v_0 = v_{h+1} = 0$, and each location is visited at most once. A solution is feasible if all requests are serviced exactly once, and all routes respect the side constraints.

For any request (p, d), if the pickup location p belongs to the route, then the delivery location d must also belong to the same route. Additionally, the pickup must always precede the delivery in the path. These are known as pairing and precedence constraints, respectively.

Accumulated load carried by a vehicle up to the k-th location is given by $a_{v_k} = \sum_{i=0}^{k} q_{v_i}$. Capacity constraints state that $a_{v_k} \leq Q, \forall k = 1, \ldots, h$. A vehicle leaves and returns to the depot empty, so $a_{v_0} = a_{v_{h+1}} = 0$.

Time window constraints are defined in terms of b_{v_k}, the time a vehicle starts service at location v_k. The start of service at the next location in route is given by $b_{v_{k+1}} = \max(e_{v_{k+1}}, b_{v_k} + w_{v_k} + t_{v_k, v_{k+1}})$. In other words, vehicles may arrive before the beginning of a time window, but must wait until service can start. It is required that $b_{v_k} \leq l_{v_k}, \forall k = 0, \ldots, h + 1$. By definition, $b_{v_0} = 0$, that is, routes start as soon as possible.

The objective function to be minimized is evaluated according to the lexicographic order of its terms and is given by:

$$f(s) = (\ |s|, \sum_{r \in s} C(r)\ )$$

The first term minimizes the number of routes or vehicles used. Ties are broken in the evaluation by the second term, which is the accumulated cost of all routes in solution s. The cost of a route r is the sum of all edges from the input graph that belong to r and is denoted by $C(r)$.

3 Related Works

A survey of both exact and heuristic solution methods for pickup and delivery problems can be found in the work of Parragh et al. [16]. The authors explain how pickup and delivery problems can be studied as two different groups. One is the non-paired variant, where vehicles must deliver products from the depot to some customers, and collect products from customers to the depot. Another variant is the paired, where products must be transported between locations, which is precisely how the PDPTW is defined.

Due to its difficulty, the most common approach applied to solve the problem is metaheuristic. Nanry and Barnes [15] proposed the first metaheuristic to address the multiple vehicle version of the PDPTW. Their method was a Reactive Tabu Search, and defined three standard local search neighborhood movements: shift a request between routes, swap two requests between routes, and relocate a request within its route.

Li and Lim [11] proposed the current standard benchmark set of instances for the problem. The authors applied a Tabu-Embedded Simulated Annealing procedure to solve the problem, with the same neighborhoods of [15].

However, a method that has been successfully applied to the PDPTW is *Large Neighborhood Search* (LNS). Bent and Van Hentenryck [4] proposed a two-stage algorithm to solve the problem, where the first stage used *Simulated Annealing* to reduce the number of vehicles, and the second stage performed LNS to reduce the cost. Likewise, Ropke and Pisinger [19] applied a two-stage *Adaptive Large Neighborhood Search*, where both phases used the same LNS algorithm with different weights on the objective function. The latter improved almost half of the best-known solutions of the standard benchmark instances [11].

A particular case of the PDPTW, where only vehicle minimization was taken into account, was studied by Nagata and Kobayashi [14]. The method was called *Guided Ejection Search* (GES), and is based on the idea of randomly removing a route of a solution, and trying to reinsert all its requests into the remaining routes, possibly ejecting requests to make space for them. Their results showed the method was efficient in reducing the number of routes, obtaining new results for the standard benchmark set.

Curtois et al. [7] proposed a hybridization of LNS and a modification of the GES heuristic, called *Adaptive Guided Ejection Search* (AGES). The method was able to improve many instances of the standard benchmark set, and can be considered the current *state-of-the-art* algorithm to solve the PDPTW. This algorithm is the base for the method presented in this paper.

Among exact approaches, branch-and-price, and branch-and-cut are the main used algorithms. Applications of such methods include [3,8,18]. The algorithm of Baldacci et al. [3] was able to solve to optimality all instances from the set previously proposed by [18]. Although, none of the exact methods is able to cope with medium and large sized instances, such as the ones from [11] benchmark set, corroborating the use of heuristics.

Regarding matheuristics, it has been shown that they can be an effective method to solve combinatorial optimization problems [5]. Applications to VRPs include the variations with time windows [1], cross-docking [9], and to solve a number of routing problems [24]. Additionally, a similar method to the one presented in this work has been proposed to the dial-a-ride problem [17], although the heuristic components are not the same, and the objective of the problems differs slightly. A survey on matheuristics applied to VRPs can be found in [2].

All cited matheuristic methods are SPP based, which works as a column generation, where each column is a valid VRP route. Routes are generated by running a metaheuristic and storing the routes of new local optimum solutions on a pool to be used by the SPP model. Even though solutions for many of the referred works are of high quality, there is no guarantee of optimality.

4   Proposed Algorithm

Our algorithm is based on that of Curtois et al. [7]. It combines AGES and LNS to generate solutions as proposed in [7] and extends the approach to iteratively recombine routes of previous local minima through a SPP model.

Those components are embedded into an ILS framework, to allow perturbation of solutions, and acceptance of different local minima to continue the search. It is henceforth referred as IGLS and described in Algorithm 1. The current solution is denoted by s, and the best found by s^*.

In line 1 an initial solution is generated by a greedy constructive algorithm. A pool $\mathcal{P}$ of routes is initialized in line 2 and is used to create the associated SPP model. Line 3 initializes the perturbation size z_p, which changes dynamically during the execution. Then, the main loop (lines 4–19) is repeated until a given stopping condition is reached. The best solution s^* is returned in line 21.

Algorithm 1 IGLS: ILS+GES+LNS+SPP

Input: Instance with graph $G = (V, E)$; set $R \subset V$ is the set of requests
Parameters: α, A, M_g, Z_g, p_{shift}, K, M_l, L, p_{shaw}
1: $s, s^* \leftarrow$ initial_solution()
2: $\mathcal{P} \leftarrow$ initialize_pool($s, \mathcal{P}$)
3: $z_p \leftarrow \lceil \alpha \cdot |R| \rceil$
4: **repeat**
5:　　$s \leftarrow$ AGES(s, M_g, Z_g, p_{shift})
6:　　$s \leftarrow$ LNS(s, K, M_l, L, p_{shaw})
7:　　$\mathcal{P} \leftarrow$ update_pool($s, \mathcal{P}$)
8:　　$s^p \leftarrow$ solve_SPP($\mathcal{P}$)
9:　　**if** $f(s^p) < f(s^*)$ **then**
10:　　　　$s^* \leftarrow s^p$
11:　　　　$z_p \leftarrow \lceil \alpha \cdot |R| \rceil$
12:　　　　$count \leftarrow 0$
13:　　**else**
14:　　　　$s \leftarrow$ accept_solution($s, s^*, count, iter$)
15:　　　　$s \leftarrow$ perturb(s, z_p, p_{shift})
16:　　　　$z_p \leftarrow \min(count \cdot \lceil \alpha \cdot |R| \rceil, \lceil A \cdot |R| \rceil)$
17:　　　　$count \leftarrow count + 1$
18:　　**end if**
19:　　$iter \leftarrow iter + 1$
20: **until** stopping condition
21: **return** s^*

At each iteration *iter* of the loop, the AGES is executed to reduce the number of routes in solution s. Next, the LNS heuristic is used to reduce the cost of the solution s. Lines 5–6 update the pool of routes, and call the SPP model to solve the problem over the new pool $\mathcal{P}$, returning the best combination of routes s^p. If the objective function $f(s^p)$ is *strictly* better than $f(s^*)$, solution s^p is accepted in line 10, else a perturbation is performed. In line 14, Solution s^* is chosen to be perturbed with probability $(iter - count)/iter$, otherwise solution s is chosen. The intention is to intensify the search when few iterations without improvement have passed but to diversify the search when too many iterations have been executed without any update of the solution s^*.

When a new best solution is found, the counter *count* of iterations without improvement, and the size of the perturbation z_p, are reset (lines 11–12). If s^p does not improve s^*, *count* is incremented, and z_p is increased to $count \cdot \lceil \alpha \cdot |R| \rceil$, up to a maximum value $\lceil A \cdot |R| \rceil$ (lines 16–17). That is, larger perturbations are performed whenever IGLS seems trapped in a local minimum.

While the two-stage algorithms of [4,19] first execute one vehicle minimization phase, and then a cost reduction phase, the proposed method, IGLS, and that of [7], perform both stages at every iteration. This procedure leads to a more intensive search, especially regarding vehicle minimization, since after reducing the total cost there may be more opportunities to reduce the number of routes. Results obtained by the two latter methods corroborate this hypothesis.

4.1 Greedy Constructive Algorithm

In order to create an initial solution, a simple greedy constructive algorithm based on [23] is used in line 1 of IGLS. The procedure creates one route at a time, and at each iteration, it tries to insert all unrouted requests into the current route, into all possible positions within it. The request and position, which are both feasible and minimize the total cost are chosen. If no request can be inserted in the current route, a new one is created. This continues until there are no more unrouted requests, or a request cannot be inserted anywhere.

This constructive algorithm does not take into account a maximum number of available vehicles, since the next steps of the algorithm may significantly reduce the number of routes. If a feasible solution cannot be built with this procedure, even when every request is serviced by a separated route, the instance is not feasible due to capacity or time window constraints.

4.2 Adaptive Guided Ejection Search (AGES)

The AGES heuristic [7], in line 5 of IGLS, is the component most responsible for vehicle minimization. At first, a random route r is removed from the current solution s. All requests that belong to r are inserted into a stack, for Last-In-First-Out (LIFO) ordering. This generates a partial solution s'.

Next, a request u is removed from the stack and inserted in a random position in s', from every possible insertion of u in the partial solution. If request u was successfully reinserted in s', then the algorithm moves to the next request in the stack. Otherwise, if there is no possible insertion position for u in s', the procedure tries to open space for u in s' by ejecting some requests.

Requests are selected to be ejected from s' based on a simple heuristic. For every request $x \in R$, a penalty counter is associated, denoted ρ_x. Whenever a request u is not able to be reinserted into the partial solution, requiring others to be ejected, ρ_u is increased. This means requests that are hard to be reinserted will probably have a high penalty counter. Thus, it seems reasonable to assume that such requests should remain where they currently are.

Then, the ejection heuristic chooses a number k of requests to remove, minimizing the sum of their penalties. In other words, it chooses k requests that are more likely to be reinserted later. Due to the large number of possibilities to choose when k grows, the heuristic only tries for $k = 1$, and if no space could be opened for request u, for $k = 2$. If it still cannot reinsert u, the current AGES iteration is aborted. The k removed requests are inserted in the stack.

After an ejection procedure and the following reinsertion of u, the resulting solution is perturbed using the procedure detailed in Sect. 4.4. The number of AGES perturbation movements is fixed, and denoted by Z_g. A counter $count_g$ is increased for every perturbation movement, and reset whenever a new partial solution with a smaller number of unassigned requests is found.

If at a given iteration the stack is empty, a route has been successfully removed from the original solution s. Then, s' is a full solution and $f(s') < f(s)$. In such case, the heuristic tries to remove another route. The procedure terminates when $count_g > M_g$ perturbation movements were executed without finding s' with less unrouted requests. This stopping condition tries to keep AGES running for as long as it appears to make progress.

4.3 Large Neighborhood Search (LNS)

Large Neighborhood Search has been shown to perform well on VRPs in general. It differs from the standard local search because it changes large portions of the solution, instead of a few components. In the PDPTW, for example, it has been used to remove a possibly large number of requests and reinsert them [19], while local search moves only moved one or two requests at a time [11,15].

The LNS used in line 6 of IGLS is the same of [7]. It has two removal heuristics: random removal [19], and shaw removal [21]. In the first, requests are randomly removed from solution s. In the second, requests are removed according to a relatedness measure, which takes into account the distance, demand and current service time of the requests. This relatedness between two requests is estimated with the same function proposed in [19], including the weights of each term. Requests considered related to already removed requests are more likely to be chosen. Though, the first request is selected at random.

In every iteration one of the two heuristics is selected, the shaw with probability p_{shaw}. The number of removed requests is chosen randomly in the interval defined by $[K, |R|/2]$, where K is a parameter. It is reasonable to expect that the more intelligent shaw removal has a higher probability of being chosen.

In order to reinsert the requests removed, the regret heuristic is used [19]. This heuristic uses a look-ahead and estimates how much would be lost if a request was not inserted in its best route, by considering its k best routes for insertion in the current solution. This way, requests that can only be inserted in a few routes will be inserted first, minimizing regrets on the decision. The look-ahead size k is chosen randomly from $\{1, 2, 3, 4, 5, |s|\}$.

A partial solution s' is generated from the original solution s when requests are removed. If some requests cannot be reinserted into s', the procedure is aborted and it continues with s, whereas if all requests were reinserted generating new solution s'', two options arise. When $f(s'') < f(s)$, s'' is accepted and the search continues from it, while if $f(s'') \geq f(s)$, an acceptance criterion based on *Late Acceptance Hill Climbing* (LAHC) [6] is applied. A list of size L is kept with values of visited solutions, and when to verify if a given solution s'' should be accepted, the algorithm checks previously stored values to decide. With LAHC, LNS may accept non-improving solutions to continue the search, allowing it to move out from certain basin of attractions more easily.

The LNS stopping conditions differ from the ones used in [7]. It ends after a number M_l of iterations without improvement, or after $10,000$ iterations overall, whichever happens first. A strict maximum number of iterations has been imposed because, in larger instances, the method spends too much time in the LNS when it would be preferable to continue and further reduce the number of vehicles. Thus, it avoids long unnecessary computation times, and as the results seem to indicate it has no negative influence on the final solution quality.

4.4 Perturbation

Solutions are perturbed by two neighborhood movements: random shift and random swap. A number of maximum moves is given as a parameter for the perturbation. Just as in [14], if a given movement cannot be applied, because would lead to an infeasible solution, it is aborted, and the next one is to be tried. Each movement selects one of two neighborhoods with a given probability. The probability of selecting random shift is given by p_{shift}.

In the random shift, two routes r_1 and r_2 are randomly picked from s. A random request u is removed from r_1 and inserted at a random feasible position in route r_2. While in the random swap, two routes r_1 and r_2 are randomly picked from s. A random request u_1 is removed from r_1, and another random request u_2 is removed from r_2. Then, u_1 is inserted in its best position in route r_2, and the same follows for u_2 in route r_1.

This procedure is used for both AGES and as the perturbation of the overall ILS in line 15 of IGLS. Although, perturbation sizes are different.

4.5 Set Partitioning Problem (SPP)

Vehicle Routing Problems can be formulated as a SPP in Integer Linear Programming form, as follows. Let $\mathcal{R}$ be the set of all feasible routes in a given VRP problem. Denote by a binary value $\lambda_{ir}, i \in V, r \in \mathcal{R}$ whether node i belongs to route r. Binary variable y_r assumes 1 when route r is used in solution, and zero otherwise. Thus,

$$\min \sum_{r \in \mathcal{R}} C(r) y_r$$
$$\text{s.t.} \sum_{r \in \mathcal{R}} \lambda_{ir} y_r = 1, \, i = 1, ..., n - 1$$
$$y_r \in \{0, 1\}, \quad \forall r \in \mathcal{R}$$

The objective function is to minimize the summed costs of all selected routes. The only constraint is that each request should be visited only once. Given that, the PDPTW can be formulated as a SPP as long as the routes in $\mathcal{R}$ respect all constraints defined in Sect. 2. In order to minimize the number of routes, a large constant value is assigned to every route, as the cost to schedule a vehicle.

However, the size of set $\mathcal{R}$ is exponential on the instance size, and so impractical to use. A common procedure is to somehow generate only routes that are of interest to find reasonable solutions [2]. In the case of matheuristics, this is done employing a (meta)heuristic. Although, there is no guarantee of optimality.

In this work, we denote the set $\mathcal{R}$ as $\mathcal{P}$, also called *pool of routes*. It contains routes that belong to local minima found during the algorithm search. At each iteration of IGLS, routes of the new local minimum solution are added to $\mathcal{P}$ in line 7, and the model is solved in line 8.

Computationally, the set $\mathcal{P}$ is a map, and the key is a set of requests attended by the mapped route. If two routes r_1 and r_2 have the same set of serviced requests, regardless of their visit order, and $C(r_1) < C(r_2)$, the pool will only store route r_1, since it serves the same requests with lower cost. This way the number of stored routes is reduced, and so is memory consumption. In fact, the pool of routes remained small enough for the SPP to be solved up to optimality at every iteration with little impact on computation times.

5   Computational Experiments

This section presents and discusses the computational results of our proposed algorithm. A replication of the method of [7] is also analyzed. All experiments have been done using the standard set of benchmark instances of [11], which is available online and maintained by SINTEF [22].

There are 354 instances separated into six groups according to the number of locations in the input graph. These sizes are: 100, 200, 400, 600, 800, and 1000. Each size is divided into groups depending on how locations are distributed, they can be: clustered (LC), random (LR), or a combination of both (LRC). Instances are also grouped based on the size of the planning horizon into two types: (1) with a shorter horizon; and (2) with a longer horizon.

For each instance, we have run the algorithms multiple times, due to their stochastic components. Instances of sizes 100, 200, 400, and 600 were executed ten times each, with maximum running time per execution of 300, 900, 900, and 1800 seconds, respectively. Larger instances of sizes 800 and 1000 were executed only five times, with 3600 seconds each. These running times were used by [7]. Separate runs used different random seeds.

All implementations were done in C++, and compiled using g++ with -O3 optimization flag. The associated SPP is solved using CPLEX 12.6.2 API. A computer equipped with an Intel i7 930 @ 2.8 GHz processor, 12 GB of RAM, and Ubuntu 16.04 LTS operating system was used to perform the experiments. The algorithms and CPLEX were run in single thread mode.

Due to the limited space, we were not able to provide full results for each instance. Though, complete tables are available upon request.

5.1   Reimplementation of State-of-the-Art

A replication of the algorithm proposed in [7] was needed to better support the findings in our work because the IGLS is based on the main components of the former, i.e., AGES and LNS. Additionally, results reported by [7] contained information only for a single execution of the algorithm. As the method comprises many stochastic components, results from a single execution can be misleading for certain conclusions, and hard to compare with, since there is usually a deviation in results between two separate runs.

Henceforth, the method of [7] is denoted by the acronym CLSQL[1]. Its approach has been implemented following the original article, and the same set of parameter values has been used. We denote our reimplementation as R-CLSQL.

Table 1 presents the results from our replication. It is usual in the literature of the PDPTW to present results as the accumulated values of all instances for

---

[1] Initials from the last name of the authors.

Table 1. Results of the replication. CLSQL is one run per instance, and R-CLSQL is the average of 10 or 5 runs as previously informed.

Inst. Size	CLSQL			R-CLSQL			
	#V	Cost	t(s)[a]	#V	σ_v	Cost	t(s)
100	402	58,163.27	300	*402.0*	± 0.00	*58,089.89*	310
200	601	186,158.61	900	601.3	± 1.12	192,649.45	929
400	1142	447,627.43	900	1152.4	± 8.59	470,814.57	959
600	1643	935,948.36	1800	1653.5	±11.14	975,073.23	1909
800	2146	1,551,495.36	3600	2146.2	±12.73	1,617,631.40	3783
1000	2634	2,310,830.27	3600	*2629.6*	±14.98	*2,374,610.91*	3883

[a]: results reported in [7] using an Intel Xeon E5-1620 @ 3.5 GHz

each size to summarize the final results. We present in the same way. Column *Inst. Size* contains the six sizes from the instances of [11]. For each size and algorithm, column #V presents the accumulated number of vehicles, σ_v the standard deviation of the number of vehicles between runs, and *Cost* the cost of the solutions. Computational times used are given in column *t(s)*, in seconds.

Results show that, on average, the solutions of CLSQL and our replication are similar. In fact, for sizes 100 and 1000 (in italic), our average results are better than the original work. Remaining sizes are worse, but can still be considered equivalent due to the standard deviation on the number of vehicles. For example, size 100 has zero σ_v, meaning the method always reached the best number of vehicles, however for size 600 the deviation is about 11 vehicles, which accounts for the 10 vehicles of difference reached by our replication compared to CLSQL. Moreover, the best results reported in Table 4 show that R-CLSQL reached solutions equivalent to the ones reported by the original work at least once.

We intended to verify if R-CLSQL achieved similar results to the original work. Given the reported results and deviations, we consider the replication successful and continue to use R-CLSQL in comparisons with IGLS.

5.2   Component Selection and Parameter Tuning of IGLS

In order to better decide the components to be used by the IGLS, as well as its parameter values, a tuning procedure has been performed using the tool irace [12]. It relies on statistical tests to choose values for each parameter, obtaining a configuration with good results, on average.

One of the main questions to arise when proposing a matheuristic method such as the IGLS is whether the mathematical programming component is statistically significant. Even though it is possible to consider simple tests comparing a version with, and another without the SPP module, it could be the case that such component is only actually useful when combined with other values of the parameters. Thus, the tuning procedure is used to both verify if the SPP is significant, and with which configuration it seems to work well.

Table 2. Tuned parameters and their respective values

	Parameter		Values	
Notation	Description		Range	Best
spp	use SPP model		{true,false}	true
ls	use Local Search		{true,false}	false
gen-sol	how to generate initial solution		{LS, greedy}	greedy
α	initial ILS perturbation multiplier		[0.0,5.0]	2.47
A	maximum ILS perturbation multiplier		[0.0,5.0]	4.13
p_{shift}	probability of performing shift in perturbation		[0.0,1.0]	0.58
M_g	maximum number of AGES iterations		$\{10^4, 10^5, 10^6\}$	10^6
Z_g	perturbation size for AGES		{10,100,1000}	100
K	minimum number of requests to remove in LNS		[1,4]	3
M_l	number of iterations with no improvement in LNS		[600,1500]	928
L	LAHC list size		[500,2500]	1309
p_{shaw}	probability of performing shaw removal in LNS		[0.0,1.0]	0.71

Table 2 presents the parameters tuned with irace. The first three rows are components to be used by the algorithm, rather than actual values. These components include the SPP, the use of local search method (implemented as proposed in [7]) after AGES, and the initial solution generator (where LS refers to the already mentioned local search). Column *Notation* is the expression to denote the parameter, and column *Description* briefly explains where each parameter is used. Columns *Range* and *Best* are the range of values available for each parameter, and the best value found by irace, respectively.

A total of 2000 experiments were performed by irace, using a randomly selected subset of instances. It is worth mentioning that the five elite configurations returned by irace, which have no statistical difference, had the parameter spp set to true. Then, it is reasonable to say that the SPP has a positive impact on solution quality, and it surpasses any computational cost of being used. The best values reported are used in all experiments of IGLS.

5.3 Numerical Results of IGLS

For the PDPTW, comparing solely with the best-known solutions (BKS) published in [22] can be misleading. These results are from various methods, many from proprietary algorithms, or with no peer-reviewed publication, to which there is neither information on how they were obtained, nor on how much time it was needed. Because of that, the proposed algorithm is compared to the method of CLSQL, which can be considered the current *state-of-the-art*.

Table 3 compares results of our replication R-CLSQL and the IGLS regarding average solution quality. Columns are the same as in Table 1. Improved solutions are highlighted in bold.

On average, results of R-CLSQL and IGLS are the same for instances of size 100, with slightly smaller solution cost for IGLS. For instances of sizes 200, 400, 600, and 800, the average solution was better for IGLS, with differences as

Table 3. Comparison between average results of R-CLSQL and IGLS

Inst. Size	R-CLSQL				IGLS			
	#V	σ_v	Cost	t(s)	#V	σ_v	Cost	t(s)
100	402.0	± 0.00	58,089.89	310	402.0	± 0.00	**58,080.54**	302
200	601.3	± 1.12	192,649.45	929	600.6	± 0.52	**189,189.25**	905
400	1152.4	± 8.59	470,814.57	959	1143.8	± 4.43	**464,123.57**	919
600	1653.5	±11.14	975,073.23	1909	1647.2	± 7.31	**967,684.34**	1861
800	2146.2	±12.73	1,588,410.73	3783	2143.5	± 9.58	**1,597,189.41**	3692
1000	**2629.6**	±14.98	**2,374,610.91**	3883	2631.6	±16.18	2,374,504.33	3784

little as 0.7 (200), up to 8.6 vehicles (400). However, for the set of 1000 requests, R-CLSQL had a better performance, with 2.0 fewer vehicles. It indicates that on average our approach works better for instances of small and medium sizes. Although, according to the deviation in the number of vehicles, IGLS can be considered competitive even for the largest set of instances.

Note that if comparisons had been made using the original information for a single execution of CLSQL, the conclusions would be different. Our proposed method would have had better results for sizes 100, 200, 800 and 1000, which is only partially true if we consider the average behavior of the algorithms.

Nevertheless, we compare the best results found by each algorithm among their multiple executions. Also, for informative purposes, we compare these results to the BKS values available at SINTEF[2] [22]. Table 4 compares such results. Values are no longer the accumulated average, but the accumulated value of the best solutions found.

Table 4. Comparison between best solutions of R-CLSQL and IGLS in 10 or 5 runs

Inst. Size	BKS		R-CLSQL			IGLS		
	#V	Cost	#V	Cost	t(s)	#V	Cost	t(s)
100	402	58,059.55	402	58,059.55	310	402	58,059.55	302
200	600	183,848.47	600	186,300.36	927	600	184,557.16	905
400	1133	438,718.64	1142	449,091.70	960	1138	447,767.02	913
600	1628	897,494.59	1642	939,861.14	1917	1639	930,144.54	1856
800	2116	1,519,213.78	2134	1,576,713.11	3772	2133	1,563,062.74	3704
1000	2572	2,231,988.31	2615	2,327,553.52	3885	2617	2,315,764.64	3803

Both algorithms, R-CLSQL and IGLS, were able to reach the best-known solutions for all instances of size 100 at least once. Although, for all the other instance sizes, some solutions could not be achieved by any of the methods. Moreover, the comparison between the best solutions found by the two algorithms agrees with the comparison of the average values. IGLS has better solutions for sizes 200, 400, 600 and 800, while R-CLSQL dominates 1000. The only

---

[2] According to the solutions published until April 16, 2018.

Table 5. New best-known solutions

Instance		BKS		CLSQL		IGLS			
Name	Size	#V	Cost	#V	Cost	#V	Cost	gap(%)	t(s)
LC1_2_8	200	19	3367.48	19	3397.65	19	3354.27	−0.39	900
LR1_2_4	200	10	3030.03	10	3044.69	10	3027.06	−0.10	902
LR1_2_7	200	12	3543.69	12	3550.61	12	3543.36	−0.01	900
LR1_2_8	200	9	2759.44	9	2814.32	9	2759.32	−0.01	901
LR2_2_8	200	2	2455.87	2	2586.42	2	2450.47	−0.22	902
LR2_2_9	200	3	3924.82	3	3924.82	3	3922.11	−0.07	901
LR2_2_10	200	3	3274.96	3	3274.96	3	3254.83	−0.61	904
LRC2_2_7	200	4	3018.05	4	3057.23	4	3016.53	−0.05	903
LC2_4_3	400	12	4412.75	12	4418.88	12	4407.71	−0.11	911
LRC1_4_3	400	24	7828.75	24	7856.72	24	7819.90	−0.11	900
LRC1_4_4	400	19	5806.20	19	5841.95	19	5804.47	−0.03	902
LRC2_4_2	400	10	7308.24	10	7605.61	10	7214.99	−1.28	902
LRC2_4_3	400	8	6505.71	8	6576.48	8	6483.48	−0.34	902
LRC2_4_5	400	10	7416.87	10	7462.66	10	7404.23	−0.17	901
LR2_6_2	600	9	23255.40	9	23255.40	9	22310.56	−4.06	1808
LR2_6_3	600	7	19183.41	7	19183.41	7	18337.46	−4.41	1806
LRC1_6_1	600	52	18293.94	52	18312.60	52	18293.72	−0.01	1805
LRC1_6_2	600	43	16624.01	43	17063.21	43	16576.53	−0.26	1825
LRC1_6_3	600	36	14041.72	36	14115.00	36	13987.02	−0.39	1802
LRC1_6_8	600	33	15862.32	33	15919.78	33	15812.61	−0.31	2000
LRC2_6_1	600	16	14782.39	16	14892.18	16	14665.50	−0.79	1810
LRC2_6_6	600	12	15315.05	12	17149.19	12	15200.75	−0.75	1800
LRC1_8_7	800	50	29948.00	51	28705.17	50	29378.53	−1.90	3609

exception is size 100, where they reach the same solution. On the other hand, comparing the best solutions of IGLS to the single execution of CLSQL leads to the conclusion that our method outperforms the latter in all instance sizes.

A total of 23 new BKS solutions have been found by the IGLS method, using a diversified set of parameters. These solutions have been validated and published by SINTEF [22]. Table 5 presents for each new solution, the name of the instance, and its size. Also, for each result presented, columns #V and *Cost* are the number of vehicles and the cost of the solution, respectively. We present the previous BKS published at SINTEF, the best solution reported by CLSQL [7], and the new result found by IGLS. Because all new solutions improve the cost, but not the number of vehicles, we present the perceptual improvement of the cost in column *gap(%)*, computed as $100 \cdot (S - BKS)/BKS$, where S is our new solution cost, and BKS the previous one.

The improvements achieved with the new BKS range from as little as 0.1%, up to more than 4.0% in the total cost of a solution. It further confirms that the use of a SPP model can help a method reach good solutions, even for problems with many side-constraints as the PDPTW.

6   Conclusion and Future Work

This work proposed a matheuristic approach based on Set Partitioning to solve the Pickup and Delivery Problem with Time Windows. The method combines Adaptive Guided Ejection Search, Large Neighborhood search and exact solutions of the SPP, in an Iterated Local Search framework. Both AGES and LNS had already been shown to work very well for the PDPTW, but their combination with a SPP solver had not been previously tested.

The study showed the addition of SPP can boost the results, and it seems to work very well on small and medium-sized instances, while it remains competitive for the largest ones. Comparisons with a *state-of-the-art* algorithm further indicate this, and a number of new best-known solutions has been found for a well-known benchmark set.

However, certain research paths need more investigation. It has been noted that in many instances the AGES seemed to use too much unnecessary computational time at every iteration when it could no longer reduce the number of routes. A more refined adaptive system could allow for this time to be spent by the LNS phase instead, trying to reduce the total cost. As for the LNS, many iterations where a single request could not be reinserted were wasted. Allowing some degree of infeasibility in LNS could help avoid these cases and to explore certain regions of the search space that would otherwise remain untouched. Also, the current use of the SPP is quite simple. Although it has performed well, it could be interesting to add routes of every local minimum of the LNS, but it would require an efficient pool size management since much more routes would be added per ILS iteration. At last, the component selection used during the tuning is far from exhaustive, and using an automatic algorithm design [12] for this problem could lead to even better results.

Acknowledgment. This work was partially supported by CNPq (Conselho Nacional de Desenvolvimento Científico e Tecnológico) and FAPERGS (Fundação de Amparo à Pesquisa do Estado do Rio Grande do Sul). In addition, the authors acknowledge the valuable contributions of the two anonymous reviewers.

References

1. Alvarenga, G.B., Mateus, G.R., De Tomi, G.: A genetic and set partitioning two-phase approach for the vehicle routing problem with time windows. Comput. Oper. Res. **34**(6), 1561–1584 (2007)
2. Archetti, C., Speranza, M.G.: A survey on matheuristics for routing problems. EURO J. Comput. Optim. **2**(4), 223–246 (2014)
3. Baldacci, R., Bartolini, E., Mingozzi, A.: An exact algorithm for the pickup and delivery problem with time windows. Oper. Res. **59**(2), 414–426 (2011)
4. Bent, R., Van Hentenryck, P.: A two-stage hybrid algorithm for pickup and delivery vehicle routing problems with time windows. Comput. Oper. Res. **33**(4), 875–893 (2006)

5. Boschetti, M.A., Maniezzo, V., Roffilli, M., Bolufé Röhler, A.: Matheuristics: optimization, simulation and control. In: Blesa, M.J., Blum, C., Di Gaspero, L., Roli, A., Sampels, M., Schaerf, A. (eds.) HM 2009. LNCS, vol. 5818, pp. 171–177. Springer, Heidelberg (2009). https://doi.org/10.1007/978-3-642-04918-7_13
6. Burke, E.K., Bykov, Y.: A late acceptance strategy in hill-climbing for exam timetabling problems. In: PATAT 2008 Conference, Montreal, Canada (2008)
7. Curtois, T., Landa-Silva, D., Qu, Y., Laesanklang, W.: Large neighbourhood search with adaptive guided ejection search for the pickup and delivery problem with time windows. EURO J. Transp. Logist., pp. 1–42 (2017)
8. Dumas, Y., Desrosiers, J., Soumis, F.: The pickup and delivery problem with time windows. Eur. J. Oper. Res. 54(1), 7–22 (1991)
9. Grangier, P., Gendreau, M., Lehuédé, F., Rousseau, L.M.: A matheuristic based on large neighborhood search for the vehicle routing problem with cross-docking. Comput. Oper. Res. 84, 116–126 (2017)
10. Laporte, G.: What you should know about the vehicle routing problem. Nav. Res. Logist. (NRL) 54(8), 811–819 (2007)
11. Li, H., Lim, A.: A metaheuristic for the pickup and delivery problem with time windows. Int. J. Artif. Intell. Tools 12(02), 173–186 (2003)
12. López-Ibáñez, M., Dubois-Lacoste, J., Cáceres, L.P., Birattari, M., Stützle, T.: The irace package: Iterated racing for automatic algorithm configuration. Oper. Res. Perspect. 3, 43–58 (2016)
13. Lourenço, H.R., Martin, O.C., Stützle, T.: Iterated local search: framework and applications. In: Gendreau, M., Potvin, J.Y. (eds.) Handbook of Metaheuristics. International Series in Operations Research & Management Science, vol. 146, pp. 363–397. Springer, Boston (2010). https://doi.org/10.1007/978-1-4419-1665-5_12
14. Nalepa, J., Blocho, M.: Enhanced guided ejection search for the pickup and delivery problem with time windows. In: Nguyen, N.T., Trawiński, B., Fujita, H., Hong, T.-P. (eds.) ACIIDS 2016. LNCS (LNAI), vol. 9621, pp. 388–398. Springer, Heidelberg (2016). https://doi.org/10.1007/978-3-662-49381-6_37
15. Nanry, W.P., Barnes, J.W.: Solving the pickup and delivery problem with time windows using reactive tabu search. Transp. Res. Part B Methodol. 34(2), 107–121 (2000)
16. Parragh, S.N., Doerner, K.F., Hartl, R.F.: A survey on pickup and delivery problems. J. für Betriebswirtschaft 58(1), 21–51 (2008)
17. Parragh, S.N., Schmid, V.: Hybrid column generation and large neighborhood search for the dial-a-ride problem. Comput. Oper. Res. 40(1), 490–497 (2013)
18. Ropke, S., Cordeau, J.F.: Branch-and-cut-and-price for the pickup and delivery problem with time windows. Transp. Sci. 43(3), 267–286 (2009)
19. Ropke, S., Pisinger, D.: An adaptive large neighborhood search heuristic for the pickup and delivery problem with time windows. Transp. Sci. 40(4), 455–472 (2006)
20. Savelsbergh, M.W., Sol, M.: The general pickup and delivery problem. Transp. Sci. 29(1), 17–29 (1995)
21. Shaw, P.: Using constraint programming and local search methods to solve vehicle routing problems. In: Maher, M., Puget, J.-F. (eds.) CP 1998. LNCS, vol. 1520, pp. 417–431. Springer, Heidelberg (1998). https://doi.org/10.1007/3-540-49481-2_30
22. SINTEF: Li & lim benchmark instances (2008). https://www.sintef.no/projectweb/top/pdptw/li-lim-benchmark/. Accessed 18 April 2018
23. Solomon, M.M.: Algorithms for the vehicle routing and scheduling problems with time window constraints. Oper. Res. 35(2), 254–265 (1987)
24. Subramanian, A., Uchoa, E., Ochi, L.S.: A hybrid algorithm for a class of vehicle routing problems. Comput. Oper. Res. 40(10), 2519–2531 (2013)

Towards Asymptotically Optimal One-to-One PDP Algorithms for Capacity 2+ Vehicles

Lars Nørvang Andersen[1] and Martin Olsen[2]([envelope])

[1] Department of Mathematics, Aarhus University, Aarhus, Denmark
larsa@math.au.dk
[2] Department of Business Development and Technology,
Aarhus University, Aarhus, Denmark
martino@btech.au.dk

Abstract. We consider the one-to-one Pickup and Delivery Problem (PDP) in Euclidean Space with arbitrary dimension d, where n transportation requests are picked i.i.d. with a separate origin-destination pair for each object to be moved. First, we consider the problem from the customer perspective, where the objective is to compute a plan for transporting the objects such that the Euclidean distance traveled by the vehicles *when carrying objects* is minimized. We develop a polynomial time asymptotically optimal algorithm for vehicles with capacity $o(\sqrt[2d]{n})$ for this case including the realistic setting where the capacity of the vehicles is a fixed constant and $d = 2$. This result also holds imposing LIFO constraints for loading and unloading objects. Secondly, we extend our algorithm to the classical single-vehicle PDP, where the objective is to minimize the total distance traveled by the vehicle and we present results indicating that the extended algorithm is asymptotically optimal for a fixed vehicle capacity, if the origins and destinations are picked i.i.d. using the same distribution.

1 Introduction

The challenge of computing optimal or near-optimal plans for transporting goods or people is a core problem within logistics. This problem has received a huge amount of attention from the operations research community. A generic term for this class of problems is *Vehicle Routing Problems* (VRP). Vehicle routing problems come in many flavors depending among other things on the properties of the vehicles used, the characteristics of the terrain, and the type of transportation requests considered.

In this paper, we consider the variant of the problem, where the terrain is the Euclidean space of an arbitrary dimension d and where we measure the distance using the Euclidean distance. We have one vehicle with limited capacity at our disposal, but the vehicle can carry more than one object at a time. Every object to be transported has a separate origin and destination. The objective is to

R. Cerulli et al. (Eds.): ICCL 2018, LNCS 11184, pp. 268–278, 2018.
https://doi.org/10.1007/978-3-030-00898-7_17

compute a plan for transporting the objects with a minimum distance traveled by the vehicle. We look at the non-preemptive version meaning that an object has to stay on the vehicle until it is delivered. Our version is static (offline) in the sense that all information on the transportation requests is available to us before we compute the optimal route for the vehicle.

The origins and destinations are picked using a stochastic process, and we measure the performance of an algorithm by considering the *approximation ratio*, i.e., the value of the solution computed by the algorithm divided by the value of the optimal solution. The main aim of the paper is to present polynomial time algorithms that are *asymptotically optimal* in the sense that the approximation ratio converges to 1 with probability 1 (almost surely) as the number of transportation requests goes to infinity. To the best of our knowledge, we are the first to present asymptotically optimal algorithms for the realistic setting, where the vehicles have a limited capacity greater than one. Our algorithms are easy to implement and may be useful in practice.

1.1 Related Work

In the literature, the problem considered is often referred to as the One-to-One Pickup and Delivery Problem (PDP) or the Vehicle Routing Problem with Paired Pickups and Deliveries. The single vehicle case that we look at in this paper is also known as the Traveling Salesman Problem with Pickups and Deliveries, because this case can be viewed as a Traveling Salesman Problem (TSP) with precedence constraints, where the origin of an object must be visited before the corresponding destination. If the single vehicle has capacity 1, then this problem is known as the Stacker Crane Problem (SCP). We refer the reader to the excellent surveys [3, 10, 11, 13] for an overview on vehicle routing research.

The PDP problem in focus in this paper is defined as follows:

Definition 1. *An instance of the* PDP *problem for dimension d and capacity c is a set of n requests $R = \{r_1, r_2, \ldots, r_n\}$ with $r_i \in [0,1]^d \times [0,1]^d$ for $i \in \{1, 2, \ldots, n\}$. A request $r = (s, t)$ corresponds to a transportation job for an object with origin s and destination t. The solution is a plan for transporting all objects from their origin to their destination minimizing the total Euclidean distance traveled using a single vehicle with capacity c. The vehicle shall start and end in the same point.*

As noted by many authors, the PDP problem generalizes the classical TSP problem and is thus NP-complete for $d \geq 2$. Guan [7] has shown that the PDP problem is NP-complete for $d = 1$ and $c \geq 2$, and Guan also shows how to solve this version in linear time, if we allow temporarily dropping objects. Treleaven et al. [15] present asymptotically optimal algorithms for $c = 1$ (SCP) and $d \geq 2$, where the origins are picked i.i.d. and the destinations are picked i.i.d. from separate distributions.

Stein [14] also conducts a probabilistic analysis, but he looks at the variant where $d = 2$ and $c = \infty$, and where n origins and n destinations are picked independently using a uniform distribution on some planar region. Stein shows that the value of the optimal solution divided by $\sqrt{n}$ converges almost surely to a constant times the square root of the area of the region. Stein also shows how to solve the problem he considers by concatenating two TSP tours on the origins and destinations respectively. This way of solving the problem yields a solution which is roughly 6% higher than the optimal solution (in the limit).

Psaraftis [12] has developed an $O(n^2)$ heuristic guaranteeing an approximation ratio of at most 4 for any instance for the case $d = 2$ and $c = \infty$. Haimovich and Rinnooy Kan [8] have constructed asymptotically optimal PDP algorithms for the case $d = 2$, where all transportation requests have the same depot as destination. Haimovich and Rinnooy Kan [8] also presented a PTAS for this case for $c = O(\log \log n)$, and this result has later been extended to cover cases with multiple depots and arbitrary values of d [9] or larger values of c [5].

In some cases, the customers only pay for a kilometer driven by the vehicle, if the vehicle carries an object when driving that kilometer. As an example, this is the case when the vehicle is a taxi. This leads us to the PDP-C problem that does not seem to have received much attention:

Definition 2. *The* PDP-C *problem is identical to the PDP problem with the exception that distance traveled carrying no objects is excluded.*

The PDP-C problem covers any situation where carrying objects is very expensive compared to carrying no objects. Possible areas for application are the development of taxi sharing or ride sharing schemes. The PDP-C problem also comes into play when we want to minimize the time spent for an elevator (or robot arm) that moves slowly carrying passengers (objects), but moves very fast carrying no passengers (objects).

An efficient and near-optimal subroutine solving the PDP-C problem might also be useful in the case, where we have multiple vehicles at our disposal and the objective is to minimize the completion time (the time when the last object has been delivered). A first step to solve this problem could involve partitioning objects into groups that share a vehicle using a PDP-C subroutine.

1.2   Contribution and Outline

In Sect. 2, we present an adaptive asymptotically optimal polynomial time algorithm for the PDP-C problem for $c = o(\sqrt[2d]{n})$ for any dimension d under the assumption that the transportation requests are picked independently and by identical distributions (i.i.d.). As explained above, there are many real-world problems where a PDP-C algorithm is useful. We note that we use the standard definition of the "little-o" notation. In other words, $c = o(\sqrt[2d]{n})$ means that $\frac{c}{\sqrt[2d]{n}}$ converges to 0 as n tends to infinity. It is very important to stress that this includes the realistic case where c is a fixed constant (for example $c = 4$).

A PDP algorithm is presented in Sect. 3 accompanied by what we consider to be good reasons to believe that this algorithm is asymptotically optimal for a constant capacity c, if the origins and destinations for the transportation requests are picked i.i.d. using the same distribution. The PDP algorithm is an extension of our PDP-C algorithm.

The key idea for our approach is that we solve a TSP with precedence constraints in d-dimensional Euclidean space – the PDP problem – by solving a classical TSP defined by the requests in Euclidean space with the double dimension $2d$. Neighbors in the solution to the classical TSP correspond to objects that have similar requests, so we let such neighbors share a vehicle.

As mentioned earlier, we believe that we are the first to present results on asymptotic optimality for the realistic case $1 < c < \infty$. Again, we emphazise that our contribution also covers the realistic setting with constant capacity vehicles and $d = 2$. Our results even hold if we allow temporarily dropping objects (the preemptive variant) or if we impose LIFO constraints for loading and unloading objects.

2   An Asymptotically Optimal PDP-C Algorithm

In this section, we present our PDP-C algorithm. We begin by listing the pseudocode consisting of 4 steps. We also exemplify how the steps work using the instance shown in Fig. 1 that has $d = 1$ and $c = 2$.

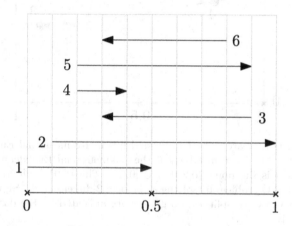

Fig. 1. An instance of a 1-dimensional PDP-C problem consisting of 6 requests, i.e., $n = 6$. As an example, request 4 shows that an object has to be picked up at 0.2 and delivered at 0.4. We assume that vehicles with capacity 2 are used.

2.1   The Pseudocode

Throughout the paper, we assume that $n = mc$ for some integer m. Otherwise, we can serve the extra objects one by one, implying an extra $O(c)$ cost that does not affect our results on asymptotic optimality.

Step 1. Use a polynomial time constant factor approximation algorithm [1,4] to compute a feasible solution T for the $2d$-dimensional Euclidean TSP problem defined by R (σ is a permutation on $\{1, 2, \ldots, n\}$):

$$T = r_{\sigma(1)} \rightarrow r_{\sigma(2)} \rightarrow \cdots \rightarrow r_{\sigma(n)} \rightarrow r_{\sigma(1)}.$$

The tour T is shown in Fig. 2.

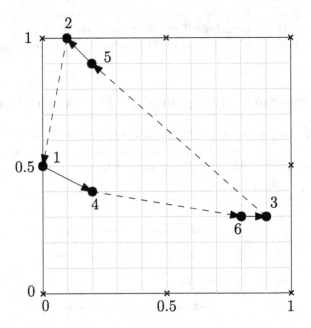

Fig. 2. The objects that are to be transported live in 1-dimensional Euclidean space. The requests from Fig. 1 are members of the 2-dimensional Euclidean space. As an example, request 4 is the point $(0.2, 0.4) \in [0,1]^2$. The PDP-C algorithm attacks a 1-dimensional PDP-C problem by solving a classical 2-dimensional TSP defined by the requests. The two ways to split T into groups are indicated by the dashed and solid arrows respectively.

Step 2. We now split T into m groups such that each group contains c consecutive requests $r_{\sigma(i)}, r_{\sigma(i+1)}, \ldots, r_{\sigma(i+c-1)}$. One possible way of partitioning the requests into groups is as follows:

$$\{r_{\sigma(1)}, r_{\sigma(2)}, \ldots, r_{\sigma(c)}\},$$

$$\{r_{\sigma(c+1)}, r_{\sigma(c+2)}, \ldots, r_{\sigma(2c)}\},$$

$$\cdots$$

$$\{r_{\sigma(n-c+1)}, r_{\sigma(n-c+2)}, \ldots, r_{\sigma(n)}\}.$$

There are c ways to do the split up (see Fig. 2). For each of the possibilities, we repeat Step 3 and obtain c candidate solutions for the PDP-C problem:

Step 3 (repeated for each possible splitting of T). The objects in a group share a vehicle. The objects for a group of requests $\{r_{\sigma(i)}, r_{\sigma(i+1)}, \ldots, r_{\sigma(i+c-1)}\}$ are picked up in the order $\sigma(i)$, $\sigma(i+1)$, $\ldots$, $\sigma(i+c-1)$ and dropped off in reverse order (LIFO). The plan corresponding to one of the ways to split T into groups is shown in Fig. 3.

Step 4. Finally, we pick the best of the c candidate solutions produced in Step 3. The plan computed by the algorithm is shown in Fig. 3.

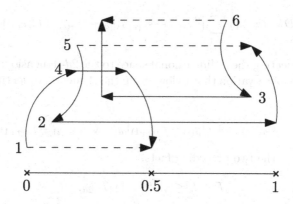

Fig. 3. The feasible solution for the PDP-C instance computed by the PDP-C algorithm. The solution is based on the splitting of T indicated by solid arrows in Fig. 2. The objects are picked up and dropped in LIFO order.

2.2   Analysis of the PDP-C Algorithm

We let SOL denote the value of the plan computed by the PDP-C algorithm. We kindly remind the reader that the value is the Euclidean distance traveled, where we only measure the distance traveled carrying objects. The value of the optimal solution is denoted by OPT. The length of the tour T is $\|T\|_{2d}$, where the subscript indicates the dimension of the underlying Euclidean space.

We now present a key lemma that links the TSP in $2d$-space to the value of the solution computed by the PDP-C algorithm:

Lemma 1

$$SOL \leq \sum_{i=1}^{n} \frac{\|s_i - t_i\|_d}{c} + \sqrt{2}\left(\frac{c-1}{c}\right)\|T\|_{2d}. \tag{1}$$

Proof. Let SOL_i, $i \in \{1, 2, \ldots, c\}$, denote the Euclidean distance covered for the i'th plan computed by Step 3. The sum $\sum_{i=1}^{c} SOL_i$ can be broken down into three terms:

$$\sum_{i=1}^{c} SOL_i = P + F + D, \tag{2}$$

where P is a term for picking up objects, F is a term for driving with a full vehicle, and D is a term for dropping off objects. Every object tries to be the final object to be picked up in precisely one of the plans:

$$F = \sum_{i=1}^{n} \|s_i - t_i\|_d. \tag{3}$$

The segments $s_{\sigma(i)} \rightarrow s_{\sigma(i+1)}$ and $t_{\sigma(i+1)} \rightarrow t_{\sigma(i)}$ are traversed in exactly $c - 1$ of the plans, $i \in \{1, 2, \ldots, n\}$ (Addition is performed cyclically: $n + 1 = 1$):

$$P + D = (c-1) \sum_{i=1}^{n} \left[\|s_{\sigma(i)} - s_{\sigma(i+1)}\|_d + \|t_{\sigma(i)} - t_{\sigma(i+1)}\|_d \right]. \tag{4}$$

The key to connecting the d-dimensional space to the $2d$-dimensional space is the following simple observation that follows from the elementary identity $\sqrt{x} + \sqrt{y} \leq \sqrt{2}\sqrt{x+y}$:

$$\|s_{\sigma(i)} - s_{\sigma(i+1)}\|_d + \|t_{\sigma(i)} - t_{\sigma(i+1)}\|_d \leq \sqrt{2}\|r_{\sigma(i)} - r_{\sigma(i+1)}\|_{2d}. \tag{5}$$

We now combine the two preceding facts:

$$P + D \leq \sqrt{2}(c-1)\|T\|_{2d}. \tag{6}$$

The lemma now follows from:

$$SOL \leq \frac{1}{c} \sum_{i=1}^{c} SOL_i. \tag{7}$$

$\square$

We are now ready to prove that our PDP-C algorithm is asymptotically optimal:

Theorem 1. *Let an infinite sequence of requests (s_i, t_i) be picked i.i.d. in $[0,1]^d \times [0,1]^d$ using a distribution satisfying that $E[\|s_i - t_i\|_d] = \mu > 0$. Let SOL_n denote the value of the plan computed by the PDP-C algorithm and let OPT_n denote the value of the optimal plan for the first n requests. If $c = o(\sqrt[2d]{n})$, then we have the following:*

$$\lim_{n \to \infty} \frac{SOL_n}{OPT_n} = 1 \ a.s. \tag{8}$$

Proof. The objects could share the bill of traveling by equally sharing the cost for each segment. This sharing scheme leads to the following lower bound on OPT_n:

$$OPT_n \geq \sum_{i=1}^{n} \frac{\|s_i - t_i\|_d}{c}. \tag{9}$$

We now combine the lower bound on OPT_n with Lemma 1:

$$\frac{SOL_n}{OPT_n} \le 1 + \sqrt{2}(c-1) \left(\sum_{i=1}^{n} \|s_i - t_i\|_d \right)^{-1} \|T\|_{2d}. \tag{10}$$

The inequality (10) is rewritten slightly:

$$\frac{SOL_n}{OPT_n} \le 1 + \sqrt{2}(c-1) \left(\frac{\sum_{i=1}^{n} \|s_i - t_i\|_d}{n} \right)^{-1} \frac{\|T\|_{2d}}{n}. \tag{11}$$

We use a constant factor approximation algorithm for solving the TSP in $2d$-dimensional space implying this upper bound on the length of T [6]:

$$\|T\|_{2d} = O(n^{\frac{2d-1}{2d}}). \tag{12}$$

Using $c = o(\sqrt[2d]{n})$, we now get the following:

$$c\|T\|_{2d} = o(n). \tag{13}$$

According to the Strong Law of Large Numbers, we have the following:

$$\lim_{n \to \infty} \left(\frac{\sum_{i=1}^{n} \|s_i - t_i\|_d}{n} \right)^{-1} = \mu^{-1} \text{ a.s.} \tag{14}$$

The lemma now follows from (11), (13) and (14). $\qquad \square$

A few comments on the convergence rate might be suitable at this point. According to [2], the limit of $\frac{\|T\|_{2d}}{n} \sqrt[2d]{n}$ is almost surely a constant, where the constant depends on the distribution of the requests with maximum value for the uniform distribution. In other words, the algorithm is *adaptive* in the sense that the right hand side of (11) tends to be smaller for big instances for the non-uniform case.

Even for relatively small values of n, we might experience a right hand side of (11) that is relatively close to 1. As an example, we consider the case $d = 1$, where we admittedly have the best conditions for convergence. Few [6] has shown that $\|T\|_2 \le \sqrt{2n} + \frac{7}{4}$ implying that the right hand side of (11) converges relatively quickly to 1 for moderate c for the case $d = 1$, if μ is not too small.

3   A PDP Algorithm

We now turn our attention to the PDP problem and present a polynomial time algorithm that can be viewed as a generalization of the *Iterated Tour Partition Heuristic* that Haimovich and Rinnooy Kan [8] presented for the case, where $d = 2$ and all the destinations are identical. The tour that we consider is a tour in $2d$-dimensional Euclidean space, where the requests are members, and we allow different destinations for arbitrary d.

Our PDP algorithm is an extension of our PDP-C algorithm: First, we figure out what objects should share the vehicle and establish a LIFO order for pickups and deliveries (PDP-C with $1 < c < \infty$). Secondly, we set up a route for the vehicle focusing on the segments when it carries no objects (PDP with $c = 1$. SCP in other words). We exemplify our PDP algorithm by adding two more figures to the PDP-C example. The pseudocode for the PDP algorithm consists of the following 6 steps:

Steps 1–4. Use the steps from the PDP-C Algorithm and compute a PDP-C solution.

Step 5. Use an algorithm from the SPLICE class [15] to compute a feasible solution S for the SCP instance defined by a pickup at the origin and a delivery at the destination for every object that was the first to be picked up by a vehicle (and, consequently, the last object to be dropped off) in the PDP-C solution. Let S_0 denote the set of segments that go <u>from</u> a delivery <u>to</u> a pickup from the solution S to the Stacker Crane Problem. See Fig. 4.

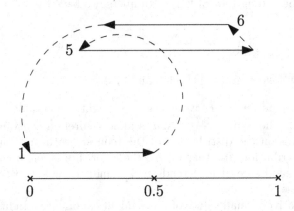

Fig. 4. The SCP and the solution S computed in Step 5. The dashed segments are the segments in S_0. The objects for the requests 1, 5 and 6 where the first to be picked up in the PDP-C solution in Fig. 3.

Step 6. A PDP solution can now be produced by combining the PDP-C solution with the segments S_0 from the Stacker Crane Plan, where no objects were carried. See Fig. 5.

We now let SOL denote the total Euclidean distance covered by the vehicle for the plan proposed by the PDP algorithm. The optimal solution is denoted by OPT. The total length of the delivery-to-pickup segments from S that we use in Step 6 is $\|S_0\|_d$, where d refers to the dimension of the Euclidean space.

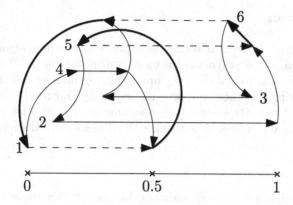

Fig. 5. The solution for the PDP problem computed by the PDP algorithm.

Lemma 2

$$SOL \leq \sum_{i=1}^{n} \frac{\|s_i - t_i\|_d}{c} + \sqrt{2}\left(\frac{c-1}{c}\right)\|T\|_{2d} + \|S_0\|_d. \tag{15}$$

Proof. Compared to Lemma 1, the extra distance driven is $\|S_0\|_d$.   □

Observation 1. *If the following conditions are met*

$$c\|S_0\|_d = o(n), \tag{16}$$

$$c\|T\|_{2d} = o(n), \text{ and} \tag{17}$$

$$\sum_{i=1}^{n} \|s_i - t_i\|_d = \Omega(n), \tag{18}$$

then the PDP algorithm is asymptotically optimal: $\lim_{n\to\infty} \frac{SOL_n}{OPT_n} = 1$.

It follows from [15] that $\|S_0\|_d$ is $o(n)$ almost surely[1] for $d \geq 3$ if S is the Stacker Crane tour computed by a SPLICE algorithm on n requests in $[0,1]^d \times [0,1]^d$ with the origins and destinations picked i.i.d. using the same distribution. The Stacker Crane tour S from our PDP algorithm consists of n/c requests, but the corresponding points are not picked independently. Informally speaking, these n/c requests seem to be spread evenly on R so we are optimistic with respect to proving that (16) holds, but more work has to be done to look into the details and conditions for convergence. Observation 1 gives us reason to believe that our PDP algorithm is asymptotically optimal for fixed capacity c, if the origins and destinations are picked i.i.d. from the same distribution, since (17) and (18) hold in this case and (16) seems plausible.

---

[1] The $o(n)$ result follows from (2) and the unnumbered equation in the proof of Theorem 4.5 in [15].

4   Conclusion

We have presented a polynomial time asymptotically optimal algorithm for the PDP-C problem and a polynomial time algorithm for the PDP problem that we have good reasons to believe is asymptotically optimal as well (under certain assumptions for picking the transportation requests). Our results deal with vehicles with limited capacity greater than one. One obvious idea for future work is incorporating time windows by extending the dimension of the request space.

References

1. Arora, S.: Polynomial time approximation schemes for Euclidean traveling salesman and other geometric problems. J. ACM **45**(5), 753–782 (1998). https://doi.org/10.1145/290179.290180
2. Beardwood, J., Halton, J.H., Hammersley, J.M.: The shortest path through many points. Math. Proc. Camb. Philosoph. Soc. **55**(4), 299–327 (1959). https://doi.org/10.1017/S0305004100034095
3. Berbeglia, G., Cordeau, J.F., Gribkovskaia, I., Laporte, G.: Static pickup and delivery problems: a classification scheme and survey. TOP: Off. J. Span. Soc. Stat. Oper. Res. **15**(1), 1–31 (2007)
4. Christofides, N.: Worst-case analysis of a new heuristic for the travelling salesman problem. Technical report 388, Graduate School of Industrial Administration, Carnegie Mellon University (1976)
5. Das, A., Mathieu, C.: A quasipolynomial time approximation scheme for Euclidean capacitated vehicle routing. Algorithmica **73**(1), 115–142 (2015)
6. Few, L.: The shortest path and the shortest road through n points. Mathematika **2**(2), 141–144 (1955). https://doi.org/10.1112/S0025579300000784
7. Guan, D.: Routing a vehicle of capacity greater than one. Disc. Appl. Math. **81**(1), 41–57 (1998). https://doi.org/10.1016/S0166-218X(97)00074-7
8. Haimovich, M., Rinnooy Kan, A.H.G.: Bounds and heuristics for capacitated routing problems. Math. Oper. Res. **10**(4), 527–542 (1985)
9. Khachay, M., Dubinin, R.: PTAS for the euclidean capacitated vehicle routing problem in R^d. In: Kochetov, Y., Khachay, M., Beresnev, V., Nurminski, E., Pardalos, P. (eds.) DOOR 2016. LNCS, vol. 9869, pp. 193–205. Springer, Cham (2016). https://doi.org/10.1007/978-3-319-44914-2_16
10. Parragh, S.N., Doerner, K.F., Hartl, R.F.: A survey on pickup and delivery problems (part I). J. Betriebswirtschaft **58**(1), 21–51 (2008)
11. Parragh, S.N., Doerner, K.F., Hartl, R.F.: A survey on pickup and delivery problems (part II). J. Betriebswirtschaft **58**(2), 81–117 (2008)
12. Psaraftis, H.: Analysis of an o(n) heuristic for the single vehicle many-to-many Euclidean dial-a-ride problem. Transp. Res. Part B: Methodol. **17**, 133–145 (1981)
13. Savelsbergh, M.W.P., Sol, M.: The general pickup and delivery problem. Transp. Sci. **29**, 17–29 (1995)
14. Stein, D.M.: An asymptotic, probabilistic analysis of a routing problem. Math. Oper. Res. **3**(2), 89–101 (1978). https://doi.org/10.1287/moor.3.2.89
15. Treleaven, K., Pavone, M., Frazzoli, E.: Asymptotically optimal algorithms for one-to-one pickup and delivery problems with applications to transportation systems. IEEE Trans. Autom. Control **58**(9), 2261–2276 (2013). https://doi.org/10.1109/TAC.2013.2259993

A Many-to-One Algorithm to Solve a Many-to-Many Matching Problem for Routing

Wenjing Guo[⊠], Wouter Beelaerts van Blokland, and Rudy R. Negenborn

Delft University of Technology, Delft, The Netherlands
{W.Guo-2,W.W.A.BeelaertsvanBlokland,R.R.Negenborn}@tudelft.nl

Abstract. This paper investigates the multi-commodity multi-service matching problem of synchromodal hinterland container transportation. To improve the computational efficiency, this paper proposes a many-to-one algorithm to solve the many-to-many matching problem. We assess the performance of the proposed method with 51 instances of the problem, and perform sensitivity analysis to analyze the influence of different demand patterns. The computational results indicate that the algorithm is suitable for large-scale instances of the problem.

Keywords: Hinterland container transportation · Synchromodality
Multi-commodity · Matching · Many-to-one algorithm

1 Introduction

With the increasing competition of market share in global trade, deep-sea ports' operators have extended their attention to hinterland container transportation. Unlike the port-to-port global transportation, hinterland transportation is the transportation from deep-sea ports (such as Rotterdam Port) to inland terminals (such as Venlo terminal), or vice versa. Therefore, the transport modalities mainly include truck, barge and train.

Compared with truck transportation, intermodal transportation benefits from the reduction of transport cost and carbon emissions, and the improvement of reliability [10]. However, intermodal transportation faces several challenges recently, including low modal split and less capacity utilization. According to statistics, road transportation occupies 76.4% of inland transportation within EU-28 (European Union's 28 countries) in 2016 [3]. The capacity utilization of intermodal services (barge and train) is just around 50% [9].

Synchromodal transportation is a concept aimed at improving the modal split and the capacity utilization of intermodal services [4]. It mainly consists of differentiated service transportation, modal-free booking, horizontal collaboration, integrated planning and real-time switching. Differentiated service transportation means decision makers (network operators) provide differentiated prices to shippers for different services. For example, if the lead time of commodities

© Springer Nature Switzerland AG 2018
R. Cerulli et al. (Eds.): ICCL 2018, LNCS 11184, pp. 279–294, 2018.
https://doi.org/10.1007/978-3-030-00898-7_18

requires to be 12 h, the service price will be 35 €/TEU; if the lead time is 1 day, the price will be 20 €/TEU. Regarding modal-free booking, it means that shippers leave the modality choice to network operators. Horizontal collaboration is the collaboration between carriers for a better utilization of intermodal services. For instance, a barge company cooperates with a train company by sharing the capacity of barge and train. In terms of integrated planning, it means decision makers make decisions for all the commodities and services to achieve a better solution. Real-time switching refers to the re-planning of decisions when disturbances happen (such as newly incoming requests and service delay).

This paper focuses on the operational integrated planning problem of synchromodal hinterland container transportation, specifically, the multi-commodity multimodal routing choice problem. This problem is defined as network operators make routing decisions for all the commodities from different shippers by considering all the services from different carriers [8]. The services can be classified into three groups: truck services (with flexible time window, no capacity limitation), self-operated intermodal services (owned by network operators, service-based cost, fixed time schedule and limited capacity) and subcontracted intermodal services (owned by third party companies, container-based cost, fixed time schedule and limited capacity). Thus, the characteristics of this problem include: time window requirements of commodities and services, transfers between different services with time-space compatibility, capacity limitations of intermodal services, and cost sharing among commodities using the same self-operated intermodal services [5].

Due to the computational complexity, we propose in [5] to formulate the multi-commodity synchromodal routing choice problem from a matching perspective. Based on the multi-commodity multi-service matching design, the number of binary variables has largely reduced compared with arc-node based formulations. In turn, the computational complexity has been proved to be much lower. As the multi-commodity multi-service matching problem is a mixed-integer nonlinear problem, we design in [5] a mixed-integer linear programming to solve the problem. The experiment results show that the mixed-integer linear programming is suitable for medium-sized instances of this problem. However, in practice, the daily container throughput in deep-sea ports is very large. In the port of Rotterdam, the daily throughput of containers is around 30,000 TEU; the number of daily inland vessels is around 280. Therefore, an efficient solution approach for large-scale instances of the problem is required.

1.1 Literature Review

In the literature, few papers investigate the solution approaches for large-scale demand-service matching problems [1,2,6,7,11]. The demand-service matching problems can be divided into two groups: multi-demand single service (MDSS) matching problem and multi-demand multi-service (MDMS) matching problem. In terms of MDSS matching, each demand can be matched with only one service. By comparison, in MDMS matching, each demand can be matched with multiple services by transferring between them. In [1,6,11], the authors investigate solution approaches for the MDSS matching problem; in [2,7], the solution methods to solve the MDMS matching problem are studied.

In [11], the authors propose to solve a multi-passenger single driver matching problem by making combinations for passengers at first and then matching the passenger combinations with drivers. Based on the time-space compatibility and drivers' capacity limitation, the number of feasible combinations and matches can be largely reduced. In [1,6], the same approach is proposed to solve a multi-task single driver matching problem and a multi-passenger single driver problem, respectively. Their experiment results show that the demand combination algorithm is suitable for the large-scale multi-demand single driver (MDSD) matching problem which has limited capacity of private cars. For services with large capacity, the number of feasible demand combinations will be very large and even results in combinatorial explosion problem. In the multi-commodity multi-service (MCMS) matching problem, the average capacity of intermodal services is very large. Thus, the demand combination algorithm is unsuitable for the MCMS matching problem.

In [2], the authors propose to make a list for demands and then use a first come first serve (FCFS) algorithm to solve a multi-parcel multi-driver problem. However, for parcels that are arriving at the same time, they does not explain the priority of requests that the decision makers should consider. In [7], to solve a multi-rider multi-driver problem, the authors propose the FCFS algorithm at first step and then design a dynamic programming algorithm to improve the matching performance. However, this approach is still unsuitable for the MCMS matching problem. The reason is that the cost of using self-operated intermodal services is calculated by service-based cost. Thus, for each commodity, the cost of matching with a self-operated intermodal service is dependent on the total volume of containers from different shippers that assigned to this service. In turn, the cost of a match between a commodity and a self-operated intermodal service is uncertain. The FCFS algorithm cannot be used for a network with uncertain weight.

Inspired from these solution approaches, this paper proposes a many-to-one algorithm to solve the multi-commodity multi-service (MCMS) matching problem. The many-to-one algorithm consists of three steps: the preprocessing of path generation, the preprocessing of feasible matches, and the binary integer linear programming. The preprocessing of path generation aims to reduce the number of feasible service combinations. The preprocessing of feasible matches is designed to reduce the number of feasible matches between commodities and feasible service combinations. The binary integer linear programming aims to linearize the binary integer nonlinear problem.

1.2 Contribution and Outline

The contribution of this paper is the design of a many-to-one algorithm to solve the multi-commodity multi-service matching problem under synchromodal hinterland transportation network. In the rest of this paper, we provide the problem description in Sect. 2. We design the many-to-one algorithm in Sect. 3. After that, we conduct numerical experiments to test the algorithm performance in Sect. 4. Finally, we conclude this paper in Sect. 5.

2   Problem Description

The multi-commodity multi-service matching problem defined in this paper contains two sets of participants. The first set is shippers that require transportation services to transport their commodities D. Commodity $d \in D$ is characterized by container volume V_d, origin terminal O_d and destination terminal D_d, the earliest departure time TD_d at the origin terminal and the latest arrival time TA_d at the destination terminal. Another set represents the carriers that provide subcontracted intermodal services S, self-operated intermodal services Q and truck services R. Subcontracted intermodal service $s \in S$ is characterized by origin terminal O_s and destination terminal D_s, capacity V_s, departure time TD_s, arrival time TA_s, transit cost c_s, transit time t_s, transit distance d_s and carbon emissions e_s. Self-operated intermodal service $q \in Q$ is characterized by origin terminal O_q and destination terminal D_q, capacity V_q, departure time TD_q, arrival time TA_q, transit cost c_q, transit time t_q, transit distance d_q and carbon emissions e_q. Truck service is flexible, which is assumed to be no capacity restrictions and no time window limitations. Truck service $r \in R$ can be characterized by origin O_r and destination D_r, transit cost c_r, transit time t_r, transit distance d_r and carbon emissions e_r.

The decision makers aim to match multiple commodities with multiple services for centralized optimization. Each commodity can be matched with multiple services by transferring at terminals; each service can be matched with multiple commodities within limited capacity. The objective consists of transit cost, transfer cost, storage cost and carbon tax. We define N as the set of terminals. Without loss of generality, we assume that the loading/unloading cost coefficient c_f, loading/unloading time t_f and storage cost coefficient c_w at each terminal are the same. The carbon tax coefficient is set as c_{tax}. We also assume the time window of commodities to be a hard time window. In addition, commodities are assumed to be unsplittable. Thus, for each commodity, only one path will be assigned. A path might consist of several services.

3   Solution Approach

As illustrated in Sect. 1, the mixed integer linear programming (MILP) proposed in [5] is only suitable for medium-sized instances. For large-scale instances of the MCMS matching problem, an efficient solution approach is required to reduce the computational complexity. Next, we propose a many-to-one algorithm to solve the problem. The many-to-one (M2O) algorithm consists of three steps: the preprocessing of path generation, the preprocessing of feasible matches and the binary integer linear programming.

A path is a combination of services. For example, a path p consists of a self-operated intermodal service q and a truck service r, thus, $p = \{q, r\}$. A service combination (path) is feasible only if the services within the collection meet time-space compatibility. The preprocessing of path generation is therefore designed to find feasible service combinations. After that, each commodity will be assigned to a path rather than multiple services. A match between a commodity and a path is feasible only if they meet time-space compatibility. Thus, the preprocessing of feasible matches aims to reduce the number of feasible matches between commodities and feasible paths. Based on the first two steps, the feasible path collection for each commodity will be generated. The original multi-commodity multi-service matching problem will be transformed to a multi-commodity single path problem. Furthermore, the time window compatibility have already considered in the preprocessing procedures. Thus, we do not need to design time constraints in the multi-commodity single path matching formulation.

However, as the cost of self-operated intermodal services is service-based cost, the matching cost between a commodity and a path including self-operated intermodal services is uncertain. In turn, the multi-commodity single path matching problem is a binary integer nonlinear problem. Therefore, the third step aims to linearize the nonlinear formulations.

Based on above three steps, a multi-commodity multi-service matching problem can be solved. For example, an instance of the problem consists of three commodities k_1, k_2, k_3 and three services s_1, q_1, r_1, as shown in Fig. 1. Here, $s1 : \{1, 2, 30, 7, 11, 2.5, 13.7\}$ represents that subcontracted intermodal service s_1 will depart from node 1 at 7:00 and arrives node 2 at 11:00 with capacity 30 TEU. Its transit cost is 2.5 €/TEU, and the carbon emissions is 13.7 kg/TEU. $q1 : \{1, 3, 100, 9, 13, 2111, 44\}$ represents that self-operated intermodal service q_1 will depart from node 1 at 9:00 and arrives node 3 at 13:00 with capacity 100 TEU. Its transit cost is 2111 €, and the carbon emissions is 44 kg/TEU. $r1 : \{2, 3, 1, 31, 53\}$ represents that truck service r_1 depart from node 2 to node 3 without time window schedule and capacity limitations. Its transit time is 1 h, transit cost is 31 €/TEU, and the carbon emissions is 53 kg/TEU. $k3 : \{1, 3, 25, 8, 24\}$ means that commodity k_3 needs to be transported from node 1 after 8:00 to node 3 before 24:00 with 25 TEU containers.

By using the preprocessing of path generation, all the feasible paths can be found. The feasible path collection include $s_1, q_1, r_1, \{s_1, r_1\}$. On the basis of the preprocessing of feasible matches, the feasible matches between commodities and feasible paths can be figured out. The feasible match collection consists of $(k_1, s_1), (k_2, q_1), (k_2, \{s_1, r_1\}), (k_3, q_1)$. Based on the binary integer linear programing, the optimal matches can be found.

3.1   Preprocessing of Path Generation

A path p can consist of a single service or multiple services. Thus, for an instance of the problem with n services, the number of possible service combinations is $C_n^1 + C_n^2 + ... + C_n^n = 2^n - 1$. Here, C_n^i means the number of i-combinations of a set with n elements. Due to the transfer limitation and time-space compatibility

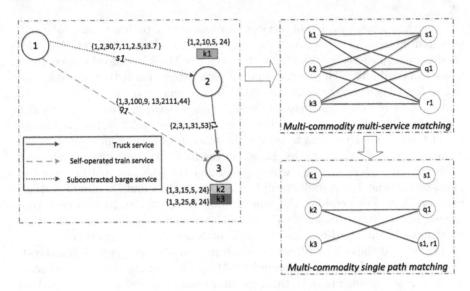

Fig. 1. An example of multi-commodity single path matching

constraints, some of the service combinations are infeasible. Next, we propose a preprocessing algorithm to generate the feasible service combinations, as shown in Algorithm 1. We limit the maximum transfer to 1. As in practice, transfer cost is sometimes higher than the transit cost of truck services. Thus, for each service combination, the maximum number of services is two. The set P denotes the collection of all feasible paths. For each hinterland terminal $n \in T = \{2, 3, ..., N\}$, the set of feasible paths with one service $P_{n,1}$ and two services $P_{n,2}$ will be generated, respectively. The index of the deep-sea port is set to 1. In addition, a path p is feasible only if it satisfies the following constraints:

Spatial Constraints. In a service combination with two services, the origin of the second service should be the destination of the first service.

Time Window Constraints. In a service combination with two intermodal services, the departure time of the second service should be greater than the arrival time of the first service plus loading and unloading time.

3.2   Preprocessing of Feasible Matches

By using the preprocessing of path generation, the set of feasible paths are produced. For an instance of the problem with N_D commodities and N_P feasible paths, the number of possible matches between commodities and feasible paths is $N_D N_P$. Therefore, for large-scale instances, the number of possible matches is very large. The preprocessing of feasible matches is designed to generate feasible matches, as shown in Algorithm 2. For each commodity $d \in D$, the feasible match sets $P_{d,1}$ between d and path $p \in P_{D_d,1}$, and $P_{d,2}$ between d and path $p \in P_{D_d,2}$

Algorithm 1. Path generation algorithm

Input: The origin, destination, scheduled timetable of all the intermodal services $S \cap Q$. The origin, destination of all the truck services R. The loading/unloading time t_f.
Output: Sets of the feasible paths for each hinterland terminal $P_{n,1} \cap P_{n,2}$.
Initialize: Let $P_{n,1} \leftarrow \emptyset$ and $P_{n,2} \leftarrow \emptyset$.

```
 1: for all destination n ∈ T do
 2:     for all service j ∈ S ∪ Q ∪ R do
 3:         if Oⱼ = 1 and Dⱼ = n then
 4:             Pₙ,₁ ← Pₙ,₁ ∪ {j}
 5:         end if
 6:     end for
 7: end for
 8: for all destination n ∈ T do
 9:     for all intermodal service j ∈ S ∪ Q do
10:         if Oⱼ ≠ 1 and Dⱼ = n then
11:             for all p ∈ P_{Oⱼ,1} do
12:                 if p ∈ S ∪ Q and TAₚ + 2t_f ≤ TDⱼ then
13:                     Pₙ,₂ ← Pₙ,₂ ∪ {{p,j}}
14:                 else if p ∈ R then
15:                     Pₙ,₂ ← Pₙ,₂ ∪ {{p,j}}
16:                 end if
17:             end for
18:         end if
19:     end for
20:     for all truck service j ∈ R do
21:         if Oⱼ ≠ 1 and Dⱼ = n then
22:             for all p ∈ P_{Oⱼ,1} do
23:                 Pₙ,₂ ← Pₙ,₂ ∪ {{p,j}}
24:             end for
25:         end if
26:     end for
27: end for
```

---

will be generated, respectively. A match (d, p) between commodity d and path p is feasible if it satisfies the following constraints:

Spatial Constraints. The destination terminal of commodity d should be the same destination terminal of path p.

Time Window Constraints. The earliest departure time of commodity d should be earlier than the departure time of path p plus loading time t_f. The arrival time of path p plus unloading time t_f should be earlier than the latest arrival time of commodity d.

3.3   Binary Integer Linear Programming

Based on above preprocessing procedure, the multi-commodity multi-service matching problem has been changed to a multi-commodity single path problem.

Algorithm 2. Feasible match generation algorithm

Input: Set of the feasible paths for each hinterland terminal. The destination terminal D_d, earliest departure time TD_d and latest arrival time TA_d of all the commodities D.
Output: Set of the feasible paths for commodity d, $P_{d,1} \cap P_{d,2}$.
Initialize: Let $P_{d,1} \leftarrow \emptyset$ and $P_{d,2} \leftarrow \emptyset$.

```
 1: for all destination d ∈ D do
 2:     for all path p ∈ P_{D_d,1} do
 3:         if p ∈ S ∪ Q then
 4:             if TD_d + t_f ≤ TD_p and TA_d ≥ TA_p + t_f then
 5:                 P_{d,1} ← P_{d,1} ∪ {p}
 6:             end if
 7:         else
 8:             P_{d,1} ← P_{d,1} ∪ {p}
 9:         end if
10:     end for
11: end for
12: for all destination d ∈ D do
13:     for all path p ∈ P_{D_d,2} do
14:         if p{1} ∈ S ∪ Q then
15:             if p{2} ∈ S ∪ Q then
16:                 if TD_d + t_f ≤ TD_{p{1}} and TA_d ≥ TA_{p{2}} + t_f then
17:                     P_{d,2} ← P_{d,2} ∪ {p}
18:                 end if
19:             else
20:                 if TD_d + t_f ≤ TD_{p{1}} and TA_d ≥ TA_{p{1}} + t_f + t_{p{2}} then
21:                     P_{d,2} ← P_{d,2} ∪ {p}
22:                 end if
23:             end if
24:         else
25:             if p{2} ∈ S ∪ Q then
26:                 if TD_d + t_{p{1}} + t_f ≤ TD_{p{2}} and TA_d ≥ TA_{p{2}} + t_f then
27:                     P_{d,2} ← P_{d,2} ∪ {p}
28:                 end if
29:             else
30:                 if TD_d + t_{p{1}} + t_{p{2}} ≤ TA_d then
31:                     P_{d,2} ← P_{d,2} ∪ {p}
32:                 end if
33:             end if
34:         end if
35:     end for
36: end for
```

The objective is to minimize the total cost for matches between commodities and the related feasible paths:

    Minimize

$$J = \sum_{d \in D} \sum_{p \in P_d} x_{dp} c_{dp} \tag{1}$$

where $P_d = P_{d,1} \cap P_{d,2}$ is the collection of feasible paths for commodity d, x_{dp} is a binary variable equal to 1 if commodity $d \in D$ is matched with path $p \in P_d$. Furthermore, c_{dp} is the cost of match (d,p), $c_{dp} = c1_{dp} + c2_{dp} + c3_{dp} + c4_{dp}$. Here, $c1_{dp}$ represents the transit cost, $c2_{dp}$ is transfer cost, $c3_{dp}$ is storage cost, and $c4_{dp}$ represents the carbon tax.

The cost of all the matches are deterministic parameters, except the matches with paths including self-operated intermodal services, i.e. $p \cap Q \neq \emptyset$. Because the transit cost of self-operated intermodal services is service-based cost, the transit cost of match $(d,p), q = p \cap Q$ is depends on the total volume assigned to the self-operated intermodal service q. We define P_{dq} as the subset of set P_d which only consists of paths including service q. Thus, $P_{dq} = \{p | d \in D, p \in P_d, q \in p \cap Q\}$. The transit cost of match $(d,p), q = p \cap Q$ is represented as: $c1_{dp} = c1_{dq} + c1_{dq^-}, q = p \cap Q, q^- = p \backslash \{q\}$, and q, q^- might be $\emptyset$. The transit cost of commodity d using self-operated intermodal service q is presented as:

$$c1_{dq} = \frac{c_q}{\sum_{d \in D} \sum_{p \in P_{dq}} x_{dp} V_d} V_d \qquad \forall d \in D, q \in Q \qquad (2)$$

According to Eq. (2), the expression of transit cost for all the commodities using all the self-operated intermodal services changes to:

$$\sum_{d \in D} \sum_{q \in Q} \sum_{p \in P_{dq}} x_{dp} c1_{dq} = \sum_{d \in D} \sum_{q \in Q} \sum_{p \in P_{dq}} x_{dp} \frac{c_q}{\sum_{d \in D} \sum_{p \in P_{dq}} x_{dp} V_d} V_d$$

$$= \sum_{q \in Q} c_q y_q \qquad (3)$$

We define $y_q = \frac{\sum_{d \in D} \sum_{p \in P_{dq}} x_{dp} V_d}{\sum_{d \in D} \sum_{p \in P_{dq}} x_{dp} V_d}$. If $\sum_{d \in D} \sum_{p \in P_{dq}} x_{dp} = 0$, $y_q = 0$. If $\sum_{d \in D} \sum_{p \in P_{dq}} x_{dp} \geq 1$, $y_q = 1$. Thus, $y_q = \min\{1, \sum_{d \in D} \sum_{p \in P_{dq}} x_{dp}\}$, where y_q is a binary variable. We define c_{dp}^* as the total cost of match (d,p) except the transit cost of Q. Thus, c_{dp}^* is deterministic parameters. Therefore, the optimization problem changes to:

$$J = \sum_{d \in D} \sum_{p \in P_d} x_{dp} c_{dp}$$

$$= \sum_{d \in D} \sum_{p \in P_d} x_{dp} (c1_{dq^-} + c2_{dp} + c3_{dp} + c4_{dp}) + \sum_{d \in D} \sum_{q \in Q} \sum_{p \in P_{dq}} x_{dp} c1_{dq} \qquad (4)$$

$$= \sum_{d \in D} \sum_{p \in P_d} x_{dp} c_{dp}^* + \sum_{q \in Q} c_q y_q$$

Subject to

$$\sum_{p \in P_d} x_{dp} = 1 \qquad \forall d \in D \qquad (5)$$

Equation (5) ensures that only one feasible path will be assigned to each commodity.

$$\sum_{d \in D} \sum_{p \in P_{ds}} x_{dp} V_d \leq V_s \qquad \forall s \in S \tag{6}$$

where $P_{ds} = \{p | d \in D, p \in P_d, s \in p \cap S\}$

Equation (6) ensures that the total volumes of commodities which is assigned to subcontracted intermodal service s do not exceed the capacity of s.

$$\sum_{d \in D} \sum_{p \in P_{dq}} x_{dp} V_d \leq V_q \qquad \forall q \in Q \tag{7}$$

where $P_{dq} = \{p | d \in D, p \in P_d, q \in p \cap Q\}$

Equation (7) ensures that the total volume of commodities which is assigned to self-operated intermodal service q do not exceed the capacity of q.

$$y_q = \min\{1, \sum_{d \in D} \sum_{p \in P_{dq}} x_{dp}\} \qquad \forall q \in Q \tag{8}$$

Equation (8) ensures that the cost of self-operated intermodal service q will be calculated only once no matter how many commodities are assigned to q. If no commodity is assigned to q, $y_q = 0$ ensures that the cost of self-operated intermodal service q will not be calculated. As Eq. (8) is nonlinear equation. The linearization of this equation is presented as:

$$y_q \geq x_{dp} \qquad \forall q \in Q, d \in D, p \in P_{dq} \tag{9}$$

4   Numerical Experiment

To evaluate the performance of the proposed many-to-one algorithm, we design 51 random instances of the multi-commodity multi-service matching problem. Each instance has a different size in terms of the number of subcontracted intermodal services, the number of self-operated intermodal services, the number of commodities, ratio between total commodity volume and total intermodal service capacity, ratio between average commodity volume and average intermodal service capacity, lead time of commodities, departure time of commodities, departure time of intermodal services. These instances are generated based on the instances proposed by [5]. We copy the instances with coordinates of network topology, transit cost coefficient and carbon emission coefficient.

The multi-commodity multi-service instances are solved on a desktop compute with Intel Core i5 2 GHz processor and 8 GB of RAM. The optimization problems are coded in MATLAB, and solved using CPLEX 12.6.3 with standard tuning.

4.1   Preprocessing Performance

In this section, we analyze the computational performance of preprocessing procedures. It consists of two steps: the preprocessing of path generation and the preprocessing of feasible matches.

Preprocessing of Path Generation. The first step is the preprocessing of path generation to generate feasible service combinations for each hinterland terminal. Because for intermodal services the departure timetables are scheduled in advance, this procedure is realized before commodities arriving. We design 9 instances to test the computational performance of the preprocessing of path generation. All of these instances are designed with 9 truck services, but with different number of intermodal services. From Fig. 2, we can see that for a network with 96 intermodal services, the computational time of preprocessing of path generation is no more than 0.03 s.

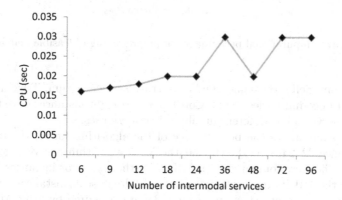

Fig. 2. Computational performance of preprocessing of path generation

Preprocessing of Feasible Matches. The second step is the preprocessing of feasible matches. In this paper, we assume that all the commodities arrived in advance, which we call static commodity-service matching. For each commodity, based on the time-space compatibility, the feasible paths can be find. Here, we design another 12 instances to test the computational performance of preprocessing of feasible matches. All of these instances are designed with the same number of services, but with different number of commodities. From Fig. 3, we can see that for the instance with 960 commodities (total container volume is 19156 TEU), the computational time of preprocessing of feasible matches is no more than 0.6 s.

4.2 Algorithm Performance

Next, we design another 11 instances to compare the computational performance of the MILP proposed in [5] and the M2O algorithm proposed in this paper, as shown in Table 1. We consider two performance measurements: the total transportation cost and the computational time. The first four instances have the same number of inland terminals (Num. InlTer), the same number of subcontracted intermodal services (Num. Sub), the same number of self-operated intermodal

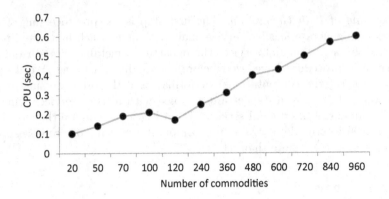

Fig. 3. Computational performance of preprocessing of feasible matches

services (Num. Self), the same number of truck services (Num. Tru) and different number of commodities (Num. Com). The last eight instances have the same number of services but different number of commodities.

Table 1 summarizes the performance of the algorithm for all the instances. From instance 22–24, it is shown that the M2O algorithm has very good computational efficiency compared with MILP. Furthermore, for instances 25–32, it shows that the MILP programming cannot solve large-scale instances by running out of memory. In contrast, no more than 3 min is required by using M2O algorithm. If we use CPLEX directly to solve the MILP programming rather than using MATLAB calling CPLEX, the MILP can solve instance 25–30, as shown in Table 1. However, the computational time of MILP is much higher than the M2O algorithm.

As we limit the maximum transfer to 1, the solution obtained by M2O algorithm is suboptimal. Nevertheless, the results obtained from M2O algorithm is quite close to optimal solutions. For instance 22–32, the gaps between suboptimal solutions and optimal solutions are no more than 0.008%.

4.3 Sensitivity Analysis

To perform sensitivity analysis over some of the parameters of multi-commodity multi-service matching problem, we design another 19 instances with the same number of services. The total capacity of intermodal services is 7200 TEU (3600 TEU for subcontracted intermodal services and 3600 TEU for self-operated intermodal services). The average capacity of intermodal services is 100 TEU. To measure the performance of matching solutions, we consider four different indicators: computation time (CPU), total cost, modal split and capacity utilization (CapUti).

We design instances 33–36 to test the influence of the ratio between the average volume of commodities and the average capacity of intermodal services (Ratio 1). For instance 33–36, Ratio 1 are increasing from 0.2 to 3.6. The ratio

Table 1. Algorithm performance

Instances	Num. InlTer	Num. Sub	Num. Self	Num. Tru	Num. Com	CPU		Cost		CPLEX	
						MILP	M2O	MILP	M2O	CPU (sec)	Cost
22	5	6	6	6	20	3.8	0.11	18500	18500		
23	5	6	6	6	50	47.89	0.18	57033	57039		
24	5	6	6	6	70	136.51	0.29	81705	81847		
25	5	48	24	9	120	Out of memory	0.82		124230	6.8	124220
26	5	48	24	9	240	Out of memory	2.38		257450	26.38	257390
27	5	48	24	9	360	Out of memory	6.75		417920	217.08	417910
28	5	48	24	9	480	Out of memory	11.76		566570	239.87	566550
29	5	48	24	9	600	Out of memory	21.31		730020	285.77	729970
30	5	48	24	9	720	Out of memory	35.86		889080	301.21	889060
31	5	48	24	9	840	Out of memory	56.59		1037500		
32	5	48	24	9	960	Out of memory	147.03		1206800		

between total volume of commodities and the total capacity of the intermodal services (Ratio 2), the lead time of commodities (LeTim), the departure time of commodities (DepTime. Com), the departure time of intermodal services (DepTime.Int) are the same. The results of the sensitivity analysis are shown in Table 2. It indicates that the higher the average volume of commodities, the lower the modal split and the capacity utilization of intermodal services (Int). It is reasonable because lager value of average volume of commodities means more containers need to be bundled together by choosing the same routes. In turn, limited the number of feasible paths including intermodal services.

Regarding instances 36–39, we design different ratio between total volume of commodities and the total capacity of the intermodal services (Ratio 2), and keep other parameters as the same. The results suggest that the larger the total volume of commodities, the lower the modal split of intermodal services. However, the capacity utilization of intermodal services increases with the increasing of Ratio 2. The reason is that the more commodities matching with intermodal services, the more capacity of intermodal services will be utilized. However, due to the capacity limitation of intermodal services, the larger the volume of the commodities, the more the commodities that will not be matched with intermodal services.

In terms of instance 40–42, we design different lead time of commodities (LeTim. Com) while keep other parameters as the same. The matching results indicate that the modal split of intermodal services increases with the increasing of leading time of commodities. The capacity utilization of intermodal services has the same trend.

Table 2. Sensitivity analysis over 19 instances

Instances	Ratio 1	Ratio 2	LeTim. Com	DepTim. Com	DepTim.Int	CPU (sec)	Cost	CapUti (%)	Modal split (%)	
									Int	Tru
33	0.20	1.00	24	Whole day	Morning	2.15	456330	38.96	32.35	67.65
34	0.60	1.00	24	Whole day	Morning	0.55	464340	38.51	31.93	68.07
35	1.44	1.00	24	Whole day	Morning	0.29	474460	28.14	21.47	78.53
36	3.60	1.00	24	Whole day	Morning	0.16	536160	0	0	1
37	0.20	0.33	24	Whole day	Morning	0.5	148320	13.81	35.21	64.79
38	0.20	1.00	24	Whole day	Morning	2.25	456330	38.96	32.35	67.65
39	0.20	1.67	24	Whole day	Morning	12.5	774290	57.79	29.37	70.63
40	0.60	1.00	12	Whole day	Evening	1.29	475430	42.74	33.66	66.34
41	0.60	1.00	18	Whole day	Evening	0.8	430320	54.72	47.16	52.84
42	0.60	1.00	24	Whole day	Evening	2.97	410640	61.21	54.31	45.69
43	0.60	1.00	18	Morning	Morning	0.89	411840	62.78	52.46	47.54
44	0.60	1.00	18	Whole day	Morning	0.59	464340	38.51	31.93	68.07
45	0.60	1.00	18	Evening	Morning	0.16	549640	0	0	1
46	0.60	1.00	18	Morning	Whole day	1.61	411590	64.71	55.14	44.86
47	0.60	1.00	18	Whole day	Whole day	0.92	398900	69.64	58.7	41.3
48	0.60	1.00	18	Evening	Whole day	0.7	448350	42.54	35.86	64.14
49	0.60	1.00	18	Morning	Evening	0.68	471480	38.64	34.8	65.2
50	0.60	1.00	18	Whole day	Evening	0.71	430320	54.72	47.16	52.84
51	0.60	1.00	18	Evening	Evening	1.2	407220	64.67	54.98	45.02

Besides, we design instances 43–51 to identify the relationship between the time window of commodities (DepTime. Com) and the time window of inter-modal services (DepTime.Int). These instances have the same average volume of commodities, the same total volume of commodities and the same lead time of commodities. However, the departure time of commodities and intermodal services are different. For instance 43–45, the departure time of intermodal services are scheduled in the morning from 1:00 to 12:00. The results show that instance 43, which departure time of commodities is also in the morning, has best performance of modal split of intermodal services. For instance 46–48, the departure time of intermodal services that are scheduled in the whole day from 1:00 to 24:00. Instance 47, which departure time of commodities is also in the whole day, has best performance of modal split of intermodal services. For instance 49–51, intermodal services are scheduled to depart in the evening from 13:00 to 24:00. Instance 51, which departs in the evening, has best performance of modal split of intermodal services. It shows that the instance which the time window of commodities is consistent with the time schedule of intermodal services has best performance.

5 Conclusion and Future Research

This paper proposes a many-to-one algorithm to solve the multi-commodity multi-service matching problem of routing in synchromodal hinterland container transportation. The algorithm includes three steps: (1) The preprocessing of path generation is designed to reduce the number of service combinations; (2) The preprocessing of feasible matches is designed to reduce the number of matches between commodities and feasible service combinations; (3) The binary integer linear programming is investigated to solve the multi-commodity single path matching problem. In total, 51 instances of the problem are used to assess the computational performance of the many-to-one algorithm. The experiments results indicate that this algorithm is suitable for large-scale instances of the problem. Compared with an exact algorithm, the solution gaps between optimal solutions and suboptimal solutions are no more than 0.008%. Based on the sensitivity analysis, it is illustrated that the average volume, total volume, lead time, and departure time of commodities have a considerable influence on the modal split and the capacity utilization of intermodal services.

In this paper, we assume the demand information is known in advance. However, in practice, requests arrive in real-time. Future research will investigate the real-time multi-commodity multi-service matching problem of routing in synchromodal hinterland container transportation.

Acknowledgments. This research is financially supported by the China Scholarship Council under Grant 201606950003.

References

1. Arslan, A., Agatz, N., Kroon, L.G., Zuidwijk, R.A.: Crowdsourced delivery - a pickup and delivery problem with ad-hoc drivers. SSRN Electron. J. 1–26 (2016)
2. Chen, W., Mes, M., Schutten, M.: Multi-hop driver-parcel matching problem with time windows. Flex. Serv. Manuf. J. 1–37 (2017)
3. Eurostat: Modal split of freight transport (2018). http://ec.europa.eu/eurostat/tgm/refreshTableAction.do?tab=table&plugin=1&pcode=t2020_rk320&language=en
4. Guo, W., Beelaerts van Blokland, W., Lodewijks, G.: Survey on characteristics and challenges of synchromodal transportation in global cold chains. In: Bektaş, T., Coniglio, S., Martinez-Sykora, A., Voß, S. (eds.) Computational Logistics. LNCS, vol. 10572, pp. 420–434. Springer, Cham (2017). https://doi.org/10.1007/978-3-319-68496-3_28
5. Guo, W., Beelaerts van Blokland, W., Negenborn, R.R.: Multi-commodity multi-service matching design for hinterland container transportation. Delft University of Technology, Technical report (2018)

6. Mahmoudi, M., Zhou, X.: Finding optimal solutions for vehicle routing problem with pickup and delivery services with time windows: a dynamic programming approach based on state–space–time network representations. Transp. Res. Part B Methodol. **89**, 19–42 (2016)

7. Masoud, N., Jayakrishnan, R.: A real-time algorithm to solve the peer-to-peer ride-matching problem in a flexible ridesharing system. Transp. Res. Part B Methodol. **106**, 218–236 (2017)

8. van Riessen, B., Negenborn, R.R., Dekker, R.: Real-time container transport planning with decision trees based on offline obtained optimal solutions. Decis. Support Syst. **89**, 1–16 (2016)

9. van Riessen, B., Negenborn, R.R., Lodewijks, G., Dekker, R.: Impact and relevance of transit disturbances on planning in intermodal container networks using disturbance cost analysis. Maritime Econ. Logist. **17**(4), 440–463 (2014)

10. SteadieSeifi, M., Dellaert, N., Nuijten, W., Woensel, T.V., Raoufi, R.: Multimodal freight transportation planning: a literature review. Eur. J. Oper. Res. **233**(1), 1–15 (2014)

11. Stiglic, M., Agatz, N., Savelsbergh, M., Gradisar, M.: The benefits of meeting points in ride-sharing systems. Transp. Res. Part B Methodol. **82**, 36–53 (2015)

A Heuristic Approach to the Driver and Vehicle Routing Problem

Bencomo Domínguez-Martín, Inmaculada Rodríguez-Martín(ID),
and Juan-José Salazar-González(✉)(ID)

DMEIO, Facultad de Ciencias, Universidad de La Laguna, Tenerife, Spain
{bdomingu,irguez,jjsalaza}@ull.es

Abstract. This paper presents a heuristic approach for a routing problem with two depots and involving vehicles and drivers. Vehicles must leave from one depot and arrive to the other, while drivers should leave and return to the same depot and their routes can not exceed a given duration. With these conditions, drivers must change vehicles in order to go back to their base depots. These changes can only take place at some particular nodes known as exchange locations. The heuristic approach in this paper is a two-phase method for the problem with one exchange location. In the first phase it solves a mathematical model to obtain drivers' routes. In the second phase it builds vehicles' routes over the given drivers' routes. Computational results show that the proposed approach can find high quality solutions for instances with up to 50 nodes.

Keywords: Vehicle routing · Multi depot · Heuristics

1 Introduction

The *Driver and Vehicle Routing Problem* (DVRP) addressed in this paper is defined as follows. We are given two depots, where a given number of vehicles and drivers are based, and a set of customers. Each customer must be served by a vehicle and a driver. Vehicles start their routes at their base depot and end at the other depot, while drivers must start and end their routes at their base depot. The vehicles have to be always led by a driver, and drivers need a vehicle to move from one location to another, either driving themselves or as passengers. When there are more than one driver in a vehicle, any of them can lead the vehicle. The duration of a driver route is the time between the departure from and the arrival to the depot, and it includes the time driving and traveling as passenger. Moreover, drivers' routes cannot exceed a given time duration. Drivers can switch vehicles only at some given points known as exchange locations, which are the only customer locations that can be visited by more than one vehicle. To make feasible the interaction between drivers and vehicles, their routes must be time

Supported by the research projects MTM2015-63680-R (MINECO/FEDER) and ProID2017010132 (Gobierno de Canarias).

R. Cerulli et al. (Eds.): ICCL 2018, LNCS 11184, pp. 295–305, 2018.
https://doi.org/10.1007/978-3-030-00898-7_19

synchronized. The objective is to design the routes of the vehicles and the drivers in order to minimize the total drivers' cost.

Figure 1 shows the optimal DVRP solution for an instance with 5 customers (nodes 1 to 5). Nodes 0 and 6 are the two depots, and vehicles' exchanges can take place only at node 5. There are four drivers and four vehicles available at each depot, the duration of all the travels between nodes is set to 1, and the maximum duration of the drivers' routes is set to 4. The figure shows that only two vehicles and two drivers from each depot are used in the optimal solution. The drivers make circular routes (dashed lines) starting and ending at the same depot, and vehicles' routes start at a depot and end at the other one (solid lines). We see that drivers exchange vehicles at node 5. For example, one of the drivers leaving from depot 0 makes the route 0-5-4-0. At node 5 he moves from one vehicle going from 0 to 6 to another going from 6 to 0 in order to be able to return to his base. The second driver leaving from depot 0 makes the route 0-5-2-0. The two vehicles leaving from 0 make the routes 0-5-1-6 and 0-5-3-6, respectively. As for depot 6, the drivers leaving it make the routes 6-5-1-6 and 6-5-3-6, while its vehicles' routes are 6-5-2-0 and 6-5-4-0. This solution has a unique driver in each vehicle at all time.

This problem was inspired by the characteristics of the air transportation in the Canary Islands, where a set of flights must be served by crews and aircrafts under some conditions (see Salazar-González [10]). In particular, there are two main airports in Canary Islands (Tenerife North and Las Palmas); while crews must return each day to their base airport (to avoid overnight in hotels), aircrafts start and end their daily routes at different bases due to short maintenance operations (aircraft maintenance can only be performed in Las Palmas, and each aircraft must be checked every other day). There are other situations like,

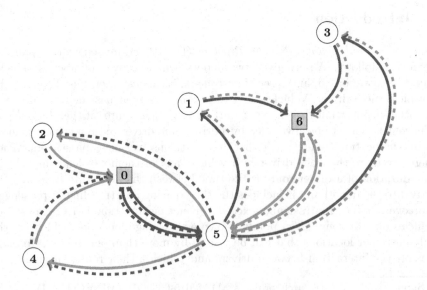

Fig. 1. Solution example

for example, long-distance ground transportation, where the DVRP might have applications. In those cases, vehicles traveling long distances are not always led by the same driver along the entire route, and drivers return to their home bases at the end of their work shifts. All routes must pass through locations where drivers can switch vehicles. In the airline context, the flights are the customers requiring service, and the switches are performed by crews on airports. This is a major difference with respect to ground transportation, where some customer locations may act as exchange locations.

The DVRP was introduced by Domínguez-Martín et al. [3]. In that paper the authors propose a mathematical programming formulation for the problem, and develop an exact branch-and-cut algorithm. A related vehicle and driver scheduling problem had been previously addressed in [2]. In fact, most of the literature on problems that involve vehicles and drivers fall within the field of the Vehicle-and-Crew Scheduling (see, for example, [1,5–9]). Usually in those problems the aim is to find the minimum-cost assignment of crews and vehicles to a given set of trips with fixed starting and ending times. The DVRP can also be seen as a vehicle routing problem with multiple synchronization constraints (see the classification scheme by Drexl [4]). If so considered, it would fit into the category called *movement synchronization en route*, that gathers problems with non-autonomous vehicles that require autonomous vehicles to move in space, and that may join and separate at locations, different from the depot, that they visit during their routes. Practical applications of the DVRP could also include time-windows for serving each customer, but we adopt the same assumptions that in [3] and do not consider them.

In this paper we present a two-phase heuristic method for the DVRP with one exchange location, that provides high quality solutions for the instances considered. The first phase of the algorithm generates driver's routes and the second phase builds suitable vehicles' routes to end up with a feasible DVRP solution. The remainder of the paper is organized as follows. Section 2 gives the details of the heuristic algorithm. Section 3 shows the computational results, and finally Sect. 4 is devoted to conclusions and future research.

2   Heuristic Approach

As said before, the heuristic method we propose for solving the DVRP with one exchange location is a two-phase method. The first phase finds drivers' routes, while the second phase builds vehicles' routes for those drivers' routes. The key point of the algorithm is the generation of the drivers' routes. We will do it by solving a mathematical model that imposes to the drivers all the problem requirements, and some additional constrains that will ensure that the obtained routes will be compatible with some vehicles' routes. These procedures are described in more detail in the following subsections.

2.1   First Phase: Generation of Drivers' Routes

To mathematically model the driver's routes problem we need some notation. We use the general notation in [3], although our approach is designed to solve instances with one exchange location. Let $V_c = \{1, \ldots, n\}$ be the set of customer locations and $D = \{0, n + 1\}$ be the set of depots. The n customer locations are partitioned in two subsets, so that $V_c = V_r \cup V_e$. The set V_e contains the *exchange* location, that is, the place where drivers can exchange vehicles, and $V_r = V_c \setminus V_e$ is the set of *regular* customer locations. Both types of customers must be visited by at least one driver. Let $G = (V, A)$ be a complete directed graph with vertex set $V = D \cup V_r \cup V_e$ and arc set $A = \{(i, j) : i, j \in V, i \neq j\}$. To refer to the set of arcs with tail in a set $S \subseteq V$ and head in $V \setminus S$, we use $\delta^+(S)$ instead of $\{(i, j) \in A : i \in S, j \notin S\}$, and we use $\delta^-(S)$ instead of $\delta^+(V \setminus S)$. Given two subsets of vertices S and S', $(S : S')$ represents the arc set $\{(i, j) \in A : i \in S, j \in S'\}$. The set of drivers available at each depot $d \in D$ is denoted by K_d, and $K = \cup_{d \in D} K_d$. The time needed to traverse an arc (i, j) is denoted by t_{ij}. Drivers' routes cannot exceed a given time limit T. There is a known cost c_{ij} to pay when a driver traverses the arc $(i, j) \in A$. This cost may represent the salary of the driver and be related to the distance or the time needed to go from i to j. Each driver must start and end its route at the same depot. Our objective is to find in G minimum cost routes for the drivers that will then be compatible with vehicles routes to form a feasible DVRP solution. To model the problem we use two sets of decision variables. Variables x_{ij}^k take value 1 if the driver k traverses the arc (i, j), and 0 otherwise. Variables u_i^k take value 1 if the driver k visits the customer i, and 0 otherwise. We write $x^k(A')$ instead of $\sum_{(i,j) \in A'} x_{ij}^k$, for each $A' \subset A$ and each $k \in K$.

Then the drivers' routes problem can be formulated as follows:

$$\min \sum_{k \in K} \sum_{(i,j) \in A} c_{ij} x_{ij}^k \tag{1}$$

subject to

$$x^k(\delta^+(i)) = x^k(\delta^-(i)) = u_i^k \qquad i \in V_c, k \in K \tag{2}$$

$$\sum_{k \in K} u_i^k \geq 1 \qquad i \in V_c \tag{3}$$

$$x^k(\delta^+(n+1)) = x^k(\delta^-(n+1)) = 0 \qquad k \in K_0 \tag{4}$$

$$x^k(\delta^+(0)) = x^k(\delta^-(0)) = 0 \qquad k \in K_{n+1} \tag{5}$$

$$\sum_{(i,j) \in A} t_{ij} x_{ij}^k \leq T \qquad k \in K \tag{6}$$

$$\sum_{i \in V_e} u_i^k \geq x^k(\delta^+(d)) \qquad k \in K_d, d \in D \tag{7}$$

$$x^k(\delta^-(S)) \geq u_i^k \qquad k \in K, S \subseteq V_c, i \in S \tag{8}$$

$$x_{ij}^k + x_{i'j}^{k'} \leq 1 \qquad j \in V_r, \; i, i' \in V, \; k, k' \in K : \tag{9}$$
$$i \neq i' \qquad k \neq k'$$

$$\sum_{k \in K_0} x^k(\delta^+(0)) = \sum_{k \in K_{n+1}} x^k(\delta^+(n+1)) \tag{10}$$

$$x_{ij}^k \in \{0, 1\} \qquad k \in K, (i, j) \in A \tag{11}$$

$$u_i^k \in \{0, 1\} \qquad i \in V, k \in K. \tag{12}$$

The objective function (1) minimizes the total cost of the drivers' routes. Equation (2) determine if a driver k visits the customer i. Constraints (3) establish that at least one driver must visit each customer. Equalities (4) and (5) guarantee that a driver from one depot will not end up at the other depot. Inequalities (6) limit to T the maximum time that a driver can spend on a route. Inequalities (7) ensure that all the drivers used in the solution visit the exchange location. Constrains (8) eliminate subtours involving only customers. Inequalities (9) ensure that, if two drivers k and k' arrive to a regular customer j, then they must come together from the same location i. Recall that regular customers must be served by one vehicle, but several drivers are allowed to go in the vehicle. Constraint (10) imposes that the same number of drivers start their routes at depots 0 and $n+1$. This helps to find vehicles' routes compatible with the obtained drivers' routes. Finally, constraints (11) and (12) define the domain of the variables.

The linear programming relaxation of this model can be strengthened using three families of valid inequalities already presented in [3]. The first family ensures that at least one driver leaves each depot:

$$\sum_{k \in K_d} x^k(\delta^+(d)) \geq 1 \qquad d \in D. \tag{13}$$

The second family helps to avoid some symmetries in the solutions, due to permutations of the drivers:

$$x^k(\delta^+(d)) \geq x^{k+1}(\delta^+(d)) \qquad d \in D, k = 1,\ldots,|K_d| - 1. \tag{14}$$

The third family forbids a driver route to return to its depot after having visited regular customers only:

$$x^k(S : V_c \setminus S) \geq x^k(d : S) \qquad d \in D, k \in K_d, S \subseteq V_r. \tag{15}$$

Constraints (9) can be strengthened by considering all arcs in $\delta^-(j)$ instead of only the arcs (i,j) and (i',j). To this end, consider any partition of $\delta^-(j)$ in E and F with $(i,j) \in E$ and $(i',j) \in F$. Then, a stronger inequality is $x^k(E) + x^{k'}(F) \leq 1$, as a driver k cannot arrive to j through an arc in E when another driver k' has arrived to j through an arc in F.

From the assumption that $|V_e| = 1$, if a vehicle arrives to the regular customer before passing through the exchange location, all drivers in that vehicle are from the depot where it started; otherwise, all the drivers in a vehicle are from the depot where that vehicle will arrive. This observation allows us to insert a new family of inequalities ensuring that all the drivers visiting a regular customer must belong to the same depot:

$$u_j^k + u_j^{k'} \leq 1 \qquad j \in V_r, k \in K_d, k' \in K_{d'}, d, d' \in D : d \neq d', \tag{16}$$

and reduce (9) to only drivers k and k' from the same depot d, for all $d \in D$.

To solve the driver's routes problem we design a branch-and-cut algorithm based on the mathematical model and valid inequalities just presented. A branch-and-cut algorithm combines a branch-and-bound method for exploring a decision tree and a cutting plane method for computing bounds. At each node of the search tree, the cutting plane method improves a linear relaxation of the problem. When this is not further possible, the branch-and-bound algorithm proceeds. In our algorithm, the computation begins by solving the linear relaxation of the model (1)–(7), (10)–(12), plus (13)–(14). Then, in each cutting plane iteration, violated inequalities (9), (16), (8) and (15) are added, if found, in this sequence. The violation of inequalities (9) and (16) can be checked in polynomial time by simple enumeration. The separation procedures for (8) and (15) are described in [3].

2.2 Second Phase: Generation of Vehicles' Routes

We have designed an iterative procedure to build vehicles' routes compatible with the drivers' routes obtained in the first phase of the heuristic. The procedure works as follows. If an arc $(d, i) \in \delta^+(d)$ is traversed by one or more drivers $k \in K_d$ (i.e., the solution x^* from the first phase has a variable $x_{di}^{*k} = 1$), then a vehicle departs from d using that arc, and follows the route of the driver until the exchange location is reached. From that point to the other depot, d', the vehicle follows the route of a driver $k' \in K_{d'}$. When $|V_e| = 1$ all drivers' routes pass through the exchange location and Equation (10) guarantees that a matching between drivers of different depots is possible at the exchange location.

When all arcs in $\delta^+(d)$ have been processed this way, we check whether the number of vehicles departing from depot d coincides with the number of drivers arriving to depot d' through different arcs. If yes, we already have a feasible DVRP solution. If not, it means that there is some driver $k' \in K_{d'}$ going from the exchange location to d' without a vehicle. To correct this situation, an extra vehicle is taken from d to that exchange location following an existing driver route, and then from there to d' following the route of driver k'.

3 Computational Results

The algorithm was coded in C++ and run on a desktop computer with an Intel(R) Core(TM) i3-3240 CPU @ 4.40 GHz, 8 GB RAM, and Windows 7 Professional. We used CPLEX 12.7 as MIP solver. The computational experiments were carried on the set of instances used in [3]. These are randomly generated instances with sizes $n+2 \in \{10, 15, 20, 25, 30\}$ and node coordinates in the square $[0, 100] \times [0, 100]$. The first and the last nodes generated are the depots 0 and $n + 1$. There are three vehicles and three drivers available at each depot. One customer location is chosen randomly to act as exchange location. The costs c_{ij} are defined as the Euclidean distances between i and j, and the times needed to traverse the arcs are defined as $t_{ij} = c_{ij}/60 + 0.5$. The parameter T that limits the duration of the drivers' routes takes four increasing values denoted by T_A, T_B, T_C, and T_D, being T_A the tightest value of T that allows to find a feasible solution. There are five instances for each problem size, resulting in a set of 25 instances.

Table 1. Results for small and medium sized instances with tight T values

name	$n+2$	T_A	timeE	timeH	gapH	T_B	timeE	timeH	gapH
n10-1	10	6	7.22	1.15	0.00	7	0.41	0.51	0.00
n10-2	10	5	17.43	0.37	0.00	6	0.06	0.12	0.00
n10-3	10	5	272.77	30.72	0.00	6	327.38	0.87	0.00
n10-4	10	5	4.52	1.95	0.00	6	8.00	0.90	0.00
n10-5	10	5	6.91	1.28	0.00	6	0.59	0.64	0.00
n15-1	15	6	34.43	8.38	0.00	8	0.83	1.68	0.00
n15-2	15	6	60.59	3.06	0.00	8	1.06	1.58	0.00
n15-3	15	6	50.44	4.24	0.00	8	9.72	3.49	0.00
n15-4	15	6	7200.00	46.33	0.00	8	473.82	7.97	0.00
n15-5	15	6	83.09	14.71	0.00	8	7200.00	35.40	0.00
n20-1	20	7	7200.00	600.00	−4.28	9	7200.00	368.12	−0.35
n20-2	20	7	4071.75	13.37	0.00	9	4.84	5.57	0.00
n20-3	20	7	6576.33	73.84	0.00	9	7200.00	194.61	0.00
n20-4	20	7	997.44	20.37	0.00	9	6305.20	104.47	0.00
n20-5	20	7	3623.64	140.71	0.00	9	1574.83	85.96	0.00
n25-1	25	8	2002.73	155.10	0.00	10	7200.00	600.00	−0.56
n25-2	25	8	7200.00	600.00	−3.39	10	7200.00	600.00	0.56
n25-3	25	8	3463.28	151.01	0.00	10	4473.84	207.22	0.00
n25-4	25	8	3530.49	168.98	0.00	10	7200.00	600.00	0.00
n25-5	25	8	3720.39	41.04	0.00	10	7200.00	600.00	−1.99
n30-1	30	9	7200.00	600.00	1.48	12	7200.00	600.00	−4.63
n30-2	30	9	7200.00	600.00	−7.66	12	7200.00	600.00	−4.29
n30-3	30	9	7200.00	600.00	0.00	12	7200.00	600.00	−2.81
n30-4	30	9	7200.00	222.50	−4.89	12	7200.00	600.00	−2.07
n30-5	30	9	7200.00	600.00	1.46	12	7200.00	600.00	−1.35

We have also created sixteen new larger instances, with 50 nodes (i.e. $n = 48$). In these instances, one depot is randomly placed in $[0, 20] \times [0, 100]$ and the other one in $[80, 100] \times [0, 100]$, the exchange location is located in $[40, 60] \times [0, 100]$, and the remaining customers in $[0, 100] \times [0, 100]$. This configuration intends to represent a situation that usually appears in ground transportation applications, with the depots far from each other and the exchange point placed at some central location between them. For these instances we set $T = 18$. The other parameters are defined as in the smaller cases. The whole set of instances is available from the authors upon request.

We ran the heuristic algorithm with a time limit of 600 s on each instance. Tables 1 and 2 show the results obtained for the instances with up to 30 nodes, with tight and loose T values respectively. We compare the heuristic results with

Table 2. Results for small and medium sized instances with looser T values

name	$n+2$	T_C	timeE	timeH	gapH	T_D	timeE	timeH	gapH
n10-1	10	8	0.73	0.42	0.00	10	0.05	0.08	0.00
n10-2	10	7	0.08	0.09	0.00	10	0.05	0.09	0.00
n10-3	10	7	0.14	0.67	0.00	10	0.06	0.03	0.00
n10-4	10	7	0.27	0.14	0.00	10	0.06	0.09	0.00
n10-5	10	7	0.12	0.27	0.00	10	0.06	0.06	0.00
n15-1	15	9	0.97	1.68	0.00	12	0.23	0.36	0.00
n15-2	15	9	0.23	0.22	0.00	12	0.25	0.23	0.00
n15-3	15	9	0.19	0.41	0.00	12	0.20	0.28	0.00
n15-4	15	9	0.69	1.23	0.00	12	1.37	1.50	0.00
n15-5	15	9	0.59	0.55	0.00	12	0.31	0.53	0.00
n20-1	20	10	10.41	4.79	0.00	14	2.48	2.82	0.00
n20-2	20	10	9.63	2.06	0.00	14	2.06	3.43	0.00
n20-3	20	10	32.84	3.39	0.00	14	2.82	3.10	0.00
n20-4	20	10	9.66	4.74	0.00	14	0.44	0.67	0.00
n20-5	20	10	7.78	5.85	0.00	14	13.71	6.02	0.00
n25-1	25	11	110.18	6.96	0.00	16	4.57	2.61	0.00
n25-2	25	11	167.98	12.28	0.00	16	52.12	12.89	0.00
n25-3	25	11	9.87	5.82	0.00	16	1.31	1.98	0.00
n25-4	25	11	22.28	5.38	0.00	16	9.50	3.51	0.00
n25-5	25	11	58.72	4.74	0.00	16	2.87	2.37	0.00
n30-1	30	13	135.27	38.06	0.00	18	57.35	17.14	0.00
n30-2	30	13	62.06	25.26	0.00	18	22.12	7.69	0.00
n30-3	30	13	38.78	12.45	0.00	18	10.41	5.34	0.00
n30-4	30	13	146.89	21.82	0.00	18	121.43	17.96	0.00
n30-5	30	13	18.35	7.11	0.00	18	35.32	8.38	0.00

the results of the exact method described in [3]. The exact method was run with a time limit of two hours, and it gives the optimal solution if that time limit is not reached (the optimality gaps for the unsolved instances are given in [3]). Columns' headings stand for: instance name (*name*), problem size ($n+2$), time limit on the drivers' routes (T_A, T_B, T_C, or T_D), computing time in seconds taken by the exact method (*timeE*), computing time in seconds taken by the heuristic algorithm (*timeH*), and percentage deviation between the objective values of the best solutions found by the exact and heuristic approaches (*gapH*).

Negative figures in Columns *gapH* mean that the heuristic algorithm outperforms the exact method and, therefore, new best known solutions have been found for the corresponding instances. A zero deviation percentage means that the heuristic result equals the exact method result. Note that for all but three of the instances in Table 1, those with tight T values, the heuristic solution either coincides with the optimal solution, when it is known, or it is better than the result reported in [3]. Moreover, the heuristic running times are usually much smaller than that of the exact method, and this is more noticeable for the hardest instances. For example, the exact method takes generally 1 h to solve the instances with 25 nodes and $T = T_A$, while the heuristic takes usually less than 4 min. Regarding the instances with looser T values, in Table 2, the heuristic manages to find the optimal solution for all of them.

Table 3. Results for instances with $n + 2 = 50$ and $T = 18$

name	ubE	timeE	lbE	gapE	ubH	timeH	gapH
n50-1	606	788.62	594.30	1.93	606	142.37	0.00
n50-2	–	7200.00	–	–	592	177.45	–
n50-3	581	1943.24	567.13	2.39	581	302.14	0.00
n50-4	593	2822.73	584.36	1.46	593	134.04	0.00
n50-5	933	7200.00	599.28	35.77	613	226.69	−34.30
n50-6	584	675.78	573.64	1.77	584	107.17	0.00
n50-7	548	1166.34	539.96	1.47	548	132.30	0.00
n50-8	588	939.03	578.92	1.54	588	94.79	0.00
n50-9	603	6081.71	587.14	2.63	603	325.28	0.00
n50-10	597	5218.73	586.11	1.82	597	107.34	0.00
n50-11	607	7200.00	579.67	4.50	592	511.14	−2.47
n50-12	1269	7200.00	593.23	53.25	825	600.00	−34.99
n50-13	–	7200.00	–	–	780	600.00	–
n50-14	–	7200.00	–	–	613	143.08	–
n50-15	867	7200.00	621.59	28.31	821	600.00	−5.31
n50-16	613	3616.71	603.30	1.58	613	206.25	0.00

Finally, Table 3 shows the results for the set of instances with 50 nodes. Since these instances are new in the literature, the table shows the objective value of the best solution found by the exact and the heuristic algorithms in columns ubE and ubH, respectively, the best lower bound computed by the exact algorithm in lbE, and the optimality gap $gapE$ between the best upper and lower bounds found by the exact algorithm. It is worth noting that these instances are difficult to be solved by the exact approach described in [3], which was unable to find even feasible solutions for three instances after two hours of computation. By contrast, the heuristic approach described in this paper found feasible solutions for all the instances. These heuristic solutions are optimal for the nine instances that were successfully solved by the exact approach, and are better than the best solutions found by the exact approach for the others instances.

4   Conclusions and Future Research

We have presented in this paper a first heuristic approach for solving the DVRP, a complex two-depot routing problem that aims to couple vehicles and drivers that perform different types of routes (circular and time-constrained for the drivers, and non-circular for the vehicles). The proposed approach is a two-phase method for solving instances with one exchange location. It first finds drivers' routes, and then builds vehicles' routes compatible with them. The algorithm is tested on a set of instances from the literature, with up to 30 nodes and different degrees of difficulty, and on larger instances with 50 nodes. The results are very promising, since the algorithm finds, in very competitive computing times, the optimal solution in most cases when it is known, and provides a new best solution in many other cases.

However, the algorithm has its limitations when tested on instances with more than 50 nodes. We tried to solve instances with 100 and 150 nodes, and on those instances the algorithm did not find feasible solutions. The main drawback is that an integer linear programming model has to be solved to find the drivers' routes. This model involves a large number of variables and constraints, and the branch-and-cut used to solve it takes too long when the size of the problem increases. This opens several lines for future research. One of them is to try to improve the performance of the branch-and-cut method used in the first phase, by adding new families of valid inequalities. We could also put a time limit to this part of the heuristic approach. If the time invested in the first phase could be kept into acceptable limits, we could go further and embed the actual two-phase method into some metaheuristic scheme such as multistart. A way of doing it would be to start from random assignments of customers to depots, thus reducing the complexity of the drivers' problem, and then proceed with the two-phase method. Other alternative lines of research would be to replace the mathematical model in the first phase by a combinatorial algorithm, to make a deeper analysis on instances with several exchange locations, or to add time-window requirements on customer services. Moreover, the extension of the proposed heuristic to tackle the general case, with multiple exchange locations, also remains as future work.

References

1. Cordeau, J.-F., Stojković, G., Soumis, F., Desrosiers, J.: Benders decomposition for simultaneous aircraft routing and crew scheduling. Transp. Sci. **35**, 375–388 (2001)
2. Domínguez-Martín, B., Rodríguez-Martín, I., Salazar-González, J.-J.: An exact algorithm for a Vehicle-and-Driver Scheduling Problem. Comput. Oper. Res. **81**, 247–256 (2017)
3. Domínguez-Martín, B., Rodríguez-Martín, I., Salazar-González, J.-J.: The driver and vehicle routing problem. Comput. Oper. Res. **92**, 56–64 (2018)
4. Drexl, M.: Synchronization in vehicle routing - a survey of VRPs with multiple synchronization constraints. Transp. Sci. **46**, 297–316 (2012)

5. Freling, R., Huisman, D., Wagelmans, A.P.M.: Models and algorithms for integration of Vehicle and Crew Scheduling. J. Sched. **6**, 63–85 (2003)
6. Huisman, D., Freling, R., Wagelmans, A.P.M.: Multiple-depot integrated Vehicle and Crew Scheduling. Transp. Sci. **39**, 491–502 (2005)
7. Lam, E., Van Hentenryck, P., Kilby, P.: Joint vehicle and crew routing and scheduling. In: Pesant, G. (ed.) CP 2015. LNCS, vol. 9255, pp. 654–670. Springer, Cham (2015). https://doi.org/10.1007/978-3-319-23219-5_45
8. Mesquita, M., Paias, A.: Set partitioning/covering-based approaches for the integrated vehicle and crew scheduling problem. Comput. Oper. Res. **35**, 1562–1575 (2008)
9. Papadakos, N.: Integrated airline scheduling. Comput. Oper. Res. **36**, 176–195 (2009)
10. Salazar-González, J.J.: Approaches to solve the fleet-assignment, aircraft-routing, crew-pairing and crew-rostering problems of a regional carrier. Omega **43**, 71–82 (2014)

Solving Full-Vehicle-Mode Vehicle Routing Problems Using ACO

Yahui Liu[✉] and Buyang Cao

School of Software Engineering, Tongji University, Shanghai, China
{liuyahui, caobuyang}@tongji.edu.cn

Abstract. The capacitated vehicle routing problem is a very classic but simple type of vehicle routing problem (VRP). There are variants of the VRP in practice based on different constraints which are called rich VRP (RVRP). In this article, variants of the VRP, including fixed vehicle types and dynamic vehicle type combinations are analyzed. An improved ant colony optimization (ACO) algorithm is designed to resolve this group of VRPs. The fixed vehicle type VRP, homogenous fleet VRP and heterogeneous fleet VRP are defined by one or multiple vehicle types in RVRP. Because of the evolution of transportation equipment, some new vehicle types such as truck and full trailer as well as tractor and semitrailer are introduced. The static and dynamic usages of different vehicle types vary with the business operations. We define this kind of VRPs as full-vehicle-mode (FVM) VRP in this paper. The associated ACO algorithm is developed to solve FVM-VRP problems. Computational experiments are performed and the results are presented to demonstrate the efficiency of the proposed algorithm.

Keywords: Vehicle routing problem · Ant colony optimization
Dynamic vehicle mode · Multi vehicle type

1 Introduction

The vehicle routing problem (VRP) is a complex combinatorial optimization problem. In practice a VRP includes a variety of business constraints. In this paper we consider various vehicle types involved in the fleet. According to [1, 2], in rich vehicle routing problems (RVRP) homogeneous and heterogeneous fleets may need to be dealt with. Nowadays, to meet the evolution of business models in logistics and improve the efficiency and turnover of vehicles, there is a new class of VRPs in which vehicle types and their configurations including truck and full trailer mode [3–7] as well as tractor and semitrailer mode [8–10] need to be considered dynamically during the solution process. In this case, the truck or the tractor just needs quickly exchange a new full trailer or semitrailer before continuing the operation without waiting a long time for unloading and loading goods. Due to the combination of the full trailer or semitrailer, the final model of a vehicle will be different in terms of the load capacity, mileage limit, travel speed and costs. In addition, different depots or customer sites can also have restrictions on the vehicle type based on the business logic. In this paper, we study the case where the vehicle types can be combined dynamically. We define this type of VRP

© Springer Nature Switzerland AG 2018
R. Cerulli et al. (Eds.): ICCL 2018, LNCS 11184, pp. 306–315, 2018.
https://doi.org/10.1007/978-3-030-00898-7_20

as full-vehicle-mode VRP (FVM-VRP). A solver based on ant colony optimization (ACO) is designed to solve this problem.

ACO is a metaheuristic based on ants foraging behavior to embody the foraging intelligence of an ant colony. ACO was applied to solve the traveling salesman problem in its first publication [11]. ACO and its improved versions are widely used to solve VRPs and improved results have been obtained [12–16].

Aiming at the FVM-VRP in this article, an improved ACO algorithm with multiple pheromones is developed. To solve the FVM-VRP effectively, multiple pheromones are set for different vehicle types to guide the associated vehicles. At the same time, in order to verify the effect of the related algorithms, two practical logistics projects are selected to be implemented and the results are convincing.

2 Vehicle Modes in the VRP

2.1 Full Vehicle Mode

It is well known that the VRP is designed to solve the problem of using a specific vehicle or vehicles to transport orders for the predefined sites with minimal or close to minimal costs. The vehicle type is a critical factor as it imposes the restrictions on capacities of load weight, volume, mileage and available time. Different vehicle types also have different speeds and transportation costs. Furthermore, not all vehicle types can service all customer site because of physical or other constraints. Sometimes, we also need to consider the vehicle type combination dynamically besides those predefined vehicle type usages. All of these make FVM-VRP more difficult to be solved.

In the RVRP literature, the fleet to be considered is either homogeneous or heterogeneous. However, the vehicle types, vehicles' attributes, and the numbers of the corresponding vehicles are usually given before solving a RVRP.

Because of the rapid development of logistics business practice, some more economical and more efficient vehicle usages or configurations have emerged, e.g., the dynamic vehicle coupling where truck and full trailer as well as tractor and semitrailer are involved. In general, full trailers and semitrailers cost much less than trucks or tractors do. In this case, a logistics company will purchase full trailers or semitrailers in large quantities to increase the flexibility of the logistics operation. It introduces the dynamic vehicle type configurations in the business and yet increases the difficulty in solving the VRP. Which full trailer or semitrailer will be hauled is determined by the load, volume, and costs dynamically.

The difference between a dynamic and a static vehicle mode in a VRP is in the initial vehicle mode setting: in a static vehicle mode, all vehicle properties are known while in a dynamic one all vehicle properties are unknown until a full trailer or semitrailer is hooked. While the flexibility and efficiency of an FVM-VRP are increased, the decision making complexity for this type of problems increases dramatically.

We can summarize the business constraints regarding the static and dynamic vehicle modes as follows:

- The dynamic vehicle mode means only the number of vehicles is known a-priori, but a vehicle property of capacities can be varied based on the configuration of a full trailer or semitrailer.
- The customer site imposes the restriction on which type of vehicles can visit for both static and dynamic vehicle mode VRP.
- Different vehicle types impact the loading capacities, transportation costs, moving speed and it influences the solution of a VRP. However, the capacity and transportation costs are not confirmed a priori in the dynamic vehicle mode of an FVM-VRP.
- When starting from the depot, for the static vehicle mode VRP the specific vehicles can be selected while the dynamic vehicle mode VRP has to consider which trailer should be hauled based on a particular scenario (cost, physical restrictions, etc.) before other decisions can be made.
- The customer sites can have time window constraints for all vehicle types to visit.
- The customer requirements can have one of or both pickup and delivery demand.
- The optimization goal is to find the routes with the minimum cost while meeting all constraints.

The FVM-VRP deals with both the static and dynamic vehicle mode VRP. The above seven constraints include most constraints regarding vehicle types; that is, why it is called full-vehicle-mode VRP and these make the FVM-VRP very different from a classic VRP. In order to solve the problem from practice, a corresponding improved version of the classical ACO algorithm is developed to tackle the problem.

2.2   The Pheromone Setting for Vehicle Types in ACO

In the ACO algorithm, the ant is guided to select the next candidate site to visit by pheromone which are set up a priori by other ants. The positive feedback of pheromone leads the ant to find a better solution. In a regular VRP, an ant corresponds to a vehicle. If there is only one type of vehicle, then there is only one kind of ant and pheromone.

Since pheromones guide ants, multiple pheromones are needed in order to deal with multiple vehicle types. Pheromones are distributed in road networks, and the concept of multiple lanes of a road in real life is introduced to the improved version of ACO. In this ACO model, the multi-lane stands for different vehicle types and the pheromone in each lane for different types are different. Otherwise, a single pheromone might not provide the correct guidance and could even mislead the selection of different vehicle types for the candidate site to be visited next. The reason is that the pheromone only reflects the route guidance information that is corresponding to the previous feasible solution for the same vehicle type, and the pheromone does not have the guidance for other vehicle types.

In the ACO algorithm, there is an issue of pheromone initialization. At the beginning of a solution procedure the initial values of the multiple pheromones are the same which is similar to the classic ACO.

2.3   The Selection Mechanism for Different Vehicle Types in ACO

In ACO, according to the pheromone and the attraction factor of the current selected vehicle type, we can use formula (1) which is the standard formula to compute the possibility for visiting the next site. The initial selection of vehicle type at the depot will be determined according to certain rules derived from the pheromones and attraction values (defined by formula (2)) for multi vehicle types, or still using formula (1) for a single vehicle type. In formula (2), the pheromones for selecting both, vehicle type and the next site to be visited, are calculated, because for the FVM-VRP, the vehicle type should be selected first before its next route selection depending on its current vehicle type. The abstractive factor remains the same as the one in the classic ACO, but the different pheromones for all vehicle types obtained will support the next site to be visited and the associated vehicle types meeting the restriction imposed by the next site, where the pheromone is computed based on formula (2). For a non-depot site to be considered as the next candidate site to be visited, the selection mechanism applies the classic formula (1) since the vehicle type has already been selected. Different vehicle types should have the same initial selection probability to ensure the diversity in the later solution stages. At the beginning, different vehicle types have the same probability to be selected because the pheromones of different vehicle types possess the same initial pheromone. During the solution process, the pheromones of different vehicle types will be updated according to the quality and the corresponding constraints of the previous feasible solution. The pheromone of the underlying vehicle type in the best solution will be obviously higher than those of other vehicle types. At the same time, its pheromone attractions to certain candidate points will be updated to become larger. Therefore, the pheromone is used to guide the selection of the most suitable vehicle type for the transportation task.

$$p_{ij}^{k}(t) = \frac{[\tau_{ij}(t)]^{\alpha}[\eta_{ij}(t)]^{\beta}}{\sum_{l \in N_i^k}[\tau_{il}(t)]^{\alpha}[\eta_{il}(t)]^{\beta}} \quad if \ j \in N_i^k \tag{1}$$

$$p_{ijm}^{k}(t) = \frac{[\tau_{ijm}(t)]^{\alpha}[\eta_{ij}(t)]^{\beta}}{\sum_{m \in [0,k]} \sum_{l \in N_i^k}[\tau_{ilm}(t)]^{\alpha}[\eta_{il}(t)]^{\beta}} \quad if \ j \in N_i^k \tag{2}$$

The algorithm for solving the FVM-VRP when selecting a candidate site is described as follows:

Candidate Selection algorithm for FMV VRP:

Procedure Candidate Selection Algorithm (

τ_{ijm} // The pheromone from site i to site j for vehicle type m;

η_{ij} // The abstractive factor from site i to site j; this is not related with vehicle type;

p_{ijm}^k // Probability to select next site j from site i to site j with vehicle type m;

1. If current site is the depot
2.   Calculate values $[\tau_{ijm}(t)]^\alpha [\eta_{ij}(t)]^\beta$ for every candidate site j for every vehicle type m based on their pheromone τ and abstractive factor η;
3.   Generate a random rate between 0 and 1;
4.   Select next site based on the random rate and probabilities of candidates using formula (2);
5.   Get the Vehicle type and Next Site based on the selection;
6. Else
7.   Calculate values $[\tau_{ij}(t)]^\alpha [\eta_{ij}(t)]^\beta$ for every candidate site j based on their pheromone τ and abstractive factor η; remember the pheromone is only for the current vehicle type which is already selected from the depot and cannot be changed after that.
8.   Generate a random rate between 0 and 1;
9.   Select next site based on formula (1);
10.   Get the Next Site based on the selection;
11. Endif

In the above procedure, when starting from the depot, we should calculate all candidates with all available vehicle types through their abstract factor of site distance, and the pheromones of all available vehicle types, and randomly select the next site and its vehicle type by formula (2), and the small difference between static and dynamic vehicle mode is that the vehicle number of each vehicle type is variable for the dynamic one but is fixed for the static one. If starting from the non-depot site, the vehicle type is already selected, so, the classic probability selection using formula (1) will be used only after filtering the pheromone only for the current vehicle type.

3 Computational Experiments

3.1 Project Background

The dataset for the computational experiments come from a real project conducted jointly with a new energy automobile manufactory. Its logistics business can be described as follows:

1. There are three types of full trailers, 10.5M, 12.5M and 17.5M(M for meters), which have different loading capacities of weight and volume, travel speeds, and transportation costs.
2. Logistics operation of this company is to transport automobile parts between its depot (factory) and its vendors in 25 nearby cities. All vendors have their own time windows for the delivery service. Some vendor sites cannot fit 17.5 m full trailer, and only the other two full trailers are allowed.
3. The transportation uses container boards, which can simplify the capacity calculation.
4. The logistics operation includes pickup and delivery, i.e., hauling full trailers filled container board with parts to the factory from vendor sites and returning the empty container board back to the vendors.
5. There are different numbers of container boards needed to be transported accord to the daily product plans.
6. The objective is to find the transportation (routes) plan with the lowest cost while meeting the business logic mentioned above.

3.2 Benchmarks

The road network data includes street segments and distances. All the distances and travel times between any pair of nodes (depot, vendors locations) are calculated to form an origin-destination (OD) matrix. The locations of the depot and vendors are geocoded based upon their addresses.

Currently all transportation plans are conducted by some experienced staff members. They try to create the best plan upon the transportation manifests. In addition to the datasets for the logistics operation we also collect the transportation plans created manually by the experienced staff members to conduct the comparisons. The computational results are listed in Table 1.

Table 1. Optimization comparison result between experienced staff and the ACO algorithm

	Before optimization	After optimization	Savings	Savings percentage
Number of vehicles	12 * 12.5M	2 * 17.5M + 6 * 12.5M	4	33.33%
Total mileage (km)	9186.858	7612.748	1574.11	17.13%
Total cost (RMB)	58612.15404	51926.32709	6685.82695	11.41%

The properties of the full trailers are listed in Table 2, which include vehicle type name, driving speed, load capacities of plates, cost of distance. Different vehicle types have different properties, and all this will affect the vehicle type selection. There are site park constraints listed Table 3 for which vehicle types can park for the sites, and this makes it more difficult for the ants to select a next candidate if not all vehicle types are compatible for the next candidate sites.

Table 2. Full Trailer Properties

Full trailer type	Driving speed (KM/H)	Load capacity (Plates)	Cost (RMB/KM)
17.5M	50	34	6.93
12.5M	60	24	6.38
10.5M	70	20	6.16

Table 3. Site properties (A is the depot)

ID	Service window	Park type	ID	Service window	Park type	ID	Service window	Park type
A	00:00–24:00	10.5/12.5/17.5	J	09:00–19:00	10.5/12.5/17.5	S	00:00–24:00	10.5/12.5
B	08:00–18:00	10.5/12.5/17.5	K	08:00–22:00	10.5/12.5/17.5	T	08:30–19:30	10.5/12.5/17.5
C	08:30–20:00	10.5/12.5/17.5	L	08:30–18:30	10.5/12.5	U	00:00–24:00	10.5/12.5/17.5
D	08:30–18:30	10.5/12.5	M	00:00–24:00	10.5/12.5/17.5	V	07:00–19:00	10.5/12.5
E	08:30–18:30	10.5/12.5/17.5	N	09:00–20:00	10.5/12.5/17.5	W	08:00–22:00	10.5/12.5/17.5
F	07:00–21:00	10.5/12.5/17.5	O	08:30–20:00	10.5/12.5/17.5	X	08:30–19:30	10.5/12.5
G	00:00–24:00	10.5/12.5/17.5	P	09:30–19:30	10.5/12.5	Y	00:00–24:00	10.5/12.5/17.5
H	08:30–18:30	10.5/12.5/17.5	Q	08:30–18:30	10.5/12.5/17.5	Z	09:00–20:00	10.5/12.5/17.5
I	00:00–24:00	10.5/12.5	R	07:00–21:00	10.5/12.5/17.5			

Table 3 describes site properties, including site ID, service windows, park vehicle type. Note: the transport is a whole day running from 00:00 to 24:00, and most sites have hard time windows for service, which means the vehicle cannot be serviced after the windows and needs to wait if it arrives before the window. The transport time between any two sites is calculated by an electronic map like Google map based on the GPS (Global Position System) coordinate of the sites. Based on that the arrival time and departure time for the candidate site are calculated. There is a fixed half an hour for the vehicle to load/unload plates which also needs be calculated.

Table 4 lists transportation requirements in terms of container plates between depot site A and the vendor sites. Since some empty container boards need to be returned to the vendor sites, our FVM-VRP needs to consider pickup and delivery services.

Table 4. Transport requirement between sites

From–To	Plates	From–To	Plates	From–To	Plates	From–To	Plates
A–B	5	I–A	8	A–W	10	R–A	10
A–D	7	J–A	12	A–Z	12	S–A	15
A–G	10	K–A	10	B–A	8	T–A	12
A–K	4	L–A	12	C–A	8	U–A	8
A–M	8	M–A	7	D–A	10	V–A	5
A–V	6	N–A	10	E–A	15	W–A	8
A–U	6	O–A	12	F–A	12	X–A	15
A–Q	3	P–A	8	G–A	8	Y–A	12
A–S	7	Q—A	8	H–A	18	Z–A	10

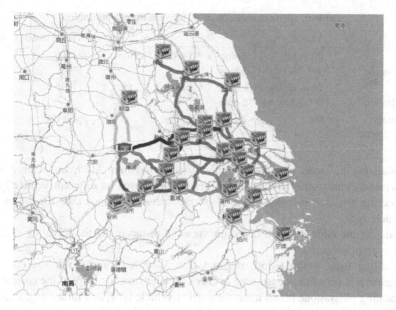

Fig. 1. Scheduling result in electronic map

Our proposed ACO algorithm is able to solve this real VRP successfully while considering static and dynamic vehicle configurations, pickup and delivery services, time windows, site physical size restrictions on vehicle types, and vehicle capacity, driving speed and transport cost at same time. The resulting route is depicted on the map (Fig. 1) for the dispatchers to evaluate and update upon their experience. In the diagram different colors represent different vehicles.

Another static vehicle mode project using the improved ACO for the FVM-VRP is a cold chain transportation project for food material, with constraints including:

- There are 80 clients in 69 buildings, and distances are calculated by an eletronic map by their GPS coordinates.
- Every client has loose time windows; most of them are between 6:00 to 18:00
- There are four vehicle types, including 1 ton, 2 tons, 3 tons and 5 tons for loading capacities.
- Every vehicle is almost fully loaded, so it is hard to reduce the vehicle number, but the total distance or total cost for transport reduces by 27.4% with the improved ACO.

The result is listed in Table 5 for details.

The source code of the ACO is modified based on max-min ACO [17], and the modifications include:

- Adding a flag to switch between static or dynamic vehicle mode, other parameters such as α, β, ρ are the same as in the original max-min ACO where the number of ants is the same as the number of sites.

Table 5. Performance for cold chain transport optimzation

Item	Vehicles	Total distance (km)
Before optimization	20 vehicles, including: 1 * 1 ton, 5 * 2 tons, 9 * 3 tons, 1 * 5 tons	2503
After optimization	20 vehicles, including: 1 * 1 ton, 5 * 2 tons, 9 * 3 tons, 1 * 5 tons	1815
Percentage improvement	0%	27.4%

- Initializing the road netwok is similar to the max-min ACO, but setting multi pheromones instead of single ones for each edge based on different vehicle types.
- Updating the cost calculation based on the specific distance cost for the current vehicle type.
- Updating the edge duration based on the edge distance and the travaling speed of the current vehicle type.
- Disabling local optimization of 3-opt if a VRP has time window constraints, because it is very time-consuming to check all later sites for 3-opt of the current site.
- Updating pheromone based upon the current vehicle type, and intializing, updating and evaporating logic is the same as fort he max-min ACO.
- Updating max-min ACO source code from C to C++ 11.

The running enviroment is ubuntu 16.04 in a computer with 32G memory.

4   Conclusions

The FVM-VRP is a complex VRP which can be found in real logistics applications. This paper analyzed the static and dynamic vehicle mode considerations in VRPs and proposes the FVM-VRP to deal with the requirements from real logistics applications, and an improved version of an ACO algorithm was proposed to solve the corresponding problem.

The dataset from a real project of an automobile manufacturer with dynamic vehicle mode and consideration of multi vehicle types, time windows and restrictions imposed by customer sites on vehicle types was used to conduct computational experiments; another static vehicle mode of cold chain transportation with multi vehicle types is also listed. The results obtained by our approach were compared to the manual ones. The results demonstrated that our algorithm is able to achieve more satisfactory outcomes and seems promising for the real applications.

Future research should combine more constraints in the current FVM-VRP, such as multi-trip FVM-VRP which means return-and-reload to be considered, Multiple time windows of a customer site should be included in the study, too.

Acknowledgements. We are indebted to Prof. Stefan Voss and three anonymous reviewers for insightful observations and suggestions that have helped to improve our paper This work was partially supported by NSFC of China project [grant number 41771410], CIUC and TJAD [grant number CIUC20150011].

References

1. Caceres-Cruz, J., Arias, P., Guimarans, D., Riera, D., Juan, A.A.: Rich vehicle routing problem: survey. ACM Comput. Surv. (CSUR) **47**(2), article #32 (2015)
2. Lahyani, R., Khemakhem, M., Semet, F.: Rich vehicle routing problems: from a taxonomy to a definition. Eur. J. Oper. Res. **241**(1), 1–14 (2015)
3. Lum, O., Chen, P., Wang, X., Golden, B., Wasil, E.: A heuristic approach for the swap-body vehicle routing problem. In: 14th INFORMS Computing Society Conference, pp. 172–187 (2015)
4. Parragh, S.N., Cordeau, J.F.: Branch-and-price and adaptive large neighborhood search for the truck and trailer routing problem with time windows. Comput. Oper. Res. **83**, 28–44 (2017)
5. Drexl, M.: Branch-and-cut algorithms for the vehicle routing problem with trailers and transshipments. Networks **63**(1), 119–133 (2014)
6. Rothenbächer, A.K., Drexl, M., Irnich, S.: Branch-and-price-and-cut for the truck-and-trailer routing problem with time windows. Transport. Sci. (2018). Online available
7. Torres, I., Cruz, C., Verdegay, J.L.: Solving the truck and trailer routing problem with fuzzy constraints. Int. J. Comput. Intell. Syst. **8**(4), 713–724 (2015)
8. Li, H., Lv, T., Li, Y.: The tractor and semitrailer routing problem with many-to-many demand considering carbon dioxide emissions. Transp. Res. Part D: Transp. Environ. **34**, 68–82 (2015)
9. Pollaris, H., Braekers, K., Caris, A., Janssens, G., Limbourg, S.: The fleet size and mix vehicle routing problem with sequence-based pallet loading and axle weight constraints. In: Proceedings of the BIVEC-GIBET Transport Research Days 2017: Towards an Autonomous and Interconnected Transport Future, pp. 162–176 (2017)
10. Li, H., Lv, T., Lu, Y.: The combination truck routing problem: a survey. Procedia Eng. **137**, 639–648 (2016)
11. Ariyasingha, I.D.I.D., Fernando, T.G.I.: Performance analysis of the multi-objective ant colony optimization algorithms for the traveling salesman problem. Swarm Evol. Comput. **23**, 11–26 (2015)
12. Gambardella, L.M., Taillard, É., Agazzi, G.: MACS-VRPTW: a multiple colony system for vehicle routing problems with time windows. In: New Ideas in Optimization, pp. 63–76. McGraw-Hill (1999)
13. Reed, M., Yiannakou, A., Evering, R.: An ant colony algorithm for the multi-compartment vehicle routing problem. Appl. Soft Comput. **15**, 169–176 (2014)
14. Rajappa, G.P., Wilck, J.H., Bell, J.E.: An ant colony optimization and hybrid metaheuristics algorithm to solve the split delivery vehicle routing problem. Int. J. Appl. Indust. Eng. (IJAIE) **3**(1), 55–73 (2016)
15. Kalayci, C.B., Kaya, C.: An ant colony system empowered variable neighborhood search algorithm for the vehicle routing problem with simultaneous pickup and delivery. Expert Syst. Appl. **66**, 163–175 (2016)
16. Wang, X., Choi, T.M., Liu, H., Yue, X.: Novel ant colony optimization methods for simplifying solution construction in vehicle routing problems. IEEE Trans. Intell. Transp. Syst. **17**(11), 3132–3141 (2016)
17. http://www.aco-metaheuristic.org/aco-code/

Optimising Routing in an Agent-Centric Synchromodal Network with Shared Information

Myrte A. M. De Juncker[1,3], Frank Phillipson[1(✉)], Lianne A. M. Bruijns[1,2], and Alex Sangers[1]

[1] TNO, PO Box 96800, 2509 JE The Hague, The Netherlands
frank.phillipson@tno.nl
[2] Delft University of Technology, Delft, The Netherlands
[3] Eindhoven University of Technology, Eindhoven, The Netherlands

Abstract. Our research focuses on synchromodal planning problems in which information is shared between all agents in the system and they choose their routes based on an individual optimisation objective. We show the effect of the information availability by developing three different methods to determine the optimal paths, to motivate logistic players to cooperate in a synchromodal system.

Keywords: Synchromodal logistics · Agent centric network
User equilibrium

1 Introduction

Freight transportation is growing and so is the need for an efficient organisation of hinterland transport services. The main element of a synchromodal transportation [21] is the integration of transport service on different modalities with real-time availability of information. Changes have to be made to the network in order to create a synchromodal system. Among others, there is need for an integrated network and service design, an integrated operation and control, contracts that allow synchronised transport, a stronger collaboration and a mind shift in planning and control.

In [17], synchromodal planning problems are classified in two directions: available information and the degree of control and optimisation. Both can take either a local view, where only own information is known and optimisation is for an individual objective, or a global view, where information is available for the entire network and the optimisation is aimed at a shared goal. If the information is available globally but every agent only optimises their own objective, the approach is called *selfish*. Information is a broad term, some of the information is public, which means every agent can get this information. Other information is private and has to be shared between different stakeholders in the network. This sharing can be difficult to achieve, since the stakeholders need to be willing to

R. Cerulli et al. (Eds.): ICCL 2018, LNCS 11184, pp. 316–330, 2018.
https://doi.org/10.1007/978-3-030-00898-7_21

share their private information to competitors and clients. In this paper, we focus on agent-centric synchromodal networks, where each agent is selfish and wants to optimise its own objective function. The public information we encounter is information about the occupancy of different links in the network. This means that we know, at any point in time, how many agents are on every link of the network. One can already see examples of this public information being used for road networks. For example: route guiding systems already have an option to recalculate your shortest path based on information about current congestion in the network. Private information on the other hand, does require different stakeholders to cooperate. The private information we encounter is information about upcoming orders. Logistic service providers normally know what orders are going to arrive in the near future. If these logistic service providers would be willing to share this information with all stakeholders, all stakeholders can react upon this information. This means that agents also have information about the, probable, future occupancy of certain links. Note that much research is known on agent based road traffic, looking for a user equilibrium. In these networks centrally controlled optimization is not possible. In logistics we assume both approaches are possible and therefor worthwhile investing.

This paper investigates the effect of different information availability in these kind of networks to show the logistic players the value of cooperating. We assume cases where only public information is available and cases where agents have access to both public and private information. We present three models for the analysis, using different methods to generate paths through the network for the agents:

- *Model 1*: Naive implementation with public information. We assume that each agent checks the public information at departure and will react accordingly. This means they will take the shortest path for the current state of the network. After the choice is made, they will not deviate.
- *Model 2*: Very similar to the first, except the fact that agents *do* switch routes before reaching their destination. This means that at each decision point, i.e., an intermediate node in the network, they again check the state of the network and reroute if necessary.
- *Model 3*: Assumes full information. Each agents knows the future destinations and routes of other agents for a certain planning horizon. This includes public as well as private information. With this information, the algorithm seeks to find the optimal routes for all agents that arrive somewhere in the planning horizon.

In Sect. 2 we give an overview of the literature on Dynamic Traffic Assignment models and on the effect of information in road networks. In Sect. 3 we describe the used models for the analysis and the underlying simulation techniques. Next, in Sect. 4 we discuss the results we found. In the final section, Sect. 5, we derive the conclusions of our work and mention recommended further research. Note that this work is based on the thesis [7].

2   Literature Review

In Dynamic Traffic Assignment (DTA) problems one can take one of two approaches in choosing objectives:

1. *System optimal (SO)*: in this case one wants to optimise a system objective; e.g. congestion or average travel time. This also means that all vehicles are controlled by a central controller.
2. *User equilibrium (UE)*: here every vehicle in the system wants to optimise an individual performance measure; e.g. travel time or costs.

There exist models for these DTA problems, but there is no model that provides a universal solution for general networks [16]. In DTA models, there is a trade-off between traffic realism and the theoretical guarantee of properties such as existence, uniqueness and stability.

Peeta and Ziliaskopoulos [16] give an overview of the different modelling approaches to DTA problems. These approaches can be divided into four categories: *mathematical programming, optimal control, variational inequality* and *simulation-based*. We mention interesting literature on all four approaches. We also describe the advantages and disadvantages of using certain approaches.

Mathematical programming DTA models aim to formulate the problem as a mathematical discrete time program. The first formulation was by Merchant and Nemhauser [13]. Birge and Ho [4] extended this model to the stochastic case by allowing for random demand desires. Jansen [10] describes the UE DTA problem as a mathematical program. However, all formulations are non-convex because of the FIFO requirement. While there is enough literature on non-convex optimisation, in a DTA context analytical and computational tractability are lost for general networks. Together with the difficulty to prohibit holding-back of traffic, mathematical programming formulations lack efficient solutions for realistic instances. Carey and Subrahmanian [5] illustrate some of the issues that arise because of the FIFO requirement and the holding-back of traffic in mathematical programming formulations. An overview on dynamic dispatching problems with stochastic requests can be found in [22].

Optimal control theory DTA formulations are continuous time problems. Here the origin-destination rates are assumed to be known continuous functions of time. For the formulations we refer to [9,19,20]. The main issue with optimal control formulations is that there is no efficient solution algorithm. As mentioned before, these formulations also have no explicit constraints for the FIFO requirement and the holding back of traffic. Therefore, new research has been done in variational inequality formulations.

Variational inequality formulations were introduced by Dafermos [6]. Variational inequality formulations are used in equilibrium problems. One defines an inequality involving a functional, which has to be solved for all possible values of a variable. Nagurney [14] provides a summary of variational inequality formulations and addresses various equilibrium problems in network economics, under which the traffic network equilibrium. Variational inequality formulations can handle more realistic traffic scenarios, but the approaches are computationally intensive. Also, the problems with the FIFO requirement and the holding-back of traffic remain.

Simulation-based DTA models use a traffic simulator in order to handle realistic traffic scenarios. The main issue with simulation-based models is that theoretical insights cannot be gathered from the models. The solution methods in these simulation-based models often use the traffic simulator as part of the solution. This is called the *predictive-iterative method*, where the simulator is used in each iteration to predict future traffic conditions given a certain route assignment. Based on these predictions a new route assignment is determined and so on. One of these iterative models is described by Peeta and Mahmassani [15]. A similar iterative approach for a UE DTA is taken by Kaufman, Smith and Wunderlich [11]. These models are much more realistic than the analytic ones and therefore widely used in analyses. However, deployment in real life is only feasible if the algorithms are computational efficient. Ben-Akiva et al. propose DynaMIT in [3] (and its route guidance in [2]), which uses a demand and supply simulator to generate UE route guidance under a rolling horizon framework.

Next to DTA models there is also literature available on the effect of information in road networks, see for example the papers by Mahmassani and Jayakrishnan [12] and Dia [8], both describes a modelling framework to analyse the effect of in-vehicle real-time information. The framework consists of a *simulation component* and a *user decisions component*.

3   Models

In this section we describe the models used for the analysis. Firstly, in Sect. 3.1, we state the assumptions for all models. Then, Sect. 3.2 describes the simulation we developed for the agents moving through a network. How the routing is down in the simulation is decided using one of the three models. Section 3.3 describes models 1 and 2. These heuristics assume that the agents in the network only know the occupancy of the links in the network up to the current time. Section 3.4 describes model 3. This model relies on the assumption that there is perfect knowledge. This means that all agents also know how many agents are due to arrive in the future.

3.1   Assumptions

First, we assume to have a transportation network, where nodes are locations and links are connections between locations which can be various modalities, such as trucks (roads), trains and barges. In the remainder of the paper we assume by agents containers in the network, that has to be transported from a certain origin to a certain destination using one or several modalities. Next, in the models we assume certain properties as stated below.

Assumption 1. *All nodes can be reached by truck.*

Assumption 2. *The costs of travelling links are non-negative.*

Assumption 3. *All information is available to all agents.*

For the first two models, Assumption 3 means that the occupancy on all links is known up to the departure time of the agent. For model 3, we assume full information. Therefore Assumption 3 means that the occupancy on all links is known as well as future orders, i.e., containers that want to travel from an origin to a destination within the network, and their routes.

In Sect. 3.2 we mention how we calculate time- and state-dependent travel times for our network. There are also some specific assumptions on the travel times.

Assumption 4. *The travel times of roads only depend on the occupancy of the link.*

Assumption 5. *The departure times of the trains and barges are known.*

Assumption 6. *The capacities of all trains/barges are known.*

Assumption 7. *The travel times of trains and barges over a certain link are constant.*

3.2 Description of Simulation

The simulation used in all models is an event-driven simulation, where each new event triggers a change in the network. Possible events are:

- Request route: a new agent asks for a route from an origin to a destination.
- Enter link: an agent will traverse a certain link in the network.
- Leaving link: when an agent reaches its destination or an intermediate node, he leaves the link.

The simulation handles the events one by one until a certain end time. The entire duration of the simulation is referred to as the planning horizon. In this simulation one can keep track of all kinds of performance measures: individual travel times, average travel times, occupancy on roads, etc.

The travel times in the simulation are time- and state-dependent. However, we assume that we know how many agents occupy certain links and at what time they want to traverse the link. This means that the time and state is fixed for the calculation of the travel times.

The travel times for roads are calculated using the approach of Akçelik [1]. He describes Davidson's function, which is a general-purpose travel-time formula for transport planning purposes. To overcome some issues, Akçelik proposes an alternative formulation, described in Eq. (1).

$$t = t_0 + 0.25T_f \left(z + \left(z^2 + \frac{8J_A \cdot x}{C \cdot T_f} \right)^{0.5} \right) \tag{1}$$

where

t = average travel time per unit distance,
t_0 = min. (free-flow) travel time per unit distance,
T_f = flow/analysis period,
J_A = a delay parameter,
z = $x - 1$,
x = $\frac{v}{C}$, : degree of saturation,
v = demand flow rate,
C = capacity.

This function assumes a constant demand pattern and no initial queue at the start of the flow period. The travel time is defined as experienced by all vehicles *arriving* during the specified flow period.

The author also proposes some parameters for this travel time function representing various road classes. We adjust these parameters for trucks using freeways in the Netherlands. Since the maximum velocity for trucks on freeways is 80 km/h, we choose the following parameters: $v_0 = 80$ and $J_A = 0.1$. The capacity is chosen with respect to the other parameters in our simulation.

For trains and barges the travel time is calculated differently. As they will leave at certain departure times and have a specified capacity, we have to calculate the time it has to wait for the next available departure. We assume fixed departure times, capacities and travel times.

Let us assume we have a container at time t that wants to travel over a link representing rail or waterway. Given departure time t_D and travel time t_T, the total travel time for the container at time t is given by $t_D - t + t_T$. However, the capacity of the train or barge, denoted by C, is not yet accounted for. The next C containers have to wait for the next departure, t_{D+1} and so on.

3.3   Public Information Models

The first two models are loosely based on how drivers in a road network can adjust their route in current traffic. We assume that the knowledge of the network is available to all agents, but only up to the current time.

In the first model, all agents will know the state of the network and act upon this information. However, once this choice has been made, it will not be altered anymore. In the second heuristic, agents can decide to switch routes as conditions in the network change.

Model 1: Minimum-Cost Routing Without Rerouting. The first model tries to find the minimum-cost route in a greedy way. For all containers that request a route, the shortest path is calculated with a dynamic shortest-path algorithm. Here we use the algorithm described in the paper by Ramalingam and Reps [18]. They obtain a new dynamic single-source shortest-path problem, which can be extended to a dynamic all-pairs shortest-path problem. All containers that request a route, are thus given a route based on the current conditions.

The containers will follow this route and do not adjust on a later time. The first model is described in Algorithm 1. Note that this is the entire algorithm, including the simulation.

Algorithm 1. Heuristic 1: Minimum-cost routing without rerouting

1: $t = 0$
2: **while** $t <$ planning horizon **do**
3:     $event =$ first event from event queue
4:     $t =$ time of event
5:     **if** $event$ is an $request\ route$ event **then**
6:         Calculate shortest path under current conditions
7:         Give this route to the current agent
8:         Add $enter\ link$ event for time t on first link
9:     **else if** $event$ is a $enter\ link$ event **then**
10:         Calculate travel time, t_T
11:         Add $leave\ link$ event for time $t + t_T$
12:         Increase occupancy on this link by 1
13:         **if** agent is not yet at its destination **then**
14:             Add $enter\ link$ event for the next link for time $t + t_T$
15:         **end if**
16:     **else if** $event$ is an $leave\ link$ event **then**
17:         Decrease occupancy on this link by 1
18:     **end if**
19: **end while**

Model 2: Minimum-Cost Routing with Rerouting. An improvement on the previous model is the rerouting of agents. As each agent traverses the network, it encounters intermediate nodes. However, for the simulation it does not matter if the agent arriving at that node arrived from outside the network or from another node. Therefore, we can recalculate the shortest path for each agent. Since the conditions in the network have likely changed, so may have the shortest path. This model is described in Algorithm 2. Again this algorithm includes the simulation.

3.4   Full Information Model

The third model calculates the optimal route for each agent in a certain planning horizon. For a given planning horizon with full information, a user equilibrium is reached between all agents.

Model 3: Full Infomation, User Equilibrium Routing. For all classes of agents in the network, i.e., all origin-destination pairs, we know all arrivals in the planning horizon. For each of the links we should know the time- and state-dependent travel time function. These travel time functions are already discussed

Algorithm 2. Model 2: Minimum-cost routing with rerouting

1: $t = 0$
2: **while** $t <$ planning horizon **do**
3:     $event =$ first event from event queue
4:     $t =$ time of event
5:     **if** $event$ is an $request\ route$ event **then**
6:         Calculate shortest path under current conditions
7:         Give this route to the current agent
8:         Add $enter\ link$ event for time t on first link
9:     **else if** $event$ is a $enter\ link$ event **then**
10:         Calculate travel time, t_T
11:         Add $leave\ link$ event for time $t + t_T$
12:         Increase occupancy on this link by 1
13:         **if** agent is not yet at its destination **then**
14:             Create an $request\ route$ event for the next node at time $t + t_T$
15:         **end if**
16:     **else if** $event$ is an $leave\ link$ event **then**
17:         Decrease occupancy on this link by 1
18:     **end if**
19: **end while**

---

in Sect. 3.2. We should also know the costs of traversing a link. We focus on the user equilibrium case which uses Wardrop's User Equilibrium Condition [23]: the system has a *user equilibirium* if no agent can improve his/her experienced travel time by unilaterally switching routes (for a given departure time).

This model is an iterative procedure which switches between a simulation of events under given route assignments and calculating new route assignments for differences in the travel times and costs. By rerouting the agents in each iteration we want to reach a fixed point in this system. In this fixed point the travel times and costs are not altered anymore and thus also the shortest paths are not altered anymore. Kaufman et al. [11] describe that this fixed point solution satisfies the UE condition.

In Fig. 1 one can see the global idea of our solution. The steps of the algorithm are described in further detail below. The algorithm is based on the algorithm for traffic networks described by Peeta and Mahmassani [15].

- **Step 1.** The iteration counter i is set to 0. For all origin-destination pairs and all time steps we calculate feasible paths. We used Dijkstra's algorithm to create these paths with the free-flow travel times and costs. The first route assignment is sending all incoming agents over their shortest path. This route assignment is denoted by R_0
- **Step 2.** Perform simulation of the planning horizon by sending the agents over their path assignments R_i. In this simulation we log the changes throughout time, obtaining the number of agents on each link on each time. We also keep track of the individual travel times of the agents.

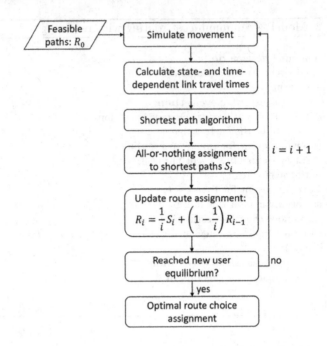

Fig. 1. Flow chart describing the general solution algorithm.

– **Step 3.** Compute the new time- and state-dependent travel times with the information gathered from the simulation. How we calculate these travel times is described in Sect. 3.2.
– **Step 4.** Compute the new time-dependent shortest paths for all origin-destination pairs, using the algorithm proposed by Ziliaskopoulos and Mahmassani [24]. This algorithm calculates the time-dependent shortest paths from all nodes in a network to a given destination node (denoted by N). Note that this is a discrete-time algorithm, thus it calculates the shortest paths for each time step over a given time horizon. It is based on Bellman's principle of optimality.
– **Step 5.** We create an auxiliary route assignment by the *all-or-nothing assignment*. The all-or-nothing assignment basically sends all agents from a certain origin-destination pair that want to depart at a certain time step over the same route. The auxiliary route assignment we create, denoted by S_i, is sending all agents on their shortest path calculated in the previous step.
– **Step 6.** Then we calculate the new route assignment. To reach convergence, the new route assignment is a combination between the old route assignment and the auxiliary route assignment calculated in the previous step. A new route assignment is calculated with the use of the Method of Successive Averages (MSA):

$$R_i = \frac{1}{i}S_i + (1 - \frac{1}{i})R_{i-1}. \tag{2}$$

- **Step 7.** The convergence criterion is based upon how much the occupancy on the links changes from one simulation to the next. We keep the log from the previous simulation and the one from this simulation and check the difference in occupancy on each link. If this difference is less than 5%, the convergence criterion is met.
- **Step 8.** If the convergence criterion is met, terminate the algorithm. Otherwise repeat from step 2 with the new route assignments.

It is important to know that the shortest-path algorithm in step 4 is discrete and therefore the planning horizon is divided in multiple time steps. At the end of the algorithm we know the shortest paths for all agents at each time step. To cope with this fact in step 2, we group all arrivals in the same time step and handle them as if they all occur at the beginning of this time step.

4   Results

We developed three different models: two based on public information and one based on full information. We elaborate on some small networks that highlight differences between and the performance of those models. The first network can be found in Fig. 2. Here one can see the length of the links and the capacity of each link (length/capacity). For each of these links the costs increase with the amount of agents in the network. One can understand that when a link is over-utilised, i.e., there are more agents on the link than the capacity allows, these costs increase steeply. When there is enough capacity for the amount of agents, the costs are similar to the free flow travel costs. The travel times follow from Eq. (1), using $t_0 = 1/6.666$ and $J_a = 0.1$.

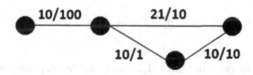

Fig. 2. Example 1, with length and capacity of nodes (notation: length/capacity).

In this first example we want agents to travel from the left of this network to the right. This means they all first need to traverse a link that has a length of 10 and a capacity of 100. Afterwards, they can choose between one route with length 21 and capacity 10 or a route comprised of two links: one with length 10 and capacity 1 and one with length 10 and capacity 10.

In the second example, Fig. 3, agents also need to travel from left to right. There are two routes available, both comprised of two links. Both routes start with a link with length 50 and capacity 100. The first route then has a link with length 20 and capacity 100, while the other route has a link with length 5 and capacity 1.

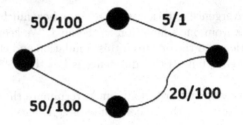

Fig. 3. Example 2, with length and capacity of nodes (notation: length/capacity).

In both examples we assume that 10 agents want to traverse the network. Firstly, we assume that all these agents want to traverse the network at the same time. For example 1, these results can be found in Fig. 4 and for example 2 in Fig. 5.

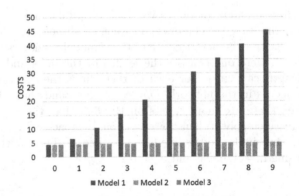

Fig. 4. Costs for each agent 0,...,9 in example 1 with clustered travelling.

In Fig. 4, we can see that model 1 performs very poorly compared to the other two. The reason for this behaviour is that it looks at the network and sees the route via the link with capacity 1 as the shortest path, since this path has length 30 and the other path has length 31. All agents depart at the same time, therefore no agents are on the link with the small capacity and thus the shortest route will seem to be the same for all agents. Therefore, all agents will travel via the lower route. However, when the agents do arrive at the second link in their route, all the other agents also want to traverse that route, which means there is an enormous extra cost for this link. Model 2 does not show this behaviour, since the shortest path is recalculated at the first intermediate node. Here the first agent will travel the route with length 30, but the next agent sees that the link is already in use and will therefore use the link with length 21. This recalculation of the shortest path is not done in heuristic 1. The solution algorithm also divides the agents, such that they all have a low cost of travelling.

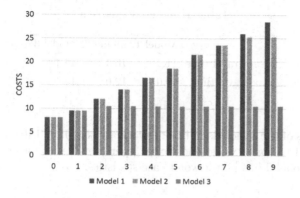

Fig. 5. Costs for each agent in example 2 with clustered travelling.

Figure 5 shows the costs for example 2. Here both models with public information perform badly. The reason for this is that both models see the route on the right side of the network as the shortest path at the departure time of the agents. Therefore, both models send all agents on the right link with length 50. Model 1 does not have a possibility to reroute and therefore all agents will be stuck on this route. This creates a situation where all agents want to traverse a route of capacity 1. This leads to extra costs. Model 2 *does* have a possibility to reroute, but only at the intermediate node. This means that at this point agents would have to travel back in order to avoid the link with capacity 1. But this means extra costs of at least a link with length 50. Therefore, most agents will still be routed on the link with capacity 1. However, one can see that agents 8 and 9 have a lower cost for model 2. This means that at this point it is actually beneficial to go back to the first node and take the left route. The solution algorithm has full information and thus knows that sending all agents over the right route will lead to trouble. Therefore, the solution algorithm already divides the agents over the left and right route from the start.

Now we show the results for a somewhat larger example as depicted in Fig. 6. For this network, given a number of 80 departs per time step (of 5 min) travelling from left to right, repeated for 10 time steps. Again, arriving at a certain arc, the travel time on that arc is calculated using Eq. (1).

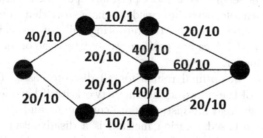

Fig. 6. Example 3.

Table 1. Results Example 3

	Model 1	Model 2	Model 3
Mean travel time	11.4	10.4	10.1
Max travel time	19.0	12.0	13.1

The results of the 3 models in this network are shown in Table 1. We see that indeed model 1 performs worst and model 3 performs best on average. However, we see that model 2 has better worst case statistics.

5   Conclusions

In this paper we investigated the effect of different information availability in agent-centric synchromodal networks. For the optimal routing with only public information we developed our two models. Obviously, model 2, that allows for rerouting, outperforms model 1. The reason for this is that agents (containers) are able to change their route when new information enters the network. Model 2 makes use of all public information that is available. The optimal routing with full information could be determined with the third model. An iterative process is used to find optimal routes for the entire planning horizon. The models were tested on small examples and on a larger one. We mention the advantages and disadvantages of each model below.

- *Model 1*: performs well if the arrivals of the orders are spread out over the planning horizon. It needs time to see the congestion of the network build up. The model performs poorly on networks where orders need to travel a large distance. For all orders, the routes are determined at their departure time. That means that the more time it takes for an order to move through the network, the more can change in the network. This will result in sub-optimal routes.
- *Model 2*: performs better when the orders are spread out over the planning horizon. Although it does have the availability to reroute certain containers, the availability of alternative routes plays a huge role on its performance. If one of the routes seems like a short path at departure time, the other routes need to be reachable from intermediate nodes on that route. If there are no alternative routes available, the rerouting will not result in smaller costs. However, this model does not perform worse when the orders in the network need to travel longer.
- *Model 3*: results in optimal routes for all discrete cases. This means that the parameters of the network and the network itself have less effect on this method. However, real life instances are continuous and the discrete nature of the simulation-based solution method is a disadvantage. Note that this problem can be solved by taking smaller time steps within the algorithm.

In further research the models should be applied to real life data to show the economic effect of information sharing in synchromodal transportation networks. Next, research should be done on ways to reach the global optimum in an agent based (user equilibrium reaching) network, i.e., by introducing tolls or other kind of (artificial) road pricing.

Acknowledgement. This work has been carried out within the project 'Complexity Methods for Predictive Synchromodality' (Comet-PS), supported by NWO (the Netherlands Organisation for Scientific Research), TKI-Dinalog (Top Consortium Knowledge and Innovation) and the Early Research Program 'Grip on Complexity' of TNO (The Netherlands Organisation for Applied Scientific Research).

References

1. Akcelik, R.: Travel time functions for transport planning purposes: Davidson's function, its time dependent form and alternative travel time function. Aust. Road Res. **21**(3) (1991)
2. Ben-Akiva, M., Bierlaire, M., Bottom, J., Koutsopoulos, H., Mishalani, R.: Development of a route guidance generation system for real-time application. In: 8th IFAC Symposium on Transportation Systems. No. TRANSP-OR-CONF-2006-063 (1997)
3. Ben-Akiva, M., Bierlaire, M., Koutsopoulos, H., Mishalani, R.: DynaMIT: a simulation-based system for traffic prediction. In: DACCORD Short Term Forecasting Workshop, pp. 1–12 (1998)
4. Birge, J., Ho, J.: Optimal flows in stochastic dynamic networks with congestion. Oper. Res. **41**(1), 203–216 (1993)
5. Carey, M., Subrahmanian, E.: An approach to modelling time-varying flows on congested networks. Transp. Res. Part B: Methodol. **34**(3), 157–183 (2000)
6. Dafermos, S.: Traffic equilibrium and variational inequalities. Transp. Sci. **14**(1), 42–54 (1980)
7. De Juncker, M.: Optimising routing in an agent-centric synchromodal network with shared information. Master's thesis, Eindhoven University of Technology (2017)
8. Dia, H.: An agent-based approach to modelling driver route choice behaviour under the influence of real-time information. Transp. Res. Part C Emerg. Technol. **10**(5), 331–349 (2002)
9. Friesz, T., Luque, J., Tobin, R., Wie, B.: Dynamic network traffic assignment considered as a continuous time optimal control problem. Oper. Res. **37**(6), 893–901 (1989)
10. Janson, B.: Convergent algorithm for dynamic traffic assignment. Transportation Research Record (1328) (1991)
11. Kaufman, D., Smith, R., Wunderlich, K.: User-equilibrium properties of fixed points in dynamic traffic assignment. Transp. Res. Part C Emerg. Technol. **6**(1), 1–16 (1998)
12. Mahmassani, H., Jayakrishnan, R.: System performance and user response under real-time information in a congested traffic corridor. Transp. Res. Part A Gen. **25**(5), 293–307 (1991)
13. Merchant, D., Nemhauser, G.: A model and an algorithm for the dynamic traffic assignment problems. Transp. Sci. **12**(3), 183–199 (1978)

14. Nagurney, A.: Network Economics: A Variational Inequality Approach, vol. 10. Springer Science & Business Media (2013)
15. Peeta, S., Mahmassani, H.: System optimal and user equilibrium time-dependent traffic assignment in congested networks. Ann. Oper. Res. **60**(1), 81–113 (1995)
16. Peeta, S., Ziliaskopoulos, A.: Foundations of dynamic traffic assignment: The past, the present and the future. Netw. Spat. Econ. **1**(3), 233–265 (2001)
17. Phillipson, F.: A thought on optimisation and self-organisation in synchromodal logistics. Technical report TNO (2017)
18. Ramalingam, G., Reps, T.: An incremental algorithm for a generalization of the shortest-path problem. J. Algorithms **21**(2), 267–305 (1996)
19. Ran, B., Boyce, D., LeBlanc, L.: A new class of instantaneous dynamic user-optimal traffic assignment models. Oper. Res. **41**(1), 192–202 (1993)
20. Ran, B., Shimazaki, T.: A general model and algorithm for the dynamic traffic assignment problems. In: Transport Policy, Management & Technology Towards 2001: Selected Proceedings of the Fifth World Conference on Transport Research, vol. 4 (1989)
21. Tavasszy, L., Behdani, B., Konings, R.: Intermodality and synchromodality. SSRN Online (2015)
22. Ulmer, M.W., Heilig, L., Voß, S.: On the value and challenge of real-time information in dynamic dispatching of service vehicles. Bus. Inf. Syst. Eng. **59**(3), 161–171 (2017)
23. Wardrop, J.: Some theoretical aspects of road traffic research. In: Proceedings of the Institute of Civil Engineers, London (1900)
24. Ziliaskopoulos, A., Mahmassani, H.: Time-dependent, shortest-path algorithm for real-time intelligent vehicle highway system applications. Transportation research record, pp. 94–94 (1993)

Adapting the A* Algorithm to Increase Vehicular Crowd-Sensing Coverage

Sergio Di Martino[1(✉)], Paola Festa[2], and Dario Asprone[1]

[1] Department of Electrical and Telecommunications Engineering,
University of Naples Federico II, Naples, Italy
sergio.dimartino@unina.it, dario.asprone@gmail.com
[2] Department of Mathematics and Applications "R. Caccioppoli",
University of Naples Federico II, Naples, Italy
paola.festa@unina.it

Abstract. Current vehicles are incorporating an even wider number of environmental sensors, mainly needed to improve safety, efficiency and quality of life for passengers. These sensors bring a high potential to significantly contribute also to urban surveillance for Smart Cities by leveraging opportunistic crowd-sensing approaches. In this context, the achievable spatio-temporal sensing coverage is an issue that requires more investigations, since usually vehicles are not uniformly distributed over the road network, as drivers mostly select a shortest time path to destination. In this paper we present an evolution of the standard **A*** algorithm to enhance vehicular crowd-sensing coverage. In particular, with our solution, the route is chosen in a probabilistic way, among all those satisfying a constraint on the total length of the path. The proposed algorithm has been empirically evaluated by means of a public dataset of real taxi trajectories, showing promising performances in terms of achievable sensing coverage.

1 Introduction

The features and services offered by vehicles are changing significantly in the last years, due to the application of Information and Communication Technologies. On one hand, novel *Advanced Driver Assistance Systems* (ADAS) requires awareness of the environment surrounding the vehicle, like relative position of pedestrians, bikes, and other vehicles, which is obtained by means of sensors like frontal and backward radars, surrounding cameras, ultrasonic sensors, and so on. On the other hand, to improve the quality of life for passengers, many comfort services also require advanced contextual sensors, like sun or rain intensity, amount of pollutants in the air, and so on.

The integration of this wide array of environmental sensors together with communication technologies is opening a whole new scenario for location-based applications [1], where modern connected vehicles can be exploited as probes in a context of Smart Cities, sensing in real-time what is happening on the streets. Many exciting services could be developed on top of the information collected

© Springer Nature Switzerland AG 2018
R. Cerulli et al. (Eds.): ICCL 2018, LNCS 11184, pp. 331–343, 2018.
https://doi.org/10.1007/978-3-030-00898-7_22

from probe vehicles, like more accurate traffic and parking predictions, better surveillance of urban scenarios, and so on [2]. One of key performance indicators (KPI) of a crowd-sensing solution is the achievable sensing coverage. In case of vehicular crowd-sensing, this is mostly influenced by the numbers of connected cars and by their trajectories. Oddly there is a number of studies in the literature addressing the required amount of probe vehicles to offer service, (e.g.: [3–6]), but only one work regarding the optimization of their trajectories [7]. Indeed, the huge literature on routing applied to smart mobility has mainly focused other problems, like eco-routing (e.g.: [8]) or dynamic car-pooling (e.g.:[9]). Nevertheless, it is known that one of the current limitations with probe vehicles comes from the non-uniform distribution of cars over the road network, as drivers normally prefer a route leading to a shortest time path to a destination [7], which is an efficient delivery solution for logistics or passenger vehicles.

To address this issue, in [10] we preliminary presented a new probabilistic routing algorithm, corresponding to a variant of the standard $\mathbf{A}^*$ procedure, where, given the same source, destination, and road network, we computed potentially different routes, as long as the total travel distance is within a selectable threshold (in percentage over the total route). This is, in our opinion, a viable solution for improving the sensing coverage in the next future, without requiring a global coordination of the paths, which would require a complex and costly shared infrastructure. To the best of our knowledge, this formulation of the problem (i.e. a routing algorithm to maximise the sensing coverage of a fleet of vehicles without a central coordination) is new and has never been investigated before, since the solution proposed in [7] is based on a global coordination of the vehicles.

In this paper we extend the findings presented in [10], by providing a much deeper description of the algorithm, and by adding a qualitative evaluation of the coverage of the road network, that could be achieved by using our proposal, by exploiting a dataset of real taxi trajectories collected in San Francisco (USA) over five weeks. After cleansing the dataset, we selected more than 320,000 routes, and computed the potential improvements in terms of sensing coverage achievable by following the routes proposed by our algorithm.

The remainder of the paper is organized as follows. In Sect. 2 we describe in details the concept of Vehicular Crowd-Sensing. In Sect. 3, after some background information, the algorithm we propose to improve sensing coverage is presented. In Sect. 4 we describe the way we preliminary assessed the solution, while in Sect. 5 we present the preliminary computational results obtained by applying our strategy on a real scenario. Conclusions and final remarks are given in Sect. 6.

2   Vehicular Crowd-Sensing

Currently, the average number of sensors in a vehicle is around 60–100, able to record a myriad of physical phenomena [11], but it expected that, in the near future, this number might rise up to 200 sensors per vehicle [12]. A partial list of the typology of these sensors is shown in Fig. 1.

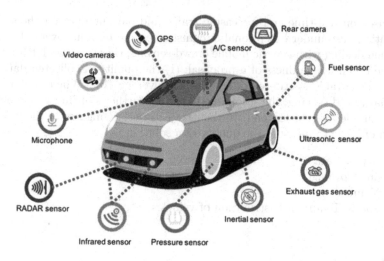

Fig. 1. Some of current vehicle's sensors, from [12].

The integration of this wide array of environmental sensors together with communication technologies is opening a whole new scenario for location-based applications [1]. Indeed, in the last years, a hot research trend is the study on how to exploit these sensors to opportunistically crowd-sense contextual information around the vehicle [13], since connected vehicles may give rise to one of the biggest and most pervasive sensor networks around the world, making possible to develop a collective intelligence, or contextual awareness, in a detail never experienced before [2].

More in details, the general idea is that, while a vehicle drives by a road segment, its sensors are constantly scanning the surrounding environment. The location of each information of interest (e.g.: a free parking space, the amount of pollutants over a certain threshold, the presence of rain, etc.) can be sent by the vehicle to a back-end server via the cellular network [14], where it is aggregated with data coming from other connected vehicles, and processed to generate new contextual knowledge.

As an example, a significant fraction of modern vehicles has sensors for air temperature, pressure, sun intensity and direction, amount of rain and quality of the air. Thus, each car is a kind of itinerant weather station. The collection and aggregation of this information from multiple vehicles could provide better weather forecasts than traditional services, due to the vehicular pervasivity over the global surface [15].

Mobility is another field where opportunistic crowd-sensing performed by vehicles is of particular relevance [16]. Indeed, having a deeper global insight on the current state of mobility situation, drivers could be supported by smarter Intelligent Transportation Systems (ITS). For example, drivers could be guided towards roads with less traffic and/or higher chances to find free parking spaces

[17], based on real-time and forecasted traffic and parking data for the selected destination, also under exceptional circumstances and extreme events [18,19].

When dealing with generic mobile crowd-sensing, there are two KPIs to measure the potential goodness of a proposal: (I) the quality of collected data, and (II) the achievable spatio-temporal sensing coverage [13]. When dealing with vehicular crowd-sensing, the second point can be further split in two factors, i.e. (a) the number of vehicles, and (b) their trajectories, giving rise to the following schema of KPIs to measure the quality of a solution [7]:

1. Quality of Data
2. Sensing Coverage
   (a) Number of vehicles
   (b) Spatio-Temporal Distribution of vehicles

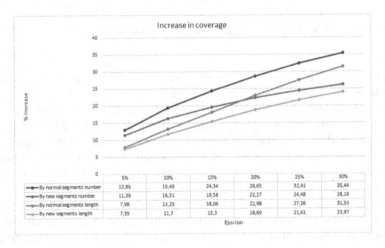

Fig. 2. An example of spatio-temporal coverage, achievable with 486 taxis in San Francisco (from [20]).

The KPI (1) mainly depends on the quality of the sensors. In the case of vehicular crowd-sensing, this quality highly varies across different car manufacturers, but is in general good enough for most of the application (especially when exploiting sensors regarding safety-relevant services). As for KPI (2), the problem with the spatio-temporal coverage is well represented in Fig. 2 [20], which shows the average daily time gaps between consecutive passing of probe vehicles, over some road segments in the Financial District of San Francisco, considering real GPS traces from a fleet of 486 taxis.

From the figure it is clearly visible that some streets are traversed by a sensing vehicle very often, with an average temporal gap smaller than 10 min between two consecutive vehicles. On some adjacent street, this interval goes significantly up, even up to hours between the consecutive passages of probe vehicles. Clearly, with such a variability, the potential use cases implementable by means of probe vehicles may have significant limitations.

As a consequence, there is the need for new optimization techniques, to maximise this spatio-temporal sensing coverage. Two main strategies can be envisioned to address this issue:

- Centralized Routing
- Ego Routing

In the former case, a centralized back-end infrastructure is needed, which has real-time information about source and destination of all connected vehicles, plus a knowledge of the sensing demand. This approach has the advantage to be the best possible solution, asking a driver to do a (minor) detour to monitor something on a given road segment only when there is an actual sensing demand on that specific segment. The problem of this solution mainly regards the costs and feasibility, requiring that all the drivers are connected to the same shared back-end and willing to accept its proposed path planning. In the latter case, no central coordination is present, so each driver of a sensing vehicle is free to follow the route he/she prefers. The main advantage is the minor cost and an easier feasibility, even if the sensing coverage might be far from being optimal.

In this paper we propose an adaptation of the $\mathbf{A^*}$ algorithm, to deal with the second case.

3   The Proposed Routing Algorithm

In this section we start by quickly recalling some basic concepts of the $\mathbf{A^*}$ algorithm, then we present the proposed solution.

3.1   The A* Algorithm

An $\mathbf{A^*}$ search algorithm for the *single source-single destination* shortest path problem with nonnegative edge lengths (see among others [21]) focuses an "informed" search of a shortest path from a given source node s to a given destination node d in a weighted di-graph $G = (V, E)$ through the use of a heuristic function h. In more detail, it works with "modified" labels $l(v)$, $\forall\ v \in V$ given by $l(v) = d(v) + h(v, d)$, where $d(v)$ represents the classical label associated to v as in any classical shortest path problem method, while $h(v, d)$ expresses an estimate of the shortest distance from v to the destination node d.

An $\mathbf{A^*}$ algorithm is guaranteed to produce an optimal solution path (whenever a path from the source node to the destination node exists) if the heuristic function h is monotone, i.e., if $h(d, d) = 0$ and for all nodes x and y, it results that $h(x, d) \leq W(x, y) + h(y, d)$, where $W(x, y)$ is the shortest distance from x to y. It can be easily seen that if the heuristic function h evaluates to zero at each node, then $\mathbf{A^*}$ reduces to the Dijkstra's algorithm.

3.2   The Proposed Solution

The proposed work is based on the $\mathbf{A}^*_\varepsilon$ algorithm [22], presented by Judea Perl in 1982. Using not one, but two different heuristic functions to choose the expanding node, this routing algorithm could expand fewer nodes than the classical $\mathbf{A}^*$, at the cost of relaxing the optimality of the result.

The breakthrough of this $\mathbf{A}^*$ variant was that it gave a bound on the optimality relaxation, tied to a parameter ε, in which it guaranteed that the returned path would never be longer more than ε times the minimal one.

As the focus of the proposed work was identifying a route in a somewhat random manner, while limiting the maximum possible returned path length, $\mathbf{A}^*_\varepsilon$ proved to be the perfect base, even if not used as its authors originally intended.

The proposed algorithm, given the required source and destination of the route and a multiplicative coefficient ε, first finds in some way, for example using a standard implementation of $\mathbf{A}^*$, the length of the shortest path between the provided endpoints. If no such path exists, then the algorithm terminates, and returns an empty path itself. Otherwise, to better distribute the length of the different paths returned over time, a candidate lower bound $lBound$ is chosen randomly among m evenly spaced values comprised between the shortest path length l_P and $l_P \cdot \varepsilon$, the latter not included. We experimentally found that the value of m is not significantly affecting the general performance of the algorithm. The actual routing algorithm is now called and given as parameters the two endpoints, the lower and upper bounds ($lBound$ and $l_P \cdot \varepsilon$), and ε. If no path could be found that met the required bounds, a smaller lower bound is chosen, until either a path is found or $lBound = l_P$, which assures that at least the shortest path will be acceptable.

The inner path-finding algorithm, being based on $\mathbf{A}^*_\varepsilon$, works by creating, at every iteration, a list of nodes which are extremes to the best paths found until now, and, as such, are candidates for expansion: given a feasible heuristic function $h : V \rightarrow \mathbb{R}$ that estimates the distance of a node from the destination, a function $d : V \rightarrow \mathbb{R}$ that returns the length of the shortest path found until now between a node and the source, and their sum function $f(n) = h(n) + d(n)$, the list of candidates is comprised of all the nodes m such as

$$m \in \left\{ v \in V \mid f(v) \leq \min_{n \in V} f(n) \cdot \varepsilon \right\}. \tag{1}$$

After having done this, in $\mathbf{A}^*_\varepsilon$ a second heuristic $t : V \rightarrow \mathbb{R}$ is evaluated for every candidate node, and the one with the smallest image is chosen as the expanding node.

The basic version of the proposed algorithm differs from $\mathbf{A}^*_\varepsilon$ in this last operation, as it chooses the node to be expanded in a random manner. The different ways in which a node can be chosen randomly lead to different efficiency and effectiveness. In the preliminary experiments that were carried out, reported in this paper, every node in the list had a probability of being chosen equal to $\frac{2i}{N*(N+1)}$, where i is the index of the node in the total ascending ordering of the list by the value of f, and $N = |V|$. Other possible strategies, such as giving the same probability of being chosen to every node, resulted in worse outcomes. Anyhow, the identification of other strategies deserves more investigations.

This first version, while completely working, was not the most efficient possible, and did not naturally offer the possibility of introducing a lower bound on returned paths. For these reasons, a bidirectional version was implemented and used for the experiments described later on in the paper. While being a pretty straightforward bidirectional version of an A^* variant, it has been possible to simplify the termination criteria usually used in the latter because we were not interested in the shortest path, but in one of the many paths whose length was comprised between a lower and an upper bound: as soon as the algorithm encounters a fitting path, it terminates.

4   Experimental Design

In this section we describe the experimental protocol we adopted for the evaluation of the proposed algorithm, in terms of employed data, procedure and measures.

4.1   The Dataset

The taxi dataset was collected within the *Cabspotting* project [23], which aimed at the extraction of socio-economic properties of regions from the taxi patterns. Each taxi periodically provided information on its latitude and longitude, timestamp, and occupancy (1 = occupied, 0 = free) to a central server. The resulting dataset contains 11,219,955 GPS coordinates, collected from 536 vehicles of the *Yellow Cab* company, over 25 days in the San Francisco Bay Area, from 2008/05/17 until 2008/06/10.

The first challenging task was to obtain map matched trajectories, by aligning the sequence of GPS points contained in the FCD with the road network provided by *OpenStreetMap*. The median time gap between two consecutive GPS measures is 60 s, with time gaps ranging between 30 and 120 s, for 86% of all observations. Such a low frequency of the FCD collection, together with the intrinsic noise of the GPS, required using advanced map matching techniques. In more detail, since for each taxi we had FCD covering more than 3 weeks, the first step was to segment each taxi's data flow into a set of independent *trajectories*. A taxi trajectory T_r can be defined as a sequence of GPS points corresponding to a trip with the same passenger occupancy state. Each GPS point $p_i = (x_i, y_i, t_i)$ has a longitude x_i, latitude y_i, and a timestamp t_i. A trajectory T_r is thus a sequence of points $p_1 \rightarrow p_2 \rightarrow \ldots \rightarrow p_n$, where the state of passenger occupancy of the taxi is the same from p_1 to p_n. Thus, differently from other similar works (e.g. [24]), we are also considering taxi trajectories where the vehicle is not occupied, since clearly the sensing of parking spaces can be done also in these cases. We also split the sequence of points every time there was a time gap longer than 3 min between consecutive GPS points, on the assumption that the taxi was not operating in that time frame. Finally, from this set of trajectories, we discarded all those having less than 5 points and/or implausible speed between two consecutive points.

On these selected taxi trajectories, we applied the map matching algorithm described in [25], which is based on [26]. For each GPS point in a trajectory, candidate street segments are identified from the *OpenStreetMap* road network. Road segments between two consecutive GPS points projected on the map are identified by a shortest path search. In a following global optimization step, the sequence of candidate segments achieving the highest score, based on spatial and temporal criteria, is selected. As a result of this task, 3,371,552 GPS points were successfully matched to the *OpenStreetMap* road network.

Then, all trajectories containing a loop were removed, as well as all the trajectories containing 3 points or less. As a result, 420,790 potential taxi runs were identified.

4.2 Experimental Procedure

The experiments are thus executed on the previously described dataset. We checked each of these 420,790 traces, by re-running the standard $\mathbf{A}^*$ algorithm, given the respective sources and destinations of the considered paths. This allowed us on one hand to compute the results of the *baseline*, intended as the normal routes computed by the standard $\mathbf{A}^*$. On the other hand, this allowed us to further cleanse the dataset, by removing all the routes whose sequence of segments did not match the one provided by $\mathbf{A}^*$.

As a result, we kept 324,199 taxi runs, whose sources and destinations were used to evaluate the proposed algorithm. More in details, we choose six different configurations for ε, namely 1.05, 1.1, 1.15, 1.2, 1.25 and 1.3, thus permitting detours from 5% up to 30%. To minimize the bias due to the probabilistic nature of the algorithm, we run it 100 times for each route and each value of ε, and then we averaged across the 100 outcomes.

We computed the results for the following six attributes:

1. Total Length of the paths identified by the $\mathbf{A}^*$.
2. Total Length of the paths identified by the $\mathbf{A}^*_\varepsilon$.
3. Total Number of Segments covered by the $\mathbf{A}^*$.
4. Total Number of Segments covered by the $\mathbf{A}^*_\varepsilon$.
5. Road Segments covered only by the $\mathbf{A}^*$.
6. Road Segments covered only by the $\mathbf{A}^*_\varepsilon$.

5   Results and Discussion

In this section we present and discuss the obtained preliminary results we conducted. In particular, in Table 1 we report the summary statistics of the results obtained by the standard $\mathbf{A}^*$ algorithm and by the new $\mathbf{A}^*_\varepsilon$, on the couples (*Source, Destination*) on the taxi trips described in the previous section, with the six values of ε.

Table 1. Summary statistics of the evaluation

Epsilon	1.05	1.1	1.15	1.2	1.25	1.3
A* Length (km)	8,578					
A_ε^* Length (km)	9,262.45	9,714.80	10,127.56	10,549.33	10,942.40	11,282.39
A* Segments	224,835					
A_ε^* Segments	253,739.6	268,653.3	279,566.8	289,261.4	297,708.1	304,507.2
Segments only A*	5,599.53	6,866.58	8,181.95	9,338.1	10,451.37	11,462.82
Segments only A_ε^*	34,504.14	50,684.93	62,913.83	73,764.49	83,324.54	91,135.04

Let us note that the results of A_ε^* are intended as the average over the 100 runs. While the total distance covered by all the taxis using the A* algorithm is 8,578 km (224,835 segments), this distance clearly varies with A_ε^* according to the value of ε. The same consideration holds also for the number of covered road segments. We can observe that the number of segments covered only by the A* increases while changing ε. This is due to the fact that the bigger is ε, the more the A_ε^* looks for routes far from the standard version of the algorithm.

The most relevant information in the results is the number of new segments, covered only by A_ε^*. Indeed, this number represents the amount of new streets that might be potentially sensed by probe vehicles using our routing algorithm.

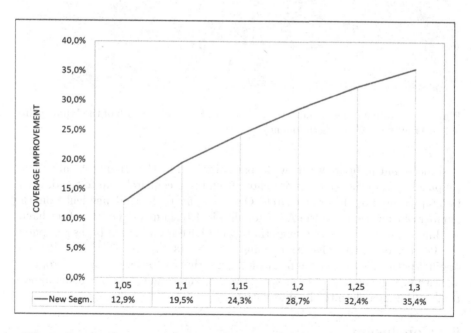

Fig. 3. Coverage increment, in %, due to the proposed algorithm, for different ε.

We also computed the relative improvement in terms of covered segments, defined as:

$$\frac{Segments\ Only\ \boldsymbol{A}^*_{\varepsilon} - Segments\ Only\ \boldsymbol{A}^*}{\boldsymbol{A}^*\ Segments} \tag{2}$$

In Fig. 3 we report both in textual and graphical form these results. From this figure we can see an interesting behavior, that, given an ε, the achievable relative improvement over the standard $\boldsymbol{A}^*$ is always greater than ε, in terms of new explored segments, and thus in sensing coverage of the probe vehicles.

Finally, we performed also a preliminary qualitative assessment of the coverage, which is visible in Fig. 4.

Fig. 4. Qualitative Assessment of the coverage of the road network of the Municipality of San Francisco. (Color figure online)

Indeed, in this figure we show the path obtained by the original $\boldsymbol{A}^*$ algorithm (in purple), from the source represented from the green circle, up to the destination represented by the orange circle. Of course this path is independent from the number of executions, as the $\boldsymbol{A}^*$ algorithm is deterministic. On the other hand, in blue we represent the segments that would have been part of paths computed by 100 runs of the $\boldsymbol{A}^*_{\varepsilon}$ algorithm, using a $\varepsilon$of 0.2 (so up to 20% longer than the original path). As we can see from the figure, thanks to the use of the proposed algorithm, the coverage of the road network is significantly higher, being almost total in the area close to the destination.

6   Conclusion

A lot of research is currently being focused on exploiting the information that could be opportunistically crowd-sensed by the multitude of sensors equipping modern connected cars, with the goal to design new services, especially useful in the context of Smart Cities. Anyhow, an open issue with probe vehicles is the achievable spatio-temporal sensing coverage, since cars are not uniformly distributed over the road network, because drivers mostly select a shortest time path to destination.

In this paper we have presented an adaptation of the $\mathbf{A^*}$ routing algorithm to increase the spatio-temporal sensing coverage achievable by probe vehicles, without requiring a central coordination. Indeed, the proposed algorithm, named $\mathbf{A^*_\varepsilon}$, search the graph of the road network to find a route in a probabilistic way, being at most longer than ε times with respect to the standard route identified by the traditional $\mathbf{A^*}$. Since the route selection is random, each probe vehicle might follow a different sequence of road segments, given the same origin, destination and road network, thus increasing the number of monitored streets.

The proposed algorithm has been empirically evaluated by means of a public dataset of real taxi traces, collected over five weeks in San Francisco. Starting from 11 millions of GPS points, we identified over 320,000 trajectories of 486 taxis whose routes corresponded to the ones computed by the standard $\mathbf{A^*}$ algorithm. For these traces, we computed the performances of the proposed $\mathbf{A^*_\varepsilon}$ algorithm, with six different values of ε.

Preliminary results show that the 320,000 routes computed by the $\mathbf{A^*_\varepsilon}$ algorithm cover a significant number of new road segments (up to 91,135), thus greatly improving the spatio-temporal coverage achievable by the fleet of probe vehicles. Also a qualitative investigation on the spatial distribution of the covered segments shows significant benefits for the given problem.

Future research directions will require a probabilistic complexity analysis of the algorithm, as well as evaluation over road network having a different topology than the one in San Francisco, which is based on a standard grid.

Another interesting evolution would be to compare the proposed probabilistic approach with the results obtainable by a centralized solution, with a priori knowledge of the sensing coverage over the road network, like the one in [7].

References

1. Vegni, A.M., Loscri, V.: A survey on vehicular social networks. IEEE Commun. Surv. Tutor. **17**(4), 2397–2419 (2015)
2. Di Martino, S.: Discovering information from spatial big data for ITS. Intelligent Transport Systems (ITS): Past, Present and Future Directions, pp. 109–130 (2017)
3. Bock, F., Martino, S.D., Sester, M.: What are the potentialities of crowdsourcing for dynamic maps of on-street parking spaces? In: Proceedings of the 9th ACM SIGSPATIAL International Workshop on Computational Transportation Science, pp. 19–24. ACM (2016)
4. Bock, F., Attanasio, Y., Di Martino, S.: Spatio-temporal road coverage of probe vehicles: a case study on crowd-sensing of parking availability with taxis. In: Bregt, A., Sarjakoski, T., van Lammeren, R., Rip, F. (eds.) GIScience 2017. LNGC, pp. 165–184. Springer, Cham (2017). https://doi.org/10.1007/978-3-319-56759-4_10

5. Mathur, S., et al.: ParkNet: drive-by sensing of road-side parking statistics. In: Proceedings of 8th International Conference on Mobile Systems, Applications, and Services, pp. 123–136. ACM, New York (2010)

6. Bock, F., Di Martino, S.: How many probe vehicles do we need to collect on-street parking information? In: 2017 5th IEEE International Conference on Models and Technologies for Intelligent Transportation Systems (MT-ITS), pp. 538–543. IEEE (2017)

7. Masutani, O.: A sensing coverage analysis of a route control method for vehicular crowd sensing. In: 2015 IEEE International Conference on Pervasive Computing and Communication Workshops (PerCom Workshops), pp. 396–401. IEEE (2015)

8. Boriboonsomsin, K., Barth, M.J., Zhu, W., Vu, A.: Eco-routing navigation system based on multisource historical and real-time traffic information. IEEE Trans. Intell. Transp. Syst. **13**(4), 1694–1704 (2012)

9. Di Martino, S., Giorio, C., Galiero, R.: A rich cloud application to improve sustainable mobility. In: Tanaka, K., Fröhlich, P., Kim, K.-S. (eds.) W2GIS 2011. LNCS, vol. 6574, pp. 109–123. Springer, Heidelberg (2011). https://doi.org/10.1007/978-3-642-19173-2_10

10. Asprone, D., Di Martino, S., Festa, P.: Improving sensing coverage of probe vehicles with probabilistic routing. In: R. Luaces, M., Karimipour, F. (eds.) W2GIS 2018. LNCS, vol. 10819, pp. 1–10. Springer, Cham (2018). https://doi.org/10.1007/978-3-319-90053-7_1

11. Lee, E., Lee, E.-K., Gerla, M., Oh, S.Y.: Vehicular cloud networking: architecture and design principles. IEEE Commun. Mag. **52**(2), 148–155 (2014)

12. Guerrero-Ibáñez, J., Zeadally, S., Contreras-Castillo, J.: Sensor technologies for intelligent transportation systems. Sensors **18**(4), 1212 (2018)

13. Ganti, R.K., Ye, F., Lei, H.: Mobile crowdsensing: current state and future challenges. IEEE Commun. Mag. **49**(11), 32–39 (2011)

14. Robert Bosch GmbH: Bosch Community-based parking (2016). http://www.bosch-mobility-solutions.com/en/connected-mobility/community-based-parking/. Accessed 27 June 2016

15. Petty, K., Mahoney III, W.: Enhancing road weather information through vehicle infrastructure integration. Transp. Res. Rec. J. Transp. Res. Board **2015**, 132–140 (2007)

16. Wang, X., Zheng, X., Zhang, Q., Wang, T., Shen, D.: Crowdsourcing in ITS: the state of the work and the networking. IEEE Trans. Intell. Transp. Syst. **17**(6), 1596–1605 (2016)

17. Di Martino, S., Rossi, S.: An architecture for a mobility recommender system in smart cities. Procedia Comput. Sci. **98**, 425–430 (2016)

18. Kwoczek, S., Di Martino, S., Nejdl, W.: Stuck around the stadium? An approach to identify road segments affected by planned special events. In: 2015 IEEE 18th International Conference on Intelligent Transportation Systems (ITSC), pp. 1255–1260. IEEE (2015)

19. Kwoczek, S., Di Martino, S., Nejdl, W.: Predicting and visualizing traffic congestion in the presence of planned special events. J. Visual Lang. Comput. **25**(6), 973–980 (2014)

20. Bock, F., Attanasio, Y., Di Martino, S.: Spatio-temporal road coverage of probe vehicles: a case study on crowd-sensing of parking availability with taxis. In: Bregt, A., Sarjakoski, T., van Lammeren, R., Rip, F. (eds.) Societal Geo-innovation. Lecture Notes in Geoinformation and Cartography, pp. 165–184. Springer, UK (2017). https://doi.org/10.1007/978-3-319-56759-4_10

21. Lugar, G.F., Stubblefield, W.A.: Artificial Intelligence Structures and Strategies for Complex Problem Solving. Addison Wesley Longman, Harlow (1998)
22. Pearl, J., Kim, J.H.: Studies in semi-admissible heuristics. IEEE Trans. Pattern Anal. Mach. Intell. **4**, 392–399 (1982)
23. Piorkowski, M., Sarafijanovic-Djukic, N., Grossglauser, M.: CRAWDAD dataset EPFL/mobility, 24 February 2009. Downloaded from http://crawdad.org/epfl/mobility/20090224. Accessed Feb 2009
24. Chen, C., et al.: Crowddeliver: planning city-wide package delivery paths leveraging the crowd of taxis. IEEE Trans. Intell. Transp. Syst. **18**(6), 1478–1496 (2017)
25. Axer, S., Pascucci, F., Friedrich, B.: Estimation of traffic signal timing data and total delay for urban intersections based on low-frequency floating car data. In: Proceedings of the 6th Mobil.TUM 2015 (2015)
26. Lou, Y., Zhang, C., Zheng, Y., Xie, X., Wang, W., Huang, Y.: Map-matching for low-sampling-rate GPS trajectories. In: ACM SIGSPATIAL GIS 2009. ACM, November 2009. https://www.microsoft.com/en-us/research/publication/map-matching-for-low-sampling-rate-gps-trajectories/

Pricing Car-Sharing Services in Multi-Modal Transportation Systems: An Analysis of the Cases of Copenhagen and Milan

Rebecca Grüner Hansen and Giovanni Pantuso$^{(\boxtimes)}$ ⓘ

Department of Mathematical Sciences, University of Copenhagen,
Universitetsparken 5, 2100 Copenhagen, Denmark
gp@math.ku.dk

Abstract. In this article we study the problem of pricing car-sharing services in multi-modal urban transportation systems. The pricing problem takes into account the competition of alternative mobility services such as public transportation and bicycles and incorporates customer preferences by means of utility functions. The problem is formulated as a linear demand-based discrete optimization problem. A case study based on the cities of Copenhagen and Milan suggests that cycling habits and the efficiency of public transportation services have a significant effect on the viability of car-sharing services.

1 Introduction

During the past decade, car-sharing systems have become an attractive means of urban mobility in several cities around the world and dozens of companies have been built to provide such novel mobility services. In car-sharing services, customers share the use of a fleet of cars that is owned, maintained, and managed by a *Car-sharing Operator* (CSO). The customers are typically able to access shared cars without interacting directly with the CSO as reservations, pick ups, and returns are often self-serviced through the internet. Car-sharing services can be divided into two categories, namely *free-floating systems* and *station-based systems*. Free-floating systems enable users to pick up and return shared cars at any parking spot within a specified business area. In station-based systems, cars are assigned to dedicated stations and users must pick up and return cars at the specified stations. In this case we distinguish *two-way* systems, requiring the user to return the car at the pick up station, and *one-way* systems, allowing the user to return the car at a different station. Users generally pay based on their use of the car in addition to a possible subscription fee, while all vehicle costs are born by the CSO (e.g., fuel, insurance and maintenance).

CSOs face novel challenges at different planning levels which have attracted the interest of the scientific community in recent years. At the strategic level the CSO must decide the fleet size and business area [1,2], the trip booking

ⓒ Springer Nature Switzerland AG 2018
R. Cerulli et al. (Eds.): ICCL 2018, LNCS 11184, pp. 344–359, 2018.
https://doi.org/10.1007/978-3-030-00898-7_23

scheme [3,4] and, in station-based systems, the location, number and capacity of stations [5–8]. At the operating level, CSOs face planning problems such as the repositioning of vehicles [4,5,9–16], maintenance [11,17], charging and refueling [2,8,17,18].

In this paper we focus on the problem of pricing car-sharing services. Particularly, we look at car-sharing services within the context of multi-modal transportation systems. Classical urban transportation means such as bus, subway, and bicycles, can in fact be seen as competitors of car-sharing services in the market of urban transportation services. Therefore, CSOs need to take into account the alternative transportation means within a city, as well as customer preferences, when deciding about pricing schemes. The preferences of customers are often formalized using specific models such as logit models. However, the resulting integrated models are typically computationally difficult due to the non-linear interaction between the decision variables. In addition, convexification and linearization of such models (see, e.g., [19,20]) might not help to solve real-life instances (see [21]). Therefore, we propose a linear demand-based discrete optimization model in the spirit of [22]. The model explicitly takes into account that customers demand for transportation depends on the price set by the CSO as well as on the characteristics and price of the alternative transportation services. Customers preferences are included in the optimization model by means of a utility function which can be adapted to the specific market. When the utility function is linear in the price, the optimization model can be formulated as a MILP, thus avoiding the non-linearity typically generated by classical choice models.

The contribution of this paper is twofold. First, we provide a novel optimization model for pricing car-sharing services in multi-modal transportation systems which explicitly takes into account customers preferences and the competition of alternative transportation means. Second, we offer an analysis of car-sharing services in Copenhagen and Milan which investigates the influence of different characteristics of public transportation services. Similarly, [23] addressed the effects of relocation in a car-sharing service in Hamburg, [24] provided an empirical analysis of car-sharing usage in Munich and Berlin, and [25] studied the elements driving satisfaction for bike-sharing users in Milan.

In Sect. 2 we describe the pricing problem and in Sect. 3 we introduce the corresponding mathematical model. In Sect. 4 we use the model to study the cases of Copenhagen and Milan, while in Sect. 5 we draw final conclusions.

2   Problem Description

We consider a CSO operating in a city which offers a number of (private or public) transportation services (e.g., buses, metro, cycling lines). The CSO must determine the price of car-sharing rides. CSOs typically charge a per-minute fee plus a constant drop-off fee which depends on the zone of the city where the car is returned. For instance Car2Go (www.car2go.com, a CSO operating in several cities around the globe) divides Milan in zone A (comprising the city center and its surroundings) and B (comprising the outskirts of the city) and charges €4.90 when returning the car in zone B (no extra charge for zone A)[1]. Consistently with common practice, we assume a pricing scheme made of a per-minute fee and a drop-off fee. However, we generalize such pricing scheme by assuming the drop-off fee depends on the customer's origin and destination (O-D) pair, while the per-minute fee is common to all O-D pairs. Such pricing scheme allows the CSO to consider the city's specific transportation means at a higher level of granularity and price car-sharing rides according to the specific O-D pair, thus taking into account the competition on individual routes. In addition, it provides the CSO an instrument to offer customers incentives for moving the cars in accordance with some ideal distribution plan, and thus reducing the need for staff-based repositioning of cars. However, this requires that, upon booking, the CSO is able to inform their users about the drop-off fees based on their current location and all possible destinations.

Given an O-D pair, customers can choose between a number of transportation services. The set of available transportation services depends on the specific O-D pair. The demand for car-sharing rides between an O-D pair depends on the customers personal preferences and on the characteristics of the available transportation services, such as price, travel time, and waiting time. Specifically, a customer's choice depends on the utility obtained by choosing a service, and each customer chooses the service that gives them the highest utility.

Therefore, given an O-D pair within the city, the available transportation services, their prices and characteristics, the set of customer types characterized by their utility functions, the CSO's problem of pricing car-sharing services consists of deciding (i) whether to offer car-sharing services between the given O-D pair and (ii) the O-D pair specific drop-off fee in order to maximize its profit.

3   Mathematical Model

We formulate the problem using the demand-based discrete optimization framework proposed by [22] which entails modeling customers response to pricing decisions by means of a utility function. We begin by clarifying the necessary modeling assumptions in Sect. 3.1 and, following, we introduce the notation and the mathematical model in Sect. 3.2.

3.1   Modeling Assumption

We assume that the market for urban transportation between an O-D pair within the city consists of a finite number of customers or, alternatively, of a finite number of groups of customers with homogeneous behavior. We also assume that, for the given O-D pair, the set of transportation services, their prices and a list of their features (e.g., travel time and waiting time) is known to the CSO and to the customers, that price and characteristics are identical for all customers,

---

[1] Source: www.car2go.com, accessed on January 6th 2018.

and that all transportation services are available to all customers. However, the CSO might decide not to offer car-sharing services between a given O-D pair if unprofitable. Furthermore, we assume that the market is closed, meaning that every customer must choose exactly one transportation service.

We assume that each (group of) customer(s) is characterized by a utility function. The utility function is a real-valued function of the characteristics of the transportation services. Each customer values each characteristic differently according to their utility function. We assume that each customer chooses the available service which gives them the highest utility. In practice, the utility function is not fully known to the CSO. Therefore, we assume that the actual utility for a customer is a random variable for the CSO. An example of utility function will be given in Sect. 4.1.

We assume that the CSO offers a pricing scheme consisting of a per-minute fee common to all O-D pairs, plus a drop-off fee which is O-D specific and must be decided by the CSO. We assume that the drop-off fee is known by the customers upon reserving a shared car. Finally, for the sake of simplicity, we assume that users drive directly from the origin to the destination. This assumption can be easily relaxed by assuming user-specific paths trough the city.

3.2   Notation and Model

In this section we first introduce the notation and then the optimization model.

Sets

$\mathcal{C}$ the set of customers or groups of customers

$\mathcal{S}$ the set of all transportation services

$\mathcal{S}^{CS} \subseteq \mathcal{S}$ the set of transportation services offered by the CSO, such as different models of shared cars

$\mathcal{R}$ the set of utility scenarios

$\mathcal{L}_s$ the set of possible drop-off fee levels for service $s \in \mathcal{S}^{CS}$

Parameters

P_s^M the price-per-minute of car-sharing service $s \in \mathcal{S}^{CS}$

P_{sl}^D the drop-off fee at level $l \in \mathcal{L}_s$ for car-sharing service $s \in \mathcal{S}^{CS}$

P_s the price of transportation service $s \in \mathcal{S} \setminus \mathcal{S}^{CS}$

T_s^{CS} the travel time between the given O-D using car-sharing service $s \in \mathcal{S}^{CS}$

C_{sc} the cost of offering car-sharing service $s \in \mathcal{S}^{CS}$ to customer $c \in \mathcal{C}$ on the given O-D pair

ϵ_{scr} realization of the random utility error for service $s \in \mathcal{S}$ and customer $c \in \mathcal{C}$ under scenario $r \in \mathcal{R}$

M_{cr} upper bound on the difference in utility between two services for customer $c \in \mathcal{C}$ in scenario $r \in \mathcal{R}$

$\pi_s^1, \ldots, \pi_s^N$ a list of N attributes for transportation service $s \in \mathcal{S}$

$f_c : \mathbb{R}^{N+1} \to \mathbb{R}$ the utility function for customer $c \in \mathcal{C}$

Variables

p_s the price for service $s \in \mathcal{S}$

u_{scr} the utility obtained by customer $c \in \mathcal{C}$ for service $s \in \mathcal{S}$ under scenario $r \in \mathcal{R}$

y_{sc} a binary variable taking value 1 if service $s \in \mathcal{S}$ is offered to customer $c \in \mathcal{C}$, 0 otherwise

y_{scr} a binary variable taking value 1 if service $s \in \mathcal{S}$ is offered to customer $c \in \mathcal{C}$ under scenario $r \in \mathcal{R}$, 0 otherwise

w_{scr} a binary variable taking value 1, if service $s \in \mathcal{S}$ is chosen by customer $c \in \mathcal{C}$ under scenario $r \in \mathcal{R}$, 0 otherwise

λ_{sl} a binary variable taking value 1, if price level $l \in \mathcal{L}_s$ is chosen for service $s \in \mathcal{S}^{CS}$, 0 otherwise

μ_{szcr} a binary variable taking value 1 if customer $c \in \mathcal{C}$ obtains a higher utility by choosing service $s \in \mathcal{S}$ over service $z \in \mathcal{S}$ under scenario $r \in \mathcal{R}$, 0 otherwise

η_{szcr} a binary variable taking value 1 if both service $s \in \mathcal{S}$ and $z \in \mathcal{S}$ are available to customer $c \in \mathcal{C}$ under scenario $r \in \mathcal{R}$, 0 otherwise

α_{scrl} a binary variable taking value 1 if service $s \in \mathcal{S}^{CS}$ is chosen by customer $c \in \mathcal{C}$ under scenario $r \in \mathcal{R}$ at price level $l \in \mathcal{L}_s$, 0 otherwise.

The problem of pricing car-sharing services between a given O-D pain can thus be stated as follows.

$$\max \sum_{s \in \mathcal{S}^{CS}} \left(P_s^M T_s^{CS} + \frac{1}{|\mathcal{R}|} \sum_{c \in \mathcal{C}} \sum_{r \in \mathcal{R}} \sum_{l \in \mathcal{L}_s} P_{sl}^D \alpha_{scrl} \right) - \sum_{s \in \mathcal{S}} \sum_{c \in \mathcal{C}} C_{sc} y_{sc} \tag{1a}$$

$$\text{s.t.} \quad u_{scr} = f_c(p_s, \pi_s^1, \ldots, \pi_s^N) + \epsilon_{scr} \qquad c \in \mathcal{C}, s \in \mathcal{S}, r \in \mathcal{R}, \tag{1b}$$

$$p_s = P_s^M T_s^{CS} + \sum_{l \in \mathcal{L}_s} P_{sl}^D \lambda_{sl} \qquad\qquad s \in \mathcal{S}^{CS}, \tag{1c}$$

$$p_s = P_s \qquad\qquad s \in \mathcal{S} \setminus \mathcal{S}^{CS}, \tag{1d}$$

$$M_{cr} \eta_{szcr} - 2 M_{cr} \leq u_{scr} - u_{zcr} - M_{cr} \mu_{sznr} \tag{1e}$$
$$c \in \mathcal{C}, s \neq z \in \mathcal{S}, r \in \mathcal{R},$$

$$u_{scr} - u_{zcr} - M_{cr} \mu_{szcr} \leq (1 - \eta_{szcr}) M_{cr} \tag{1f}$$
$$c \in \mathcal{C}, s \neq z \in \mathcal{S}, r \in \mathcal{R},$$

$$\mu_{szcr} + \mu_{szcr} \leq 1 \qquad\qquad c \in \mathcal{C}, s \neq z \in \mathcal{S}, r \in \mathcal{R}, \tag{1g}$$

$$y_{scr} + y_{zcr} \leq 1 + \eta_{szcr} \qquad\qquad c \in \mathcal{C}, s \neq z \in \mathcal{S}, r \in \mathcal{R}, \tag{1h}$$

$$\eta_{szcr} \leq y_{scr} \qquad\qquad c \in \mathcal{C}, s \neq z \in \mathcal{S}, r \in \mathcal{R}, \tag{1i}$$

$$\eta_{szcr} \leq y_{zcr} \qquad\qquad c \in \mathcal{C}, s \neq z \in \mathcal{S}, r \in \mathcal{R}, \tag{1j}$$

$$\mu_{szcr} \leq y_{scr} \qquad\qquad c \in \mathcal{C}, s \neq z \in \mathcal{S}, r \in \mathcal{R}, \tag{1k}$$

$$w_{scr} \le \mu_{szcr} \qquad\qquad c \in \mathcal{C}, s \ne z \in \mathcal{S}, r \in \mathcal{R}, \qquad (1l)$$

$$\sum_{s \in \mathcal{S}} w_{scr} = 1 \qquad\qquad c \in \mathcal{C}, r \in \mathcal{R}, \qquad (1m)$$

$$\lambda_{sl} + w_{scr} \le 1 + \alpha_{scrl} \qquad\qquad c \in \mathcal{C}, s \in \mathcal{S}^{CS}, r \in \mathcal{R}, l \in \mathcal{L}_s, \qquad (1n)$$

$$\alpha_{scrl} \le \lambda_{sl} \qquad\qquad c \in \mathcal{C}, s \in \mathcal{S}^{CS}, r \in \mathcal{R}, l \in \mathcal{L}_s, \qquad (1o)$$

$$\alpha_{scrl} \le w_{scr} \qquad\qquad c \in \mathcal{C}, s \in \mathcal{S}^{CS}, r \in \mathcal{R}, l \in \mathcal{L}_s, \qquad (1p)$$

$$\sum_{l \in \mathcal{L}_s} \lambda_{sl} = 1 \qquad\qquad s \in \mathcal{S}^{CS}, \qquad (1q)$$

$$y_{scr} \le y_{sc} \qquad\qquad c \in \mathcal{C}, s \in \mathcal{S}^{CS}, r \in \mathcal{R}, \qquad (1r)$$

$$y_{sc} = 1 \qquad\qquad c \in \mathcal{C}, s \in \mathcal{S} \setminus \mathcal{S}^{CS}, \qquad (1s)$$

$$y_{scr} = 1 \qquad\qquad c \in \mathcal{C}, s \in \mathcal{S} \setminus \mathcal{S}^{CS}, r \in \mathcal{R}, \qquad (1t)$$

$$p_s \ge 0 \qquad\qquad s \in \mathcal{S}, \qquad (1u)$$

$$y_{sc} \in \{0,1\} \qquad\qquad c \in \mathcal{C}, s \in \mathcal{S}, \qquad (1v)$$

$$y_{scr}, w_{scr} \in \{0,1\} \qquad\qquad c \in \mathcal{C}, s \in \mathcal{S}, r \in \mathcal{R}, \qquad (1w)$$

$$\lambda_{sl} \in \{0,1\} \qquad\qquad s \in \mathcal{S}^{CS}, l \in \mathcal{L}_s, \qquad (1x)$$

$$\mu_{szcr}, \eta_{szcr} \in \{0,1\} \qquad\qquad c \in \mathcal{C}, s \ne z \in \mathcal{S}, r \in \mathcal{R}, \qquad (1y)$$

$$\alpha_{scrl} \in \{0,1\} \qquad\qquad c \in \mathcal{C}, s \in \mathcal{S}^{CS}, r \in \mathcal{R}, l \in \mathcal{L}_s. \qquad (1z)$$

Objective function (1a) represents the expected profit generated on the given O-D pair. Constraints (1b) define the utility as the sum of a customer-specific utility dependent on the attributes of the transportation systems (the part of the utility the CSO can explain) and a random term ϵ_{scr} which plays the twofold role of describing the component of the utility that the CSO cannot explain as well as possible irrational customer choices. When $f_c(p_s, \pi_s^1, \ldots, \pi_s^N)$ is linear in p_s model (1a)–(1z) is a MILP. However, it is not required that $f_c(\cdot)$ is linear in the remaining attributes $\pi_s^1, \ldots, \pi_s^N$. In Sect. 4.1 we introduce a specific utility function based on the available literature. Constraints (1c) and (1d) set the price for the transportation services offered by the CSO (the sum of per-minute and drop-off fee) and by other parties, respectively. Constraints (1e) and (1f) ensure that, among two services a customer always chooses the one with the highest utility. Constraints (1g) ensure that, given services s and z, either s has a higher utility than z or viceversa. Constraints (1h) ensure that η_{szcr} takes value 1 if both service s and z are offered to customer c under scenario r. Consistently, constraints (1i) and (1j) ensure that variable η_{szcr} takes value 0 if either service s or z are not offered to customer c under scenario r. Constraints (1k) state that service s cannot be preferred to service z by customer c under scenario r if the service is not offered to the customer. Constraints (1l) state that customer c can choose service s only if its utility is the highest in scenario r. Constraints (1m) ensure that each customer chooses exactly one service. Constraints (1n)–(1p) are required in order to obtain a linear objective function. Constraints (1n) ensure that α_{scrl} takes value 1 if price level l has been chosen for service s and customer

c has chosen service s under scenario r. Constraints (1o) and (1p) ensure that α_{scrl} takes value 0 if price level l has not been chosen and if customer c has not chosen service s, respectively. Constraints (1q) ensure that only one price level is selected. Constraints (1r) ensure that if a service is not offered to customer c it is not offered in any of the scenarios. Constraints (1s) and (1t) ensure that the transportation services other than car-sharing are always available to all users. Finally, constraints (1u)–(1z) define the domain for the decision variables.

4   The Cases of Copenhagen and Milan

In this section we use model (1a)–(1z) to investigate the profitability of car-sharing services in the cities of Copenhagen, Denmark, and Milan, Italy. Particularly, the scope of the computational study is to analyze the price a CSO is able to set between different zones of the cities, and the corresponding market response. Model (1a)–(1z) has been implemented in GAMS 24.4.6 and solved using CPLEX on a machine with 4 GB RAM and a 2.3 GHz CPU.

Car-sharing services have been adopted in both cities. To our knowledge only one free-floating car-sharing service is operating in Copenhagen as of January 2018, while at least four can be counted in Milan. In both cities there exists a public transportation provider offering services such as buses, metro lines, and surface/underground trains. Cycling trails reach a higher level of capillarity in Copenhagen, where bicycles are a common transportation option. According to [26] nine out of ten Danes own a bicycle and in 2016 the number of bicycles crossing the city center of Copenhagen exceeded the number of cars. On the contrary, cycling is not as popular in Milan to the extent that the municipality is seeking economic incentives to improve cycling mobility [27]. Therefore, for the city of Copenhagen we consider three transportation services, namely car-sharing, public transportation, and bicycles while for Milan we consider car-sharing and public transportation. In both cities, public transportation between a given O-D pair may include commuting and, for the sake of simplicity, we assume bicycles cannot be taken on board public transportation.

In Sect. 4.1 we describe the utility function used in the computational study and the groups of customers considered. In Sect. 4.2 we describe the attributes of the transportation services. Finally, in Sect. 4.3 we discuss the results obtained.

4.1   Utility Function

We use the utility function provided by [28] with minor adjustments to our specific case. The function is linear in the price p_s rendering model (1a)–(1z) a MILP. For each $s \in \mathcal{S}$ and $c \in \mathcal{C}$ the utility can be stated as (2).

$$f_c(p_s, T_s^{CS}, T_s^{PT}, T_s^B, T_s^W, T_s^{Wait}) = \beta_c^P p_s + \beta_c^{CS} T_s^{CS} + \beta_c^{PT} T_s^{PT}$$
$$+ \tau(T_s^B)\beta_c^B T_s^B + \tau(T_s^W)\beta_c^W T_s^W + \beta_c^{Wait} T_s^{Wait} \qquad (2)$$

Here, T_s^{CS} represents the time spent riding a shared car, T_s^{PT} the total time spent in public transportation, T_s^B the time spent riding a bicycle, T_s^W the walking time which includes the walking time to the nearest transportation service (such as a shared car or bus stop), between public transportation means, and to the final destination and, finally, T_s^{Wait} the total waiting time. The β coefficients of (2) are quantified following the procedure illustrated by [28] (after converting in Euro the values provided in Italian Liras when necessary). Two customer segments are introduced, namely *lower-middle class* (LMC) and *upper-middle class* (UMC), thus $\mathcal{C} = \{LMC, UMC\}$. We obtain $\beta_c^P = -188.33$ and $\beta_c^P = -70.63$, for $c = LMC$ and UMC, respectively. Furthermore, we set $\beta_c^{CS} = -1$, $\beta_c^{PT} = -2$, $\beta_c^B = -2.5$, $\beta_c^W = -3$ and $\beta_c^{Wait} = -6$ for all $c \in \mathcal{C}$. The function $\tau : \mathbb{R} \to \mathbb{R}$ is defined as $\tau(t) = \lceil \frac{t}{10} \rceil$ and allows us to model the utility of cycling and walking as a piece-wise linear function representing the fact that the utility of walking and cycling decreases faster as the walking and cycling time increases.

Finally, uncertainty in the preferences of customers is considered by creating $|\mathcal{R}| = 100$ utility scenarios. Each scenario consists of a realization of the error term $\epsilon_{scr} = \xi_{scr} f_c(p_s, T_s^{CS}, T_s^{PT}, T_s^B, T_s^W, T_s^{Wait})$, where ξ_{scr} is an i.i.d $\mathcal{N}(0, 0.1)$ sample. This corresponds to assuming a normally distributed error with a 10% standard deviation.

4.2   Characteristics of the Cities

We consider a base case which includes car-sharing, public transportation, and bicycle for Copenhagen and car-sharing and public transportation for Milan. However, the influence of cycling habits in both cities is investigated in Sect. 4.4. Copenhagen and Milan have been divided into eight and ten evenly spread zones, respectively. For each zone a central point acts as origin/destination. For each city, O-D pair, and transportation service $s \in \mathcal{S}$, the values of the attributes p_s, T_s^{CS}, T_s^{PT}, T_s^B, T_s^W, T_s^{Wait} are calculated based on the actual transportation services and distances. For each transportation service, we assume customers always choose the fastest option (e.g., driving route or public transportation connection). The fastest driving and cycling routes are found through *Google Maps*. The fastest public transportation connections are found through *Rejseplanen* (www.rejseplanen.dk) for Copenhagen and *Google Maps* and *ATM* (www.atm. it) for Milan. We assume a cycling speed of 16 km/h, which includes stops at traffic lights and a walking speed of 5 km/h. Furthermore, we assume shared cars are always available within 500 m from the origin. The impact of a reduced distance from shared cars is investigated in Sect. 4.4. All the time-related attributes for each O-D pair and transportation services are provided in Appendix A.

The price for bicycle rides is always zero, while the prices of public transportation services are taken from the local providers and are 1.60 € for all O-D pairs in Copenhagen[2] and 1.5 € for each O-D pair in Milan. Finally, the price of car-sharing services is set according to current market prices. Particularly, we register that in Milan the per-minute fee offered as of January 2018 varies between 0.24 and 0.29 €/min between the different CSO. We adopt a lower per-minute fee, namely 0.20 €/min, in order to assess the opportunity of including an

---

[2] Assuming the usage of a widely available transportation card named *rejsekort*.

O-D specific drop-off fee. We consider four possible drop-off fees, namely $0, 1, 2$ and $3 \in$. In Sect. 4.4 the influence of different per-minute fees is investigated. Finally, for the sake of simplicity, the cost of car-sharing services is ignored, i.e., $C_{sc} = 0$ for all $s \in \mathcal{S}^{CS}$ and $c \in \mathcal{C}$ so that we consider the maximization of the revenue, and we assume a trip from O to D has the same characteristics as a trip from D to O.

4.3   Results for the Base Case

Tables 1 and 2 report, for each O-D pair in Copenhagen and Milan, respectively, the CSO's expected revenue (assuming one customer for each segment), the chosen drop-off fee, and the distribution of customers among transportation services (alternatively the probability that the customer chooses a transportation service). Based on the results in Tables 1 and 2, car-sharing appears much more competitive in Milan than in Copenhagen. In Copenhagen, the CSO makes a positive revenue only on one O-D pair, while in Milan the CSO makes a positive revenue on almost all the O-D pairs. In Copenhagen, the great majority of the customers is attracted by the possibility of cycling (inexpensive and relatively easy due to the short distances). It can be noticed that the O-D pair Østerbro-Ørestad, the only O-D pair for which the CSO makes a profit in Copenhagen, is also the only one with a cycling distance longer than 30 min. On the other hand, Table 2 shows that in Milan, despite public transportation services are a serious competitor (especially for the LMC customers), car-sharing services can attract a fair percentage of customers. However, the results show that the CSO does not have enough market power to charge a drop-off fee. The competitiveness of car-sharing services is highly price-sensitive, and the viability of car-sharing services depends on the cost or running the service.

The lower competitiveness of car-sharing services in Copenhagen is consistent, for example, with a statement released by Car2Go upon closing their service in Copenhagen (reported by [29, 30]): *"Car2Go has not reached the critical mass in demand necessary to establish a successful, viable and robust business in Denmark"*. Our analysis suggests that cycling habits might be one of the main reasons behind the different successes of car-sharing services in Copenhagen and Milan. This is further investigated in Sect. 4.4. However, the necessary simplification made in our analysis might also influence the results. Particularly, we categorized customers based only on their price sensitivity while further discrimination by e.g., age and health conditions, might provide additional insights.

4.4   Factors Influencing Car-Sharing Services

We investigate the influence of cycling habits by assessing the profitability of car-sharing services in Copenhagen after excluding the possibility of cycling, and in Milan after including the possibility of cycling. The results show that the CSO makes a positive revenue in 24 out of 28 O-D pairs in Copenhagen when

Table 1. Results for Copenhagen. The expected revenue assumes one customer for each customer group. %CS, %PT and %B indicate the percentage of customers choosing car-sharing, public transportation and bicycle, respectively.

Origin	Destination	Expected revenue [€]	P_{il}^D [€]	% CS		% PT		% B	
				LMC	UMC	LMC	UMC	LMC	UMC
Østerbro	København K	0	0	0	0	0	0	100	100
Østerbro	Nørrebro	0	0	0	0	0	0	100	100
Østerbro	Fredriksberg C	0	0	0	0	0	0	100	100
Østerbro	Frederiksberg	0	0	0	0	0	0	100	100
Østerbro	Vesterbro	0	0	0	0	0	7	100	93
Østerbro	Ørestad	0.352	0	0	8	0	72	100	20
Østerbro	Øst Amager	0	0	0	0	0	34	100	66
København K	Nørrebro	0	0	0	0	0	0	100	100
København K	Fredriksberg C	0	0	0	0	0	0	100	100
København K	Frederiksberg	0	0	0	0	0	0	100	100
København K	Vesterbro	0	0	0	0	0	0	100	100
København K	Ørestad	0	0	0	0	0	0	100	100
København K	Øst Amager	0	0	0	0	0	0	100	100
Nørrebro	Fredriksberg C	0	0	0	0	0	0	100	100
Nørrebro	Freder iksberg	0	0	0	0	0	0	100	100
Nørrebro	Vesterbro	0	0	0	0	0	0	100	100
Nørrebro	Ørestad	0	0	0	0	0	3	100	97
Nørrebro	Øst Amager	0	0	0	0	0	28	100	72
Fredriksberg C	Frederiksberg	0	0	0	0	0	0	100	100
Fredriksberg C	Vesterbro	0	0	0	0	0	0	100	100
Fredriksberg C	Ørestad	0	0	0	0	0	0	100	100
Fredriksberg C	Øst Amager	0	0	0	0	0	0	100	100
Frederiksberg	Vesterbro	0	0	0	0	0	0	100	100
Frederiksberg	Ørestad	0	0	0	0	0	3	100	97
Frederiksberg	Øst Amager	0	0	0	0	0	25	100	75
Vesterbro	Ørestad	0	0	0	0	0	0	100	100
Vesterbro	Øst Amager	0	0	0	0	0	1	100	99
Ørestad	Øst Amager	0	0	0	0	0	0	100	100

the possibility of cycling is excluded. For these O-D pairs, a fair amount of (particularly UMC) customers chooses car-sharing services and, in a number of O-D pairs, car-sharing services are selected more than public transportation, especially when public transportation connections require commuting and waiting. However, also in this case the CSO does not have market power to charge a drop-off fee. In the city of Milan, a dramatic migration of customers from car-sharing and public transportation towards bicycles can be observed. For each O-D pair considered, almost all customers choose to move by bicycle. These results are certainly influenced by the simplifications in the utility function which does not include elements such as the purpose of the trip, weather conditions and carry-on items. However, the results clearly illustrate a trend towards bicycles should they become an actually viable transportation system. Thus, it emerges that cycling represents a though competitor to take into account when setting up and pricing car-sharing services. Furthermore, it emerges that CSOs can define better pricing by looking at the configuration of the public transportation systems and particularly at O-D pairs with inefficient connections due to, e.g., long waiting time.

In the cases discussed so far the per-minute fee was 0.20 €/min, a tariff lower than current market prices in order to assess the possibility to set an O-D-specific drop-off fee. We assess three alternative per-minute fees, namely 0.30 (just above market prices), 0.25 (about average market price), and 0.15 €/min (significantly lower than market prices). As intuition suggests, the results show that customers, of both customers classes, shift towards car-sharing services as the per-minute fee decreases. For the case of Milan, the total expected revenue decreases by 67.62% (with respect to the base case discussed in Sect. 4.3) with a per-minute fee of 0.30 €, and by 39.43% with a per-minute fee of 0.25 €, but increases by 53.11% with a per-minute fee of 0.15 € due to the significant increase in car-sharing demand. These results show that the per-minute fee is a crucial parameter to influence the penetration of car-sharing services in a city. However, the possibility to impose a drop-off fee remains limited even with a very low per-minute fee.

CSOs determine the proximity of shared cars to users by adjusting the size and distribution of the fleet. In order to assess how the proximity to a shared car influences customers choices and pricing decisions, we consider the base case of Milan and we assume a (possibly unrealistic) zero distance to shared cars. Similar scenarios may be obtained for example with a very large fleet of cars. The results illustrate that, with respect to Table 2 (where the distance to the nearest car is 500 meters), the percentage of customers choosing car-sharing services generally increases and, consequently, the total expected revenue. However, car-sharing does not attract customers on the four O-D pairs where it was never selected in the base case, illustrating that when public transportation connection are particularly advantageous, car-sharing has little room for gaining market shares. Also in this case the drop-off fee is set to zero on all O-D pairs. Thus, while increased proximity of shared cars can attract more customers and increase the revenue (by 19.63% in our case), it does not provide CSOs the possibility to

Table 2. Results for Milan. The expected revenue assumes one customer for each customer group. %CS and %PT indicate the percentage of customers choosing carsharing and public transportation, respectively.

Origin	Destination	Expected revenue [€]	P_{il}^D [€]	% CS		% PT	
				LMC	UMC	LMC	UMC
Portobello	Derganino	2.882	0	34	97	66	3
Portobello	China Town	3.060	0	70	100	30	0
Portobello	Sempione	2.880	0	63	97	37	3
Portobello	Washinghton	4.158	0	89	100	11	0
Portobello	Carrobbio	0.432	0	0	12	100	88
Portobello	Ticinese	0.378	0	0	9	100	91
Portobello	Guastalla	0.054	0	0	1	100	99
Portobello	QDM	0.294	0	0	7	100	93
Portobello	Central Station	0.324	0	0	9	100	91
Derganino	China Town	2.912	0	82	100	18	0
Derganino	Sempione	2.496	0	11	85	89	15
Derganino	Washington	2.210	0	0	65	100	35
Derganino	Carrobbio	0	0	0	0	100	100
Derganino	Ticinese	0	0	0	0	100	100
Derganino	Guastalla	0.044	0	0	1	100	99
Derganino	QDM	0	0	0	0	100	100
Derganino	Central Station	0.858	0	0	33	100	67
China Town	Sempione	3.008	0	88	100	12	0
China Town	Washinghton	2.990	0	18	97	82	3
China Town	Carrobbio	0.224	0	0	7	100	93
China Town	Ticinese	0.360	0	0	9	100	91
China Town	Guastalla	0.528	0	0	12	100	88
China Town	QDM	0.324	0	0	9	100	91
China Town	Central Station	0.810	0	0	27	100	73
Sempione	Washinghton	2.744	0	96	100	4	0
Sempione	Carrobbio	3.132	0	74	100	26	0
Sempione	Ticinese	2.928	0	25	97	75	3
Sempione	Guastalla	1.938	0	0	57	100	43
Sempione	QDM	0.540	0	0	18	100	82
Sempione	Central Station	0.038	0	0	1	100	99
Washinghton	Carrobbio	3.220	0	61	100	39	0
Washinghton	Ticinese	3.072	0	31	97	69	3
Washinghton	Guastalla	2.496	0	2	76	98	24
Washinghton	QDM	0.324	0	0	9	100	91
Washinghton	Central Station	0.046	0	0	1	100	99
Carrobbio	Ticinese	2.416	0	65	86	35	14
Carrobbio	Guastalla	1.876	0	2	65	98	35
Carrobbio	QDM	0.030	0	0	1	100	99
Carrobbio	Central Station	0	0	0	0	100	100
Ticinese	Guastalla	3.136	0	96	100	4	0
Ticinese	QDM	1.638	0	4	59	96	41
Ticinese	Central Station	0.304	0	0	8	100	92
Guastalla	QDM	1.680	0	6	64	94	36
Guastalla	Central Station	1.020	0	0	34	100	66
QDM	Central Station	0.676	0	0	26	100	74

replace good public transportation connections, nor enough market power to set a drop-off fee.

Finally, in order to study the effect of public transportation frequency we study the base case of Milan with waiting times increased by 50%. The results show that some LMC customers choose car-sharing in 22 O-D pairs against the 19 of the basic case. For 11 out of 45 O-D pairs all UMC customers choose car-sharing services, against the 8 of the basic case. As a consequence, the total expected revenue increases by 21.41%. Cities with inefficient public transportation services appear therefore a better environment for car-sharing services. This also illustrates the potential of defining pricing strategies which vary with the frequency and configuration of public transportation services.

5   Conclusions

This paper presented novel optimization model for pricing car-sharing services taking into account alternative transportation means as well as customers preferences via a utility function. When the utility function is linear in the price of car-sharing services the model can be formulated as a MILP. The model is amenable to further characterizations and enhancements, and to be integrated into broader analytic tools for car-sharing services.

The model is used to illustrate the viability of car-sharing services in Copenhagen and Milan. The study shows that cycling habits have a crucial impact on the market response to car-sharing. Furthermore, it emerges that companies have little margins to increase prices, mainly due to the competition of classical transportation services. However, a richer characterizations of customers preferences might illustrate market power which was not captured by our study. Furthermore, our results show that inefficiency in public transportation services such as long waiting times (due to e.g., low frequency) can be exploited by CSOs to gain market shares.

A   Attributes of the Origin-Destination Pairs Considered

See Table 3.

Table 3. Time-related attributes of car-sharing (CS), public transportation (PT), and bicycle (B) for the O-D pairs of interest in Copenhagen and Milan.

City	Origin	Destination	Service	Attributes (min) T_s^{CS}	T_s^{PT}	T_s^{W}	T_s^{B}	T_s^{Wait}
Copenhagen								
C	Østerbro	København K	CS	12	0	5.95	0	0
C	Østerbro	København K	PT	0	12	3.27	0	5
C	Østerbro	København K	B	0	0	0	13.50	0
C	Østerbro	Nørrebro	CS	11	0	5.95	0	0
C	Østerbro	Nørrebro	PT	0	10	6.01	0	6
C	Østerbro	Nørrebro	B	0	0	0	10.50	0
C	Østerbro	Fredriksberg C	CS	16	0	5.95	0	0
C	Østerbro	Fredriksberg C	PT	0	20	4.00	0	13
C	Østerbro	Fredriksberg C	B	0	0	0	16.50	0
C	Østerbro	Frederiksberg	CS	18	0	5.95	0	0
C	Østerbro	Frederiksberg	PT	0	24	3.05	0	9
C	Østerbro	Frederiksberg	B	0	0	0	21	0
C	Østerbro	Vesterbro	CS	21	0	5.95	0	0
C	Østerbro	Vesterbro	PT	0	30	5.15	0	4
C	Østerbro	Vesterbro	B	0	0	0	23.63	0
C	Østerbro	Ørestad	CS	22	0	5.95	0	0
C	Østerbro	Ørestad	PT	0	23	13.31	0	9
C	Østerbro	Ørestad	B	0	0	0	33	0
C	Østerbro	Øst Amager	CS	25	0	5.95	0	0
C	Østerbro	Øst Amager	PT	0	21	7.73	0	9
C	Østerbro	Øst Amager	B	0	0	0	29.25	0
C	København K	Nørrebro	CS	11	0	5.95	0	0
C	København K	Nørrebro	PT	0	12	7.25	0	15
C	København K	Nørrebro	B	0	0	0	12.75	0
C	København K	Fredriksberg C	CS	11	0	5.95	0	0
C	København K	Fredriksberg C	PT	0	7	9.05	0	6
C	København K	Fredriksberg C	B	0	0	0	13.50	0
C	København K	Frederiksberg	CS	17	0	5.95	0	0
C	København K	Frederiksberg	PT	0	8	12.33	0	3
C	København K	Frederiksberg	B	0	0	0	19.88	0
C	København K	Vesterbro	CS	11	0	5.95	0	0
C	København K	Vesterbro	PT	0	18	2.17	0	4
C	København K	Vesterbro	B	0	0	0	14.25	0
C	København K	Ørestad	CS	12	0	5.95	0	0
C	København K	Ørestad	PT	0	8	15.39	0	5
C	København K	Ørestad	B	0	0	0	19.50	0
C	København K	Øst Amager	CS	12	0	5.95	0	0
C	København K	Øst Amager	PT	0	9	4.74	0	8
C	København K	Øst Amager	B	0	0	0	15.75	0
C	Nørrebro	Fredriksberg C	CS	9	0	5.95	0	0
C	Nørrebro	Fredriksberg C	PT	0	13	2.93	0	14
C	Nørrebro	Fredriksberg C	B	0	0	0	9	0
C	Nørrebro	Frederiksberg	CS	10	0	5.95	0	0
C	Nørrebro	Frederiksberg	PT	0	12	5.54	0	10
C	Nørrebro	Frederiksberg	B	0	0	0	11.63	0
C	Nørrebro	Vesterbro	CS	15	0	5.95	0	0
C	Nørrebro	Vesterbro	PT	0	23	4.29	0	9
C	Nørrebro	Vesterbro	B	0	0	0	16.13	0
C	Nørrebro	Ørestad	CS	18	0	5.95	0	0
C	Nørrebro	Ørestad	PT	0	21	11.07	0	10
C	Nørrebro	Ørestad	B	0	0	0	28.13	0
C	Nørrebro	Øst Amager	CS	21	0	5.95	0	0
C	Nørrebro	Øst Amager	PT	0	18	5.49	0	10
C	Nørrebro	Øst Amager	B	0	0	0	27.75	0
C	Fredriksberg C	Frederiksberg	CS	8	0	5.95	0	0
C	Fredriksberg C	Frederiksberg	PT	0	7	5.61	0	8
C	Fredriksberg C	Frederiksberg	B	0	0	0	8.25	0
C	Fredriksberg C	Vesterbro	CS	8	0	5.95	0	0
C	Fredriksberg C	Vesterbro	PT	0	6	6.32	0	10
C	Fredriksberg C	Vesterbro	B	0	0	0	6.75	0
C	Fredriksberg C	Ørestad	CS	15	0	5.95	0	0
C	Fredriksberg C	Ørestad	PT	0	13	18.49	0	6
C	Fredriksberg C	Ørestad	B	0	0	0	22.13	0
C	Fredriksberg C	Øst Amager	CS	18	0	5.95	0	0
C	Fredriksberg C	Øst Amager	PT	0	10	12.90	0	6
C	Fredriksberg C	Øst Amager	B	0	0	0	22.50	0
C	Frederiksberg	Vesterbro	CS	11	0	5.95	0	0
C	Frederiksberg	Vesterbro	PT	0	14	6.46	0	11
C	Frederiksberg	Vesterbro	B	0	0	0	12.75	0
C	Frederiksberg	Ørestad	CS	20	0	5.95	0	0
C	Frederiksberg	Ørestad	PT	0	16	16.73	0	6
C	Frederiksberg	Ørestad	B	0	0	0	28.13	0
C	Frederiksberg	Øst Amager	CS	25	0	5.95	0	0
C	Frederiksberg	Øst Amager	PT	0	13	11.14	0	6
C	Frederiksberg	Øst Amager	B	0	0	0	29.25	0
C	Vesterbro	Ørestad	CS	15	0	5.95	0	0
C	Vesterbro	Ørestad	PT	0	19	12.95	0	11
C	Vesterbro	Ørestad	B	0	0	0	17.25	0
C	Vesterbro	Øst Amager	CS	19	0	5.95	0	0
C	Vesterbro	Øst Amager	PT	0	17	6.56	0	11
C	Vesterbro	Øst Amager	B	0	0	0	22.50	0
C	Ørestad	Øst Amager	CS	8	0	5.95	0	0
C	Ørestad	Øst Amager	PT	0	13	11.67	0	12
C	Ørestad	Øst Amager	B	0	0	0	13.88	0
M	Washinghton	Carrobbio	CS	10	0	5.95	0	0
M	Washinghton	Carrobbio	PT	0	10	12.86	0	6
M	Washinghton	Carrobbio	B	0	0	0	9.38	0
M	Washinghton	Ticinese	CS	12	0	5.95	0	0
M	Washinghton	Ticinese	PT	0	16	11.31	0	10
M	Washinghton	Ticinese	B	0	0	0	12.38	0
M	Washinghton	Guastalla	CS	16	0	5.95	0	0
M	Washinghton	Guastalla	PT	0	12	19.64	0	8
M	Washinghton	Guastalla	B	0	0	0	18.75	0
M	Washinghton	QDM	CS	18	0	5.95	0	0
M	Washinghton	QDM	PT	0	9	17.26	0	3
M	Washinghton	QDM	B	0	0	0	16.88	0
M	Washinghton	Central Station	CS	23	0	5.95	0	0
M	Washinghton	Central Station	PT	0	12	18.45	0	3
M	Washinghton	Central Station	B	0	0	0	24.38	0
M	Ticinese	Guastalla	CS	8	0	5.95	0	0
M	Ticinese	Guastalla	PT	0	9	11.90	0	6
M	Ticinese	Guastalla	B	0	0	0	6.75	0
M	Ticinese	QDM	CS	13	0	5.95	0	0
M	Ticinese	QDM	PT	0	4	14.29	0	4
M	Ticinese	QDM	B	0	0	0	12.00	0
M	Ticinese	Central Station	CS	19	0	5.95	0	0
M	Ticinese	Central Station	PT	0	9	16.07	0	4
M	Ticinese	Central Station	B	0	0	0	18.00	0

City	Origin	Destination	Service	Attributes (min) T_s^{CS}	T_s^{PT}	T_s^{W}	T_s^{B}	T_s^{Wait}
Milan								
M	Portobello	Derganino	CS	11	0	5.95	0	0
M	Portobello	Derganino	PT	0	12	11.31	0	6
M	Portobello	Derganino	B	0	0	0	13.13	0
M	Portobello	China Town	CS	9	0	5.95	0	0
M	Portobello	China Town	PT	0	6	14.29	0	3
M	Portobello	Sempione	CS	9	0	5.95	0	0
M	Portobello	Sempione	PT	0	12	6.90	0	9
M	Portobello	Sempione	B	0	0	0	10.88	0
M	Portobello	Washinghton	CS	11	0	5.95	0	0
M	Portobello	Washinghton	PT	0	5	23.81	0	5
M	Portobello	Washinghton	B	0	0	0	13.50	0
M	Portobello	Carrobbio	CS	18	0	5.95	0	0
M	Portobello	Carrobbio	PT	0	13	13.69	0	6
M	Portobello	Carrobbio	B	0	0	0	21.75	0
M	Portobello	Ticinese	CS	21	0	5.95	0	0
M	Portobello	Ticinese	PT	0	19	11.31	0	10
M	Portobello	Ticinese	B	0	0	0	17.63	0
M	Portobello	Guastalla	CS	27	0	5.95	0	0
M	Portobello	Guastalla	PT	0	15	19.64	0	8
M	Portobello	Guastalla	B	0	0	0	27.38	0
M	Portobello	QDM	CS	21	0	5.95	0	0
M	Portobello	QDM	PT	0	13	17.26	0	5
M	Portobello	QDM	B	0	0	0	19.88	0
M	Portobello	Central Station	CS	18	0	5.95	0	0
M	Portobello	Central Station	PT	0	11	10.95	0	8
M	Portobello	Central Station	B	0	0	0	20.63	0
M	Derganino	China Town	CS	8	0	5.95	0	0
M	Derganino	China Town	PT	0	11	7.74	0	8
M	Derganino	China Town	B	0	0	0	10.88	0
M	Derganino	Sempione	CS	13	0	5.95	0	0
M	Derganino	Sempione	PT	0	7	12.50	0	10
M	Derganino	Sempione	B	0	0	0	17.63	0
M	Derganino	Washinghton	CS	17	0	5.95	0	0
M	Derganino	Washinghton	PT	0	9	19.05	0	10
M	Derganino	Washinghton	B	0	0	0	22.50	0
M	Derganino	Carrobbio	CS	22	0	5.95	0	0
M	Derganino	Carrobbio	PT	0	31	2.38	0	8
M	Derganino	Carrobbio	B	0	0	0	24.38	0
M	Derganino	Ticinese	CS	25	0	5.95	0	0
M	Derganino	Ticinese	PT	0	11	12.50	0	10
M	Derganino	Ticinese	B	0	0	0	28.50	0
M	Derganino	Guastalla	CS	22	0	5.95	0	0
M	Derganino	Guastalla	PT	0	12	10.12	0	10
M	Derganino	Guastalla	B	0	0	0	27.38	0
M	Derganino	QDM	CS	19	0	5.95	0	0
M	Derganino	QDM	PT	0	6	6.55	0	10
M	Derganino	QDM	B	0	0	0	23.25	0
M	Derganino	Central Station	CS	13	0	5.95	0	0
M	Derganino	Central Station	PT	0	15	4.76	0	9
M	Derganino	Central Station	B	0	0	0	13.50	0
M	China Town	Sempione	CS	8	0	5.95	0	0
M	China Town	Sempione	PT	0	5	11.07	0	5
M	China Town	Sempione	B	0	0	0	8.25	0
M	China Town	Washinghton	CS	13	0	5.95	0	0
M	China Town	Washinghton	PT	0	20	13.69	0	9
M	China Town	Washinghton	B	0	0	0	13.50	0
M	China Town	Carrobbio	CS	16	0	5.95	0	0
M	China Town	Carrobbio	PT	0	20	4.05	0	8
M	China Town	Carrobbio	B	0	0	0	15.38	0
M	China Town	Ticinese	CS	20	0	5.95	0	0
M	China Town	Ticinese	PT	0	20	13.10	0	6
M	China Town	Ticinese	B	0	0	0	19.50	0
M	China Town	Guastalla	CS	22	0	5.95	0	0
M	China Town	Guastalla	PT	0	16	17.26	0	10
M	China Town	Guastalla	B	0	0	0	20.63	0
M	China Town	QDM	CS	18	0	5.95	0	0
M	China Town	QDM	PT	0	10	12.50	0	6
M	China Town	QDM	B	0	0	0	15.38	0
M	China Town	Central Station	CS	15	0	5.95	0	0
M	China Town	Central Station	PT	0	4	16.31	0	3
M	China Town	Central Station	B	0	0	0	16.88	0
M	Sempione	Washinghton	CS	7	0	5.95	0	0
M	Sempione	Washinghton	PT	0	10	5.48	0	7
M	Sempione	Washinghton	B	0	0	0	6.75	0
M	Sempione	Carrobbio	CS	9	0	5.95	0	0
M	Sempione	Carrobbio	PT	0	1	17.26	0	3
M	Sempione	Carrobbio	B	0	0	0	7.50	0
M	Sempione	Ticinese	CS	12	0	5.95	0	0
M	Sempione	Ticinese	PT	0	13	10.83	0	10
M	Sempione	Ticinese	B	0	0	0	11.63	0
M	Sempione	Guastalla	CS	17	0	5.95	0	0
M	Sempione	Guastalla	PT	0	10	16.07	0	10
M	Sempione	Guastalla	B	0	0	0	18.00	0
M	Sempione	QDM	CS	15	0	5.95	0	0
M	Sempione	QDM	PT	0	8	12.50	0	3
M	Sempione	QDM	B	0	0	0	10.50	0
M	Sempione	Central Station	CS	19	0	5.95	0	0
M	Sempione	Central Station	PT	0	9	13.33	0	3
M	Sempione	Central Station	B	0	0	0	18.38	0
M	Carrobbio	Ticinese	CS	8	0	5.95	0	0
M	Carrobbio	Ticinese	PT	0	4	7.62	0	5
M	Carrobbio	Ticinese	B	0	0	0	7.88	0
M	Carrobbio	Guastalla	CS	14	0	5.95	0	0
M	Carrobbio	Guastalla	PT	0	3	19.64	0	3
M	Carrobbio	Guastalla	B	0	0	0	14.25	0
M	Carrobbio	QDM	CS	15	0	5.95	0	0
M	Carrobbio	QDM	PT	0	15	7.62	0	4
M	Carrobbio	QDM	B	0	0	0	12.75	0
M	Carrobbio	Central Station	CS	21	0	5.95	0	0
M	Carrobbio	Central Station	PT	0	11	10.36	0	3
M	Carrobbio	Central Station	B	0	0	0	21.38	0
M	Guastalla	QDM	CS	12	0	5.95	0	0
M	Guastalla	QDM	PT	0	6	11.90	0	4
M	Guastalla	QDM	B	0	0	0	10.50	0
M	Guastalla	Central Station	CS	15	0	5.95	0	0
M	Guastalla	Central Station	PT	0	11	13.69	0	4
M	Guastalla	Central Station	B	0	0	0	13.50	0
M	QDM	Central Station	CS	13	0	5.95	0	0
M	QDM	Central Station	PT	0	5	10.12	0	4
M	QDM	Central Station	B	0	0	0	9.38	0

References

1. George, D.K., Xia, C.H.: Fleet-sizing and service availability for a vehicle rental system via closed queueing networks. Eur. J. Oper. Res. **211**(1), 198–207 (2011)
2. Cepolina, E.M., Farina, A.: A new shared vehicle system for urban areas. Transp. Res. Part C Emerg. Technol. **21**(1), 230–243 (2012)
3. Correia, G.H.D.A., Jorge, D.R., Antunes, D.M.: The added value of accounting for users' flexibility and information on the potential of a station-based one-way carsharing system: an application in Lisbon, Portugal. J. Intell. Transp. Syst. **18**(3), 299–308 (2014)
4. Kaspi, M., Raviv, T., Tzur, M.: Parking reservation policies in one-way vehicle sharing systems. Transp. Res. Part B Methodol. **62**, 35–50 (2014)
5. de Almeida Correia, G.H., Antunes, A.P.: Optimization approach to depot location and trip selection in one-way carsharing systems. Transp. Res. Part E Logist. Transp. Rev. **48**(1), 233–247 (2012)
6. Kumar, P., Bierlaire, M.: Optimizing Locations for a Vehicle Sharing System (2012). https://infoscience.epfl.ch/record/195890
7. Boyaci, B., Zografos, K.G., Geroliminis, N.: An optimization framework for the development of efficient one-way car-sharing systems. Eur. J. Oper. Res. **240**(3), 718–733 (2015)
8. Li, X., Ma, J., Cui, J., Ghiasi, A., Zhou, F.: Design framework of large-scale one-way electric vehicle sharing systems: a continuum approximation model. Transp. Res. Part B Methodol. **88**, 21–45 (2016)
9. Kek, A., Cheu, R., Chor, M.: Relocation simulation model for multiple-station shared-use vehicle systems. Transp. Res. Rec. J. Transp. Res. Board **1986**, 81–88 (2006)
10. Fan, W., Machemehl, R., Lownes, N.: Carsharing: dynamic decision-making problem for vehicle allocation. Transp. Res. Rec. J. Transp. Res. Board **2063**, 97–104 (2008)
11. Kek, A.G., Cheu, R.L., Meng, Q., Fung, C.H.: A decision support system for vehicle relocation operations in carsharing systems. Transp. Res. Part E Logist. Transp. Rev. **45**(1), 149–158 (2009)
12. Nair, R., Miller-Hooks, E.: Fleet management for vehicle sharing operations. Transp. Sci. **45**(4), 524–540 (2011)
13. Weikl, S., Bogenberger, K.: Relocation strategies and algorithms for free-floating car sharing systems. IEEE Intell. Transp. Syst. Mag. **5**(4), 100–111 (2013)
14. Jorge, D., Correia, G.H.A., Barnhart, C.: Comparing optimal relocation operations with simulated relocation policies in one-way carsharing systems. IEEE Trans. Intell. Transp. Syst. **15**(4), 1667–1675 (2014)
15. Nourinejad, M., Zhu, S., Bahrami, S., Roorda, M.J.: Vehicle relocation and staff rebalancing in one-way carsharing systems. Transp. Res. Part E Logist. Transp. Rev. **81**, 98–113 (2015)
16. Bruglieri, M., Pezzella, F., Pisacane, O.: Heuristic algorithms for the operator-based relocation problem in one-way electric carsharing systems. Discret. Optim. **23**, 56–80 (2017)
17. Santos, G., Correia, G.: A MIP model to optimize real time maintenance and relocation operations in one-way carsharing systems. Transp. Res. Procedia **10**, 384–392 (2015)

18. Kühne, K.S., Rickenberg, T.A., Breitner, M.H.: An optimization model and a decision support system to optimize car sharing stations with electric vehicles. In: Lübbecke, M., Koster, A., Letmathe, P., Madlener, R., Peis, B., Walther, G. (eds.) Operations Research Proceedings 2014, pp. 313–320. Springer, Cham (2016). https://doi.org/10.1007/978-3-319-28697-6_44

19. Benati, S., Hansen, P.: The maximum capture problem with random utilities: problem formulation and algorithms. Eur. J. Oper. Res. **143**(3), 518–530 (2002)

20. Zhang, Y., Berman, O., Verter, V.: The impact of client choice on preventive healthcare facility network design. OR Spectr. **34**(2), 349–370 (2012)

21. Sharif Azadeh, S., Marcotte, P., Savard, G.: A non-parametric approach to demand forecasting in revenue management. Comput. Oper. Res. **63**, 23–31 (2015)

22. Bierlaire, M., Azadeh, S.S.: Demand-based discrete optimization. Technical report, Transport and Mobility Laboratory School of Architecture, Civil and Environmental Engineering Ecole Polytechnique Fédérale de Lausanne, Lausanne (2016)

23. Herrmann, S., Schulte, F., Voß, S.: Increasing acceptance of free-floating car sharing systems using smart relocation strategies: a survey based study of car2go Hamburg. In: González-Ramírez, R.G., Schulte, F., Voß, S., Ceroni Díaz, J.A. (eds.) ICCL 2014. LNCS, vol. 8760, pp. 151–162. Springer, Cham (2014). https://doi.org/10.1007/978-3-319-11421-7_10

24. Schmöller, S., Weikl, S., Müller, J., Bogenberger, K.: Empirical analysis of free-floating carsharing usage: the Munich and Berlin case. Transp. Res. Part C Emerg. Technol. **56**, 34–51 (2015)

25. Manzi, G., Saibene, G.: Are they telling the truth? Revealing hidden traits of satisfaction with a public bike-sharing service. Int. J. Sustain. Transp. **12**(4), 253–270 (2018)

26. Cycling Embassy of Denmark: Facts About Cycling in Denmark - Cycling Embassy of Denmark (2014). http://www.cycling-embassy.dk/facts-about-cycling-in-denmark/statistics/

27. The Guardian: Cash for cycling: polluted Milan might pay commuters to bike to work—Cities—The Guardian (2016). https://www.theguardian.com/cities/2016/feb/29/cash-cycling-polluted-milan-italy-pay-commuters-bike-to-work

28. Modesti, P., Sciomachen, A.: A utility measure for finding multiobjective shortest paths in urban multimodal transportation networks. Eur. J. Oper. Res. **111**(3), 495–508 (1998)

29. CPH Post: Car2go shutting down in Copenhagen (2016). http://cphpost.dk/news/car2go-shutting-down-in-copenhagen.html

30. The Local: Car2go drives out of Copenhagen after flop - The Local (2016). https://www.thelocal.dk/20160120/car2go-on-copenhagen-road-to-nowhere

Network Design and Scheduling

Exact Methods and Heuristics for the Liner Shipping Crew Scheduling Problem

Valerio Maria Sereno[1], Line Blander Reinhardt[2(✉)], and Stefan Guericke[3]

[1] DTU Management Engineering, The Technical University of Denmark,
Kongens Lyngby, Denmark
[2] Institute of Materials and Production, Aalborg University,
Aalborg, Denmark
lbr@mp.aau.dk
[3] Data Science and Artificial Intelligence, A.P. Moller - Maersk,
Copenhagen, Denmark
stefan.guericke@maersk.com

Abstract. In this paper the liner shipping crew scheduling problem is described and modelled. Three different models have been formulated and tested for the scheduling problem. A mixed integer formulation and a set covering formulation are constructed and solved using exact methods. A mat-heuristic based on column generation has been implemented and tested. Moreover, a simple heuristic is implemented as a benchmark value. The models and methods were tested on smaller instances of the problem. The results show that good results can be achieved within 5 min using the heuristic and around an hour using the set partitioning formulation.

1 Introduction

During the last decade the global containerized market has been oversupplied with capacity. The oversupply in the containerized fleet is the result of past investment decisions and slower-than-expected demand growth (UNCTAD [16]). Carriers are responding to this situation by trying to cut costs.

In liner shipping crew costs have been considered a secondary issue [5], since labor cost is a minor cost component compared to fuel cost, vessel expenses, terminal handling charges, etc. Nevertheless, the crew scheduling process is resource consuming from a tactical perspective. Moreover, the companies explore new opportunities for cutting cost, investigating also smaller expense accounts. The seafarers are generally divided into officers and ratings, with the ratings defined as noncommissioned sailors in the navy. The worldwide population of seafarers serving on internationally trading merchant ships was estimated at 1,647,500 in 2015 [3]. Approximately 47% of them were officers and the rest were rating. Nearly 60% of the global demand for officers was estimated to be "on board ships" with around 40% being ashore at any given time.

© Springer Nature Switzerland AG 2018
R. Cerulli et al. (Eds.): ICCL 2018, LNCS 11184, pp. 363–378, 2018.
https://doi.org/10.1007/978-3-030-00898-7_24

According to The Baltic and International Maritime Council, the world's largest international shipping association, while the global supply of officers is forecast to increase steadily, this trend is expected to be outpaced by increasing demand, with an estimated undercoverage of almost 150,000 officers in 2025. The current and future shortage of maritime officers could increase wages and decrease the availability of officers thus increasing the demand for better crew planning.

Concluding, nowadays, the maritime industry is facing many challenges. China's economic slowdown, low commodities' prices, exchange rate volatility and oversupplied capacity are negatively influencing the freight rates. Carriers are reacting trying to reduce costs and increase the efficiency of their operations. At the same time, the industry could experience a shortage of skilled personnel in near future. In this scenario, an effective planning of work force could become a crucial aspect for the maritime industry.

1.1   The Crew Scheduling Problem

In the Operational Research context, many publications deal with the crew scheduling problem. These papers often introduce solution methods tailored to the constraints and conditions prevailing in a particular industry. Among the different cases studied, it is possible to identify a set of common characteristics to crew scheduling problems in any given setting. The definition used here is a generalization of the problem presented by Beasley and Cao in [1].

A crew scheduling problem (CSP) is commonly defined by a set of employees and a set of tasks. The employees can be considered as individuals or as a team. They will generally be defined by a geographical location, from where they will start their work, a set of skills indicating if they can perform a task, a temporal availability and a wage. The tasks have usually specific time windows, a location and some skill requirements. Time windows identify the period where the tasks are carried out, locations represent the geographical place of the task, while the required skills determine a set of potential crew members that can perform the task. The aim of the problem is to allocate crew members to the tasks in order to optimize the defined objective function respecting the geographical and temporal constraints as well as the labour regulations.

In the last few decades, crew scheduling problems have been investigated widely in a variety of contexts. The increase of research attention could be motivated by economic considerations [2] and in many industries, such as production companies and hospitals, wages are one of the biggest components of direct costs. In the airline industry competition has been quite fierce the last couple of decade and in the health care sector costs are escalating resulting in and increased amount of research in both areas. However, the application of mathematical modelling techniques in the maritime setting has received little attention in the literature.

Liner shipping companies are still using manual methods to find feasible rosters, resulting in large amounts of man hours used on generating solutions. The research presented here analyzes the crew scheduling problem faced by a major liner shipping company. Several different formulations are applied to the problem: a mixed integer model and a set covering formulation. Moreover, a column generation heuristic is applied using the set covering formulation. Results show a good convergence to the optimum.

The fundamental aim of the problem is to produce a set of crew rosters, one for each vessel, which respects the safe sailing regulations and the labor laws. The "tasks" in this crew scheduling problem are defined by the crew roles (or ranks) that must be fulfilled on-board of each vessel during all the planning horizon. The required crew size for each ship is dictated by its Minimum safe manning certificate as defined in [10].

1.2 Literature Review

The crew scheduling problem was introduced for the first time by Edie [8] in 1954. In the paper, the author introduces a methodology for the toll booths' personnel scheduling at the port of New York authority. The study was subsequently reformulated by Dantzig [6], using *work patterns*. The reformulation made it possible to reconstruct the problem to a variant of the standard transportation-type linear integer problem.

One of the most studied crew scheduling problems within transportation is the airline crew scheduling problems. The airline crew scheduling problem are divided in *crew pairings* and *crew rostering* phases. Firstly, anonymous crew rotations are formed out of the flight legs and a pairing is a sequence of flights that is assigned to a generic crew member. A pairing can contain one or several *deadheads*, where a deadhead is defined as a flight where the crew is off-duty, i.e. they travel as passengers. In the crew rostering phase, the pairings are assigned to crew members individually along with other activities such as rest periods, ground duties, etc. As argued in [11], most of the published research in airline crew scheduling is related to the crew pairing problem since major cost benefits can be achieved through an effective pairing.

Several different methods have been applied to solve the CSP in the airline sector. In [4,9,15] the general rostering problem as set partitioning problem was analyzed. While Ryan in [15] solved the problem by integer programming technique, [4,9] applied a column generation approach. In [15] *line of work (LoW)* rosters, are generated by enumeration and filtered following some criteria. The LoW rosters are subsequently used to solve a set partitioning problem in a constraint branching tree algorithm.

Dohn and Mason in [7] present a general approach for staff rostering. In [7] a set partitioning master problem and a three stage subproblem modelled as constrained resource shortest path problem is presented. The model is developed around the concept of *work-stretch* defined as a combination of on-days and off-days. A sequence of work-stretch determine a roster line, i.e. a column of the master problem. In the subproblem, the shifts are first combined into a on-stretch finding the shortest path in a graph where the nodes represent the shifts and arcs between the nodes exist if two shifts can be consecutive. The first stage of the subproblem generates, for each staff member, the optimal content of every type of on-stretch, based on his/her preferences. In the second phase, the on-stretches are combined with the off-stretches to form work-stretches. Once that the set of optimal work-stretches have been determinated, a shortest path problem is solved through dynamic programming to generate a roster line with negative reduced cost. Integer solutions were obtained through a modified constraint branching technique (see [14]). A useful characteristic of this framework is that it can be set up for different particular problems; shift, on-stretch, off-stretch, work-stretch and so on, can be defined as entities and their attributes can be modified to follow a specific formulation. This approach has been inspiration for the set partitioning methods presented here.

To the best of our knowledge, there is no publication that addresses the crew scheduling problem on liner shipping vessel. Nevertheless, ships related CSP research have been completed in different maritime contexts.

In [13] Li and Womer dealt with the manpower scheduling problem on a military vessel. The problem was modelled as project scheduling problem with multi-purpose resources, where multi-skilled sailors form teams to accomplish the tasks.

Leggate presented a vessel crew scheduling problem for support services in the offshore oil industry in [12]. Among the different papers analyzed, [12] presents the highest number of analogies with the liner vessel CSP. The similarities include contracts duration, rest periods, planning horizon and some experience requirements, however, the decision about travel is not included in [12]. Leggate presents two different problem formulations in his work. A simpler *Task-based formulation*, based on some assumptions on crew contracts and working patterns, and a more complex *Time-window* formulation.

2   Models

The main aim of the liner shipping crew scheduling is to minimize the transportation cost, defined as the sum of the airfare and non-airfare costs, such as hotel and ground travel. In addition, an analysis regarding the number of seafarers used was performed including the seafarers' yearly salary as a fixed cost.

For the tests it is assumed that all the vessels are empty at the beginning of the planning horizon. This, however, does not affect the model as availability days could be included for the crew. Moreover, only regular crew members are considered and their nationality is not considered, thus visa requirements are disregarded. The on-board period lasts from the moment a seafarer boards until the seafarer leaves the vessel. The rest period starts when a seafarer is disembarked. The union of on-board and rest period is defined as a duty. The boarding and disembarkment procedures only takes place when a vessel visits a port. A port call is modelled by a single point in time. Moreover, three manning safe requirement conditions are considered: Number of employees in each rank required, Handover period from one crew to the next and minimum combined experience on different ranks, which imposes that the sum of the experience of two seafarers, working in different ranks, has to be higher than a certain amount

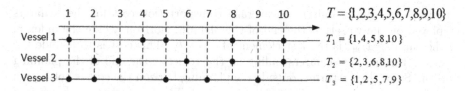

Fig. 1. Example of subsets of discrete time points for three vessels

(i.e. in order to be allowed to sail together, a Captain and a Chief Engineer have to have a combined sailing experience higher than 12 months). Two experience requirements are considered: minimum combined experience of two seafarers in their respective ranks and minimum combined experience of two seafarers in the company. Moreover, in cases where more than one seafarer is required in a certain rank, only the experience of one of them should be considered.

2.1 Integer Model

Let $v \in V$ be a vessel v in the set of vessels V, $e \in E$ a seafarer e in the set of seafarers E, and $r \in R$ a rank r in the set of ranks R, then R_v is the subset containing the ranks required on vessel v, E_r is the subset containing employees covering rank r. Since every seafarer is employed in only one rank, the subsets E_r are mutually exclusive, with $\bigcup_{\forall r} E_r = E$ and $E_r \cap E_{r'} = \emptyset$. A set U composed by pairs of ranks $r^{ex} = (r_1, r_2)$ is required to formulate the constraints about the minimum experience needed to be satisfied by two seafarers, covering different roles. Thus let U be the set of rank pairs (r_1, r_2) where $r_1, r_2 \in R$. Let P be the set of all ports visited by the different vessels and $p \in P$ is a port. Moreover, let $c \in C$ be a contract type of the seafarers. For each vessel v it is possible to define a subset of discrete time points T_v at which vessel v visits any port p and let then T be the set containing the time points from T_v for all $v \in V$. A small graphical example of this operation is reported in Fig. 1, which shows for three vessels the individual set of port calls, represented with a black dot on the line of time, and the correspondent set of discrete time points.

2.1.1 Objective Function

The objective function minimizes the total travelling cost for all the seafarers used along the whole period considered. The function is presented in (1) where the binary variable b_{evt} is one when seafarer e is boarded on vessel v at time $t-1$ and zero otherwise. The binary variable d_{evt} is one when seafarer e departs in time t from vessel v, and the binary variable x_{evt} is one when e works on vessel v at time t.

$$\text{Minimize } z = \sum_{e \in E} \sum_{v \in V} c_{ev}^{avg} \cdot x_{ev\tau} + \left(\sum_{t \in T_v} c_{evt-1}^{T} \cdot b_{evt} + c_{evt}^{T} \cdot d_{evt} \right) \tag{1}$$

The term $c_{ev}^{avg} \cdot x_{ev\tau}$ reflects an average transportation cost for the seafarers employed on the vessels at the end of the planning horizon, where the index τ, in the variable x_{evt}, is the last element of a set T_v. This cost is to include the cost of returning the seafarers sailing on the vessel at the end of the planning horizon. Since their return is outside the planning horizon the cost is not known and thus the cost is approximated as described. The parameter c_{evt}^T represents the cost of sending a seafarer from his home airport to the port visited by vessel v at time t.

2.1.2   Modelling Boarding and Departure of a Seafarer

To model the boarding and departure of a seafarer on a vessel a number of constraints are needed. The constraints (2) and (3) allow the boarding and departure variables to take positive value only if it is possible for seafarer e to travel from his home airport to the port visited by vessel v at time t and vice versa, with the parameter θ_{evt}, which is one if employee e can travel to vessel v at time t and zero otherwise, the constraints (4) ensure that the boarding variable is positive if the seafarer was not on the vessel at time $t-1$ and on the vessel at time t, the constraints (5) handle the first port call of each vessel in the case a seafarer is already working on that vessel before time 1. This condition was imposed through the use of the parameter p_{ev}, which is equal to one if seafarer e is employed on vessel v prior time 1. Constraints (6) force the departure variable to be positive if the seafarer is on vessel at time t and is not at time $t+1$. Lastly, constraints (7) assure that a seafarer can work on maximum one vessel at a time

$$b_{evt} \leq \theta_{evt} \qquad \forall e \in E, v \in V, t \in T_v \qquad (2)$$

$$d_{evt} \leq \theta_{evt} \qquad \forall e \in E, v \in V, t \in T_v \qquad (3)$$

$$b_{evt} \geq x_{evt} - x_{evt-1} \qquad \forall e \in E, v \in V, t \in T_v \setminus \{1\} \qquad (4)$$

$$b_{ev1} \geq x_{ev1} - p_{ev} \qquad \forall e \in E, v \in V \qquad (5)$$

$$d_{evt} \geq x_{evt+1} - x_{evt} \qquad \forall e \in E, v \in V, t \in \{1, .., |T_v| - 1\} \qquad (6)$$

$$\sum_{v \in V} x_{evt} \leq 1 \qquad \forall e \in E, t \in T_v. \qquad (7)$$

2.1.3   Manning and Handover Requirements

The minimum number of crew members on each role is set by inequality (8). This equation ensures that for every vessel v at every time t, the sum of all the seafarers working as rank r is at least equal to m_r, which is the minimum number of crew members required, for the rank.

$$\sum_{e \in E_r} x_{evt} \geq m_{rv} \qquad \forall r \in R_v, v \in V \qquad (8)$$

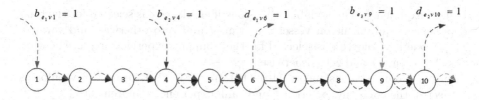

Fig. 2. Graphical example of the handover period

The sailing safe regulations impose a handover period every time a seafarer leaves a vessel. To model this for every vessel, rank and time period (v, r, t_2) define a time t_1 such that, t_1 is the time of the closest port call that precedes t_2 with at least h_r days to t_2. h_r is defined as minimum handover period for rank r on vessel v, while ν_{t_1,t_2} is equal to the number of days between t_1 and t_2. Thus formulated mathematically let $t_1 = \{t_1 \in T_v : t_1 \le t_2 \wedge \nu_{t_1,t_2} = min\{\nu_{t_1,t_2} \ge h_r\}\}$ Once defined the relation between t_2 and t_1 it was possible to formulate the handover constraints (9).

$$\sum_{e_1 \in E_r \setminus \{e_2\}} x_{e_1 v t_1} + m_r \left(1 - d_{e_2 v t_2}\right) \ge m_r,$$

$$\forall e_2 \in E_r, r \in R_v, v \in V, t_2 \in T_v, \nu_{t_1,t_2} = min\{\nu_{t_1,t_2} \ge h_r\} \tag{9}$$

The constraint imposes that every time an employee leaves a vessel at time t_2, the sum of the other employees working with the same role on that vessel at time t_1 has to be at least equal to the manning requirement. An example of the consequences of the handover constraint is reported in Fig. 2. The figure shows a partial port rotation of a vessel where the port calls are represented by the nodes. For the Figure, seafarer e_1 boards at port call 1 and leaves the vessel at port call 6. Before he leaves, seafarer e_2 boards the vessel at 4, thus the two seafarers sail together in the legs 4–5 and 5–6, respecting the handover period requirement. The same happens in the leg 9–10 between seafarers e_2 and e_3

2.1.4 Experience Requirements

The employees' experience is evaluated on two different aspects: amount of time a seafarer has worked in a specific rank and amount of time he has worked in the company. The parameter L_e^C is the number of months of experience from the employee e in the company and the parameter L_e^R is the number of months of experience of employee e in his rank. The amount of combined experience required is represented by the parameters $\lambda_{r_1 r_2}^C$ and $\lambda_{r_1 r_2}^R$, which are respectively the minimum amount of combined experience required in the company and the minimum amount of combined experience required in the rank for roles r_1 and r_2 in R_v and $(r_1, r_2) \in U$.

In order to guarantee the respect of the experience requirements, it was necessary to add two new binary variables to select only the seafarer with the highest experience. The variable α_{evt}^R is one if employee e is selected for the rank experience requirement on vessel v at time t and zero otherwise and variable α_{evt}^C is one if employee e is selected for the company experience requirement on vessel v at time t and zero otherwise.

The two groups of constraints, (10)–(12) and (13)–(15), respectively enforces respect of the company experience and rank experience requirement.

$$\alpha_{evt}^C \leq x_{evt} \qquad\qquad \forall e \in E, v \in V, t \in T_v, r \in R_v \quad (10)$$

$$\sum_{e \in E_r} \alpha_{evt}^C = 1 \qquad\qquad \forall v \in V, t \in T_v, r \in R_v \quad (11)$$

$$\sum_{e_1 \in E_{r1}} \alpha_{e_1 vt}^C \cdot L_{e_1}^C + \sum_{e_2 \in E_{r2}} \alpha_{e_2 evt}^C \cdot L_{e_2}^C \geq \lambda_{r_1 r_2}^C$$

$$\forall v \in V, t \in T_v, r_1, r_2 \in R_v, (r_1, r_2) \in U \qquad\qquad (12)$$

$$\alpha_{evt}^R \leq x_{evt} \qquad\qquad \forall e \in E, v \in V, t \in T_v, r \in R_v \quad (13)$$

$$\sum_{e \in E_r} \alpha_{evt}^R = 1 \qquad\qquad \forall v \in V, t \in T_v, r \in R_v \quad (14)$$

$$\sum_{e_1 \in E_{r1}} \alpha_{e_1 vt}^R \cdot L_{e_1}^R + \sum_{e_2 \in E_{r2}} \alpha_{e_2 evt}^R \cdot L_{e_2}^R \geq \lambda_{r_1 r_2}^R$$

$$\forall v \in V, t \in T_v, r_1, r_2 \in R_v, (r_1, r_2) \in U \qquad\qquad (15)$$

Constraints (10) and (13) impose that a seafarer can be considered for the experience requirements only if he is employed on that vessel at that moment. Constraints (11) and (14) restrict the choice to one seafarer for requirement, on every vessel, time and rank. Those four constraints allows for selecting different seafarers for the company and rank requirements in cases where more than one seafarer is employed in a rank. Constraints (12) and (15) set the minimum amount of combined experience between two ranks both covered on one vessel. Notice that, when a role is covered by multiple seafarers, the constraint should account only the seafarer with maximum experience.

2.1.5   On-Board and Off-Board Duration

The length of an embark period for a seafarer can vary between a minimum and maximum value. Moreover, for every seafarer, the duration of the rest period depends on the length of his last on-board. Let the parameter l_e^{min} be the minimum duration of on-board period for seafarer e and l_e^{max} be the maximum duration of an on-board period for seafarer e. Moreover, let ϵ_e be the ratio between minimum on-board period and minimum rest period for seafarer e.

The length of the embark is modelled with the variables l_{et} representing the number of days employee e has been boarded at time t and l_{et}^E representing the duration in days for an embark period carried out from employee e and ended in t.

The variable l_{et}, if a seafarer is embarked, keeps track of how many days he has spent on board. This calculation is shown in constraints (17), where parameter ν_{t_1,t_2} is the number of days between two time instants. l_{et}^E, instead, is used to ensure that the on-board duration is within the limits set by the seafarer's contract.

$$l_{e1} \geq q_e \qquad\qquad \forall e \in E \quad (16)$$

$$l_{et} \geq l_{et-1} + \sum_{v \in V} x_{evt} \cdot \nu_{t,t-1} - l_e^{max} \cdot \sum_{v \in V} d_{evt-1} \qquad \forall t \in T\backslash\{1\}, e \in E \quad (17)$$

$$l_{et}^E \geq l_{et} - \left(1 - \sum_{v \in V} d_{evt}\right) \cdot l_e^{max} \qquad\qquad \forall e \in E, t \in T \quad (18)$$

$$l_{et}^E \geq l_e^{min} \sum_{v \in V} d_{evt} \qquad\qquad \forall e \in E, t \in T \quad (19)$$

Constraints (16) forces the variable l_{e1}, in the cases where a seafarer is already on board of a vessel at time 1, to be greater than q_e, which is equal to the number of days a seafarer has spent on a vessel at $t = 1$. Constraint (17) imposes that, for every seafarer and time, the value of the variable l_{et} has to be greater than or equal to the variable at the previous time point plus the number of days between t and $t-1$ if the seafarer works in t. The variable is relaxed in t if the employee leaves the vessel in $t-1$. Constraints (18) forces l_{et}^E to be greater or equal to l_{et} when employee e departs from a vessel. Constraint (19) ensures that for every seafarer, an on-board period ended in t is at least equal to the minimum on board period.

After every deployment, a seafarer is entitled to a rest period r_{et}, which correspond to the number of days without work that seafarer e is entitled at time t.

Constraints (20) set the rest variable at time 1 for the employees that are already in their off-board period, with s_e equal to the amount of days of rest seafarer e is entitled to at time 1. The parameter ν_{t_1,t_2} is the number of days between the two time instants t_1 and t_2. When a seafarer leaves a vessel, r_{et} is forced to take a value at least equal to the period of rest the seafarers earned by constraints (21). This value is decreased through constraints (22) in the same way that the duration of the on-board was increased in constraints (17). Constraints (23) force an employee to not work if he is entitled of a rest period at time t.

$$r_{e1} \geq s_e \qquad\qquad \forall e \in E \quad (20)$$

$$r_{et} \geq \epsilon_e \cdot l_{et}^E \qquad\qquad \forall e \in E, t \in T \quad (21)$$

$$r_{et} \geq r_{et-1} - \left(1 - \sum_{v \in V} x_{evt}\right) \cdot \nu_{t,t-1} \qquad \forall e \in E, t \in T\backslash\{1\} \quad (22)$$

$$\epsilon_e \cdot l_e^{max} \cdot \left(1 - \sum_{v \in V} x_{evt}\right) \geq r_{et} \qquad\qquad \forall e \in E, t \in T_v \quad (23)$$

Combining the constraints from (1) to (23) results in the initial compact model.

2.1.6   Minimization of the Pool of Seafarers

The model presented in the previous section was extended to evaluate the optimal pool size of seafarers. In order to asses this aspect, a new binary variable y_e was introduced, which is one if employee e is deployed at least once and zero otherwise. Moreover, a new set of constraints (24) was needed to ensure that the variable y_e takes positive value if seafarer e is utilized, where the optimal size of the factor M is equal to the cardinallity of the set T. In the new objective function (25) a cost s_e is equal to the annual salary of the employee e and this must be payed for every seafarer utilized.

$$M \cdot y_e \geq \sum_{v \in V} \sum_{t \in T_v} x_{evt} \qquad \forall e \in E \tag{24}$$

$$\text{Minimize } z = \sum_{e \in E} y_e \cdot s_e + \sum_{v \in V} c_{ev}^{avg} \cdot x_{ev\tau} + \left(\sum_{t \in T_v} c_{et-1}^T \cdot b_{evt} + c_{et}^T \cdot d_{evt} \right) \tag{25}$$

2.2   Set Covering Model

Another model tested is the set covering formulation. With respect to the MIP formulation, the set covering formulation has the advantage of modeling a duty, representing an seafares embarkment from boarding to dispatching, through the use of a single variable, reducing the feasible space.

The objective function is reported in (26), where c_j is the cost of assigning the seafarer to duty j, i.e. the sum of the boarding and departure costs, and γ_{ej} is a binary variable that states if seafarer $e \in E$ is assigned to duty $j \in J_e$, where E is the set of seafarers. J is the set of duties and $J_e \subseteq J$ is the set of duties for seafarer e.

OBJECTIVE FUNCTION

$$\text{Minimize} \sum_{e \in E} \sum_{j \in J_e} c_j \cdot \gamma_{ej} \tag{26}$$

SUBJECT TO

$$\sum_{j \in J_e} \gamma_{ej} \cdot \sigma_{jt} \leq 1 \qquad\qquad \forall e \in E, t \in T \tag{27}$$

$$\sum_{e \in E_r} \sum_{j \in J_e} \gamma_{ej} \cdot z_{jvt} \geq m_r \qquad \forall t \in T_v \setminus H_j, r \in R_v, v \in V \tag{28}$$

$$\phi_{evt}^C \leq \sum_{j \in J_e} \gamma_{ej} \cdot z_{jvt} \qquad \forall e \in E, v \in V, t \in T_v \tag{29}$$

$$\sum_{e \in E_r} \phi_{evt}^C = 1 \qquad\qquad \forall t \in T_v, r \in R_v, v \in V \tag{30}$$

$$\sum_{e_1 \in E_{r_1}} \phi^C_{e_1 vt} \cdot L^C_{e_1} + \sum_{e_2 \in E_{r_2}} \phi^C_{e_2 vt} \cdot L^C_{e_2} \geq \lambda^C_{r_1 r_2} \quad \forall (r_1, r_2) \in U, r_1, r_2 \in R_v, v \in V, t \in T_v \quad (31)$$

$$\phi^R_{evt} \leq \sum_{j \in J_e} \gamma_{ej} \cdot z_{jvt} \quad \forall e \in E, v \in V, t \in T_v \quad (32)$$

$$\sum_{e \in E_r} \phi^R_{evt} = 1 \quad \forall t \in T_v, r \in R_v, v \in V \quad (33)$$

$$\sum_{e_1 \in E_{r_1}} \phi^R_{e_1 vt} \cdot L^R_{e_1} + \sum_{e_2 \in E_{r_2}} \phi^R_{e_2 t} \cdot L^R_{e_2} \geq \lambda^R_{r_1 r_2} \quad \forall (r_1, r_2) \in U, r_1, r_2 \in R_v, v \in V, t \in T_v \quad (34)$$

Constraints (27) ensure that only one roster j is assigned to a seafarer e at time t, where parameter σ_{jt} is positive if roster j include working or resting at time t. The handover period and the manning requirements are modelled through constraint (28). Let V be the set of vessels, $T_v \subseteq T$ the set of time points at which vessel v is at port and $H_j \subseteq T_v$ set of time points where duty j require handover. Thus, the constraint ensure that the number of seafarers assigned on each vessel v, rank r and time t is higher than the minimum number required m_r, where the binary parameter z_{jvt} determines if roster j includes working on vessel v at time t. Let the binary variable ϕ^C_{evt} be one if seafarer $e \in E$ satisfies the company experience requirement on vessel v at time point t and zero otherwise and let ϕ^R_{evt} be the corresponding binary variable for the rank requirement. Equalities (29) and (32) force the respective variables to zero if the seafarer does not work in the time period considered and equalities (30) and (33) impose that only one seafarer can be chosen to satisfy the experience requirement. At last, constraints (31) and (34) imposes that the sum of the amount of experience of the two seafarers is higher than the minimum amount required ($\lambda_{r_1 r_2}$), with L_e amount of experience of seafarer e, E_r set of seafarers available in rank r and U set of pair of ranks subject to experience requirements.

2.3   Column Generation Heuristic

Column generation was used on the set partitioning model in the root node. In this framework the Master problem is the relaxed version of the set covering model and the subproblem selects the duties to insert in the MP based on their reduced cost. All feasible duties were generated through an algorithm at the beginning of the procedure. The running time of the duty generation algorithm on the instances tested was 1.15 sec or less. However, to reduce the solution time of the master the duties are only added if they have a reduced cost. An initial feasible solution is achieved by using the solution of the greedy benchmark described in Sect. 2.4. The procedure ends when no duty with negative reduced cost are available in the root node and the integer master problem is solved using the branch-and-bound method. Thus the heuristic is basically to iteratively find all the cost columns with negative reduced cost for the relaxed master problem in the root node of the branch and bound tree and when there for the solution to the relaxed master problem can not be found any columns/duties with negative reduced cost the master problem is solved as a MIP problem.

2.4   Greedy Heuristic Used as Benchmark

Since no benchmark values were available, a simple heuristic method was developed to compare the solutions reported by the model with a solution that could be produced by the planners. The heuristic method considers only the minimization of transportation cost. This choice was based on the fact that the minimization of the pool of seafarers is not part of the planners operations. Before presenting the method, it is important to underline that the solutions reported by the heuristic method can be highly different from a solution generated by the planners. Nevertheless, the method was used to give a general evaluation of the possible savings and highlight some issues derived from a manual approach.

The method simulates, in broad terms, the process currently carried out by the planners in the company and is based on the following assumptions:

1. As company policy, the planners try to organize the departure of a seafarer when his embark duration is close to the half of the maximum allowed duration.
2. When they assign a seafarer, the planners do not know when and where he will leave the vessel.
3. The off-boards of the seafarers are decided during the weekly review of the schedule in which the planners consider the next two-weeks timetable of the vessel. If a seafarer can not leave the vessel in one of those port calls, the decision is postponed to the next week.
4. The minimum cost is always chosen by the planner.
5. If it is possible, the planners assign the seafarers in order to minimize the handover period.

The heuristic method developed assigns seafarers to vessels and ranks with a greedy approach following the temporal order. The method uses the variable $last_{vrn}$ defined as the point of time when it is necessary for vessel v to board a seafarer in rank r, where n represent the number of employees required on vessel v in rank r. This variable is initialized to the first time visit of each vessel and the method continues to assign seafarers to vessels, updating the values of $last_{vrn}$, until all of them have value equal to the last time point. For each seafarer the duties are stored in a list D_e ordered for starting time of the duties, where $s[d]$ represents the on-board time, $f[d]$ the point of time where the seafarer is available again and $h[d]$ the handover starting time.

In every iteration the method identifies the list of all possible duties starting at $last_{vrn}$. If there are no duties starting at that point of time, the algorithm moves backward to the previous point of time, increasing the handover period (Assumption 5). If there are duties starting at the point of time considered, the method selects all of them with an on-board duration $l[d]$ between the average duration of an on-board, defined by the seafarer contract and seniority, plus/minus a week (Eq. (35)). If no on-board periods are identified, the parameter i is increased by one, simulating the postponement of the decision in the next week (Assumption 3).

$$\frac{l_e^{max} + l_e^{min}}{2} - 7 \leq l[d] \leq \frac{l_e^{max} + l_e^{min}}{2} + 7 \cdot i \tag{35}$$

The list of possible on-board periods is subsequently sorted firstly by boarding cost and then by departure cost (Assumption 4). At this point, the algorithm goes through all the on-board periods selected and all the seafarers available, at that point of time, checking if the selected seafarer satisfies the minimum experience requirement. Since the data used had an average low value of rank's experience, the seafarers were ordered for increasing values of this parameter. In the first assignment of each rank on each vessel, if no seafarer was already assigned in one of the pairs of ranks that have an experience requirement, the minimum amount of experience needed was set to 3 months. This choice was made in order to not assign seafarers with really low experience that could have brought to infeasible solutions in cases with a small pool of seafarers. Once a seafarer has been assigned, the algorithm runs backwards through the time periods, identifying the closest time period ($t^* \leq t$) with a number of days between t and t^* at least equal to the handover period, and it sets the $last_{vrn}$ variable to this value (Assumption 5). If no seafarer is available to carry out one of the duties selected, the method decreases the value of the variable $last_{vrn}$ to the value of the previous vessel port call. The results of this heuristic benchmark are shown in the next section.

3   Data and Results

The models were implemented using Java and Gurobi solver. The models were tested on four data instances.

3.1   Data

The four data instances used in the testing can be seen in Table 1.
    All the data instances consider a planning horizon of 1 year.

Table 1. Data instances

Instances	Seafarers	Contracts	Vessels	Manning req.	Port calls
Instance1	443	1	1	9	87
Instance2	443	1	2	18	174
Instance3	443	1	5	48	432
Instance4	443	1	9	86	671

3.2   Results Using Integer Model

The model presented in the objective only minimizes the traveling costs of the seafarer. However, even though seafarers are paid embarked or not a company may wish to minimize their seafarer base. Therefore we have an additional test where an estimated cost of wages is included. The direct model was tested, in both the formulations, on **instance1**. As expected, the model reported low performances, reaching the time limit of 48 h, in both cases. More specifically, in the test without wages, the model found an integer solution with a gap to the lower bound of 57.0%. The test with wages reported a gap of 41.4%, as shown in Table 2.

From these tests, it was clear that a Mixed Integer Program was not the optimal approach to the problem. The Big M formulations used to model the duration of the on-boards and rest periods caused a slowing down in the branch-and-bound process. Moreover, since the l_{evt} and r_{evt} variables were modelled with greater than or equal to ($\geq$) constraints, their values, in the solution reported, were not exact in cases where the seafarers were not deployed again just after the end of their rest periods. This aspect implies that a re-optimization of the solution is necessary to obtain the exact duration of the on-boards and rest periods. For those reasons, the set covering formulation of the Liner Vessel Crew Scheduling problem was tested.

Table 2. Integer programming model results, instance1

	Without wages	With wages
Running time (Hr)	48	48
GAP	**57.0%**	**41.4%**

Due to limited space only a few test results are presented in Tables 2, 3 and 4. In Table 3 running times for the set covering formulation on 4 test instances are presented. For all instances the set covering formulation found the optimal solution. In Table 4 the results of the column generation heuristic are presented for **instance3**. In general, the column generation had better performance respect the set covering formulation with a time reduction $\sim$70%. The column generation was able to find the optimal solution for all instances tested despite the lack of a branch-and-price procedure.

The heuristic method was tested on all four instances considered, however, we here only show results for **instance3** and **instance4**. Its results were considerably higher than the optimal solution reported by the Set Covering formulation. Table 5 shows the results and the percentage change reported by the heuristic. From the Table it can be noticed that, as expected, the method assigned the seafarers to duties with a duration close to the average of the allowed on-board duration (90 days for the seniors and 150 days for the junior) resulting in a higher number of duties assigned. The average costs show two fundamental weaknesses of a manual approach. Firstly, most likely, the average transportation costs will

Table 3. SCP without wages, running times

Time (Sec)	Instance1	Instance2	Instance3	Instance4
Data reading and setting	4.82	5.10	12.76	18.91
Generation of Duties	0.34	0.43	0.70	1.15
Construction of the model	6.12	25.99	164.65	428.36
Resolution of the model	67.58	194.66	1,073.65	3,430.21
Total time	78.86	226.18	1,251.76	3,878.63

Table 4. CG without wages, instance3 Results. Add-10 is when 10 columns with negative reduced costs are added, Add-5 is when 5 are added and add-all is where all columns with negative reduced costs are added

	Add-All	Add-10	Add-5
Number of iterations	5	14	16
Total number of column insert	173,751	23,998	18,417
Total number of variables	535,239	385,486	379,905
Difference of Num. of variables with SCP (%)	−60.1%	−71.2%	−71.6%
Time to solve master problem	287.0	92.1	126,6
Total time (Sec)	**885.2**	**377.7**	**418.9**
Optimality gap	**0.0%**	**0.0%**	**0.65%**

Table 5. Heuristic, results and percentage change from set partitioning results

	Instance3	SCP change	Instance4	SCP change
Number of seafarers	122	−4.7%	208	−2.8%
Number of duties	181	+10.4%	329	+11.9%
Duty per seafarer	1.48	−35.4%	1.58	+15.1%
AVG Junior On-board duration (Days)	149.00	−0.4%	151.00	+5.1%
AVG Senior On-board duration (Days)	89.00	−14.4%	90.00	−13.6%
AVG On-board duration (Days)	114.00	−10.1%	115.00	−7,5%
Total Transport Cost	-	**88.19%**	-	**99.98%**

increase when every choice is made with a planning horizon of two weeks, because every time that a seafarer leaves a vessel, the decision is made only considering a two-weeks schedule, and therefore, cheaper transportation solutions, outside the examined period, are not be considered. Secondly, every time that a choice is made in the manual benchmark, even if it represents the best choice at that moment, it has consequences on the possible future choices and it can bring about worse solutions. Those aspects influence the performance of the heuristic method that had reported on average solutions 74.3% more expensive than the Set covering model.

4   Conclusion

A compact and a set partitioning model has been formulated for the liner shipping seafarer crew scheduling problem exact and heuristic methods have been applied and tested. Good results were achieved in a timely manner for smaller instances with the set covering problem formulations and for larger instances a Column Generation heuristic showed really good results solving all instances in less than 5 min. When this is said is should be noted that the largest problem was dealing with 9 vessels where as larger liner shipping companies often operates several hundred vessels. The column generation heuristic showed good potential for handling larger cases as all instances were solved to optimality within 5 min.

References

1. Beasley, J.E., Cao, B.: A tree search algorithm for the crew scheduling problem. Eur. J. Oper. Res. **94**(3), 517–526 (1996)
2. den Bergh, J., Beliën, J., De Bruecker, P., Demeulemeester, E., De Boeck, L.: Personnel scheduling: a literature review. Eur. J. Oper. Res. **226**(3), 367–385 (2013)
3. BIMCO and ICS: Manpower report: the global supply and demand for seafarers in 2015, Framework (2015). http://www.ics-shipping.org/docs/default-source/publications/employment-and-training/bimco-and-ics-manpower-report-2015.pdf
4. Borndörfer, R., Schelten, U., Schlechte, T., Weider, S.: A column generation approach to airline crew scheduling. In: Haasis, H.-D., Kopfer, H., Schönberger, J. (eds.) Operations Research Proceedings 2005, vol. 2005, pp. 343–348. Springer, Heidelberg (2006). https://doi.org/10.1007/3-540-32539-5_54
5. Christiansen, M., Fagerholt, K., Nygreen, B., Ronen, D.: Maritime transportation. Handb. Oper. Res. Manag. Sci. **14**, 189–284 (2007)
6. Dantzig, G.B.: Letter to the editor-a comment on traffic delays at toll booths. J. Oper. Res. Soc. Am. **3**, 229–341 (1954)
7. Dohn, A., Mason, A.: Branch-and-price for staff rostering: an efficient implementation using generic programming and nested column generation. Eur. J. Oper. Res. **230**, 157–169 (2013)
8. Edie, L.C.: Traffic delays at toll booths. J. Oper. Res. Soc. Am. **2**, 107–138 (1954)
9. Gamache, M., Soumis, F., Marquis, G., Desrosiers, J.: A column generation approach for large-scale aircrew rostering problems. Oper. Res. **47**, 247–263 (1999)
10. IMO: Principles of safe manning, Framework, A27/Res.1047 (2011). www.imo.org/en/OurWork/HumanElement/VisionPrinciplesGoals/Documents/1047(27).pdf
11. Kohl, N., Karisch, S.E.: Airline crew rostering: problem types, modeling, and optimization. Ann. Oper. Res. **127**, 223–257 (2004)
12. Leggate, A.: A vessel crew scheduling problem: formulations and solution methods. Report University of Strathclyde (2016)
13. Li, H.T., Womer, K.: A decomposition approach for shipboard manpower scheduling. Mil. Oper. Res. **14**, 67–90 (2009)
14. Ryan, D.M., Foster, B.A.: An integer programming approach to scheduling. In: Computer Scheduling of Public Transport Urban Passenger Vehicle and Crew Scheduling, pp. 269–280 (1981)
15. Ryan, D.M.: The solution of massive generalized set partitioning problems in aircrew rostering. J. Oper. Res. Soc. **4392**, 459–467 (1992)
16. UNCTAD: Review of Maritime Transport 2016, Technical report 2016. http://unctad.org/en/PublicationsLibrary/rmt2016_en.pdf

The Balanced Dispatching Problem in Passengers Transport Services on Demand

Patricio J. Araya-Córdova[1], Fabián Díaz-Nuñez[2], Javier Moraga[2],
Franco Quezada[3], Luis Rojo-González[2], and Óscar C. Vásquez[2(✉)]

[1] Department of Economics, Faculty of Economics and Business, University of Chile,
Diagonal Paraguay 257, Santiago, Chile
parayaco@fen.uchile.cl
[2] Industrial Engineering Department, Universidad de Santiago de Chile,
Av. Ecuador 3769, Estación Central, Chile
{fabian.diaz,javier.moraga,luis.rojo.g,
oscar.vasquez}@usach.cl
[3] Laboratoire d'Informatique de Paris 6, LIP6, Sorbonne Université,
4 Place Jussieu, 75005 Paris, France
franco.quezada@lip6.fr

Abstract. We introduce the balanced dispatching problem in passengers transport services on demand, such as taxi services, and propose an easy-to-implement algorithm in an online scenario, where all necessary information is only revealed with the arrival of requested transport services, seeking to guarantee quality of service for the passengers as well as balancing the income received by the drivers of the vehicles who perform the transportation services. The algorithm is based on theoretical results from the balanced incomes per worked time through the income variance minimization problem for two transport vehicles, considering the incoming service tariff, the performed services and the worked time by the transport vehicle drivers during the pay period in an online scenario. We compare our algorithm with the online dispatching algorithm currently used by Chilean companies on real instances. The numerical experiments are based on three different data sets of real instances such a labour day, one week, and one month, respectively. The obtained results show that the proposed algorithm is able to efficiently reduce the income dispersion among transport vehicle drivers within reduced running times, allowing a practical implementation into an automated dispatching system.

Keywords: Dispatching problem · Online dispatching algorithm
Passengers transport services

1 Introduction

Most companies dedicated to passenger transportation must deal with the dispatching problem for requested transport services on a daily basis, considering

© Springer Nature Switzerland AG 2018
R. Cerulli et al. (Eds.): ICCL 2018, LNCS 11184, pp. 379–387, 2018.
https://doi.org/10.1007/978-3-030-00898-7_25

different aspects related to both their external clients, i.e. passengers to be transported, as well as their internal clients, i.e. drivers of the vehicles used to transport the passengers, seeking to guarantee quality of service, as well as balancing the income received by the drivers of the vehicles who perform the transportation services.

In literature, the transport vehicles dispatching problem has been studied from different perspectives, mainly considering an *offline* scenario, in which it is assumed that all information is known at the moment when taking the dispatching decision [6]. However, in a real scenario, this assumption is not satisfied, and then the dispatching decision must be taken when the necessary information is revealed by the arrival of the requested transport service [1]. The above version of the transport vehicles dispatching problem, called *online*, is faced daily by transport companies, particularly those that must manage the dispatching of a fleet of transport vehicles, combining key company objectives such as quality of service, with the individual goals of the transport vehicle drivers.

Our Contribution. We introduce the balanced dispatching problem in passenger transport services and propose an easy-to-implement online dispatching algorithm, which aims to minimize the variance of the incomes per worked time among the transport vehicle drivers during the pay period defined by the transport company (e.g. monthly), guaranteeing that the requested transport services are carried out with some time tolerance. The algorithm is based on analytic equalities from the theoretical analysis of the incomes variance minimization problem for two transport vehicle drivers in an online scenario, considering the tariff of the performed transport services, the transport services to be performed and the worked time by transport vehicle drivers during the pay period. In order to compare the algorithm performance, we describe an online dispatching algorithm currently used in a Chilean company. The numerical results performed on real instances show that the proposed algorithm is able to efficiently reduce the income dispersion among transport vehicle drivers within reduced running times, allowing a practical implementation into an automatized dispatching system.

Related Work. The transport vehicles online dispatching problem has been studied in recent years motivated by the advances in information and communications technology (ICT). Maciejewski et al. [5] present an application of a wide-range microscopic model covering the city of Berlin and the Brandenburg region to assess the performance of a real-time dispatching strategy based on solving the taxi dispatching problem. The obtained results show improvements for both, drivers (less idle driving) and passengers (less waiting). However, computing the assignments for thousands of taxis in a huge road network turned out to be computationally demanding. Gao et al. [2] propose a new Mobile Taxi-hailing System (MTS) based on an optimal algorithm for multi-taxi dispatching, which differs from the competition modes used in traditional taxi-hailing systems, assigning vacant taxis to taxi-hailing passengers proactively. The system utility function involves the total net profits of taxis and waiting times of passengers subject to the individual net profits of taxis and the passengers' requirements for specified classes of taxis. Liu et al. [4] formulate the taxi-passenger matching as a global optimization problem by considering the pickup rate and average waiting time of passengers, proposing a parallel genetic algorithm to solve the problem. The algorithm includes initialization, crossover, mutation and a divide-and-conquer strategy for dimension reduction. The experimental results show the effectiveness and efficiency of the proposed algorithm, improving the quality of service provided by the taxi systems. Hyland and Mahmassani [3] develop and compare six dispatching strategies, that provide direct origin-to-destination service to travelers who request rides via a mobile application and expect to be picked up within a few minutes. The more sophisticated strategies significantly improved operational efficiency when fleet utilization is high (e.g. during the morning or evening peak); conversely, when fleet utilization is low, simply dispatching passengers sequentially to the nearest idle transport vehicle is comparable to more advanced strategies.

2 Statement of the Problem

We consider a company and a set $\mathcal{M}$ of transport vehicles. Let $\mathcal{K}$ be the set of transport vehicle types depending on the size and capacity. Each transport vehicles $m \in \mathcal{M}$ is classified by a subset $\mathcal{K}_m \subseteq \mathcal{K}$ of the transport vehicle classes and has a subset $\mathcal{I}_m$ of transport services, which can be performed.

Let $\mathcal{L}$ be the set of work bases distributed along the operation area of transport service. These bases host vehicles after every finished transport service, and also, start transport services from there depending on how far the pick-up from the transport service is located. Note that the transport vehicle has an assigned work base at the beginning of each time window available to be worked.

Each transport service $i \in \mathcal{I} := \cup_m \mathcal{I}_m$ needs a transport vehicle of type $k_i \in \mathcal{K}$ and has an arrival time h_i, a starting time a_i, and tariff service v_i.

The dispatching process involves seven main steps described as follows:

1. **Receive a request:** A transport service is solicited by the passenger via phone or through an online platform.
2. **Search transport vehicle availability:** The availability of a transport vehicle is searched according to the requirements of the passenger. If the transport vehicle is not available, it is necessary consulting the remaining time of the current services to communicate the new transport service conditions to the passenger, which could be accepted by the passenger and then a set of transport vehicles is available in this new scenario. In this step, a dispatching algorithm is used.
3. **Register passenger information:** Once the passenger has accepted the transport service, his information is registered together with the kind of transport vehicle who is giving the transport service.
4. **Assign the transport service:** The requested transport service is assigned to the transport vehicle driver on an online platform and wait for the confirmation.

5. **Start transport service:** Once the transport vehicle driver has accepted the assigned transport service, has to pick up the passenger and after the trip has finished the service is declared ended entering the transport service information on transport vehicle driver's application.
6. **Validate trip information on the application:** Finished the transport service, the passenger has to validate the information entered by the transport vehicle driver and, therefore, finish the transport service given.
7. **Report arrival to work base:** Once the validation of the information from the last finished transport service, the driver comes back to the nearest work base from the last finished transport service, reporting the arrival and availability. These work bases are parking points defined by the historical data of performed transport services and locations constraints. The goal is to minimize both the response time to the requested transport services and the cost related to fuel used by the transport vehicles.

The online dispatching algorithm is performed at Step 2. The algorithm goal is to assign transport services to transport vehicles according to some objective. In general, this algorithm is independently carried out on each work base once a service is requested. In this work, we adopt the algorithm goal to minimize the variance of the incomes per worked time among the transport vehicles drivers during a given pay period, guaranteeing that all requested transport services are carried out with some time tolerance.

3   The Online Dispatching Algorithm

We propose an online dispatching algorithm, where all necessary information is only revealed with the arrival of requested transport services. We need to introduce some notation in order to describe our algorithm.

Let $\mathcal{F}$ be the set of disjoint intervals for the cost transport service. Thus, each service $i \in \mathcal{I}$ belongs only to one interval $f \in \mathcal{F}$ according to the tariff transport service v_i. For convenience, we denote f_i the interval $f \in \mathcal{F}$ of transport service $i \in \mathcal{I}$. Let $r(m, f, t)$ be the ranking of transport vehicle $m \in \mathcal{M}$ ordered in decreasing order according to the waiting time on work base from last performed transport service i belongs to the interval $f \in \mathcal{F}$ at time t. We denote $\delta(t)_{m,m'}$ and $\gamma_{i,i'}$ the working time ratio between the transport vehicles m and m' at time t, $m, m' \in \mathcal{M}$ and ratio between the tariff transport services $i, i' \in \mathcal{I}$. Each transport $m \in \mathcal{M}$ has an income A_m at arrival time of the last perfomed transport service $i \in \mathcal{I}$, and a worked time d_m.

In order to develop our algorithm, we consider the case for a set $\mathcal{M}$ of two transport vehicles available at moment when a new transport service i' is solicited and propose a *worked time rule* to assign transport services. We show that the proposed rule solved the balanced dispatching problem in this setting and extend the application for $|\mathcal{M}|$ transport vehicles.

Definition 1 (*The worked time rule*). *Consider a set $\mathcal{M}$ of two transport vehicles and $r(m, f_{i'}, h_{i'}) = m$ without loss of generality, which are available to* *perform the transport service i', when it is solicited. The worked time rule defines that the new transport service i' will be performed by the transport vehicle $m = 1$ when $\delta(t)_{1,2}$ belong to the interval defined by the lower bound*

$$\frac{A_1 - A_2 - v_i}{\gamma_{i',i}v_i + 2v_i + 2A_2} - \frac{\sqrt{(v_i + A_2 - A_1)^2 + (\gamma_{i',i}v_i + 2v_i + 2A_2)(\gamma_{i',i}v_i + 2A_1)}}{\gamma_{i',i}v_i + 2v_i + 2A_2}$$

and the upper bound

$$\frac{A_1 - A_2 - v_i}{\gamma_{i',i}v_i + 2v_i + 2A_2} + \frac{\sqrt{(v_i + A_2 - A_1)^2 + (\gamma_{i',i}v_i + 2v_i + 2A_2)(\gamma_{i',i}v_i + 2A_1)}}{\gamma_{i',i}v_i + 2v_i + 2A_2},$$

where v_i is the tariff of last transport service performed by the transport vehicle $m = 2$ with $f_i = f_{i'}$.

Theorem 1. *Given a set $\mathcal{M}$ of two transport vehicles and $r(m, f_{i'}, h_{i'}) = m$ without loss of generality, which are available to perform the transport service i', when it is solicited. The worked time rule minimizes the variance of incomes per worked time of the transport vehicles at time $t \geq h_{i'}$.*

Proof. We consider a proof by cases analysis. The case (a) where the new transport service i' is assigned to the transport vehicle $m = 1$, which is the first in the ranking of the interval f'_i; and the case (b) where it is assigned to the transport vehicle $m = 2$. Note that the transport service i' is always performed by assumption.

We have that straightforward verification shows that the resulting variance of income per worked time of transport vehicles is equal to the expression (1) for case (b)

$$2\left(\frac{d_1(v_i + \gamma_{i',i}v_i + A_2) - d_2 A_1}{2d_1 d_2}\right)^2, \tag{1}$$

and equal to the expression (2) for case (a)

$$2\left(\frac{d_1(v_i + A_1) - d_2(A_1 + \gamma_{i',i}v_i)}{2d_1 d_2}\right)^2 \tag{2}$$

We compare the expressions (1) and (2) and have

$$2\left(\frac{d_1(v_i + \gamma_{i',i}v_i + A_2) - d_2 A_1}{2d_1 d_2}\right)^2 - 2\left(\frac{d_1(v_i + A_2) - d_2(A_1 + \gamma_{i',i}v_i)}{2d_1 d_2}\right)^2 = 0 \tag{3}$$

We replace $\delta(t)_{1,2} := d_1/d_2$ in expression (3) and have:

$$\delta(t)_{1,2}^2(\gamma_{i',i}v_i + 2v_i + 2A_2) + \delta(t)_{1,2}(2v_i + 2A_2 - 2A_1) - (\gamma_{i',i} + 2A_1) = 0 \tag{4}$$

We resolve the equality (4) in order to obtain the value $\delta(t)_{1,2}$ that satisfies the quadratic equation, obtaining the root values $\delta(t)_{1,2}^*$

$$\frac{A_1 - A_2 - v_i}{\gamma_{i',i}v_i + 2v_i + 2A_2} \pm \frac{\sqrt{(v_i + A_2 - A_1)^2 + (\gamma_{i',i}v_i + 2v_i + 2A_2)(\gamma_{i',i}v_i + 2A_1)}}{\gamma_{i',i}v_i + 2v_i + 2A_2},$$

$$\tag{5}$$

Thus, we observe that the case (a) has a minimum variance value when the $\delta(t)_{1,2}$ value belong to the interval defined by the expression (5), which concludes the proof.   □

The *worked time rule* guarantees that the incomes of each transport vehicle driver will be defined by considering its the effective worked time and its partial income obtained into the defined pay period (e.g. a labour day, one week, one month), avoiding arbitrariness at the moment to assign a new requested transport service. For instance, if both transport vehicle drivers are available and have the same effective worked time, i.e. $\delta(t)_{1,2} = 1$, then the requested transport service will be assigned to the driver with the partial income minimum.

We now develop our online dispatching algorithm, which is described in Algorithm 1

Algorithm 1. Online dispatching algorithm

Data: Vector $\{r(m, f, h_i)\}_{1 \leq m \leq |\mathcal{M}|, 1 \leq f \leq |\mathcal{F}|}$, transport vehicles information and information of the new transport service i.

Result: Assignment transport service i to an available transport vehicle

---

if $\mathcal{K}_i \neq \emptyset$ **then**

 | Assign the transport vehicle m which can perform the transport service i
 | and dominates the *worked time rule* among all pairs of transport vehicles in
 | the interval f_i.

else

 | Propose a starting time a_i such that some transport vehicle m can perform
 | the transport service i.
 | **if** a_i *is accepted* **then**
  | Assign the transport vehicle m which can perform the transport service
  | i and and dominates the *worked time rule* among all pairs of transport
  | vehicles in the interval f_i.
 | **else**
  | Eliminate transport service i.
 | **end**

end

The number of available transport vehicles to perform the transport service i is verified, line 1. For a number of available transport vehicles greater than or equal to one, the vehicle that dominates the worked time rule among all pair of transport vehicle in interval f_i is assigned, line 2. The *dominance* is stated by the use of the worked time rule among each pair of transport vehicles, starting with the first and the second position in the ranking for then interval f_i, considering the assigned transport vehicle by the rule in the following comparison. This procedure is repeated until comparing by pairs every transport vehicle belonging to $\mathcal{K}_i$, choosing in this manner the dominant transport vehicle. Otherwise, the algorithm offers a new starting time a_i for transport service i to be performed by the dominant transport vehicle in interval f_i, which can be accepted or rejected, lines 4–9.

4   Computational Experiments

In order to evaluate the online dispatching algorithm performance, we consider three different experiments on a set of real instances. The first, second and third experiments only assign the requested transport services during a pay period of a labour day, one week and one month, respectively. Each instance contains a set $\mathcal{M}$ of transport vehicles $\mathcal{M}$, a set $\mathcal{J}$ of time windows available to be worked and set $\mathcal{I}$ of transport services, which reach the values 67, 1,353 and 7,309, respectively. The Chilean peso (\$CL) and the minutes are considered as the monetary unit and the unit of worked time, respectively.

In addition, we describe and implement the online dispatching algorithm currently used by Chilean companies, which is applied manually and only considers a set $\mathcal{F}$ as the set of disjoint intervals for the tariff transport service and define a ranking $r(m, f, t)$ for each transport vehicle $m \in \mathcal{M}$ ordered in decreasing order according to the waiting time on work base given the last performed transport service i belongs to the interval $f \in \mathcal{F}$ at time t. The information about the effective worked time d_m or the partial income A_m of each vehicle $m \in \mathcal{M}$ at moment to arrive a new service is dismissed. The online dispatching algorithm currently used by Chilean companies is described in Algorithm 2. Note that this algorithm is carried out on each work base defined by the distance between the transport service to be performed and closest work base.

Algorithm 2. Online dispatching algorithm currently used by Chilean companies

---

Data: Vector $\{r(m, f, h_i)\}_{1 \leq m \leq |\mathcal{M}|, 1 \leq f \leq |\mathcal{F}|}$, transport vehicles partial
      information and partial information of the new transport service i.
Result: Assignment transport service to transport vehicle

---

if $\mathcal{K}_i \neq \emptyset$ **then**
   | Assign the transport vehicle m, which can perform the transport service i
   | and has the first position in the ranking for the interval f_i.
else
   | Propose a a_i such that some transport vehicle m can perform the transport
   | service i.
   | **if** a_i *is accepted* **then**
   | | Assign the transport vehicle m which can perform the transport service
   | | i and is in the first position in the ranking for the interval f_i.
   | **else**
   | | Eliminate transport service i.
   | **end**
end

The online dispatching algorithm currently used by Chilean companies is executed each time a transport service is solicited. The first operation considers the availability of the transport vehicles to perform the transport service i, line 1. If the service transport i can be performed by a number of available transport vehicles greater than or equal to one, then the transport vehicle with the first position in the ranking for interval f_i is assigned, line 2. Otherwise, the algorithm offers a new starting time a_i for transport service i to be performed the transport vehicle in the first position in the ranking for then interval f_i, which can be accepted or rejected, lines 5–9.

The online dispatching algorithms were implemented in Python 2.7 with a Notebook Intel Core i5, 2.5 GHz, 4 GB RAM. All experiment runs in at most five seconds. Table 1 shows the average of variance of incomes per worked minute among the transport vehicles in the different pay periods.

Table 1. Average of variance of incomes in Chilean peso ($CL) per worked minute among the transport vehicles in the different pay periods

Online algorithm	Experiment 1	Experiment 2	Experiment 3
Current	197, 024.99	47, 780.26	75, 166.23
Proposal	222, 305.41	40, 398.54	67, 601.23

In Experiment 1, the online dispatching algorithm currently used by Chilean companies shows that the minimum income per worked minute of transport vehicle is $CL 14.9 and the maximum is $CL 349.31, obtaining a median of $CL 112.12. Moreover, 50% of the transport vehicles earn between $CL 72.82 and $CL 143.69. Comparing with the proposed online dispatching algorithm, it shows that the 50% of the income per worked minute of the transport vehicles is concentrated between $CL 73.24 and $CL 129.47, decreasing the dispersion of values.

In Experiment 2, the online dispatching algorithm currently used by Chilean companies shows that the minimum income per worked minute of the transport vehicle is $CL 12.79 and the maximum is $CL 138.67, the average is $CL 68.29 and finally, the 50% of the transport vehicles earn between $CL 48.4 and $CL 84.42. In addition, the proposed online dispatching algorithm shows higher income per worked minute where 50% of those are between $CL 59.67 and $CL 99.84.

In Experiment 3, the online dispatching algorithm currently used by Chilean companies shows that the minimum income per worked minute of the transport vehicle is $CL 27,68 and the maximum is $CL 170.33, obtaining an average of $CL 64.44 and the 50% of the transport vehicles earn between $CL 38.18 and $CL 99.38. The proposed online dispatching algorithm concentrates the earned income per worked minute of the 50% of the transport vehicles between $CL 52.61 and $CL 102.45, decreasing the dispersion of quartiles 1 and 3.

5 Final Remarks

In this paper, we introduce the balanced dispatching problem in passenger transport services and propose an easy-to-implement online dispatching algorithm. The obtained results from the comparison with another online dispatching algorithm currently used by Chilean companies show an important difference in the average of variance of incomes per worked time among the transport vehicles for extensive pay periods (e.g. week, month), reducing the dispersion. In addition, the reduced running times allow a practical implementation into an automated dispatching system, avoiding possible arbitrariness in the manual dispatching decision.

Note that for brief pay period (e.g., a labour day) the online dispatching algorithm currently used by Chilean companies has a better performance , since initially none of the transport vehicles has assigned some transport service and then the income per worked time is zero. Therefore, assigning a new transport service do not need a further analysis, being possible for any available transport vehicle.

For future research, we propose to study the parameterization of the proposed online dispatching algorithm in order to find the work base and the disjoint interval for the tariff transport services, which improves its performance.

Finally, we leave open the question about a mathematical model under the assumption that all necessary information of the pay period is available to take the dispatching decision, which allows to compare its performance with the optimal solution.

Acknowledgement. This work is partially supported by DICYT No061817OP, Universidad de Santiago de Chile.

References

1. Borodin, A., El-Yaniv, R.: Online Computation and Competitive Analysis. Cambridge University Press, Cambridge (2005)
2. Gao, G., Xiao, M., Zhao, Z.: Optimal multi-taxi dispatch for mobile taxi-hailing systems. In: 2016 45th International Conference on Parallel Processing (ICPP), pp. 294–303. IEEE (2016)
3. Hyland, M., Mahmassani, H.S.: Dynamic autonomous vehicle fleet operations: optimization-based strategies to assign AVs to immediate traveler demand requests. Transp. Res. Part C: Emerg. Technol. **92**, 278–297 (2018)
4. Liu, Y.-W., Zhang, X.-Y., Gong, Y.-J., Chen, W.-N., Zhang, J.: A parallel genetic algorithm with region division strategy to solve taxi-passenger matching problem. In: 2017 IEEE Symposium Series on Computational Intelligence (SSCI), pp. 1–7. IEEE (2017)
5. Maciejewski, M., Bischoff, J., Nagel, K.: An assignment-based approach to efficient real-time city-scale taxi dispatching. IEEE Intell. Syst. **31**(1), 68–77 (2016)
6. Toth, P., Vigo, D.: Vehicle Routing: Problems, Methods, and Applications, vol. 18. SIAM, Philadelphia (2014)

Scheduling Assistance for Passengers with Special Needs in Large Scale Airports

Michele Mele$^{(\boxtimes)}$ and Paola Festa

Department of Mathematics and Applications,
University of Naples "Federico II", 80126 Naples, Italy
dip.matematica-app@unina.it

Abstract. In this paper we study a specific variant of the well known Fixed Job Scheduling Problem, namely the Tactical Fixed Job Scheduling Problem with Spread-Time constraints. In this problem it is required to schedule a number of jobs on non identical machines that differ each other for the set of jobs they can perform and that have constraints on the length of their duty. We present some lower bounds for the optimal value of the problem and introduce the first heuristic algorithm for solving it. We also study a specific case of interest connected with the assistance of passengers with special needs in large scale airports.

Keywords: Operations research · Combinatorial optimization
Scheduling · Heuristic algorithm · Fixed job scheduling · Accessibility

1 Introduction

In an instance of the basic Fixed Job Scheduling Problem (FJS) it is asked to find the minimum number of machines needed to process n jobs J_j with fixed release time r_j and deadline d_j $(j = 1, ..., n)$. Each machine is able to process only a job at a time and it has to complete the processing of a job after starting it with no interruption. All the machines are available along the whole (discrete) interval $[0, T]$, with $T = \max d_j$. The authors in [1] solve the problem to optimality in polynomial time with a staircase rule, dealing with the FJS as a special case of Dilworth's problem. In [2] an alternative method for finding the optimal solution of FJS is presented via a special step function. The authors in [3] propose an algorithm to compute the optimal solution of FJS using a pile.

In [4–6] some variants of the basic FJS are introduced; in particular in [4] the Fixed Job Scheduling Problem with spread-time constraints (FJSS) is presented as the basic FJS plus special real life inspired constraints. Each machine can work only for a fixed number L of time units: the spread-time is the range between the release of the first job and the deadline of the last job assigned to a machine. In [4,6] the authors introduce two procedures, one based on a greedy approach

© Springer Nature Switzerland AG 2018
R. Cerulli et al. (Eds.): ICCL 2018, LNCS 11184, pp. 388–400, 2018.
https://doi.org/10.1007/978-3-030-00898-7_26

and one based on the preemptive version of the problem, that are the main steps of the two 2-approximation algorithms they present.

The Tactical Fixed Job Scheduling Problem (TFJS) is introduced in [7]. It consists of the Basic FJS with the machines divided into classes: machines from a given class can only process jobs from a given subset of jobs. In [7,8] NP-completeness results are proved for the TFJS together with some upper and lower bounds and an exact branch-and-bound procedure.

Finally in [9] the Tactical Fixed Job Scheduling Problem with spread-time constraints (TFJSS) is introduced with a set partitioning model on which a branch-and-price approach to the problem is based.

In the next section we introduce the mathematical model for TFJSS and we prove that it is NP-complete. In Sect. 3 we define some lower bounds for the optimal value of TFJSS. In Sect. 4 we present a greedy algorithm for solving the problem and we produce computational experiments for a specific set of instances connected to a specific real world situation.

1.1   Motivation for This Research

Passengers with special needs (for example blind people, partially sighted people, people in wheelchair) suffer remarkable inconveniences when they have to take a flight, especially when they have to start a journey by their own or when they have to change flight at a transition airport. Many airports and airline companies are planning to organize assistance services for these passengers to allow them to travel in autonomy. In the last decades many rules have been stated to regulate such services. Unfortunately the road to accessibility is still long to go: in almost every international airport the assistance services use to violate some of the rules stated in the international treaties. So one may ask for the minimum number of workers needed to provide assistance and accompany passengers with special needs respecting all the stated rules. All passengers with special needs must be assisted by the same worker during their whole permanence in the airport; moreover the worker assisting a passenger has to speak a language comprehensible by him/her. Finally the basic work rules must be respected, especially the rule on the length of a day duty.

This real world problem can be seen as a TFJSS in which the workers of the airport are identified with the machines and the length of the day duty coincides with the spread-time. In this scenario the workers are not equal because they can speak different languages and so assist different passenger groups.

2   Mathematical Formulation and Complexity

An istance of TFJSS consists of n jobs J_j $(j = 1, ..., n)$ that must be processed without preemption from a fixed release time (or starting time) r_j to a fixed deadline d_j on m non identical machines that can process only one job at a time $(m < n)$. Each machine can only work for a fixed number L of consecutive time units: as already mentioned, the spread-time is defined as the range from the first job and the last job assigned to a machine. Moreover machines are divided into c classes: machines belonging to a class can only process jobs from a given subset of jobs. The goal is to minimize the number of machines required to process all jobs.

Let M^i be the set of machines of class i $(1 \leq i \leq c)$ and C_j $(1 \leq j \leq n)$ be the set of classes containing all the machines able to process J_j. Let K^i be the set of jobs that can be processed from the machines of M^i. Two jobs J_j and J_k are said to be compatible if they can be performed by the same machine. For each $J_j \in K^i$ $(1 \leq i \leq c)$ let $A_j = \{J_k \in K^i : r_k \leq r_j \leq d_k \vee r_j \leq r_k \leq d_j \vee d_k - r_j > L \vee d_j - r_k > L\}$ be the set of jobs that are not compatible with J_j. We define the decision variables y_k^i, that are equal to 1 if machine $k \in M^i$ is used to perform at least one job and equal to 0 otherwise, and the decision variables x_{jk}^i that assume value 1 if job $J_j \in K^i$ is assigned to machine $k \in M^i$ and value 0 otherwise. So we can model the TFJSS as follows:

$$z = \min \sum_{i=1}^{c} \sum_{k \in M^i} y_k^i \tag{1}$$

s. t.

$$x_{jk}^i \leq y_k^i \quad J_j \in K^i, k \in M^i, i = 1, ..., c \tag{2}$$

$$x_{jk}^i + x_{lk}^i \leq 1 \quad J_l \in A_j, J_j \in K^i, k \in M^i, i = 1, ..., c \tag{3}$$

$$\sum_{i \in C_j} \sum_{k \in M^i} x_{jk}^i = 1 \quad j = 1, ..., n \tag{4}$$

$$x_{jk}^i, y_k^i \in \{0, 1\} \quad J_j \in K^i, k \in M^i, i = 1, ..., c. \tag{5}$$

The objective functions (1) requires the minimization of the number of machines needed to perform all jobs. Constraints (2) assure that a machine is used only when at least one job is assigned to it. Constraints (3) assure that the compatibility relations are respected. Constraints (4) make a job executed once and by a unique machine.

Theorem 1. *The TFJSS is NP-complete.*

Proof. We introduce the mathematical formulation of the FJSS [4]. In an instance of this problem it is required to schedule n jobs J_j $(j = 1, ..., n)$ on m identical machines for which there is the same spread-time limit L. Jobs have fixed starting time r_j and deadline d_j and they must be processed without pre-emption. For each J_j let $A_j = \{J_k : r_k \leq r_j \leq d_k \vee r_j \leq r_k \leq d_j \vee d_k - r_j > L \vee d_j - r_k > L\}$ be the set of jobs that are not compatible with J_j. We define decision variables y_k that assume value 1 if machine k is used and 0 otherwise; and x_{jk} that assume value 1 if job J_j is processed by machine k and 0 otherwise.

We can formulate as follows:

$$z' = \min \sum_{k=1}^{m} y_k \tag{6}$$

s. t.

$$x_{jk} \le y_k \quad \forall J_j, k = 1, ..., m \tag{7}$$

$$x_{jk} + x_{lk} \le 1 \quad J_l \in A_j; \forall J_j; k = 1, ..., m \tag{8}$$

$$\sum_{k=1}^{m} x_{jk} = 1 \quad j = 1, ..., n \tag{9}$$

$$x_{jk}, y_k \in \{0, 1\} \quad \forall J_j, k = 1, ..., m. \tag{10}$$

Putting $C = 1$ in model (1)–(5) one can produce a polynomial reduction to (6)–(10); this latter is NP-complete so TFJSS is NP-complete too.   ◇

3   Lower Bounds

In this section we introduce some lower bounds for the optimal value of TFJSS. Consider a discrete time-line $[0, T]$, a fixed time $t^0 \in [0, T]$, let n^0 be the number of active jobs at t^0 (computable in polynomial time as in [1]). Put $t_1^0 = t^0 + L + 1$, $t_{-1}^0 = t^0 - L - 1$. More generally

$$t_p^0 = t_{p-1}^0 + L + 1,$$

$$t_{-p}^0 = t_{1-p}^0 - L - 1$$

when they exist in $[0, T]$. Denote with n_p^0 the number of active jobs at time t_p^0 with $p \in Z$.

Theorem 2.

$$B_1(t^0) = \sum_p n_p^0$$

is a lower bound for the optimal value of TFJSS.

Proof. In fact a machine working at time t_p^0 cannot be working at time t_{p+1}^0 or at time t_{p-1}^0. Denote with z^* the optimal value of (1)–(5). Clearly $z^* \ge B_1(t^0)$ because $B_1(t^0)$ does not take into account the jobs that have release time and deadline between some t_p^0 and t_{p+1}^0. We obtain a family of lower bounds moving $t^0 \in [0, T]$ obtainable in polynomial time.   ◇

Moreover put

$$\overline{B}_1 = \max_{t^0 \in [0, T]} B_1(t^0).$$

Clearly $\overline{B}_1$ is a lower bound for the optimal value of TFJSS.
    Take now $t^0 \in [0, T]$ and put

$$\tau_p^0 = \min\{t : t \ge t_p^0 \wedge n(t) \ge 1\},$$

$$\tau^0_{-p} = \max\{t : t \le t^0_{-p} \land n(t) \ge 1\},$$

when they exist in $[0, T]$, where $n(t)$ is the number of active jobs at time t. Note that if $n^0_p \ge 1$, then $\tau^0_p = t^0_p$.

Theorem 3.

$$B_2(t^0) = \sum_p \tau^0_p$$

is a lower bound for the optimal value of TFJSS.

Proof. In fact a machine working at time τ^0_p cannot be working at time τ^0_{p+1} or at time τ^0_{p-1}. Denote with z^* the optimal value of (1)-(5). Clearly $z^* \ge B_2(t^0)$ because $B_2(t^0)$ does not take into account the jobs that have release time and deadline between some τ^0_p and τ^0_{p+1}. We obtain a family of lower bounds for the optimal value of TFJSS moving $t^0 \in [0, T]$.   ◇

Moreover

$$\overline{B}_2 = \max_{t^0 \in [0, T]} B_2(t^0)$$

is a lower bound for the optimal value of TFJSS.

4   An Heuristic Algorithm

4.1   The Algorithm

We present an heuristic algorithm based on a greedy approach, nevertheless it allows to choose for selection criteria based on the flexibility of the machines.

The algorithm starts sorting jobs in non decreasing order of r_j. For $k = 1, ..., n$ it selects job J_k: if there is at least a machine able to process J_k among the ones to which at least one job has already been assigned and for which the spread-time constraints would be not violated, then the algorithm chooses one of those machines via a criterion C' and assigns J_k to this machine; else, it chooses a new machine among the ones that are not already used via a criterion C'' and assigns J_k to this machine.

It is possible to choose among three criteria for C' and C'' before the algorithm starts:

- Minimum flexibility (cmin): choose one of the machines (to which a job has already been assigned in the case of C', to which no job has already been assigned in the case of C'') with the minimum flexibility; i.e. able to process jobs from the minimum number of job groups and for which the spread-time constraints would be not violated;
- Maximum flexibility (cmax): choose one of the machines (to which a job has already been assigned in the case of C', to which no job has already been assigned in the case of C'') with the maximum flexibility; i.e. able to process jobs from the maximum number of job groups and for which the spread-time constraints would be not violated;
- Random: choose one of the machines able to process that job and for which the spread-time constraints would be not violated.

Algorithm 1.

Sort jobs in non decreasing order of r_j;
for $k = 1, ..., n$ **do**
    if there is at least a machine able to process J_k among the ones to which it has already been assigned at least one job and for which the spread-time constraints would be not violated, **then**
        Choose one of those machines via a criterion C' and assign J_k to this machine;
      else Choose a new machine among the ones that are not already used via a criterion C'' and assign J_k to this machine.
    end if
end for

4.2 Computational Experiments

The algorithm has been developed in Java language with Eclipse Jee Oxygen on a DELL Inspiron with 8 GB RAM, SSD hard disk and Windows 10 operative system. The instances for this paper have been created to be in accordance with real world situations that occur in large scale airports. Informal talks have been carried out with accessibility and security managers in important european international airports, detecting averages and numbers of the real world cases that they face every day.

We consider a discrete time-line of $[0, 200]$, discrete randomly chosen r_j in the whole time-line and integer duration of the jobs randomly chosen in $[5, 30]$. For the experiments we put $L = 80$ or $L = 100$, $m = \frac{13}{20}n$ and $n = 100, 500, 1000$ or 2000. The jobs are divided into four groups corresponding to the four languages spoken by the machines/workers. For simplifying the notation we identify this four languages with the Italian (basic language spoken by all the machines/workers), English, Spanish and French language. The m machines/workers are divided into five classes of equal cardinality containing respectively all the machines/workers speaking only the basic language of the airport (Italian language in our case), all machines/workers speaking only Italian and English, all machines/workers speaking only Italian, English and French, all machines/workers speaking only Italian, English and Spanish, all machines/workers speaking Italian, English, Spanish and French.

Consider an iteration of the algorithm choosing a job J_k. W. l. g. we suppose we are adopting criteria (cmin,cmin) for (C', C''), i.e. choosing the minimum flexibility criterion cmin for both C' and C'', and that there are no machines able to process J_k among the ones to which a job has already been assigned and with minimum flexibility. Suppose that the algorithm can not find a machine (of minimum flexibility) for J_k until it reaches two classes of machines/workers, call it q and q', that can process J_k and having the same cardinality. Machines from q and q' are able to process J_k and have the same flexibility, so it does not matter to J_k which one to choose, a machine/worker from q or q'. Preferring always one of the two classes could affect the final solution. In this situation the algorithm will pick a machine/worker from $q \cup q'$. The same observation is valid also for other couples of criteria for C', C''. This situation corresponds to the case in which a job/passenger asking assistance with Italian or English language cannot be allocated to a machine from the two classes with minimum flexibility: the class containing machines/workers speaking Italian, English and French and the class containing ones speaking Italian, English and Spanish have the same flexibility and can process that job. This situation can occur many times during the execution of the algorithm and the choices it makes can bring to different solutions. For this reason every instance has been solved many times.

We study 15 instances for each couple (n, L) with $n = 100, 500, 1000$ and 2000 and $L = 80$ or 100. In the left part of the tables there are the numbers of jobs/passengers requiring respectively Italian (ITA), English (ENG), French (FRA) and Spanish (SPA) language. This instances represent realistic situations in which of course the number of jobs/passengers requiring the basic language of the airport (Italian in our case) or English is greater than the number of jobs/passengers requiring the other two languages. In the right part of the tables there are the results of the solution of the instances. Each row contains the averages of the results of 20 executions of the algorithm with criteria (cmin,cmin), i.e. choosing the minimum flexibility criterion cmin for both C' and C'', for a total of 2400 executions. The reported numbers are, from left to right:

- the best value obtained for each instance (BVO)
- the average of the values obtained (AVO)
- the average of the times in ms of the initial sorting process via selection sort (SSA)
- the average of the times in ms of the main algorithm (MAA).

4.3   Conclusions

The couple of criteria (cmin,cmin) proved to be the best. All the other combinations of criteria bring to worst results and sometimes they leave some jobs/passengers unassigned, especially in the cases in which $n = 100$.

Computational study shows that the value of BVO and AVO generally depend on the distribution of the jobs during the time-line, i.e. on the number of jobs that overlap, and from the length $d_j - r_j$ of the jobs. The results are competitive with the numbers of large scale airports, as stated by talks and comparisons with accessibility and security managers of international airports. In the real cases on which the instances with $L = 80$ are modeled the number of machines/workers currently used to process $n = 500$ jobs is generally between 150 and 160. Similarly in the case in which $n = 1000$ the number of machines/workers used to accomplish all jobs is between 290 and 310; in the case of $n = 2000$ that number is generally between 570 and 600. In the real cases on which the instances with $L = 100$ are modeled the number of machines/workers currently used to process $n = 500$ jobs is generally between 110 and 120. Similarly in the case in which $n = 1000$ the number of machines/workers used to accomplish all jobs is between 220 and 240; in the case of $n = 2000$ that number is generally between 420 and 450. Apart from the cases in which $n = 100$

Table 1. Results for the instances with $n = 100$, $L = 80$

Instance	ITA	ENG	FRA	SPA	BVO	AVO	SSA (ms)	MAA (ms)
1	60	20	10	10	39	39,2	8,3	1
2	55	21	13	11	37	37	9,5	1
3	51	23	13	13	32	32,5	1,6	1,6
4	46	26	15	13	34	35,4	1,6	1,7
5	43	28	14	15	39	39,6	1,8	1,4
6	40	30	15	15	27	28,7	2,2	1,4
7	38	31	16	15	33	36,4	2	1,3
8	36	33	15	16	35	37,2	6,1	1,2
9	34	33	17	16	38	39,1	2,5	1,5
10	33	33	17	17	34	35	2,9	1,7
11	31	34	16	19	31	33,5	8,7	1
12	32	30	19	19	39	39	10,1	1
13	31	30	20	19	33	33	9,1	1
14	30	29	21	20	35	36,4	8,5	1
15	27	28	21	24	31	33,2	9	1

Table 2. Results for the instances with $n = 500$, $L = 80$

Instance	ITA	ENG	FRA	SPA	BVO	AVO	SSA (ms)	MAA (ms)
1	300	100	50	50	134	138,6	26,1	14,6
2	279	109	52	60	146	148,8	22,7	14,3
3	252	133	60	55	135	136,7	25,7	13,4
4	231	142	66	61	152	153	24,2	11,3
5	212	146	72	70	134	136,1	23,1	18,3
6	200	150	75	75	139	142,3	26,5	13,7
7	189	150	80	81	159	160,7	26,2	17,4
8	169	141	100	90	148	149,1	23,1	13,7
9	153	155	91	101	130	132,7	23	16,1
10	158	140	103	99	152	154,6	24,5	13,6
11	147	134	111	108	158	158,7	23,2	14,5
12	140	142	108	110	154	156,6	24,8	15,2
13	141	139	110	110	160	160,9	24,8	15,2
14	138	133	114	115	133	138,3	23	14,2
15	130	134	120	116	125	128	23,8	14,3

Table 3. Results for the instances with $n = 1000$, $L = 80$

Instance	ITA	ENG	FRA	SPA	BVO	AVO	SSA (ms)	MAA (ms)
1	600	200	100	100	273	275,2	43,8	21,1
2	570	212	110	108	276	278,1	48,4	18.3
3	521	241	117	121	304	304,7	46,6	20,5
4	462	262	140	136	292	295,2	45,1	19,9
5	421	289	150	140	287	289,4	46,9	20,9
6	400	300	150	150	292	294,7	41,3	21,9
7	371	330	152	147	273	276,2	52	16,7
8	363	331	150	156	277	282,1	46,5	14,1
9	351	317	165	167	264	266,2	43,3	14,3
10	324	342	171	163	277	279,9	43,6	14,7
11	330	302	190	178	285	286,3	45,5	14,3
12	310	311	179	200	289	291,7	42,9	14,2
13	303	298	204	195	288	289,8	43	13,1
14	285	287	200	228	280	284,3	40,7	13
15	272	258	240	230	278	281,2	41,1	12,3

Table 4. Results for the instances with $n = 2000$, $L = 80$

Instance	ITA	ENG	FRA	SPA	BVO	AVO	SSA (ms)	MAA (ms)
1	1200	400	200	200	522	523,9	60,6	38,5
2	1114	429	223	234	562	565,3	59,5	36,1
3	1058	457	245	240	552	555,3	61	36,6
4	950	536	258	256	557	558,4	57,3	37,1
5	859	571	291	279	526	530,1	64,1	39,1
6	800	600	300	300	560	562,3	66,1	36,5
7	772	604	319	305	592	593,6	63	38,2
8	729	580	345	346	533	535,9	69,3	41,6
9	708	572	358	362	529	532,4	62,6	41,1
10	680	553	385	382	561	563,7	63,6	37,7
11	651	546	405	398	531	534,4	61,8	38,7
12	619	539	421	421	553	556	60,3	31,6
13	605	520	439	436	539	541,4	59,4	30,3
14	582	523	450	445	558	559,5	59,5	33,1
15	558	511	462	469	567	569,3	66,4	37,3

Table 5. Results for the instances with $n = 100$, $L = 100$

Instance	ITA	ENG	FRA	SPA	BVO	AVO	SSA (ms)	MAA (ms)
1	60	20	10	10	28	28,8	2,3	1,8
2	55	22	11	12	31	31	8,1	1
3	50	24	13	13	27	27,1	3,9	1,3
4	46	25	15	14	29	29,2	2,1	1,8
5	42	29	14	15	26	26,9	2,6	1,8
6	40	30	15	15	31	31	2,4	1,7
7	38	31	16	15	23	23,9	2,1	1,5
8	36	30	18	16	28	29	3	1,6
9	34	33	17	16	30	30,2	2,7	1,6
10	33	32	18	17	29	29	2,6	1,5
11	33	30	20	17	25	26,7	2,5	1,6
12	31	31	18	20	29	29	9,5	1
13	31	30	21	18	32	33,6	9,8	1
14	29	28	23	20	24	24,8	10,3	1
15	28	27	24	21	27	28,3	9,1	1

Table 6. Results for the instances with $n = 500$, $L = 100$

Instance	ITA	ENG	FRA	SPA	BVO	AVO	SSA (ms)	MAA (ms)
1	300	100	50	50	110	111,7	20,6	14,7
2	273	111	59	57	107	108,7	21,7	15,6
3	249	120	67	64	112	114,4	21,8	12,7
4	230	129	73	68	105	106,6	24,9	14,4
5	212	140	73	75	107	109,8	22,4	14
6	200	150	75	75	102	104,9	25,2	14,2
7	186	149	83	82	104	105,9	21,4	12,7
8	165	144	99	92	99	101,9	23,7	12,9
9	153	156	90	101	107	108,7	22	15,3
10	156	140	99	105	98	100	24	13,7
11	147	133	113	107	103	105	22,3	14,8
12	143	130	116	111	101	102,4	22,6	11,7
13	136	140	111	113	108	109,1	20,5	12,9
14	135	132	124	109	97	98,9	21,9	13,3
15	133	124	127	116	98	100,3	23,3	13,7

Table 7. Results for the instances with $n = 1000$, $L = 100$

Instance	ITA	ENG	FRA	SPA	BVO	AVO	SSA (ms)	MAA (ms)
1	600	200	100	100	192	194	44,1	12
2	563	217	117	103	199	201,6	44,8	16,5
3	520	242	124	114	210	210,5	45	14,5
4	459	261	144	136	194	197,6	43,5	14,2
5	425	280	145	150	200	201,9	45	17,4
6	400	300	150	150	209	211,4	41,4	16,5
7	382	307	160	151	206	209,3	43,1	18,8
8	365	299	170	166	199	199,7	40,8	17
9	348	298	180	174	220	222,1	41,7	18,1
10	325	345	161	169	200	200,75	41,8	16,1
11	320	290	198	192	190	193,2	42,8	15,9
12	311	289	194	206	201	203,2	44,2	15,9
13	302	281	206	211	195	197,4	42,3	15,2
14	285	275	221	219	202	205,1	47,6	13,6
15	279	253	257	211	195	197,1	44,1	11,1

Table 8. Results for the instances with $n = 2000$, $L = 100$.

Instance	ITA	ENG	FRA	SPA	BVO	AVO	SSA (ms)	MAA (ms)
1	1200	400	200	200	382	385,8	62,6	30,8
2	1103	438	228	231	385	387,8	59,5	28,4
3	1026	481	253	240	371	376,3	58,6	28,4
4	948	539	262	251	387	393,6	56,9	28,8
5	852	569	288	291	386	388,5	59,4	32,9
6	800	600	300	300	370	374,1	63,2	31,1
7	783	598	302	317	376	379,5	61,6	28,8
8	756	601	327	316	395	399,3	62,5	31,1
9	730	582	348	340	380	386,9	61	34,4
10	703	580	361	356	379	381,9	65,1	36,9
11	660	681	335	324	382	385,8	65,1	37,8
12	641	555	394	410	386	388,1	59,2	25,6
13	606	530	447	417	388	391,5	62,1	29,2
14	586	521	450	443	396	396,7	67	34,2
15	553	509	481	457	380	383,1	61	28

(Tables 1 and 5), for which the numbers obtained by the algorithm are very similar to the ones recorded in airports, a comparison between the real world numbers and the results of the algorithm shows that those latter seem to be better in the majority of cases (see Tables 2, 3, 4 and Tables 6, 7, 8). This stands despite the fact that in almost every airport some of the rules stated by international conventions and represented in our mathematical model are ignored, hence the numbers recorded in real world situations arise from schedulings that are not feasible for our formulation of the problem. So the results obtained by the algorithm are generally better than the ones recorded in many real world cases.

Computational experiments show that at least the 40% of the machines/workers is saved; moreover in the majority of cases the number of machines/workers saved reaches notable levels with more than the 60 per cent of machines/workers saved in the case $L = 80$ (see Table 4, instance 1) and 71 per cent in the case $L = 100$ (see Table 8, instance 6).

References

1. Ford, L.R., Fulkerson, D.R.: Flows in Networks. Princeton University Press, Princeton, New Jersey (1962)
2. Gertsbakh, I., Stern, H.I.: Minimal resources for fixed and variable job schedules. Oper. Res. **26**(1), 68–85 (1978)
3. Gupta, U.I., Lee, D.T., Leung, J.Y.-T.: An optimal solution for the channel-assignment problem. In: IEEE Transactions on Computers C, pp. 807–810. IEEE (1979)
4. Fischetti, M., Martello, S., Toth, P.: The fixed job schedule problem with spread-time constraints. Oper. Res. **35**(6), 849–858 (1987)
5. Fischetti, M., Martello, S., Toth, P.: The fixed job schedule problem with working-time constraints. Oper. Res. **37**(3), 395–403 (1989)
6. Fischetti, M., Martello, S., Toth, P.: Approximation algorithms for fixed job schedule problems. Oper. Res. **40**(1(Suppl. 1)), S96–S108 (1992)
7. Kolen, A., Kroon, L.: License class design: complexity and algorithms. Eur. J. Oper. Res. **63**(3), 432–444 (1992)
8. Kroon, L., Salomon, M., Van Wassenhove, L.: Exact and approximation algorithms for the tactical fixed interval scheduling problem. Oper. Res. **45**(4), 624–638 (1997)
9. Zhou, S., Zhang, X., Chen, B., van de Velde, S.: Tactical fixed job scheduling with spread-time constraints. Comput. Oper. Res. **47**, 53–60 (2014)
10. Kovalyov, M.Y., Ng, C.T., Chen, T.C.E.: Fixed interval scheduling: models, applications, computational complexity and algorithms. Eur. J. Oper. Res. **178**(2), 331–342 (1992)
11. Kolen, A., Lenstra, J., Papadimitriou, C., Spieksma, F.: Interval scheduling: a survey. Nav. Res. Logist. (NRL) **54**(5), 530–543 (1992)
12. Keil, M.: On the complexity of scheduling tasks with discrete starting times. Oper. Res. Lett. **12**(5), 293–295 (1992)
13. Huang, Q., Lloyd, E.: Cost Constrained Fixed Job Scheduling. In: 8th Italian Conference on Theoretical Computer Science, pp. 111–124 (2013)
14. Williamson, D.P., Shmoys, D.B.: The Design of Approximation Algorithms. Cambridge University Press, New York (2011)

15. Scholl, A.: Balancing and Sequencing of Assembly Lines. Springer, Heidelberg (1999)
16. Martello, S., Toth, P.: A heuristic approach to the bus driver scheduling problem. Eur. J. Oper. Res. **24**(1), 106–117 (1986)
17. De Leone, R., Festa, P., Marchitto, E.: A Bus Driver Scheduling Problem: a new mathematical model and a GRASP approximate solution. J. Heuristics **17**(4), 441–466 (2011)
18. Garey, M.R., Johnson, D.S.: Computers and Intractability: A Guide to the Theory of NP-Completeness. W.H. Freeman and Company, New York (1979)
19. Cabrera, G.G., Rubio, J. M. L.: Hybrid algorithm of tabu search and integer programming for the railway crew scheduling problem. In: 2th Asia-Pacific Conference on Computational Intelligence and Industrial Applications, pp. 413–416 (2009)
20. Wren, A., Rousseau, J.M.: Bus Driver Scheduling - An Overview. School of Computer Studies Research Report Series, University of Leeds, vol. 93(31), pp. 1–14 (1993)
21. Portugal, R., Lourenco, H.R., Paixao, J.P.: Driver scheduling problem modelling. Public Transp. **1**(2), 103–120 (2008)
22. Rodrigues, M.M., de Souza, C.C., Moura, A.V.: Vehicle and crew scheduling for urban bus lines. Eur. J. Oper. Res. **170**(3), 844–862 (2006)
23. Mastelic, T., Fdhila, W., Brandic, I., Rinderle-Ma, S.: Predicting resource allocation and costs for business processes in the cloud. In: 11th World Congress on Services, pp. 47–54 (2015)

A Study on Travel Time Stochasticity in Service Network Design with Quality Targets

Giacomo Lanza[1(✉)], Teodor Gabriel Crainic[2,3], Walter Rei[2,3], and Nicoletta Ricciardi[1,2]

[1] Department of Statistical Sciences, Sapienza University of Rome, Rome, Italy
giacomo.lanza@uniroma1.it
[2] Interuniversity Research Centre on Enterprise Networks, Logistics and Transportation (CIRRELT), Montreal, Canada
[3] Department of Management and Technology, Université du Québec à Montréal, Montreal, Canada

Abstract. The scope of this paper is to advance the investigation into the importance of introducing uncertainty in service network design (SND) formulations by examining the uncertainty of travel times, a phenomenon that has been little studied up to now. The topic of our research thus is the stochastic scheduled service network design problem with service-quality targets and uncertainty on travel times, an important problem raising in the tactical planning process of consolidation-based freight carriers. Quality-service targets relate to the on-time operation of services and delivery of commodity flows to destinations. The problem is formulated as a two-stage mixed-integer linear stochastic model defined over a space-time network, with service targets modelled through penalties. Its aim is to define a cost-efficient transportation plan such that the chosen quality-service targets are respected as much as possible over time. An extensive experimental campaign is proposed using a large set of random generated instances with the scope of enhancing the understanding of the relations between the characteristics of a service network and its robustness, in terms of respect of the service schedule and delivery due dates, given business-as-usual fluctuations of travel times. Several analyses are reported identifying the features that appear in stochastic solutions to hedge against or, at least, reduce the bad effects of travel time uncertainty on the performance of a service network.

Keywords: Stochastic service network design
Stochastic travel time · Quality targets

Supported by the Ministero dell'Istruzione, dell'Università e della Ricerca (MIUR) of Italy, through its Research Projects of Relevant National Interest (PRIN) program, the Sapienza Università di Roma, Italy, through its Progetto di Ateneo La Sapienza, and the Natural Sciences and Engineering Council of Canada (NSERC), through its Discovery Grant program.

R. Cerulli et al. (Eds.): ICCL 2018, LNCS 11184, pp. 401–416, 2018.
https://doi.org/10.1007/978-3-030-00898-7_27

1   Introduction

Freight transportation operates in a highly competitive, cost and quality-of-service driven environment. In order to meet market requests and still make a profit, carriers need to minimize the costs of their services establishing sets of operating policies to perform the routing of commodity flows and the management of available resources (both human and material) in the most rational and profitable way. We focus on consolidation-based, long-haul freight transportation carriers, where the loads of different demands are grouped, loaded and moved in the same vehicle or convoy for all or part of their itineraries from origins to destinations. To achieve consolidation and servicing many different customers simultaneously with the same vehicles, carriers need to plan a set of regular transportation services, between terminals in their network, operated according to a particular schedule, which is repeated for a certain period of time, e.g., a weekly schedule repeated for six months of a so-called season. This is a rather complex tactical-planning problem that is traditionally addressed through a scheduled service network design (SSND) methodology.

SSND aims to produce the set of scheduled services, together with planned routes for the demand (services used and terminals passed through), to achieve the economic and quality targets of the carrier. The latter normally concern the reliability of service operations with respect to the published schedule and of freight deliveries with respect to promised due dates. While there is quite a body of literature on SSND models for consolidation-based transportation, few address quality-target issues, and even fewer account for fluctuations in travel times and the resulting delays and reliability breach, with monetary and possible market-loss consequences. According to our best knowledge, [12] were the first to jointly address the design of an efficient service network and the consideration of travel-time uncertainty impacting its reliability. The authors proposed a two-stage mixed-integer linear stochastic model over a space-time network. The first stage addresses the selection of services and the routing of freight flows. Service targets are modelled through penalties and addressed in the second stage, where penalties are assigned to late services and deliveries. The authors also proposed an heuristic method to address large instances.

What is still lacking, however, is a deep study focusing on the relations between the characteristics of a service network and its robustness in terms of observance of service schedules and delivery due dates, given business-as-usual fluctuations of travel times. Our objective is to fill this gap. Main questions we explore in our work: What is gained by integrating information about the stochastic nature of travel times directly into the tactical planning methodology? Are different patterns, either in the service selection or in the freight itineraries, suggested when such information is integrated into the model? Is the resulting transportation plan actually more robust with respect to travel time fluctuations? What characteristics are more important in producing such a robustness?

To perform this study, we considered a basic version of the problem in which periodic schedules are built for a number of vehicles and where only travel times vary stochastically. In order to obtain results with the lowest bias possible, we focused on optimal solutions only, for both the deterministic and the stochastic formulations. For this reason, we chose problem sizes allowing the use of standard mixed-integer software (see [12] for an approach able to address instances that cannot be directly addressed by the solver). An extensive experimental campaign was performed using a large set of random generated instances. The analysis and comparison of the stochastic and deterministic solutions provided the mean to identify characteristics that appear to hedge against or, at least, reduce the bad effects of travel time uncertainty on the performance of a service network.

The plan of the paper is as follows. We state the problem in Sect. 2. We briefly recall the stochastic formulation in Sect. 3. The experimental setting, including instance and scenario generation procedures, are reported in Sect. 4. Computational results are presented in Sect. 5. Conclusions and future research paths are discussed in Sect. 6.

2   Problem Description

We briefly recall the main elements of tactical planning and SND for consolidation-based freight carriers; for more detailed explanations we refer to [1,4,6] for rail transportation, [3] for maritime transportation, [7,8,10] for land-based long-haul transportation, and [9] for intermodal transportation.

Carriers operate over a physical network of uni or intermodal terminals connected by infrastructure (rail, road) or conceptual (navigation) links. They set up and exploit transportation services, according to a given schedule, to satisfy the regular *demand*. Each demand, or commodity, requires the transportation of a certain *quantity* of freight from an *origin* terminal, available at a certain *availability date*, to be delivered at a *destination* terminal by a required *due date*. Each service is characterized by its *origin* and *destination* terminals, its *schedule*, i.e., the departure time at origin, the departure and arrival times at intermediate stops (if any), and the arrival time at destination, as well as a number of characteristics, e.g., its capacity. To take advantage of economies of scale, the loads of different demands are *consolidated*, loaded together, into the same vehicles. Freight may thus be moved by a sequence of services between its origin and destination, undergoing consolidation (accompanied possibly by loading/unloading) and service-to-service transfer operations at intermediate terminals.

Tactical planning determines the *transportation* (or load) *plan* to be operated for a given medium-term planning horizon (typically six months to a year). The plan specifies the service network with its schedule, the itineraries of each demand within the service network, as well as the operations to be performed at each terminal. *Scheduled Service Network Design (SSND)* supports this planning phase. SSND model takes the form of a fixed-cost, capacitated, time-dependent network design formulation, whose aim is to select the services, and thus the schedule, to make up a cost-efficient service network satisfying the forecast regular demand. Besides the economic efficiency and profitability of its operations, the carrier is also aiming for service reliability in terms of performing according to the schedule and the due dates established with the customers. Carriers will thus often internally set up certain targets of on-time operations, which generally reflect trade-offs between operating costs and the estimated market impact of service performance. Simultaneously, customer contracts may carry penalties for late deliveries and carriers aim to avoid them. We model two types of *quality targets*, the *service target* and the *demand target*, as the minimum degree of conformity to the schedule and the contracted due dates for demand, respectively.

We are aware of very few contributions in the literature addressing the integration of SSND and quality targets. The vast majority of proposed SSND formulations assume deterministic travel times. Yet, it is well known that time fluctuations and delays occur even in the most tightly operated systems due to congestion conditions, adverse weather, etc. We thus proposed the Stochastic Service Network Design Problem with Service Quality Targets (SSND-QST) integrating travel-time uncertainty and service targets into a SSND model, such that targets and the extra costs of undesired delays are accounted for when selecting the service network and the demand itineraries.

The goal of this paper is to verify the worthiness of such a formulation. [13] were the first to address the problem of critically compare the performances of a deterministic and stochastic - in terms of demand - service network design, highlighting the role of consolidation not only as a powerful mean to lower costs, but also to hedge against demand uncertainty. Following their contribution, our research has the same scope of finding insights and characteristics that may define robustness for a service network. We define a schedule as being increasingly more robust, the more cost effectively it deals with varying travel times, hence the lower expected costs it leads to. Specifically, in the present case this means the ability to set up a transportation plan able to be as much as possible congruent with the quality targets desired and promised by the carrier. We address a basic version of the problem: all services are of the same type in terms of speed, priority, and capacity; service time at terminals is deterministic; services may arrive early at a stop, at no additional cost, but have to wait for service until the scheduled time; services may arrive late, in which case, terminal operations begin immediately and connections are not missed. The problem setting we consider aims to determine the "best" transportation plan given a set of possible services, with their respective *normal* travel times (that is, smooth operations without undue delays), the carrier may operate, without recourse to spot transportation.

3   Model Formulation

Similar to many SSND problems, we modelled the dynamics of the SSND-QST through a *space-time network*, discretizing the schedule length into a fixed number of time periods of equal length. Demand is represented by a set of commodities, each requiring the transport of a certain volume from an origin to a destination according to its entry and due dates. A set of potential services that the carrier may use is available. Each service has a capacity, a route in the physical network, specifying the set of consecutive terminals visited between its origin and destination, and timing information indicating the normal (ideal conditions without any delay) departure time at origin, the normal arrival time at destination, as well as the normal arrival/departure times at the other terminals visited. A *fixed* selection (operation) cost is associated to each potential service and a unit commodity *transportation* cost is associated to each commodity.

A normal travel time and a travel-time random variable are associated to each service leg (a segment between two consecutive stops) of each service. Actual travel times are observed at each period. This information must then be translated into the actual arrival times at destinations, which means that the delays incurred by services and demand-flow delays are observed only when services complete their movement. The service design and routing decisions cannot be changed at that time, but penalties, if any, have to be paid.

The model then takes the form of a two-stage stochastic optimization formulation with simple recourse. The selection of services and the routing of freight decisions are made in the first-stage. Quality targets are expressed through penalties on lateness and added to the objective function. Second-stage variables define the time instant at which a service ends its movement on a given service leg and the time instant at which a commodity arrives at its destination. Lateness of a service is considered as soon as the observed arrival time at a stop exceeds the usual arrival time for that stop; lateness of a commodity is considered as soon as the observed arrival time at destination exceeds its promised due date. Notice that, the lateness of a service at a particular stop does not necessarily imply the transported demand is also late as, e.g., the demand could have been shipped in advance with respect to its due date. Service and demand quality targets must thus be computed separately. The selection of services and the routing of freight thus aim to minimize the fixed service-selection and variable demand-routing cost, plus the expected penalty costs of the chosen plan given travel-time uncertainty. Uncertainty is approximated through a set of scenarios. The second-stage function depends on both design and routing decisions, as well as on the realizations of the random variables expressed through the scenarios. Traditional constraints are then considered, that is, commodity flow conservation, linking-capacity, non-negativity and binary constraints. The complete description and mathematical formulation of the model is available in [12].

4   Experimental Plan

We performed three sets of experiments, named *Evaluation Analysis*, *Structural Comparison*, and *Comparative Analysis*, focused on highlighting differences in reliability, costs and structural complexity between stochastic and deterministic solutions. Deterministic and stochastic mixed-integer linear programming models were implemented in OPL language. Experiments were conducted on an Intel Xeon X5675 computer with 3.07 GHz and 48 GB of RAM. *CPLEX* 12.6 (IBM ILOG, 2016) was used to obtain solutions.

4.1   Instances and Scenario Generation

We considered a physical service network inspired by the one in [5], consisting of 5 physical nodes and 10 physical arcs and shown in Fig. 1(a). The service network is defined for a schedule length of 15 periods and displays a cyclic nature [5], as illustrated in Fig. 1(b).

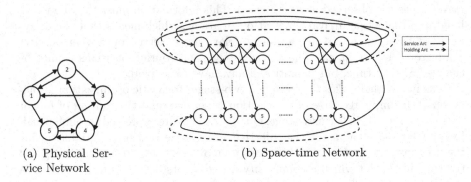

(a) Physical Ser-
vice Network

(b) Space-time Network

Fig. 1. Graph representation

We considered 6 *demand classes*, defined by number of commodities and the time available to deliver them. Three *levels of demand* are taken into account. Levels 1 to 3 consider 15, 20, and 25 commodities, respectively. Two values were considered for the *delivery-time windows*, loose (*l*), with due dates between 11 and 14 periods after the availability date (given a schedule length of 15 periods), and tight (*t*) with due dates between 9 and 12 periods after the availability dates. The demand classes are represented by $DClass(\cdot\cdot)$ with the respective values for these two attributes in the following tables.

The potential service network is the same for all instances, with 150 direct services and 7 one-stop services. We generated services with normal duration of 3, 4, and 5 periods. The fixed cost of a direct service is proportional to its normal duration, while, for a service with intermediary stop, it is 35% less than the cost of the two direct services one would need for the same path.

We modeled travel-time variations through the *Truncated Gamma* (*TG*) class of probability distributions [2], which allowed us to control the main elements defining the travel time: normal value, variability, and range, defined as the difference between the maximum and minimum travel times possible on the arc; the former stands for the worst case outside of highly hazardous major disturbances and catastrophic events, while the latter corresponds to the free running time of a service under perfect conditions. The *TG* distribution, increasing rapidly to the value of the normal travel time, followed by a gradual decrease to the maximum travel-time value (i.e., tail skewed to the right), also captures the observed phenomena of delays occurring much more frequently than early arrivals, with delay lengths generally "not too far" from the normal travel time.

Scenario generation was performed by sampling random values from a TG distribution with particular values for its mode, variance, and range of travel times. The mode of a service (service leg) is its normal duration. Twelve scenario classes, *SClass*, were generated by considering four variability levels and three ranges. The former were measured in terms of standard deviation, low for level 1, medium for level 2, and high for level 3. Level 4 considers a mixed case, where the lowest variability level is assigned to a subset of physical arcs and the highest to the remaining ones. We set the same lower bound for all cases, but varied the range by using three different upper bounds. We defined a tight range, t, computed as the mode -30% of a time-period duration, medium, m, computed as mode plus one time period, and loose, l, computed as medium plus 30% of a time period. The looser the range, the wider is the concept of "normal" travel time. Scenario classes, *SClass*($\cdot\cdot$), are thus identified by the pair *level of variability* (1, 2, 3 or 4) and *range* (t, m, or l).

Experiments were performed under three levels of increasing penalty costs for each of the two service targets, on-time arrival and maximum acceptable delay. For services, the first level of on-time-arrival penalty was set to 175% of the cost of the most expensive service, while the first level of the maximum-delay penalty was set to 215% of the same value. The second and third levels were obtained by doubling and tripling these values, respectively. A similar process was performed for the demand targets, where the on-time-arrival penalty was set to the cost of the most expensive service, while the first level of the maximum-delay penalty was set to 175% of the same value. To address the single-target formulations, we set the penalties to 0 for the target not considered in that experiment.

5   Experimental Results and Analysis

All the analyses were performed considering the 6 demand classes derived by the combined use of the 3 levels of demand and the 2 delivery-time windows. We generated 10 instances for each demand class, for a total of 60 deterministic instances. For each deterministic instance, 36 stochastic instances were constructed, combining the first 4 levels of variability, 3 ranges, and 3 penalty rates.

A solution, whether for the deterministic variant (SSND) or for the stochastic one, consists of a set of selected services and the paths used to transport commodity flows to their destinations. Solutions were found to the three stochastic formulations considering a set of 30 scenarios after having verified *in-sample* and *out-of-sample* stability [11] (more details on this analysis are available in [12]). Stochastic solutions are identified in the following as SSND-QST for the complete formulation, SSND-QST-S and SSND-QST-D when only the service or the demand targets, respectively, were considered.

5.1   Evaluation Analysis: Benefits of Stochastic Formulation

The purpose of this analysis is to quantify the benefits of explicitly considering time-stochasticity into the model rather than using a traditional deterministic assumption. It is composed of two parts. The first is a comparison of set-up costs (service selection plus routing costs) and full costs (the set-up costs plus the penalties incurred for delays) of a network. The second focuses on the estimation of the observed delay probability distributions. The latter as well as the full costs are estimated through a Monte Carlo-like simulation procedure by considering a set of 100 scenarios.

Table 1 displays the average set-up and full costs for instances belonging to the third demand class that were solved with increasing level of variability and the highest penalty level. The SSND always yields the same service network configuration, no matter how variable are travel times distributions. The stochastic SSND-QST-S set-up costs are generally similar to those of the corresponding deterministic SSND, but the structure of the service network is markedly different. As we will see in next section, in almost all cases, less services operate in SSND-QST-S than in SSND and multi-stop services are replaced by direct services. This trend is more present as the variability increases, resulting in a slight increase of activation costs. SSND-QST-D networks appear to be built to bring commodity flows as early as possible to destination, at least one period before due date. Such a behaviour requires more services to be selected and leads to an increase in set-up costs with respect to SSND. SSND-QST-S and SSND-QST-D display full costs that are always lower than the cost of the corresponding SSND. This shows that explicitly considering the stochastic nature of the travel times in the tactical planning model may hedge against, or at least reduce, the effects and consequences of uncertainty, despite an initial higher set-up cost. The general SSND-QST yields service designs that appear as a compromise between SSND-QST-D and SSND-QST-S, displaying both the lowering-service and early-freight-arrival trends. It is the service targets, however, that influence the SSND-QST the most.

Table 1. Cost of SSND, SSND-QST-S, SSND-QST-D and SSND-QST

	DClass(3t)											
	SSND		SSND-QST-S		SSND		SSND-QST-D		SSND		SSND-QST	
	Set-up	Full	Set-up	Full	Set-up	Full	Set-up	Full	Set-up	Full	Set-up	Full
SClass(1m)	6341	18221	6326	17871	6341	11068	6345	8647	6341	22949	6347	20246
SClass(2m)	6341	39440	6331	36835	6341	21332	6351	13389	6341	54431	6356	44871
SClass(3m)	6341	50362	6334	46531	6341	25200	6352	15616	6341	69221	6366	57512
	DClass(3l)											
	SSND		SSND-QST-S		SSND		SSND-QST-D		SSND		SSND-QST	
	Set-up	Full	Set-up	Full	Set-up	Full	Set-up	Full	Set-up	Full	Set-up	Full
SClass(1m)	6695	21550	6685	20636	6695	14804	6697	10186	6695	29659	6703	24390
SClass(2m)	6695	51720	6705	42864	6695	28410	6715	17215	6695	73435	6742	54049
SClass(3m)	6695	65478	6714	54481	6695	33396	6717	20736	6695	92180	6765	67839

We now focus on the delays observed over the set of 100 scenarios through simulation. We report a number of statistics to evaluate and compare the delay distribution performance of deterministic and stochastic solutions, the latter belonging to the third demand class with the highest level of variability and the highest penalty level. The average observed delay (as well as the average observed short and long delay), the minimum and maximum observed delay (bringing also to the range of the distribution), and standard deviation and 3rd quartile, as measures of dispersion, are reported in Table 2 for SSND, SSND-QST-S, SSND-QST-D and SSND-QST. For the latter case, service and commodity delays are summed together.

Given how the model was formulated, not surprisingly, the average delays of stochastic solutions are always lower than SSND. The higher delay reduction always belongs to the longest type (which is also the most penalized). SSND-QST, SSND-QST-S and SSND-QST-D always outperform SSND in all the chosen statistics, defining observed delay distributions which are consistently better than SSND in terms of range and dispersion of observations. In all cases, the third quartile of stochastic observed delay distributions is lower than the average observed delay of SSND. This defines a set of positive-skewed (that is, the mass of the distribution is concentrated on the left of the mean), shifted on the left with shorter and less dense tails stochastic observed delay distributions compared to the deterministic ones. Figure 2(a), (b) and (c) respectively show the above described distribution for SSND-QST-S, SSND-QST-D and SSND-QST compared to the observed delay probability distributions for SSND.

Table 2. Monte Carlo simulation delay analysis

	Observed average delay			Delay distribution dispersion indexes			
	Tot delay	Short	Long	Min delay	Max delay	St. dev.	3rd quartile
SSND-QST-S	8.78	4.08	4.70	2.03	19.26	2.97	10.51
SSND	13.79	7.48	6.31	3.66	25.11	3.19	11.83
SSND-QST-D	3.60	1.62	1.98	0.04	11.20	2.05	4.06
SSND	6.36	1.91	4.45	1.23	13.41	2.11	7.12
SSND-QST	12.21	5.62	6.58	2.37	25.11	4.06	14.23
SSND	17.03	6.27	10.76	5.68	36.26	5.06	18.81

5.2   Structural Analysis: Reducing Delay Risk Techniques

The purpose of this analysis is to identify the features that stochastic solutions exploit to hedge against time uncertainty.

The SSND displays, in general, characteristics typical of consolidation-based transportation networks, where different commodities share the capacity of single services for most of their journeys, passing through several intermediary stops, where they often wait idle, before arriving at destination. One also observes just-in-time arrivals, with respect to due dates, of freight at destination. Furthermore, one-stop services are usually favoured when possible, rather than no-stop services in order to lower the fixed costs.

In almost all cases, less services operate in SSND-QST-S than in SSND, even though the two solutions share part of them. The most remarkable feature relates to the decrease of multi-stop services activation, which are the most sensitive to risk of delays. Thus, if a service experiences a delay in its first leg, it will most likely arrive at destination (its second stop) later than scheduled, unless, in the second leg, the observed travel time is much lower than normal and absorbs the delay. Given the assumed distributions, complete absorption is not very likely and one-stop services have a higher risk of paying for delays. Consequently, the model would move the solution away from less-expensive multi-stop services to more expensive direct connections, lowering the risk of extra costs when the services operate. The observed trend of SSND-QST-S solutions is thus to select only the strictly necessary direct services to fulfill demand by replacing multi-stop services with direct services. As fewer services are available, commodity paths will be more tangled and involve more services and transfers, the latter implying additional idle time at intermediary terminals. SSND-QST-D networks are built to bring commodity flows as early as possible to destination, at least one period before due date. When avoiding just-in-time arrival is not possible for the total quantity of a commodity, the flow is sometimes split and a major part is shipped in advance. Such a behaviour requires, generally, a higher number of services (and higher set-up costs, as seen) compared to SSND. Table 3 displays

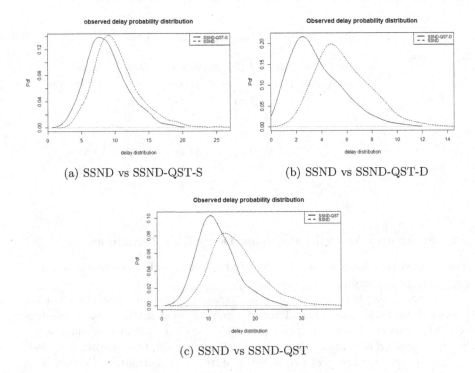

(a) SSND vs SSND-QST-S   (b) SSND vs SSND-QST-D

(c) SSND vs SSND-QST

Fig. 2. Delay probability distributions

the average number of direct and multi-stop services activated in SSND-QST-S and in SSND and the percentage amount of early and just-in-time freight arrivals in SSND-QST-D and in SSND for the same demand classes, scenario classes, and penalty level considered above.

Table 3. Trends characteristic in SSND-QST-S and SSND-QST-D

| | SSND-QST-S | | | | | | SSND-QST-D | | | | | |
| | PClass-3t | | | PClass-3l | | | PClass-3t | | | PClass-3l | | |
	Tot. serv	Direct	Not direct	Tot. serv	Direct	Not direct	Tot. serv	Early	Just-in-time	Tot. serv	Early	Just-in-time
SClass-1m	31.3	28.7	2.6	37.4	33.8	3.6	32.4	67.3	32.7	39.8	51.6	48.4
SClass-2m	31.1	28.4	2.5	37	33.7	3.3	32.9	70.5	29.5	39.7	55.3	44.7
SClass-3m	30.5	28.1	2.4	36.6	33.7	2.9	32.9	71.3	28.7	39.7	58	42
SDM	31.4	28.4	3	39.5	35.5	4	31.4	53.6	46.4	39.5	44	56

When both targets are simultaneously considered, the same not-direct-services and early-freight-arrivals oriented trends are observed. Nevertheless, the coexistence of these two components cause changing in the network at a slower rate when compared to SSND-QST-S or SSND-QST-D (see Table 4).

Table 4. Trends characteristic in SSND-QST

| | PClass-3t | | | | | PClass-3l | | | | |
	Tot. serv	Direct	Not direct	Early	Just-in-time	Tot. serv	Direct	Not direct	Early	Just-in-time
SClass-1m	31.2	28.6	2.6	66.7	33.3	38.6	35.2	3.4	53.3	46.7
SClass-2m	31.4	28.8	2.6	68.2	31.8	37.6	34.6	3	53.3	46.7
SClass-3m	31.2	28.7	2.5	70	30	36.8	33.9	2.9	55.6	44.4
SDM	31.4	28.4	3	53.6	46.4	39.5	35.5	4	44	56

It is clear that such features are fostered only by the stochastic formulation of the problem and would have never been found and exploited with a traditional time-deterministic formulation. How do these features eventually change network design configurations? Figure 3 displays an example of how SSND and SSND-QST-S differ from each other. Dashed arrows represent multi-stop services while solid arrows stand for direct services. The amount of commodity shipped is depicted on each service arc (three commodities are considered, differentiated by underlines). In the SSND two multi-stop services are activated. In SSND-QST-S the multi-stop services are avoided and replaced by either their parallel direct services or by services traveling on a complete different route. In fact, the multi-stop service traveling from hub 1 to hub 3 passing through hub 2 is replaced by two direct services, the first traveling from hub 1 to hub 5 the second from hub 5 to hub 3.

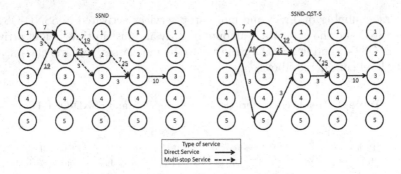

Fig. 3. SSND vs. SSND-QST-S

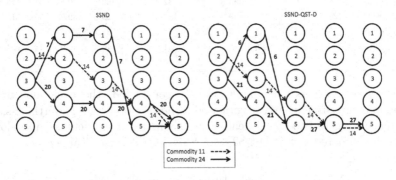

Fig. 4. SSND vs. SSND-QST-D

Figure 4 displays, instead, the paths of commodity 11 (dashed arcs) and commodity 24 (solid arcs). The amount of freight shipped on each arc is reported. In SSND-QST-D, both commodities are shipped well in advance following the same physical paths as SSND, but shifted one period before. In this specific example, if a delay is observed in the last segment of commodity paths of SSND, it would involve the 82% of its total amount, as opposed to SSND-QST-D, where it will be 0 thanks to the early-freight-arrivals trend.

5.3 Comparative Analysis: Impact of Parameters

The goal of the last analysis is to investigate how the values of the parameters of the stochastic model may change the performance of stochastic solutions. Solutions are thus obtained by varying one of the parameters at a time, keeping all the others fixed.

We first consider the impact of the *amplitude of delivery-time windows*, which plays an important role when demand-target is considered. Consider the case depicted in Fig. 5. The origin in space and time of a load is represented by vertex 1. We compare two cases; in the first, the due date is right after the availability date, while in the second, the due date is after 2 periods with respect to the availability date, respectively vertices 3 and 4. Two parallel potential services are available to ship the load, labelled service 1 and 2 with the same cost of activation. In the first case, no other possibility than service 1 may be considered in both SSND-QST-D and SSND. The commodity leaves immediately its origin and is shipped with service 1 to its destination, vertex 3, without any idle time. In case two, SSND-QST-D will always choose service 1 in order to consistently be on time and not pay additional penalty costs (taking advantage of one period of idle from 3 to 4), which is not guaranteed if service 2 would be chosen. In a deterministic setting, service 1 would never be favoured with respect to 2. Privileging service 1 instead of 2 is thus a feature displayed only by the stochastic formulation of the problem. This feature however is strictly dependent on the time amplitude between entry and due dates of commodities. The narrower are availability and due dates the less the model is able to build a robust service network.

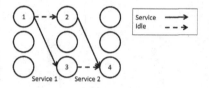

Fig. 5. Impact of delivery-time window

The *amplitude of the penalties* for late arrival also directly influences both SSND-QST-D and SSND-QST-S designs. The penalties represent the need for reliability: the higher they are, the more reliability is requested. As a consequence, the higher the level of penalties, the more the model aims to build a service network that will perform as planned when travel times vary. Table 5 displays the increase in set-up costs and decrease in total delay, in percentages, for solutions obtained with penalty levels 2 and 3, compared to solutions obtained with penalty level 1. The same demand and scenario classes considered in the previous experiments were also used here. Focusing on the SSND-QST-S, the main delay decrease concerns the long and most expensive delay. The higher the penalty, the lower are such delays. Increasing penalties threefold yields an increase in the fixed cost of the network by around 0.03%, with a decrease in the amount of total delay of around 3% for the short delays and 10% of the long ones. Similar results were observed for SSND-QST-D. The percentage of early freight arrivals is increased by 8% when the solutions based on the highest and lowest penalty levels are compared (total number of commodities is 25). The total amount of delay decreases of around 8% for the short delays and 19% of the long ones, at the expense of additional set-up cost of 0.05% only. Therefore, in response to a small increase of initial costs, high benefits may be observed in terms of reliability. Table 5 displays this analysis.

The *level of variability* influences SSND-QST-D and SSND-QST-S designs as well. Figure 6(a) show the percentage of direct and multi-stop activated services

Table 5. Penalty increase on SSND-QST-S and SSND-QST-D behaviour

	SSND-QST-S			SSND-QST-D		
	Fixed cost	Short delay	Long delay	Fixed cost	Short delay	Long delay
Penalty 1	6656.8	4046	2404.2	7323.9	916.1	799.5
Penalty 2	+0.01%	−1,06%	−5,66%	+0.01%	−3,26%	−9,99%
Penalty 3	+0.03%	−3,38%	−9,95%	+0.05%	−7,48%	−18,36%

when the level of variability increases from level 1 to 3. Figure 6(b) shows instead the percentage of commodities delivered just-in-time and at least one period before under the same condition (commodities delivered at destination through a service arc are just-in-time deliveries). The higher the variability level the more the direct-service and early-freight-arrival trends are observed.

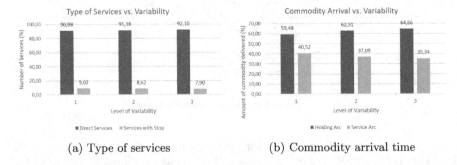

(a) Type of services   (b) Commodity arrival time

Fig. 6. Variability effects

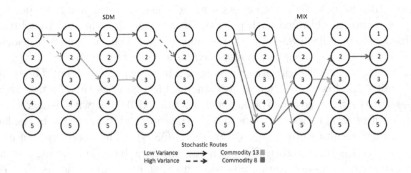

Fig. 7. SDM vs. MIX case

We also considered a mixed-level variability case. The travel time probability distributions of physical links connecting vertex 1 to vertex 2 and vice versa have a high variability (level 3), while the remaining physical arcs a low one (level 1). Figure 7 shows how the structure of the shipment plan may change.

The routes of two commodities are shown. In the SSND case, the services travelling along the more risky link 1–2 are used, since they establish a faster connection between those vertices. As opposed, in the MIX case, they are totally avoided and commodities are shipped through more tangled but also safer against high delays paths.

6 Conclusion

We proposed a study of the relations between the characteristics of a service network and its robustness in terms of respect of the service schedules and delivery due dates, given business-as-usual fluctuations of travel times. Very few papers in the literature address issues related to service network design and stochastic time, our contribution being, according to our best knowledge, the first to clearly address the issue of identifying the features a service design must have to gain in reliability. Such features define structurally different designs showing characteristics that a deterministic model would typically not produce. Several interesting research avenues are open. The introduction of uncertainty on terminal operations, an integration in a unique formulation of both demand and time uncertainty, the representation of more complex decisions/actions when delays are observed, addressing, for example, the case of missed connections are few examples of interesting paths to explore.

References

1. Assad, A.A.: Models for rail transportation. Transp. Res. Part A Policy Pract. **14**, 205–220 (1980)
2. Chapman, D.G.: Estimating the parameters of a truncated gamma distribution. Ann. Math. Stat. **27**, 498–506 (1956)
3. Christiansen, M., Fagerholt, K., Ronen, D.: Ship routing and scheduling: status and perspectives. Transp. Sci. **38**(1), 1–18 (2004)
4. Cordeau, J.F., Toth, P., Vigo, D.: A survey of optimization models for train routing and scheduling. Transp. Sci. **32**(4), 380–404 (1998)
5. Crainic, T.G., Hewitt, M., Toulouse, M., Vu, D.M.: Service network design with resource constraints. Transp. Sci. **50**(4), 1380–1393 (2014)
6. Crainic, T.G.: Rail tactical planning: issues, models and tools. In: Bianco, L., Bella, A.L. (eds.) Freight Transport Planning and Logistics, pp. 463–509. Springer, Heidelberg (1988). https://doi.org/10.1007/978-3-662-02551-2_16
7. Crainic, T.G.: Network design in freight transportation. Eur. J. Oper. Res. **122**(2), 272–288 (2000)
8. Crainic, T.G.: Long-haul freight transportation. In: Hall, R.W. (ed.) Handbook of Transportation Science, 2nd edn., pp. 451–516. Kluwer Academic Publishers, Norwell (2003)
9. Crainic, T.G., Kim, K.H.: Intermodal transportation. In: Barnhart, C., Laporte, G. (eds.) Transportation, Handbooks in Operations Research and Management Science, Chap. 8, vol. 14, pp. 467–537. North-Holland, Amsterdam (2007)
10. Crainic, T.G., Roy, J.: O.R. tools for tactical freight transportation planning. Eur. J. Oper. Res. **33**(3), 290–297 (1988)

11. Kali, P., Wallace, S.W.: Stochastic Programming. Springer, Heidelberg (1994). https://doi.org/10.1007/978-3-642-88272-2
12. Lanza, G., Crainic, T.G., Rei, W., Ricciardi, N.: Service network design problem with quality targets and stochastic travel times. CIRRELT-2017-71 (2017)
13. Lium, A.G., Crainic, T.G., Wallace, S.W.: A study of demand stochasticity in service network design. Transp. Sci. **43**(2), 144–157 (2009)

Improved Fully Polynomial Approximation Schemes for the Maximum Lateness Minimization on a Single Machine with a Fixed Operator or Machine Non-Availability Interval

Imed Kacem[1]([⊠])[ID] and Hans Kellerer[2][ID]

[1] Université de Lorraine, LCOMS EA 7306, 57000 Metz, France
imed.kacem@univ-lorraine.fr
[2] Institut für Statistik und Operations Research, University of Graz,
Universitätsstraße 15, 8010 Graz, Austria
hans.kellerer@uni-graz.at
http://lcoms.univ-lorraine.fr, http://www.uni-graz.at

Abstract. In this paper we consider the single machine scheduling problem with one non-availability interval to minimize the maximum lateness where jobs have positive tails. Two cases are considered. In the first one, the non-availability interval is due to the machine maintenance. In the second case, the non-availibility interval is related to the operator who is organizing the execution of jobs on the machine. The contribution of this paper consists in an improved FPTAS for the maintenance non-availability interval case and its extension to the operator non-availability interval case. The two FPTASs are strongly polynomial and outperform the recent ones by Kacem, Kellerer and Seifaddini presented in [12].

Keywords: Scheduling · Approximation Schemes · FPTAS
Maximum lateness minimization · Single machine
Non-availability interval · Dynamic programming

1 Introduction

In this paper we investigate some improvements to the previous work by Kacem, Kellerer and Seifaddini [12]. We consider the single machine scheduling problem with one non-availability interval to minimize the maximum lateness where jobs have positive tails. Two cases are considered. In the first one, the non-availability interval is due to the machine maintenance. In the second case, the

Supported by the LCOMS EA 7306, a research unit of the Université de Lorraine, and by the University of Graz.

non-availability interval is related to the operator who is organizing the execution of jobs on the machine. An operator non-availability period is a time interval in which no job can start, and neither can complete. The main difference between machine non-availability (MNA) and operator non-availability (ONA) consists in the fact that a job can be processed but cannot start neither finish during the ONA period. However, the machine non-availability interval is a completely forbidden period. Rapine *et al.* [18] have described the applications of this problem in the planning of a chemical experiments as follows: Each experiment is performed by an automatic system (a robot), during a specified amount of time, but a chemist is required to control its start and completion. At the beginning, the chemist launches the process (preparation step). The completion step corresponds to the experimental analysis, which is to be done in a no-wait mode to stop chemical reactions. Here, the automatic system is available all the time, where the chemists may be unavailable due to planned vacations or activities. This induces operator (chemist) non-availability intervals when experiments (jobs) can be performed by the automatic system (machine), but cannot neither start nor complete.

The study of this family of scheduling problems has been motivated by different applications in logistics. First, the production scheduling by integrating the different sources of non-availability (machine and/or operator sources) is an important application (see for example, Brauner *et al.* [1], Rapine *et al.* [18]). Moreover, the maximum lateness minimization in scheduling theory is known as equivalent to the maximum delivery date minimization, where the delivery times can be seen as transportation durations (see for example, Carlier [2], Kacem and Kellerer [9], Dessouky and Margenthaler [4]). Other online applications can be found in Kacem and Kellerer [11].

2   Related Works

The MNA case of this type of problems has been studied in the literature under various criteria (a sample of these works includes Lee [15], Kacem [8], Kacem *et al.* [13], Kubzin and Strusevich [14], Qi *et al.* [16,17], Schmidt [19], He *et al.* [6]). However, few papers studied the problem we consider in this paper. Lee [15] explored the Jackson's sequence and proved that it is a 2-approximation. Recently, Yuan *et al.* developed an interesting PTAS (Polynomial Approximation Scheme) for the studied problem [21]. Kacem [8] presented a first Fully Polynomial Time Approximation Scheme (FPTAS) for the maximum lateness minimization. It is well-known that an FPTAS is the best possible approximation scheme for an NP-hard problem, unless $P = NP$ (see for example, Kacem and Kellerer [10], Gens and Levner [5], Ibarra and Kim [7], Sahni [20]). That is why this paper is a good attempt to design more efficient approximation heuristics and approximation schemes to solve the studied problem.

For the ONA case, few works have been published. Brauner *et al.* [1] considered the problem of single machine scheduling with ONA periods. They analyzed this problem on a single machine with the makespan as a minimization criterion and they showed that the problem is NP-hard with one ONA period. They also considered the problem with K ONA periods such that the length of each ONA period is no more than $\frac{1}{\lambda}$ times the total processing time of all jobs. They introduced a worst-case ratio smaller than $1 + \frac{2K}{\lambda}$ for the so-called algorithm LS (list scheduling). They presented an approximation algorithm with a worst-case ratio close to $2 + \frac{K-1}{\lambda}$. The natural case of periods where the duration of the periods is smaller than any processing time of any job, has been considered by Rapine *et al.* [18]. They proved that the problem can be solved in polynomial time if there is only one ONA period. It was shown that the problem is NP-hard if one has $K \geq 2$ small non-availability periods and the worst-case ratio of LS is no more than $\frac{K+1}{2}$ and the problem does not admit an FPTAS for $K \geq 3$ unless P = NP.

Recently, Chen *et al.* [3] considered the single machine scheduling with one ONA period to minimize the total completion time. The problem is NP-hard even if the length of the ONA period is smaller than the processing time of any job. They have also presented an algorithm with a tight worst-case ratio of $\frac{20}{17}$. They showed that the worst-case ratio of SPT is at least $\frac{5}{3}$.

For more details, the previous paper by Kacem, Kellerer and Seifaddini [12] contains an overview on these problems.

3   Contributions

The contribution of this paper consists in an improved FPTAS for the maintenance non-availability interval case and its extension for the ONA interval case. The two FPTASs are strongly polynomial and they have reduced time complexities compared to [12]. These contributions are summarized in Table 1 for both cases.

This note is organized as follows. Section 4 recalls the exact formulation of the maintenance non-availability interval case and the improved FPTAS. Section 5 is devoted to the extension to the operator non-availability interval case and to the presentation of the associated FPTAS. Finally, Sect. 6 concludes the paper.

Table 1. Summary of results

	Result	Reference
MNA	FPTAS: $O(n log(n) + \min\{n, 1/\varepsilon\}^3/\varepsilon^2)$	Kacem et al. [12]
ONA	FPTAS: $O((n/\varepsilon)log(n) + n\min\{n, 1/\varepsilon\}^3/\varepsilon^3)$	Kacem et al. [12]
MNA	FPTAS: $O(n log(n) + \min\{n, 1/\varepsilon\}^2/\varepsilon^2)$	This paper
ONA	FPTAS: $O((n/\varepsilon)log(n) + n\min\{n, 1/\varepsilon\}^2/\varepsilon^3)$	This paper

4   Case Under MNA Interval

The considered problem $(\mathcal{P})$ can be formulated as follows. We have to schedule a set J of n jobs on a single machine, where every job j has a processing time

p_j and a tail q_j (or delivery time). The machine can process at most one job at a time and it is unavailable between T_1 and T_2 (i.e., (T_1, T_2) is a forbidden interval). Preemption of jobs is not allowed (jobs have to be performed under the non-resumable scenario). All jobs are ready to be performed at time 0. With no loss of generality, we consider that all data are integers and that jobs are indexed according to Jackson's rule [15] (i.e., jobs are indexed in nonincreasing order of tails). Therefore, we assume that $q_1 \geq q_2 \geq \ldots \geq q_n$. Let $C_j(S)$ denote the completion time of job j in a feasible schedule S for the problem and let $\varphi_S(\mathcal{P})$ be the maximum lateness (or the delivery date) yielded by schedule S for instance $\mathcal{I}$ of $(\mathcal{P})$:

$$\varphi_S(\mathcal{I}) = \max_{1 \leq j \leq n} (C_j(S) + q_j) \tag{1}$$

The aim is to find a feasible schedule S by minimizing the maximum lateness. We also denote by $\varphi^*(\mathcal{I})$ the minimal maximum lateness for instance $\mathcal{I}$. Due to the dominance of Jackson's order, an optimal schedule is composed of two sequences of jobs scheduled in nondecreasing order of their indexes [15]. If all the jobs can be inserted before T_1, the instance studied $(\mathcal{I})$ has obviously a trivial optimal solution obtained by Jackson's rule. We therefore consider only the problems in which all the jobs cannot be scheduled before T_1. Moreover, we consider that every job can be inserted before T_1 (i.e., $p_j \leq T_1$ for every $j \in J$). It is useful to recall that Lee [15] explored the Jackson's sequence JS and proved that its deviation to the optimal maximum lateness cannot exceed the largest processing time, which is equivalent to state that JS is a 2-approximation.

4.1 The Improved Procedure

The proposed FPTAS is based on the modification of the one proposed in [12] by Kacem, Kellerer and Seifaddini.

First, as described in [12], we use the simplification technique based on merging small jobs proposed in [9]. We simplify the input instance $\mathcal{I}$ as follows. Given an arbitrary $\varepsilon > 0$, with the assumption that $1/\varepsilon$ is integer, we split the interval $[0, \max_{j \in J}\{q_j\}]$ in $1/\varepsilon$ equal length intervals and we round up every tail q_j to the next multiple of $\varepsilon q_{\max}$ ($q_{\max} = \max_{j \in J}\{q_j\}$). The new instance is denoted as $\mathcal{I}'$. Then, J is divided into at most $1/\varepsilon$ subsets $J(k)$ ($1 \leq k \leq 1/\varepsilon$) where jobs in $J(k)$ have identical tails of $k\varepsilon q_{\max}$. The second modification consists in reducing the number of small jobs in every subset $J(k)$. Small jobs are those having processing times less than $\varepsilon P/2$ where $P = \sum_{j=1}^{n} p_j$. The reduction is done by merging the small jobs in each $J(k)$ so that we obtain new greater jobs having processing times between $\varepsilon P/2$ and εP. The small jobs are taken in the order of their index in this merging procedure. At most, for every subset $J(k)$, a single small job remains. We re-index jobs according to nondecreasing order of their tails. The new instance we obtain is denoted as $\mathcal{I}''$. Clearly, the number of jobs remaining in the simplified instance $\mathcal{I}''$ is less than $3/\varepsilon$. These reductions are recalled for self-consistency and their details are available in Kacem et al. [12].

It is worthy to note that such reductions cannot increase the optimal solution value of $\mathcal{I}$ too much and they can be done in linear time.

We apply a modified dynamic programming algorithm DP_ε to instance $\mathcal{I}''$ using the Jackson's sequence JS to obtain an upper bound for the maximum lateness. The main idea of DP_ε is to remove a special part of the states generated by a dynamic programming algorithm. Therefore, the modified algorithm becomes faster and yields an approximate solution instead of the optimal schedule. First, we define the following parameters:

$$\overline{n} = \min\{n, 3/\varepsilon\},$$

$$\omega_1 = \left\lceil \frac{4\overline{n}}{\varepsilon} \right\rceil,$$

$$\omega_2 = \left\lceil \frac{2}{\varepsilon} \right\rceil,$$

$$\delta_1 = \frac{\varphi_{JS}\left(\mathcal{I}''\right)}{\omega_1}$$

and

$$\delta_2 = \frac{T_1}{\omega_2}.$$

We split $[0, \varphi_{JS}\left(\mathcal{I}''\right))$ into ω_1 equal subintervals $I_m^1 = [(m-1)\delta_1, m\delta_1)_{1 \leq m \leq \omega_1}$. We also split $[0, T_1)$ into ω_2 equal subintervals $I_s^2 = [(s-1)\delta_2, s\delta_2)_{1 \leq s \leq \omega_2}$ of length δ_2. Moreover, we define the two singletons $I_{\omega_1+1}^1 = \{\varphi_{JS}\left(\mathcal{I}''\right)\}$ and $I_{\omega_2+1}^2 = \{T_1\}$. Our algorithm DP_ε generates reduced sets $\mathcal{X}_j^{\#}$ of states $[t, f]$ where t is the total processing time of jobs assigned before T_1 in the associated partial schedule and f is the maximum lateness of the same partial schedule. It is described in Algorithm 1.

4.2   Algorithm DP_ε is an Improved FPTAS

Compared to the previous FPTAS presented in [12], our new algorithm keeps two approximate states in every box $I_m^1 \times I_s^2$ ($1 \leq m \leq \omega_1+1, 1 \leq s \leq \omega_2+1$) instead of a single approximate state. As a consequence, the loss in terms of variable t will be reduced as it can be shown in the proof of the following theorem. Thus, the interval length δ_2 is taken larger compared to [12]. Moreover, in the following proof we will use a tighter recursive relation on the closeness of the approximate states and those originally generated by the standard dynamic algorithm. As a result, we will show that the new algorithm DP_ε outperforms the one provided in [12] by a linear factor in n or $1/\varepsilon$.

Theorem 1. *Given an arbitrary $\varepsilon > 0$, the modified algorithm DP_ε yields an output $\varphi_{DP_\varepsilon}\left(\mathcal{I}''\right)$ such that:*

$$\varphi_{DP_\varepsilon}\left(\mathcal{I}''\right) - \varphi^*\left(\mathcal{I}''\right) \leq \varepsilon\varphi^*\left(\mathcal{I}''\right). \tag{2}$$

Algorithm 1. The new proposed FPTAS DP_ε

The inputs of the algorithm are: ε, T_1, T_2, $\overline{n}$ and simplified instance $\mathcal{I}''$. The algorithm returns a feasible schedule with a maximum lateness value less or equal to $(1+\varepsilon)\varphi^*(\mathcal{I}'')$.

    i. set $\mathcal{X}_1^\# = \{[0, T_2 + p_1 + q_1], [p_1, p_1 + q_1]\}$.

    ii. For $j \in \{2, 3, \ldots, \overline{n}\}$,

$$\mathcal{X}_j^\# = \emptyset.$$

For every state $[t, f]$ in $\mathcal{X}_{j-1}^\#$:

1) Put $\left[t, \max\left\{f, T_2 + \sum_{i=1}^{j} p_i - t + q_j\right\}\right]$ in $\mathcal{X}_j^\#$

2) Put $[t + p_j, \max\{f, t + p_j + q_j\}]$ in $\mathcal{X}_j^\#$ if $t + p_j \leq T_1$

Remove $\mathcal{X}_{j-1}^\#$

Let $[t, f]_{m,s}$ and $[u, g]_{m,s}$ be the states in $\mathcal{X}_j^\#$ such that $f, g \in I_m^1$, $t, u \in I_s^2$ and t and u are respectively the smallest and the greatest possible values in subinterval I_s^2.

Set $\mathcal{X}_j^\# = \left\{[t, f]_{m,s}, [u, g]_{m,s} \mid 1 \leq m \leq \omega_1 + 1, 1 \leq s \leq \omega_2 + 1\right\}$.

    iii. $\varphi_{DP_\varepsilon}(\mathcal{I}'') = \min_{[t,f]\in\mathcal{X}_{\overline{n}}^\#}\{f\}$.

Proof. First, we recall the idea of the dynamic programming algorithm [8] which is necessary to explain the proof. Indeed, the problem can be optimally solved by applying the following dynamic programming algorithm DP. This algorithm generates iteratively some sets of states. At every iteration j, a set $\mathcal{X}_j$ composed of states is generated ($1 \leq j \leq \overline{n}$). Each state $[t, f]$ in $\mathcal{X}_j$ can be associated to a feasible schedule for the first j jobs. Variable t denotes the completion time of the last job scheduled before T_1 and f is the maximum lateness of the corresponding schedule. This dynamic programming is given in Algorithm 2.

Algorithm 2. The standard dynamic programming DP [8]

The inputs of the algorithm are: T_1, T_2, $\overline{n}$ and simplified instance $\mathcal{I}''$. The algorithm returns a schedule with an optimal maximum lateness value $\varphi^*(\mathcal{I}'')$.

    (i). Set $\mathcal{X}_1 = \{[0, T_2 + p_1 + q_1], [p_1, p_1 + q_1]\}$.

    (ii). For $j \in \{2, 3, \ldots, \overline{n}\}$,

$$\mathcal{X}_j = \{\}.$$

For every state $[t, f]$ in $\mathcal{X}_{j-1}$:

1) Put $\left[t, \max\left\{f, T_2 + \sum_{i=1}^{j} p_i - t + q_j\right\}\right]$ in $\mathcal{X}_j$

2) Put $[t + p_j, \max\{f, t + p_j + q_j\}]$ in $\mathcal{X}_j$ if $t + p_j \leq T_1$

Remove $\mathcal{X}_{j-1}$

    (iii). $\varphi^*(\mathcal{P}) = \min_{[t,f]\in\mathcal{X}_{\overline{n}}}\{f\}$.

Let $UB = \varphi_{JS}(\mathcal{T}'')$ be an upper bound on the optimal maximum lateness for problem $(\mathcal{T}'')$ obtained by Jackson's sequence. We add the restriction that for every state $[t, f]$ the relation $f \leq UB$ must hold.

The main idea of the FPTAS is to remove a special part of the states generated by the dynamic programming algorithm. Therefore, the modified algorithm DP_ε becomes faster and yields an approximate solution instead of the optimal schedule. The worst-case analysis of our FPTAS is based on the comparison of the execution of algorithms DP and DP_ε, which can be summarized by the following relations. For every state $[t, f]$ in $\mathcal{X}_j$ there exists a state $[t^\#, f^\#]$ in $\mathcal{X}_j^\#$ such that:

$$t - \delta_2 \leq t^\# \leq t \tag{3}$$

and

$$f^\# \leq f + \delta_2 + j\delta_1 \tag{4}$$

The two relations can be proved by induction on j.

First, for $j = 1$ we have $\mathcal{X}_1^\# = \mathcal{X}_1$. Therefore, the statement is trivial. Now, assume that the statement holds true up to level $j - 1$. Consider an arbitrary state $[t, f] \in \mathcal{X}_j$. Algorithm DP introduces this state into $\mathcal{X}_j$ when job j is added to some feasible state for the first $j-1$ jobs. Let $[t', f']$ be the above feasible state. Two cases can be distinguished: either $[t, f] = [t' + p_j, \max\{f', t' + p_j + q_j\}]$ or $[t, f] = \left[t', \max\left\{f', T_2 + \sum_{i=1}^{j} p_i - t' + q_j\right\}\right]$ must hold. For proving the statement for level j we will distinguish two cases.

Case 1: $[t, f] = [t' + p_j, \max\{f', t' + p_j + q_j\}]$
Since $[t', f'] \in \mathcal{X}_{j-1}$, there exists $[t'^\#, f'^\#] \in \mathcal{X}_{j-1}^\#$ such that $t' - \delta_2 \leq t'^\# \leq t'$ and $f'^\# \leq f' + \delta_2 + (j - 1)\delta_1$.

Consequently, the state $[t'^\# + p_j, \max\{f'^\#, t'^\# + p_j + q_j\}]$ is feasible (since $t'^\# + p_j \leq t' + p_j = t \leq T_1$) and it is generated by Algorithm DP_ε at iteration j. However it may be removed when reducing the state subset. Let $[\lambda, \mu]$ and $[\alpha, \beta]$ be the two possible states in set $\mathcal{X}_j^\#$ that remain in the same box as the state $[t'^\# + p_j, \max\{f'^\#, t'^\# + p_j + q_j\}]$ (with $\lambda \leq t'^\# + p_j \leq \alpha$). Hence, we have two situations to consider: $t' + p_j \geq \alpha$ or $t' + p_j < \alpha$.

Subcase 1.a: $t' + p_j \geq \alpha$. In this subcase, we can verify that the state $[\alpha, \beta]$ fulfills (3). Indeed, $\alpha \leq t' + p_j = t$ and by definition $\alpha \geq t'^\# + p_j \geq t' - \delta_2 + p_j = t - \delta_2$. Thus, we have

$$t - \delta_2 \leq \alpha \leq t.$$

Subcase 1.b: $t' + p_j < \alpha$. In this subcase, we can verify that the state $[\lambda, \mu]$ fulfills (3). Indeed, $\lambda \leq t'^\# + p_j \leq t' + p_j = t$ and by definition $\lambda \geq \alpha - \delta_2 > t' + p_j - \delta_2 = t - \delta_2$. Thus, we have

$$t - \delta_2 \leq \lambda \leq t.$$

On the other hand, the two values μ and β are in the same subinterval as the value $\max\{f'^\#, t'^\# + p_j + q_j\}$. Therefore, the kept state will have a maximum lateness value less or equal to $\max\{\mu, \beta\}$. Then, we conclude that

$$
\begin{aligned}
\max\{\mu, \beta\} &\leq \max\left\{f'^{\#}, t'^{\#} + p_j + q_j\right\} + \delta_1 \\
&\leq \max\left\{f' + \delta_2 + (j-1)\,\delta_1, t' + p_j + q_j\right\} + \delta_1 \\
&\leq \max\left\{f', t' + p_j + q_j\right\} + \delta_2 + j\delta_1 \\
&= f + \delta_2 + j\delta_1.
\end{aligned}
$$

Consequently, the statement holds for level j in this case.

Case 2: $[t, f] = \left[t', \max\left\{f', T_2 + \sum_{i=1}^{j} p_i - t' + q_j\right\}\right]$

Since $[t', f'] \in \mathcal{X}_{j-1}$, there exists $\left[t'^{\#}, f'^{\#}\right] \in \mathcal{X}_{j-1}^{\#}$ such that $t' - \delta_2 \leq t'^{\#} \leq t'$ and $f'^{\#} \leq f' + \delta_2 + (j-1)\,\delta_1$.

Consequently, the state $\left[t'^{\#}, \max\left\{f'^{\#}, T_2 + \sum_{i=1}^{j} p_i - t'^{\#} + q_j\right\}\right]$ is generated by Algorithm DP_ε in iteration j. However, it may be removed when reducing the state subset. Let $[\lambda', \mu']$ and $[\alpha', \beta']$ be the states in $\mathcal{X}_j^{\#}$ that are kept in the same box as $[t'^{\#}, \max\{f'^{\#}, T_2 + \sum_{i=1}^{j} p_i - t'^{\#} + q_j\}]$ and having $\lambda' \leq t'^{\#} \leq \alpha'$. Again, we have two situations to be considered: $t' \geq \alpha'$ or $t' < \alpha'$.

Subcase 2.a: $t' \geq \alpha'$. In this subcase, we can verify that the state $[\alpha', \beta']$ fulfills (3). Indeed, $\alpha' \leq t' = t$ and by definition $\alpha' \geq t'^{\#} \geq t' - \delta_2 = t - \delta_2$. Thus, we have

$$
t - \delta_2 \leq \alpha' \leq t.
$$

Subcase 2.b: $t' < \alpha'$. In this subcase, we can verify that the state $[\lambda', \mu']$ fulfills (3). Indeed, $\lambda' \leq t'^{\#} \leq t' = t$ and by definition $\lambda' \geq \alpha' - \delta_2 > t' - \delta_2 = t - \delta_2$. Thus, we have

$$
t - \delta_2 \leq \lambda' \leq t.
$$

On the other hand, the values μ' and β' are in the same subinterval as $\max\{f'^{\#}, T_2 + \sum_{i=1}^{j} p_i - t'^{\#} + q_j\}$. Therefore, the kept state will have a maximum lateness value less or equal to $\max\{f'^{\#}, T_2 + \sum_{i=1}^{j} p_i - t'^{\#} + q_j\} + \delta_1$. Moreover,

$$
\max\left\{f'^{\#}, T_2 + \sum_{i=1}^{j} p_i - t'^{\#} + q_j\right\} + \delta_1 \leq \max\{X, Y\} + \delta_1
$$

where $X = f' + \delta_2 + (j-1)\,\delta_1$ and $Y = T_2 + \sum_{i=1}^{j} p_i - t' + \delta_2 + q_j$. Thus,

$$
\max\{\mu', \beta'\} \leq \max\left\{f', T_2 + \sum_{i=1}^{j} p_i - t' + q_j\right\} + \max\{\delta_2 + j\delta_1, \delta_2 + \delta_1\}
$$

$$
\leq f + \delta_2 + j\delta_1.
$$

In conclusion, the statement holds also for level j in the second case, and this completes our inductive proof. Now, we give the proof of Eq. (2). By definition, the optimal solution can be associated to a state $[t^*, f^*]$ in $\mathcal{X}_{\bar{n}}$. From Eq. (4), there exists a state $[t^\#, f^\#]$ in $\mathcal{X}_{\bar{n}}^\#$ such that:

$$f^\# \leq f^* + \delta_2 + \bar{n}\delta_1$$
$$= f^* + \frac{T_1}{\omega_2} + \bar{n}\frac{\varphi_{JS}(\mathcal{I}'')}{\omega_1}$$
$$= f^* + \frac{T_1}{\lceil\frac{2}{\varepsilon}\rceil} + \bar{n}\frac{\varphi_{JS}(\mathcal{I}'')}{\lceil\frac{4\bar{n}}{\varepsilon}\rceil}$$
$$\leq f^* + \varepsilon\frac{T_1}{2} + \varepsilon\frac{\varphi_{JS}(\mathcal{I}'')}{4}$$
$$\leq (1+\varepsilon)\,\varphi^*(\mathcal{I}'').$$

Since $\varphi_{DP_\varepsilon}(\mathcal{I}'') \leq f^\#$, we conclude that Eq. (2) holds.

It can be easily seen that the proposed modified algorithm DP_ε can be implemented in $O\left(\bar{n}\log\bar{n} + \bar{n}^2/\varepsilon^2\right)$ time. The schedule obtained by DP_ε for instance $\mathcal{I}''$ can be easily converted into a feasible one for instance $\mathcal{I}$. This can be done in $O(n)$ time.

To summarize, we conclude that Algorithm DP_ε is an FPTAS and it can be implemented in $O\left(n\log n + \min\{n, 1/\varepsilon\}^2/\varepsilon^2\right)$ time.

5   Consequences: An Improved FPTAS for the ONA Case

Here, the studied problem (Π) can be formulated as follows. An operator has to schedule a set J of n jobs on a single machine, where every job j has a processing time p_j and a tail q_j. The machine can process at most one job at a time if the operator is available at the starting time and the completion time of such a job. The operator is unavailable during (T_1, T_2). Preemption of jobs is not allowed (jobs have to be performed under the non-resumable scenario). All jobs are ready to be performed at time 0. Without loss of generality, we consider that all data are integers and that jobs are indexed according to Jackson's rule. The aim is to find a feasible schedule S by minimizing the maximum lateness. Again, if all the jobs can be inserted before T_1, the instance studied $(\mathcal{I})$ has obviously a trivial optimal solution obtained by Jackson's rule. We therefore consider only the problems in which all the jobs cannot be scheduled before T_1. Moreover, we consider that every job can be inserted before T_1 (i.e., $p_j \leq T_1$ for every $j \in J$).

As it has been done in Kacem et $al.$ [12], an FPTAS can be established for Π. The procedure is based on guessing the so-called $straddling$ job and its starting time from a finite set of approximate values, which leads to $O(\frac{n}{\varepsilon})$ auxiliary MNA problems $(\mathcal{P})$. Thus, the application of DP_ε to all these auxiliary problems can lead to an improved FPTAS as the following theorem claims:

Theorem 2. *Problem Π admits an FPTAS and this scheme can be implemented in $O(n(\ln n)/\varepsilon + n\min\{n, 3/\varepsilon\}^2/\varepsilon^3)$ time.*

Proof. The proof is a straightforward from [12] and Theorem 1.

6   Conclusion

In this paper we consider the single machine scheduling problem with one non-availability interval to minimize the maximum lateness where jobs have positive tails. Two cases are considered. In the first one, the non-availability interval is due to the machine maintenance. In the second case, the non-availibility interval is related to the operator who is organizing the execution of jobs on the machine. The contribution of this paper consists in an improved FPTAS for the maintenance non-availability interval case and its extension to the operator non-availability interval case. The two FPTASs are strongly polynomial and outperform our previous ones published in the literature.

As a research perspective, we are extending the ideas of this paper in order to improve some existing FPTASs for other scheduling problems.

References

1. Brauner, N., et al.: Operator non-availability periods. 4OR: Q. J. Oper. Res. **7**, 239–253 (2009)
2. Carlier, J.: The one-machine sequencing problem. Eur. J. Oper. Res. **11**, 42–47 (1982)
3. Chen, Y., Zhang, A., Tan, Z.: Complexity and approximation of single machine scheduling with an operator non-availability period to minimize total completion time. Inf. Sci. **251**, 150–163 (2013)
4. Dessouky, M.I., Margenthaler, C.R.: The one-machine sequencing problem with early starts and due dates. AIIE Trans. **4**(3), 214–222 (1972)
5. Gens, G.V., Levner, E.V.: Fast approximation algorithms for job sequencing with deadlines. Discret. Appl. Math. **3**, 313–318 (1981)
6. He, Y., Zhong, W., Gu, H.: Improved algorithms for two single machine scheduling problems. Theor. Comput. Sci. **363**, 257–265 (2006)
7. Ibarra, O., Kim, C.E.: Fast approximation algorithms for the knapsack and sum of subset problems. J. ACM **22**, 463–468 (1975)
8. Kacem, I.: Approximation algorithms for the makespan minimization with positive tails on a single machine with a fixed non-availability interval. J. Comb. Optim. **17**(2), 117–133 (2009)
9. Kacem, I., Kellerer, H.: Approximation algorithms for no idle time scheduling on a single machine with release times and delivery times. Discret. Appl. Math. **164**(1), 154–160 (2014)
10. Kacem, I., Kellerer, H.: Approximation schemes for minimizing the maximum lateness on a single machine with release times under non-availability or deadline constraints. Algorithmica (2018) https://doi.org/10.1007/s00453-018-0417-6
11. Kacem, I., Kellerer, H.: Semi-online scheduling on a single machine with unexpected breakdown. Theor. Comput. Sci. **646**, 40–48 (2016)
12. Kacem, I., Kellerer, H., Seifaddini, M.: Efficient approximation schemes for the maximum lateness minimization on a single machine with a fixed operator or machine non-availability interval. J. Comb. Optim. **32**, 970–981 (2016)
13. Kacem, I., Sahnoune, M., Schmidt, G.: Strongly fully polynomial time approximation scheme for the weighted completion time minimisation problem on two-parallel capacitated machines. RAIRO - Oper. Res. **51**, 1177–1188 (2017)

14. Kubzin, M.A., Strusevich, V.A.: Planning machine maintenance in two machine shop scheduling. Oper. Res. **54**, 789–800 (2006)
15. Lee, C.Y.: Machine scheduling with an availability constraints. J. Glob. Optim. **9**, 363–384 (1996)
16. Qi, X.: A note on worst-case performance of heuristics for maintenance scheduling problems. Discret. Appl. Math. **155**, 416–422 (2007)
17. Qi, X., Chen, T., Tu, F.: Scheduling the maintenance on a single machine. J. Oper. Res. Soc. **50**, 1071–1078 (1999)
18. Rapine, C., Brauner, N., Finke, G., Lebacque, V.: Single machine scheduling with small operator-non-availability periods. J. Sched. **15**, 127–139 (2012)
19. Schmidt, G.: Scheduling with limited machine availability. Eur. J. Oper. Res. **121**, 1–15 (2000)
20. Sahni, S.: Algorithms for scheduling independent tasks. J. ACM **23**, 116–127 (1976)
21. Yuan, J.J., Shi, L., Ou, J.W.: Single machine scheduling with forbidden intervals and job delivery times. Asia-Pac. J. Oper. Res. **25**(3), 317–325 (2008)

Selected Topics in Logistics Oriented Combinatorial Optimization

Smoothing the Outflow of Stock from Picking Lines in a Distribution Centre

Le Roux Visser and Stephan E. Visagie[⊠]

Department of Logistics, Stellenbosch University, Stellenbosch, South Africa
svisagie@sun.ac.za

Abstract. In a typical retail distribution network, stock flows from factories through distribution centres (DCs) to retail outlets. An uneven flow of stock through a distribution network increases the cost of material handling and transportation. These costs increase because additional labour and vehicles must be hired on a temporary basis to handle the peak stock flows. In this paper, real data of a large South African retailer are used to model the scheduling of the order picking operation in the DC. The aim is to smooth the outflow of stock from the DC to lower labour and transport cost that arise from an uneven flow of stock. Two integer programming models are presented and both calculate solutions that can level out the peak stock flows without letting stock run too late. The model that limits the outflow to a predetermined target level, while minimising the delay of stock in the DC is recommended due to model performance and practicality of implementation.

Keywords: Distribution centre · Order picking · Smoothing stock flow

1 Introduction

In a typical distribution network of a large retailer stock flows from factories through distribution centres (DCs) to the retail outlets. Two important costs that are associated with a distribution network are the material handling cost and the transport cost. This study focuses on smoothing the stock flow through the DC over time to lower the handling cost in the DC and the transport cost between the DC and the stores of a retail chain.

This problem arises in one of the largest retail chains in South Africa. The retailer sells predominantly clothing, but also homeware and mobile phones. In the following paragraphs a brief background of the retailer's distribution network is provided to arrive at a more detailed description and scope of the problem considered in this paper.

2 The Distribution Network

The distribution network of the retailer contains the same elements as most retailers [1]. However, unlike most retailers it uses a central planning system to

© Springer Nature Switzerland AG 2018
R. Cerulli et al. (Eds.): ICCL 2018, LNCS 11184, pp. 431–445, 2018.
https://doi.org/10.1007/978-3-030-00898-7_29

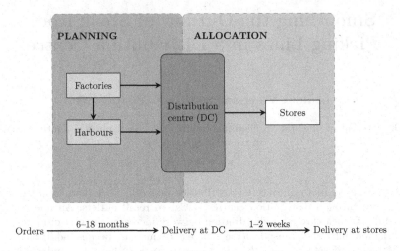

Fig. 1. A graphical representation of the distribution network and its processes.

push stock to stores. Thus, individual stores do not request (or order) stock, but the amount of stock that is pushed to each store is determined by a central office. Figure 1 contains a schematic illustration of the retailer's distribution network.

The retailer has its own factories, but most of its products are imported from the East. These products are then shipped to South Africa, mainly through the Durban harbour (at the east cost of South Africa) to the DC in Durban. In the DC stock is regrouped (by means of a picking process) for shipment to the retail stores. Two processes drive this product flow. A planning process governs the flow from the different factories to the DC. During the planning process forecasts are made about possible sales for expected fashion trends in the upcoming seasons. These plans are then used to determine the correct quantities to order from the factories. This planning process approximately covers a period of 6 to 18 months before the actual inflows into the DC take place. Once the stock arrives in bulk for the entire retailer at the DC, these plans are revised by means of an allocation process. The allocation process refines the values obtained during the forecasts in the planning phase and make an actual allocation of stock to stores. This allocation plan is put in motion by instructions known as DBNs (distributions). The central office thus issues (or releases) a DBN to the DC to execute. A DBN contains a stock keeping unit (SKU) and the number of units of that SKU that should be sent to each of the retail stores. In only one situation it may contain more than one SKU. For clothing that comes in sizes, for example, a specific style of t-shirt, the smalls of this t-shirt will have a different SKU number then the mediums. In cases such as this, there may be more that one SKU in the same DBN. However, SKUs in the same DBN must be the same clothing product that only differ in size. The reason why the different sizes are grouped together within one DBN, is that all the sizes of the same product should be processed together so that all the sizes of the same product reaches the stores at the same time.

A rough layout of the retailer's DC is shown schematically in Fig. 2. Stock arrives at the *Goods received* area. In this area the necessary quality and quantity checks are completed. From here stock can either be moved to the *Storage racks* or the *Full carton* area. Stock that can be sent out to stores in full carton quantities are moved to the *Full carton* area, whereas stock that needs to be picked in smaller quantities for the stores will be stored in the *Storage racks*. When a DBN is released to the DC, stock is moved from the storage racks to a fast picking area, called picking lines. A DBN (in other words, the SKU or SKUs listed in the DBN) is assigned to one picking line. On each picking line the store orders (constructed from the instructions in the DBNs on that picking line) are picked into cartons. The fully picked cartons are sent off to dispatch where cartons from all picking lines are consolidated into loads and then shipped by truck to the stores.

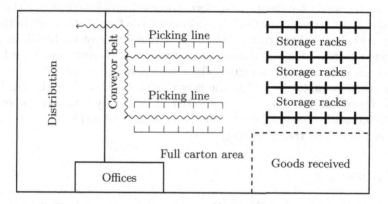

Fig. 2. A schematic representation of the layout of the DC.

Once a DBN is released to the DC, the DC has 7 days to pick (for all the stores listed in that DBN) the SKUs in that DBN and send it to the stores. The central office releases DBNs on a continuous basis. Every day the DC management groups a subset of released DBNs together on open picking lines and do the order picking for those DBNs on those picking lines. Pickers walk in the picking line and manually pick all the stock for all the associated stores on that picking line. A specific product is thus present on only one picking line. It gets picked and is send off to all stores that should receive it. If the order picking on such a line is completed, left-over stock (if any) is removed from the picking line and the picking line is repopulated with a new set of DBNs so that the picking of the new set of DBNs can commence. One such cycle of populating a picking line with stock, picking the stock and clearing the leftover stock is referred to as a wave of picking. More than one picking line operates in parallel. Thus, at any given time there are picking lines that are being populated with stock, there are others that are busy with picking and others that are being cleared out after their wave of picking.

Central planning is divided into teams that specialise in certain product types. For example, there can be a team working on boys shirts, ladies sandals, etc. These different teams send out DBNs independently of each other. This working method results in an uneven release of DBNs to the DC over time. This leads to an uneven work balance in the DC over time as well as an uneven outflow of the volume of stock that must be transported to the stores. These peaks result in additional temporary workers that must be employed to operate night shifts for the duration of the peak. Furthermore, the retailer does not own its own fleet of vehicles and thus hires the trucks to perform the distribution between the DC and the stores. Trucks can be hired on long term contracts (that are cheaper per volume-km[1]) or on short term contracts (that are more expensive per volume-km). These peaks in stock outflow from the DC result in a larger proportion of the fleet that must be hired on the short term contract to handle the peak outflows. A more constant outflow enables the retailer to hire more trucks on the cheaper long term contracts and less trucks on the more expensive short term contracts. Thus smoothing the flow of stock out of the DC will decrease both labour cost (in the form of handling cost in the DC as extra shifts are needed to handle peaks) and transport cost (because more trucks can be hired on the cheaper contracts). The main operation in the DC in terms of labour intensity and determining the flow rate of stock through the DC is the picking operation [3, 10]. The main objective of this study is thus to smooth the flow (in terms of volume) of stock over time through the picking lines in the DC to lower labour and transport cost.

3   Literature

The picking operation described in the previous section may be viewed as a type of forward picking area. Forward picking areas (also known as fast picking areas or primary picking areas) had receive much attention in literature [1–3, 15, 16]. Bartholdi and Hackman [1] devote a full chapter on forward picking areas. This chapter gives a good overview on the topic. Bartholdi and Hackman [1, p. 99] states that the two major issues when designing and operating forward pick areas are (1) which SKUs to store in the fast-pick area and (2) how much of each SKU to store in the forward area? There are numerous variants and approaches in the studies on forward picking areas, but fundamentally these studies boil down to answering these two questions. The objective of these studies is usually to show that the savings in picking costs outweigh the cost of restocking the forward picking area. This approach to forward picking areas are not useful in the picking system described here. The first question is not applicable, because all SKUs are picked in the forward picking area. Similarly, the second question is not applicable as all the units of a SKU is brought to the forward picking area and enough space should be available to take all the stock that are picked. In the system described in this paper the important decision is thus *when* should

---

[1] For the retailer in this study volume, and not weight, determine the capacity of the trucks.

a SKU be brought to the forward picking area opposed to *which* SKUs and how *much* of it, as is the case in literature.

The picking operation considered has three tiers of decisions that must be made when scheduling the waves of picking on the picking lines [10]. The three decisions can be formulated as:

1. Which set of released DBNs should be grouped together to be picked in the next wave of picking?
2. Where in the picking line should the SKUs in the selected DBNs be placed?
3. How should the pickers be routed to minimise the total walking distances?

Each of the three tiers is on its own computationally expensive to solve. All three tiers need to be solved before a wave of picking can commence, hence heuristic approaches are used solve all these tiers [9]. All three decision tiers have one primary objective, namely to minimise the walking distance (and thus time) of pickers to complete a wave of picking.

The third decision tier was the focus of a number of papers [4–6,10,11]. In these papers different approaches to route pickers with the objective to minimise the total walking distance of pickers are presented. These approaches include exact methods [10], tour construction heuristics [4], greedy heuristics [10,11], assignment heuristics [6] and metaheuristics [5].

The second decision tier of the picking system was studied in two papers [7, 13]. Counter to intuition this decision tier proved to have very little effect on the total walking distance of pickers [13]. Thus this tier could be used to achieve secondary objectives without substantially increasing the main objective of total walking distance. The first of the secondary objectives that also has a huge impact on the total picking time, is picker congestion [7]. Other secondary objectives include keeping stock from different departments (for example, men's clothes, woman's clothes etc.) grouped together in the line to limit the unpacking effort in stores. It is also important not to put different sizes of the same product in locations next to each other. It prevents picking errors (a picker may accidentally take the SKUs from the wrong location if neighbouring SKUs only differ in size).

The first decision tier of this picking system was considered in at least two papers [9,12]. The secondary objectives present in the first tier includes the minimisation of largest volume of stock on a picking line over all picking lines (to ensure work balance over all picking lines), the number of small packages coming off a picking line (a large number of small packages increase handling cost), and the total penalty incurred for DBNs scheduled after their out-of-DC date [9]. The models presented in the next section extend the secondary objectives in the first decision tier to include volume smoothing over time to decrease labour cost (to handle peaks in the DC) and the transport cost between the DC and the stores. In essence, the models in this paper can be used to split the first decision tier into two sub-decisions. This paper answer the question of which sub-set of SKUs should picked per day to smooth outflow, while models in Le Roux and Visagie [9] and Matthews and Visaige [12] can then be used to decide how a day's identified sub-set of SKUs should be organised into the picking lines to ensure work balance and minimise the number of small packages. Thereafter the second and third tier problem can be solved.

4   Model

This problem has a dynamic element because DBNs are released on a continuous basis and the selection of DBNs to process today influences the DBNs available for picking tomorrow. The problem thus needs to be solved daily to determine the course of action for that day. Thus any solution approach should be able to solve the problem dynamically. The solution of previous time periods influence the input data for the model for subsequent periods. The modelling approach should thus be able to use the solutions of previous days to incorporate in later time periods.

Two modelling approaches are introduced to smooth the DBN outflow from the DC. Both models are mixed integer programming models that can solve the problem dynamically. Before the models are introduced, the necessary assumptions and notations are introduced.

4.1   Assumptions

The mathematical models presented in this study are approximations of an underlying real life situation. In such mathematical approximations assumptions need to be made to ensure that the models properly represent the real life situation. The main assumptions are listed below:

1. DBNs have no fixed or predetermined priorities in terms of their processing date. In other words there are no DBNs that must be processed more urgently than others. The processing sequence of DBNs may thus be moved around as long as the key performance indicators (KPI) of the DC are met. If an urgent DBN does arise, it can always be handled by assigning an early out-of-DC date. The model will then schedule it as soon as possible to avoid delaying the normal out-of-DC dates.
2. An out-of-DC date can be calculated for all DBNs based on their release date. Currently the out-of-DC date is managed by the retailer to be within the released date of the DBN plus seven days.
3. Any DBNs may be scheduled together on a picking line. There are no products that may not be scheduled on the same picking line.
4. In the situation when a unique clothing product has more than one size, the DBN contains the SKUs of all these sizes (For example, the different sizes of the same style of a t-shirt). In this situation the number units that must be picked in the DBN is the sum of the units over all the SKUs contained within that DBN. Furthermore, it is assumed that the volume of each individual unit within such a DBN is the same, because it is the same product.

4.2   Sets, Parameters and Variables Used in the Models

The following necessary sets, parameters and decision variables must be defined to introduce the mathematical models. The DC plans their picking lines in daily buckets and thus the models should be run at the start of each day to determine the DBNs to process that day. Let $\mathcal{I}_t = \{1, 2, \ldots, i, \ldots, I\}$ be a set of all available DBNs on day t (the day when the model is solved). Available DBNs are all the DBNs that are released, but not yet scheduled on day t. Let $\mathcal{J}_t = \{1, 2, \ldots, j, \ldots, J\}$ be a set of the days over which the model is solved on day t. The number of days J needed to schedule all the available DBNs is determined using the target level and the number of available DBNs on day t. The model will thus find a schedule for all available DBNs at day t, but only the DBNs scheduled for day t (the first day of the solution) are scheduled for picking. The first day's scheduled DBNs are removed from $\mathcal{I}_t$ and the DBNs released on day t are added to $\mathcal{I}_t$, to form the set of available DBNs for day $t+1$, it is $\mathcal{I}_{t+1}$.

The following parameters must be calculated and are needed as input to build and solve the models. Let

m_i be the maximum number of days (remaining from time t when the model is solved) that DBN i may spend in the DC,

q_i be the number of units that must be picked in DBN i,

v_i be the volume of one unit in DBN i,

c_j be the target level for day j, and

b_j be the fraction of the picking shifts available in the DC on day j. For example, on Saturdays and Sundays fewer picking shifts are run than normal weekdays.

In addition, the following variables are used in the models. Let

z_c be the maximum deviation in volume,

z_d be the maximum deviation in number of days,

$x_{ij} = \begin{cases} 1 & \text{if DBN } i \text{ is scheduled on day } j \\ 0 & \text{otherwise,} \end{cases}$

δ_j be the under utilisation of the target level on day j,

γ_j be the over utilisation of the target level on day j,

ρ_i be the number of days DBN i is scheduled before its out-of-DC date, and

η_i be the number of days DBN i is scheduled after its out-of-DC date.

There are two aspects with a trade-off that must be handled by the mathematical models. These two aspects are the desired outflow level (target level) of the DC and the out-of-DC dates of DBNs. All the outflow volumes can be kept on (or below) the desired level but that may imply that some DBNs must be scheduled later than their out-of-DC dates. On the other hand, if all the DBNs are scheduled before their out-of-DC date, it might happen that the volume must be pushed above the target level on certain days. Two models are thus proposed. In the first model the out-of-DC date is fixed and the model then attempts to schedule all DBNs such that the maximum by which the target level is exceeded is minimised—resulting in Model 1: Limit deadline, minimise outflow above target level (LDMO). In the second model the outflow is fixed at less than or equal to the target level and the number of days a DBN is scheduled after its out-of-DC date is minimised—resulting in Model 2: Limit outflow, minimise days late after the out-of-DC date (LOMD).

A third option, where these two quantities (outflow and days delayed) are traded off against each other might also be possible in theory. In practice this trade-off should be calculated in terms of monetary value, since the main objective is to minimise the total logistics cost. Unfortunately, the data to approximate the cost of sending out DBNs after their out-of-DC date or exceeding the desired level of outflow are not available. Moreover, in practice it is easier for managers to either set the lateness and minimise the outflow or set the desired level of outflow and optimise the delay of DBNs.

4.3   Model 1: Limit Deadline, Minimise Outflow Above Target Level

The first model minimises the maximum surplus deviation from the target level determined, whilst making sure DBNs are scheduled after the release dates, but before their out-of-DC date. Mathematically this model may be formulated as

$$\text{minimise} \quad z_c \tag{1}$$

$$\text{subject to} \quad \sum_{j \in \mathcal{J}_t} j x_{ij} \le m_i \qquad i \in \mathcal{I}_t, \tag{2}$$

$$\sum_{i \in \mathcal{I}_t} v_i q_i x_{ij} + \delta_j - \gamma_j = b_j c_j \qquad j \in \mathcal{J}_t, \tag{3}$$

$$\sum_{j \in \mathcal{J}_t} x_{ij} = 1 \qquad i \in \mathcal{I}_t, \tag{4}$$

$$\gamma_j \le z_c \qquad j \in \mathcal{J}_t, \tag{5}$$

$$x_{ij} = 0/1 \qquad i \in \mathcal{I}_t, j \in \mathcal{J}_t, \tag{6}$$

$$\delta_j, \gamma_j \ge 0 \qquad j \in \mathcal{J}_t. \tag{7}$$

The objective function (1) minimises the largest value by which the target level is exceeded. The set of constraints (2) ensures that DBN i is scheduled before its out-of-DC date. The set of constraints (3) calculates the deviation from the target level on day j. The value of b_j could be set to account for days, such as Saturdays, Sundays and public holidays when the DC only functions at a fraction of its capacity. The set of constraints (4) ensures that each available DBN is scheduled exactly once in the planning period. The set of constraints (5) calculates the maximum over deviation from the target level. Constraint set (6) states that x_{ij} is a boolean variable, while constraint set (7) ensures that the continuous variables remain non-negative.

4.4 Model 2: Limit Outflow, Minimise Days Late on Out-of-DC Date

The second model minimises the maximum deviation, in days, that a DBN stays in the DC after its out-of-DC date, while ensuring that the DBNs do not exceed the target level. The objective of this model is to

$$\text{minimise} \quad z_d \tag{8}$$

$$\text{subject to} \quad \sum_{i \in \mathcal{I}_t} v_i q_i x_{ij} \leq b_j c_j \qquad j \in \mathcal{J}_t, \tag{9}$$

$$\sum_{j \in \mathcal{J}_t} j x_{ij} + \rho_i - \eta_i = m_i \qquad i \in \mathcal{I}_t, \tag{10}$$

$$\sum_{j \in \mathcal{J}_t} x_{ij} = 1 \qquad i \in \mathcal{I}_t, \tag{11}$$

$$\eta_i \leq z_d \qquad i \in \mathcal{I}_t, \tag{12}$$

$$x_{ij} = 0/1 \qquad i \in \mathcal{I}_t, j \in \mathcal{J}_t, \tag{13}$$

$$\rho_i, \eta_i \geq 0 \qquad i \in \mathcal{I}_t. \tag{14}$$

The objective function (8) minimises the maximum delay on the out-of-DC date for all DBNs. The set of constraints (9) ensures that the volume of all the DBNs scheduled for day j is less than or equal to the target level on day j. The value of b_j can be used to scale the target level on certain days – as in Model 1. The set of constraints (10) calculates the deviation in days between the scheduled date of DBN i and its out-of-DC date. The set of constraints (11) ensures that all the available DBNs are scheduled exactly once. The set of constraints (12) calculates the maximum deviation (in days) that a DBN is delayed after its out-of-DC date. The set of constraints (13) ensures that all x_{ij} are boolean, while constraint set (14) ensures that the continuous variables remain non-negative.

Both models were implemented in CPLEX [8] and tested (validated and verified) on a small data set over a time period of a month to ensure that it functions correctly and provides the correct solutions.

4.5 Data and Implementation

A raw dataset was supplied by the retailer in .csv format. The dataset contains DBN outflow data from the Durban DC, for a three year period namely from April 2015 to April 2017. The volumes of outflow was masked by multiplying it with a constant factor to not disclose any sensitive information. The data fields of interest contained in the dataset are: dbn no, completed date, created date, release date, scheduled date, product type, units per cubic metre qty, nbr of branches, nbr of skus, send qty, no of repl cycle days. These field names are self-explanatory, however, more information is needed on the product type and no of repl cycle days to understand the results. Four possible products types in a DBN may be specified, namely A,

B, C and N. The `repl cycle days` field contains the respective replenishment cycle (RC) values—but only if the product type is A. There are four possible replenishment cycles, namely RC1, RC2, RC3 and RC4.

When validating the data, the dataset proved to be clean. There were a few instances of missing dates, but could be filled with suitable dates after consulting with the DC management. Furthermore, some of the products of type A had no RC values. In this case an RC0 value was assigned to it. This does not influence the models as the RC values are not used when solving the models. They are only necessary for reporting purposes. Most of the RC data in the dataset were in days, while some were captured as weeks. Therefore, all the RC values were changed to days to ensure uniformity.

Both models were implemented in Python 3.5 [14], using the CPLEX 12.5 [8] and Pandas APIs. The CPLEX API is used to solve the linear programming problems and the Pandas API is used to extract data from MS Excel and store results in MS Excel. Both models were validated and verified on a smaller data set (over a time period of a month) to confirm that the implementation runs correctly and that the model produces accurate results. The steps in the logic flow of the Python code to solve a given model is outlined in Algorithm 1.

Algorithm 1. The logic of the python code to run models

1: **Input:** Parameters, model to use
2: **Output:** Daily schedules of DBNs
3: $t \leftarrow 1$
4: **while** $t \leq T$ **do**
5:     Read input data and build the model
6:     **Call:** CPLEX to solve the model
7:     Read solution output form CPLEX
8:     Append day t's schedule to an output file
9:     Update input data files (day t's solution and released DBNs, and update m_i)
10:     $t \leftarrow t + 1$
11: **end while**

---

5   Results

The target level changes over time because there are natural peaks (like summer holidays) that must be taken into account. After consultation with the DC management it became clear that it would be more practical for the DC to set the target level on a weekly or monthly interval, because on a daily level it changes too often and on an annual level it changes too seldom. The first year's data was thus used to determine average outflows per week (and per month) for the DC to provide estimates for possible values for the target level c_j. All the results reported here use a c_j value that is fixed for either a week or a month, but the models can handle any other interval in multiples of days. The historical outflows for the first quarter of the second year as well as the estimated weekly and monthly capacities are shown in Fig. 3.

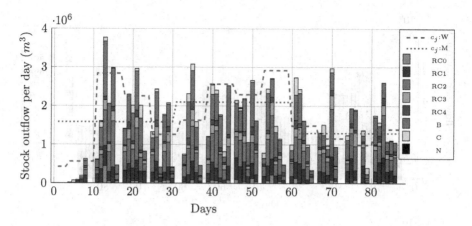

Fig. 3. The historic outflow of DBNs for the first quarter of April 2016–March 2017. RC0 to RC4, B, C and N are the different product types discussed in Sect. 4.5. The calculated weekly target level (c_j:W) and monthly target level (c_j:M) is also shown.

The models are then run to determine a schedule for all DBNs released during the second year. The schedules generated by the models can then be compared to the historical data since exactly the same set of DBNs with the same set of properties are scheduled by all three methods (Historic, Model 1 and Model 2). Although all models were solved for a full year's data, only the first quarter's results are plotted. If more then a quarter's data is plotted it becomes too cluttered and unclear. The remainder of the year's results follow the same pattern as the quarter plotted.

Figure 4 contains calculated outflows for a weekly (at the top) and monthly (at the bottom) target level when Model 1 (LOMC) was applied. On first glance it seems as though the monthly capacities generate a more even outflow than the weekly capacities. This is confirmed when actual deviations are calculated. The standard deviations from the average outflow (over the entire year) are 615 426.7, 568 380.7 and 509 193.6 units for the historical outflows, Model 1 weekly target level and Model 1 monthly target level, respectively. The standard deviation is calculated as the square root of $\sigma^2 = \frac{1}{N-1}\sum_{j=1}^{N}(x_j - \mu)^2$, where N is the number of days (one year in this case), x_j is the outflow for day j and μ is the average outflow calculated over the entire year.

The results when applying Model 2 (LCMD) can be seen in Fig. 5 with a weekly (at the top) and monthly (at the bottom) target level. The same general pattern is present in the results for Model 2 as found in Model 1, namely that the results seem to be smoother in the case of monthly capacities. For Model 2 the standard deviation from the average outflow (over the entire year) is 615 426.7, 601 471.1 and 505 201.1 for the historical, weekly and monthly cases respectively.

Both models solved fast enough to be used in practice. This means that the model can be solved in reasonable time each morning to determine which DBNs should be scheduled on that day.

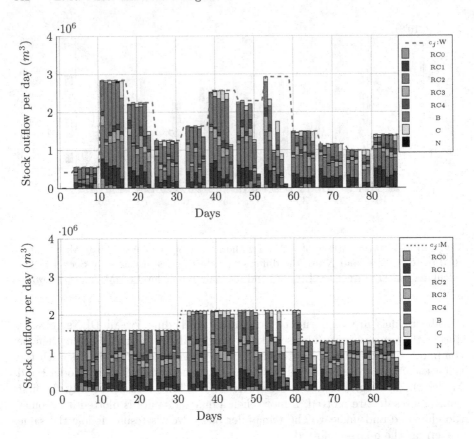

Fig. 4. The outflows calculated with Model 1 for the first quarter of April 2016–March 2017 using a weekly target level (top) and a monthly target level (bottom). RC0 to RC4, B, C and N are the different product types discussed in Sect. 4.5.

Further experiments were performed by setting up different scenarios. One scenario may be when more days (than seven) are allowed to set the out-of-DC date. This allows the models to smooth the outflow even more. For the majority of these scenarios Model 2 (LCMD) with monthly capacities performs the best in terms of standard deviation. After demonstrating these results to the DC management it became clear that Model 2 (LCMD) with monthly capacities would also be the most practical to implement. This is because the level at which the DC can operate is determined by the number of shifts (per day) scheduled for workers in the DC. The DC management can thus take historical outflow for a month to determine the number of shifts needed to set the target level such that the DC will not fall behind on their KPIs and then use Model 2 to smooth the outflow.

This result adds value to the retailer because Model 2 (LCMD) could be used to determine the subset of DBNs to schedule for a given day. Once the subset is identified the methods in Le Roux and Visagie [9] can be used to assign this

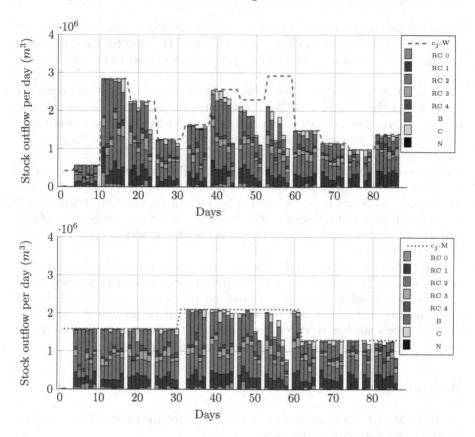

Fig. 5. The results from Model 2 for the first quarter of April 2016–March 2017 calculated for a weekly target level (top) and monthly target level (bottom). RC0 to RC4, B, C and N are the different product types discussed in Sect. 4.5.

subset of DBNs to picking lines to ensure work balance over all the picking lines and minimise the number of small packages produced by the picking lines. The classroom discipline heuristic [7] could then be used to assign SKUs to locations on each picking line to minimise congestion of pickers. Finally, the fast nearest end heuristic [10] could be used to minimise the total walking distance of the pickers. This approach would cover all three decisions tiers discussed in Sect. 3.

6   Conclusions, Recommendations and Future Work

In this paper, a problem of smoothing the outflow of stock from a DC to minimise labour and transport cost was considered. The DC uses a very specific picking line system to perform order picking. This order picking system determines the outflow of stock from the DC. Two models are presented to determine how to schedule stock on the picking lines to smooth the flow of stock over time.

The modelling approaches are to either set the level of outflow and minimise the maximum lateness of stock or to set an out-of-DC date for stock and then minimise the maximum outflow. Both approaches will smooth the outflow over time.

After studying the results and receiving feedback from the DC management it was found that Model 2 (LCMD) with monthly capacities is recommended for two reasons. The first reason is that this model delivers the best results over a wide range of scenarios. The second is that it is easier (than Model 1) to implement from a managerial perspective, because the level at which the DC can function is determined by the number of shifts workers do per day. It is relatively easy to change these shifts on a monthly interval, but not so on a weekly basis. In practice the DC runs either one, two or three eight hour shifts, which for all practical purposes, limits b_j to one of three settings. The retailer is in the process of implementing the recommended model in their warehouse management system.

The real trade-off between the cost of labour, transport and lateness of DBNs are not modelled in this paper. In future studies, approximations for these costs could be developed to determine a trade-off in terms of cost. Estimating or approximating the labour and transport cost might be possible if the correct data becomes available. However, determining the real cost of sending stock one or two days later than planned, would be a very difficult task.

References

1. Bartholdi, J.J., Hackman, S.T.: Warehouse & Distribution science (2017). Release 0.98. www.warehouse-science.com
2. Dallari, F., Marchet, G., Melacini, M.: Design of order picking system. Int. J. Adv. Manuf. Technol. **42**, 1–12 (2009)
3. De Koster, R., Le-Duc, T., Roodbergen, K.J.: Design and control of warehouse order picking: a literature review. Eur. J. Oper. Res. **182**, 481–501 (2007)
4. De Villiers, A.P., Matthews, J., Visagie, S.E.: Tour construction heuristics for an order sequencing problem. S. Afr. J. Ind. Eng. **23**(2), 56–67 (2012). https://doi.org/10.7166/23-3-511
5. De Villiers, A.P., Matthews, J., Visagie, S.E.: Metaheuristic approaches to order sequencing on a unidirectional picking line. ORiON J. Oper. Res. Soc. S. Afr. **29**(1), 31–54 (2013). https://doi.org/10.5784/29-1-152
6. De Villiers, A.P., Visagie, S.E.: Toewysingsheuristieke om die volgorde van bestellings vir 'n uitsoeklyn te bepaal. Litnet Akademies **9**(1), 1–22 (2012)
7. Hagspihl, R., Visagie, S.E.: The number of pickers and SKU arrangement on a unidirectional picking line. S. Afr. J. Ind. Eng. **25**(3), 169–183 (2014). https://doi.org/10.7166/25-3-886
8. IBM ILOG CPLEX Optimization Studio: CPLEX 12.5. http://www-03.ibm.com/software/products/en/ibmilogcpleoptistud. Accessed 5 Feb 2017
9. Le Roux, G.J., Visagie, S.E.: A multi-objective approach to the assignment of stock keeping units to unidirectional picking lines. S. Afr. J. Ind. Eng. **28**(1), 190–209 (2016). https://doi.org/10.7166/28-1-1615

10. Matthews, J., Visagie, S.E.: Order sequencing on a unidirectional cyclical picking line. Eur. J. Oper. Res. **231**(1), 79–87 (2013). https://doi.org/10.1016/j.ejor.2013. 05.011. http://www.sciencedirect.com/science/article/pii/S0377221713004104
11. Matthews, J., Visagie, S.E.: Assignment of stock keeping units to parallel unidirectional picking. S. Afr. J. Ind. Eng. **26**(1), 235–251 (2015). https://doi.org/10. 7166/26-1-907
12. Matthews, J., Visagie, S.E.: SKU assignment using correlations on unidirectional picking lines. ORiON J. Oper. Res. Soc. S. Afr. **31**(2), 61–76 (2015). https://doi. org/10.5784/31-2-531
13. Matthews, J., Visagie, S.E.: SKU arrangement on a unidirectional picking line. Int. Trans. Oper. Res. (2018, in press). https://doi.org/10.1111/itor.12550
14. Python Programming Language: Python 3.5. https://www.python.org/ downloads/release/python-350/. Accessed 15 June 2016
15. Shah, B., Khanzode, V.: Storage allocation framework for designing lean buffers in forward-reserve model: a test case. Int. J. Retail Distrib. Manag. **45**(1), 90–118 (2017)
16. Thomas, L.M., Meller, R.D.: Developing design guidelines for a case-picking warehouse. Int. J. Prod. Econ. **170**, 741–762 (2015). https://doi.org/10.1016/j.ijpe. 2015.02.011

An Effective Structural Iterative Refinement Technique for Solving the Quadratic Assignment Problem

Mehrdad Amirghasemi[1]([⊠])(ID) and Reza Zamani[2](ID)

[1] SMART Infrastructure Facility, Faculty of Engineering and Information Sciences, University of Wollongong, Wollongong, NSW, Australia
`mehrdad@uow.edu.au`
[2] School of Computing and Information Technology, Faculty of Engineering and Information Sciences, University of Wollongong, Wollongong, NSW, Australia

Abstract. The quadratic assignment problem deals with the arrangement of facilities in plants for minimizing the communication cost among the facilities. This problem is one of the focal problems both in academia and industry, absorbing the attention of researchers for more than five decades. Having a variety of applications, this problem has still no effective exact solution strategy, as the number of possible feasible solutions, even for medium-sized problems, is extremely large. This makes effective heuristics as the only viable solution strategy for this problem. In this paper, a technique is presented which aims at achieving local minimization through refining layouts structurally. For this purpose, the technique uses an efficient linear assignment technique, and enhances layouts based on the feedback provided. The results of extensive computational experiments on different benchmark instances indicate that the procedure is both robust and efficient.

Keywords: Facility location problem
Quadratic assignment problem · Iterative refinement

1 Introduction

The quadratic assignment problem (QAP) was first introduced in [17] as the problem of finding the optimal locations of a number of conjoined economical activities. These activities which can be represented with different facilities can be placed in different locations. As a mathematical model for determining the locations of interconnected facilities, the cost in this model is realistically defined as a combination of a linear and a quadratic expression.

The PhD scholarship awarded by SMART infrastructure facility, University of Wollongong, to Dr Mehrdad Amirghasemi facilitated the research reported in this paper.

R. Cerulli et al. (Eds.): ICCL 2018, LNCS 11184, pp. 446–460, 2018.
https://doi.org/10.1007/978-3-030-00898-7_30

The linear expression reflects the fact that for placing each facility at a particular location, a specified cost can be involved. The quadratic expression, on the other hand, reflects the fact that the distance between every two facilities, depending on the flow existing between those two facilities, incurs a cost obtained by the multiplication of the flow and distance.

Assuming there are n facilities to be allocated to n locations, two $n \times n$ matrices comprise the input of any procedure solving the QAP. The first matrix shows the flow between any two facilities, f_{ij}, and the second matrix represents the distance between any two locations d_{kl}. Based on such input, the output of the procedure is a vector π, in which each element shows the location of the corresponding facility, i.e., for the facility i, π_i denotes its location.

The goal is to find a vector π^* which minimizes the sum of all possible distance-flow products, as the weighted traffic volume, shown with $Z(\pi)$, and calculated as follows.

$$Z(\pi^*) = \min_{\pi} Z(\pi) \quad \& \quad Z(\pi) = \sum_{i=1}^{n} \sum_{i=1}^{n} f_{ij} d_{\pi(i)\pi(j)} \tag{1}$$

The procedure presented in this paper, called Progressive Adjusting Structural Solver (PASS) determines the locations of facilities in the QAP based on the gradual refining of arrangements. By moving from arrangement π to π', the traffic volume changes as follows.

$$Z(\pi) - Z(\pi') = \sum_{i=1}^{n} \sum_{i=1}^{n} f_{ij} (d_{\pi(i)\pi(j)} - d_{\pi'(i)\pi'(j)}) \tag{2}$$

The refinement performed by the PASS is based on a linear approximation technique which solves a linear assignment problem. The rationale behind employing a method used in solving the Linear Assignment Problem (LAP) for solving the QAP is twofold.

First, with respect to assigning facilities to locations, the QAP has some similarity with the LAP, which assigns operators to tasks. Second, whereas the best solution strategies to solve the LAP are quite efficient and can solve problems with thousands of tasks in a matter of milliseconds, there is no efficient exact solution strategy for solving the QAP, and ironically, some QAPs with only 40 facilities are still outside the current reach of exact methods.

The PASS starts with a construction method which creates an initial layout (π) and since initial layouts closer to the optimal layout (π^*) have a better chance of convergence to the optimum, the PASS further refines the solution produced by the construction method through a local search. This can bridge part of the gap between the initial and optimal layouts and make the repeated applications of the linear assignment technique more effective.

The rest of the paper is organized as follows. Section 2 presents the related work. Section 3 describe the PASS. Section 4 is devoted to the computational experiments, and the concluding remarks are provided in Sect. 5, which also outlines several directions for further research.

2 Related Work

The term "assignment problem" (AP) was originally coined in [34], and the first efficient solution methodology for the AP, known as Hungarian method, was developed in [18]. Among many of its variations, the most important variation of the AP is the QAP, which originally has been introduced in [17].

The AP is about choosing a one-to-one matching between two sets of items, each with the size of n, to minimize the total cost of the assignment. With considering pseudo items, the AP can cover cases where the sets of items have different sizes. Unlike for the QAP, which is an NP-hard problem [27], the solution methodologies for the AP are polynomial and can solve very large problems to optimality in a matter of seconds.

Advanced optimization techniques have solved some of the challenging instances of the QAP to optimality, mainly through integrating parallel processing with branch and bound algorithms [8] as well as grid computing [2]. Exact methods for the QAP have been extensively surveyed in [20], and two powerful exact methods not included in that survey are those presented in [35, 36].

Construction methods, limited enumeration techniques, metaheuristics, genetic algorithms, and hybrids are among major heuristics used to tackle the QAP. Whereas metaheuristics mainly comprise tabu searches and simulated annealing algorithms, construction methods are categorized into simple greedy algorithms, greedy randomized adaptive search mechanisms, and ant systems. Each of these categories of solution strategies has been widely explored and several techniques have been provided.

Among construction methods, ant systems can be considered as an effective strategy and several efficient ant systems presented for the problem include [6, 9, 14, 21, 29, 31, 32]. With respect to metaheurstics, the simulated annealing approaches presented in [4, 7, 23] and tabu searches presented in [3, 10, 11, 15, 24, 26, 33] are amongst the most effectual solution strategies presented. With respect to local searches, non-greedy systematic swap of facilities has proved to be very efficient [22].

Genetic algorithms, when used in conjunction with the local searches, can comprise an effective strategy in providing high quality solutions. One of these effective solution strategies is the technique used in [12]. For creating offspring, this technique is based on a special merging rule called cohesive merging that exploits the special structure of the problem in producing high-quality solutions. For this purpose, the sites are divided into two groups, with the first group taking the facilities based on the first parent and the second group taking the facilities based on the second parent.

Also, the hybrid presented in [13] uses a tabu search as a metaheuristic guiding its local search component, and has various similarities with the above procedure, employing a compounded genetic algorithm presented in [11]. Solution strategies can also be extended to cope with the situations, in which multiple facilities can be assigned to a single location based on its capacity [28].

Since the development of structural iterative refinement in the PASS has been inspired by the structural refinement of a solution in the Newton-Raphson method, Fig. 1 shows the operations of this method. This method is attributed to Isaac Newton and Joseph Raphson and an excellent description of it can be found in [16].

As is seen in Fig. 1, the method operates based on approximate linear estimation of non-linear equations, and structurally generates highly precise solutions for seemingly hard-to-solve non-linear equations. As is seen in this figure, the procedures starts with the initial point r_0 and proceeds to $r_1, r_2, r_3, \ldots$, converging to one of the roots of the equation $r \approx 1.9$. This is performed by structurally refining a solution provided by the linear approximation so that it can better fit in a non-linear equation. Larger number of refinements usually lead to results with higher quality.

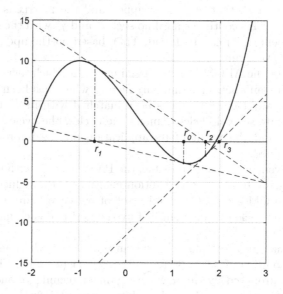

Fig. 1. Demonstrating the performance of the Newton-Raphson method in estimating the roots of the non-linear equation $2x^3 - x^2 - 8x + 5 = 0$.

3   The PASS

The development of the PASS is based on the employing of a linear assignment technique for solving the QAP. The PASS first finds an initial solution through using a construction method. Using the average distance of each location from other locations and the average flow of each facility to other facilities, the method iteratively allocates facilities to locations.

Then, a starting point is created by applying necessary swaps to the initial solution created by the construction method. Next a modified version of the Quick Match (QM) procedure [19], Arc Eliminating Quick Match (AEQM), is repeatedly applied to the starting point. The application of the AEQM changes the starting point iteratively and each new starting point is a ground for producing the next. This process is repeated as many times as needed and stops when no improvement becomes possible.

More precisely, the PASS starts with a seed which is generated by a construction method, and creates its associated relocation matrix. This matrix shows the information existing in the improvement graph [1] with each cell (i, j) of this matrix representing the improvement occurring in the case the facility i is moved to the location of the facility j. In the initial stages, each time with $O(n^2)$ operations, the relocation matrix is checked to find a row i and a column j so that $\text{cell}(i, j) + \text{cell}(j, i)$ has the minimum value in the matrix and calls such a row and a column as i^* and j^*, respectively.

If for such i^* and j^*, the value of $\text{cell}(i^*, j^*) + \text{cell}(j^*, i^*)$ is negative then the location of i^* with that of j^* is swapped and the matrix is updated with $O(n^2)$ operations. This continues until no such i^* and j^* with the negative value of $\text{cell}(i^*, j^*) + \text{cell}(j^*, i^*)$ can be found. Then based on the updated relocation matrix, the AEQM is run.

The output of the AEQM is a full assignment, in which facilities have been matched with one another, including any facility which has been matched with itself. Excluding any facility that has been matched with itself, the rest of the facilities can create several cycles. Among the cycles, the cycle with minimum cost is selected and the result of its application on the improvement of the solution is examined.

If any improvement becomes possible, the PASS restarts with this new initial solution. Otherwise, based on the criterion mentioned, the reduced costs of the last stage of the QM are changed and a set of new cycles are identified. This continues until no cycle can be identified by the QM and each facility is simply assigned to itself.

Assuming that, in a sample QAP instance of size 4, the linear assignment has led to the solution of $[(1 \to 3), (2 \to 2), (3 \to 4), (4 \to 1)]$, Fig. 2 shows the suggested cycle proposed by the linear assignment technique, and Fig. 3 shows the result of applying this cycle to the solution $\pi = \{3, 1, 2, 4\}$.

An assignment can be made based on the priority of facilities, with facilities having a higher sum of flow from all the other facilities being assigned a higher priority. Then, facilities with higher priority need to be assigned to locations which are more accessible to other locations.

The criterion of accessibility is determined based on the sum of the distances from the corresponding location to all other locations, and the locations with the lower sum of distances from all other locations are considered to be more accessible. Later, we will show how by repeating this simple routine, the solution is gradually enhanced.

The employed construction method creates an arrangement of facilities by first allocating high-flow facilities to central locations and then adding the pairs of facility-locations one at a time to minimize the traffic. The two-phase construction method employed woks as follows.

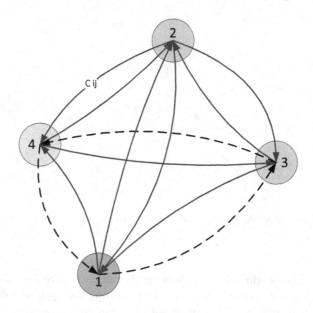

Fig. 2. Improvement graph, cycle $\{1 - 3 - 4 - 1\}$ is shown as dashed arcs

In the first phase, it determines the total distance of every location k from all other locations and the total flow of every facility i from all other facilities, denoted with TD_k, and TF_i, respectively. Since in general d_{ij} may not be equal to d_{ji}, the values of TD_k, and TF_i are calculated as follows:

$$TD_k = \sum_{l=1}^{n} d_{lk} + \sum_{l=1, l \neq k}^{n} d_{kl} \tag{3}$$

$$TF_i = \sum_{j=1}^{n} f_{ji} + \sum_{j=1, j \neq i}^{n} f_{ij} \tag{4}$$

Then the vectors of TD and TF are sorted in ascending and descending manner, respectively:

$$TD_{k_1} \leq TD_{k_2} \leq TD_{k_3} \leq \ldots \leq TD_{k_n} \tag{5}$$

$$TF_{i_1} \geq TF_{i_2} \geq TF_{i_3} \geq \ldots \geq TF_{i_n} \tag{6}$$

By using this information, high-flow facilities are allocated to the central locations. These facilities are a given percentage of facilities which have the highest flow among other facilities. On the other hand, central locations are a certain percentage of locations which have the smallest total distance from other locations. The next phase of the method simply adds the pairs of facility-locations one at a time to minimize the traffic until all facilities have been assigned.

As its input, any linear assignment technique needs a bipartite graph with a number of nodes connected by a series of arcs. Arcs in a linear assignment graph connect facilities to facilities and indicate that if the starting facility moves to

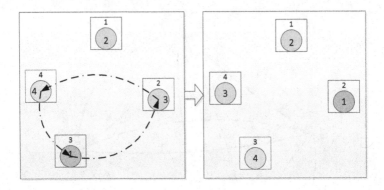

Fig. 3. Applying the cycle $L = \{1 - 3 - 4 - 1\}$ to $\pi = \{3, 1, 2, 4\}$ resulting in $\pi_L = \{2, 1, 4, 3\}$

the position of the ending facility, how much the cost increases, with negative values indicating a decrease in the cost. After finding the optimal solution in the LA, the specified arcs in this optimal solution can create between 1 and $\lfloor n/2 \rfloor$ different cycles.

The cycles with the size of 3 and more are identified and ranked based on their gains, with the gain being defined as the negative of the cost. For identifying the priority of each cycle, only its total gain, ρ_2, is considered. The cycle which has the highest value of ρ_2 is called the prioritized cycle. In the calculation of ρ_2, again ties are broken in the favor of cycles with the lower number of members. For forcing an arc to be taken out of a cycle, the AEQM increases its reduced cost to a large value and forces this cycle not to participate in the optimal solution.

The arc with the highest cost in the cycle is chosen for removal. For instance, consider the cycle $[(5), 4, 1, 8, 3, (5)]$. As is seen, the arcs comprising this cycle are $(5, 4)$, $(4, 1)$, $(1, 8)$, $(8, 3)$, and $(3, 5)$. Hence, if the costs of these five arcs in the assignment problem are $-6, 3, -7, 9$, and -2, then the cost associated with facilities 4, 1, 8, 3, and 5 are $-3, -4, 2, 7$, and -8, respectively, calculated as $(-6+3), (3-7), (-7+9), (9-2)$, and $(-2-6)$. This indicates that the facility 8 has incurred the highest cost, 7, and is the candidate facility for being removed from the cycle. Because the values of cost and gain have been defined as the opposite of one another, the highest cost is always equivalent of the lowest gain.

For computing the change of the cost when facility i goes to the location of facility j, in the symmetric case, the flow between facility i and facility j and their distance from each other is the same. However, in the asymmetric case, it is assumed that facility i is in the location of facility j and that facility j is in the location of the facility i.

Having described the concept of relocation matrix and cycles proposed by the linear assignment technique, now we can outline the major operations performed by the PASS. These operations, which are performed consecutively, include (i) generating an initial solution, (ii) calculating the relocation matrix, (iii) identifying swaps through using the relocation matrix, (iv) applying the beneficial swaps to the solution, (v) updating the relocation matrix based on the applied swap(s), (vi) converting the relocation matrix to the positive matrix, (vii) identifying cycles by using linear assignment which receives the positive matrix as its input, (viii) selecting the best cycle and applying it to the solution, (ix) removing the selected cycle through modifying reduced costs, and (x) iterative improvement of the solution while the termination criterion has not been met.

The PASS starts with generating an initial solution and then generates the relocation matrix of that solution. Next, while the relocation matrix suggests new swaps, such swaps are generated and applied to the initial solution, with the corresponding relocation matrix being updated.

Then the positive matrix is constructed based on the relocation matrix by subtracting the minimum of each row from all elements of the corresponding row. This conversion is due to the fact that AEQM, for solving the linear assignment, works only with non-negative costs. Subsequently, the linear assignment procedure uses the positive matrix as its input and identifies all of the cycles.

The identified cycles, one by one, are selected and applied to the solution, then among them the best cycle is selected. It is worth mentioning that only in cases where such a best cycle causes improvement, the corresponding solution is modified. After the possible modification of solution, the linear assignment technique is updated through modifying the relevant reduced costs, so that the previously applied cycle cannot be proposed again.

While the termination criterion has not been met, a new initial solution is generated and is progressively updated through performing all the aforementioned operations. In effect, the number of initial solutions which are generated and progressively improved depends on the allocated time-limit of the PASS. After the termination criterion is met, the best solution encountered is returned as output and the procedure is terminated.

As is seen, the PASS is aimed at enhancing the final solution through three stages. Whereas in the first stage, by the construction method, it generates initial solutions, in the second stage, improvements are made to each solution through applying the swaps suggested by the relocation matrix. The third stage starts to improve solutions which have already been enhanced in the second stage. The linear assignment technique as well as the positive matrix, which is created based on the relocation matrix, are used in this third stage. In this stage, cycles are applied to the improved solution to make further enhancement. The third stage has become efficient by modifying reduced costs and preventing extra computational effort for starting the linear assignment technique from scratch.

Figure 4 shows the best solution obtained versus iteration for the first seven start-over on the lipa60a benchmark instance. This benchmark instance is one of the instances of the QAPLIB [5], discussed in the next section.

Since in Fig. 4, with every start-over, the value of the corresponding initial solution is represented, it can easily be seen that in each of them, independent of others, a solution is found which can have a chance of improving the best obtained solution. The greater the value of the start-over, the smaller is the chance of improving the best obtained solution. Figure 5 shows the best solution

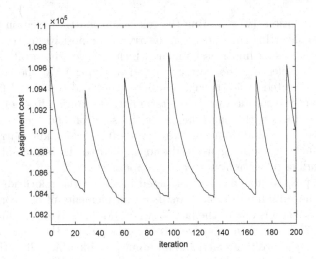

Fig. 4. The best solution obtained versus iteration for the first seven start-overs on the lipa60a instance by the PASS

obtained versus iteration on the lipa60a instance, for the first 5000 solutions evaluated.

4   Computational Experiments

The PASS has been implemented in C++ and compiled via GNU GCC compiler on a Desktop PC with 2.93 GHz speed and 8 GBs of RAM. As many as 74 representative problem instances from the QAPLIB [5] have been extracted for the procedure to be run on. The QAPLIB has a particular website and is revised frequently. Considering n as the size of each instance, in each run, a time-limit of $n/3$ min has been set for the purpose of terminating the procedure.

The instances on which the PASS has been run include 8 uniformly generated Taillard instances [30], all shown the pattern of taiXXa; 13 Skorin-Kapov instances, all shown with the pattern of skoXXX, and 53 instances shown with the different patterns of burXXX, hadXX, chrXXX, kraXXX, steXXX, escXXX and els19.

In all these patterns, X stands for a digit. Moreover, the last 53 instances are all real-life problems contributed by different authors to the QAPLIB. It is worth mentioning that among these 74 instances, the instances existing in skoXXX and taiXXa are comparatively harder than the others.

The procedure has two main parameters including (i) α, which shows the percentage of facilities to be assigned to high-grade locations and, (ii) k, the grasp cardinality [25] indicating the diversity of the initial solution at the cost of ignoring some high-quality regions. Based on employing several of these instances in testing different values of α and k we decided to set α and k to 0.45 and

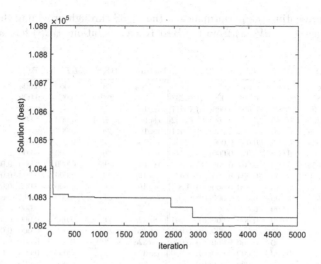

Fig. 5. The best solution obtained versus iteration on the lipa60a instance for the first 5000 solutions evaluated by the PASS.

9, respectively. Moreover, when constructing a starting solution, with a probability of $p = 0.01$ the initial solution is also generated as a uniform random permutation.

For reporting the performance measure, in line with other procedures, the PASS has been restarted 10 times with different initial seeds for each instance. Table 1 shows the results of running the procedure on all of the 74 instances. In this table, $\%DEV_{best}$ shows the best deviation percentage from the Best Known Solution (BKS) and is calculated as $100 \times (s - BKS)/BKS$, with s being defined as the best solution returned by the PASS.

In Table 1, as well as the value of $\%DEV_{best}$, the value of $\%DEV_{avg}$ has also been reported, which shows the average of deviation percentage over 10 runs. The other reported quantity is the shortest time, T_{best}, in which the best solution for each instance has been obtained, in minutes.

Since for some instances, the best solution has been obtained in several of the runs, the average time, T_{avg}, in which the best solution for each instance has been obtained, T_{avg}, has also been reported. As can be seen, the PASS obtains the best deviation percentage of 0.000 from the best known solutions of all instances except for chr20b and it is very fast in achieving such results.

Moreover, due to some similarity between our approach and the Ejection Chain [26], the results of the direct comparison between the two procedures has been shown in Table 2. The best variant of the ejection chain algorithm is named as EC3 and has been tested by its authors on SGI ALTIX with Intel Itanium processor and 1.3 GHz speed.

Because EC3 and the PASS have been run on two different computers, we provide the reported CPU FLOPS (floating-point operations per second) for the corresponding CPUs to make comparisons fair, indicating that Intel Itanium

Table 1. Representing the performance of the PASS through providing the best deviation percentage ($\%DEV_{best}$) from the Best Known Solution (BKS) for each different instance

Instance	BKS	$\%DEV_{best}$	T_{best}	$\%DEV_{avg}$	T_{avg}	Instance	BKS	$\%DEV_{best}$	T_{best}	$\%DEV_{avg}$	T_{avg}
bur26a	5426670	0.000	0.014	0.000(10)	0.106	esc16a	68	0.000	0.000	0.000(10)	0.000
bur26b	3817852	0.000	0.006	0.000(10)	0.287	esc16b	292	0.000	0.000	0.000(10)	0.000
bur26c	5426795	0.000	0.002	0.000(10)	0.017	esc16c	160	0.000	0.000	0.000(10)	0.000
bur26d	3821225	0.000	0.000	0.000(10)	0.002	esc16d	16	0.000	0.000	0.000(10)	0.000
bur26e	5386879	0.000	0.000	0.000(10)	0.011	esc16e	28	0.000	0.000	0.000(10)	0.000
bur26f	3782044	0.000	0.001	0.000(10)	0.006	esc16f	0	0.000	0.000	0.000(10)	0.000
bur26g	10117172	0.000	0.020	0.000(10)	0.340	esc16g	26	0.000	0.000	0.000(10)	0.000
bur26h	7098658	0.000	0.001	0.000(10)	0.003	esc16h	996	0.000	0.000	0.000(10)	0.000
els19	17212548	0.000	0.004	0.000(10)	0.036	esc16i	14	0.000	0.000	0.000(10)	0.000
had12	1652	0.000	0.000	0.000(10)	0.000	esc16j	8	0.000	0.000	0.000(10)	0.000
had14	2724	0.000	0.000	0.000(10)	0.000	esc32a	130	0.000	0.014	0.000(10)	0.156
had16	3720	0.000	0.000	0.000(10)	0.000	esc32b	168	0.000	0.000	0.000(10)	0.001
had18	5358	0.000	0.000	0.000(10)	0.001	esc32c	642	0.000	0.000	0.000(10)	0.000
had20	6922	0.000	0.000	0.000(10)	0.001	esc32d	200	0.000	0.000	0.000(10)	0.000
chr12a	9552	0.000	0.000	0.000(10)	0.000	esc32e	2	0.000	0.000	0.000(10)	0.000
chr12b	9742	0.000	0.000	0.000(10)	0.000	esc32g	6	0.000	0.000	0.000(10)	0.000
chr12c	11156	0.000	0.000	0.000(10)	0.000	esc32h	438	0.000	0.000	0.000(10)	0.002
chr15a	9896	0.000	0.000	0.000(10)	0.001	esc64a	116	0.000	0.000	0.000(10)	0.000
chr15b	7990	0.000	0.000	0.000(10)	0.000	esc128	64	0.000	0.000	0.000(10)	0.021
chr15c	9504	0.000	0.000	0.000(10)	0.002	tai20a	703482	0.000	0.014	0.000(10)	0.152
chr18a	11098	0.000	0.000	0.000(10)	0.011	tai25a	1167256	0.000	0.112	0.041(9)	0.746
chr18b	1534	0.000	0.000	0.000(10)	0.001	tai30a	1818146	0.016	0.054	0.181	1.604
chr20a	2192	0.000	0.130	0.839(3)	0.128	tai40a	3139370	0.378	11.866	0.851	8.351
chr20b	2298	2.089	0.121	2.794	0.239	tai50a	4938796	1.185	16.038	1.324	11.843
chr20c	14142	0.000	0.001	0.000(10)	0.004	tai60a	7208572	1.184	5.445	1.458	9.816
chr22a	6156	0.000	0.459	0.507(3)	0.419	tai80a	13499184	1.234	23.158	1.524	17.869
chr22b	6194	0.000	0.065	1.101(1)	0.571	tai100a	21052466	1.317	31.880	1.409	19.596
chr25a	3796	0.000	0.026	1.496(5)	0.437	sko42	15812	0.000	0.001	0.000(10)	1.795
kra30a	88900	0.000	0.000	0.000(10)	0.010	sko49	23386	0.000	0.694	0.007(9)	5.730
kra30b	91420	0.000	0.009	0.000(10)	0.043	sko56	34458	0.017	18.450	0.033	9.583
kra32	88700	0.000	0.001	0.000(10)	0.005	sko64	48498	0.000	8.724	0.051(1)	15.131
ste36a	9526	0.000	0.003	0.000(10)	0.173	sko72	66256	0.054	10.583	0.111	15.707
ste36b	15852	0.000	0.001	0.000(10)	0.014	sko81	90998	0.088	7.155	0.156	11.158
ste36c	8239110	0.000	0.214	0.000(10)	0.887	sko90	115534	0.090	7.725	0.146	12.925
						sko100a	152002	0.072	27.606	0.164	12.210
						sko100b	153890	0.179	2.360	0.221	17.207
						sko100c	147862	0.058	7.785	0.113	19.185
						sko100d	149576	0.126	28.513	0.227	19.511
						sko100e	149150	0.090	25.210	0.182	11.195
						sko100f	149036	0.134	30.427	0.263	14.549

1.3 GHz (SGI Altix) is reported to handle up to 4.03 to 5.20 Giga FLOPS per core[1], whereas our desktop CPU can handle between 3.05 to 3.62 Giga FLOPS per core[2]. This signifies that the computer on which EC3 was run is faster than our computer. As is seen, even with the assumption that both computer have the same speed, the PASS has outperformed the EC3 on Skorin-Kapov instances both in term of solution quality and the running time; however, at the expense of larger running times, the EC3 has produced better results on Taillard instances.

---

[1] http://icl.cs.utk.edu/hpcc/hpcc_record.cgi?id=41.

[2] https://asteroidsathome.net/boinc/cpu_list.php (Intel(R) Core(TM) i7 CPU 870 @ 2.93 GHz).

Table 2. Comparison with Rego et al. (2010) [26]

Instance	BKS	MNRP		EC3	
		%DEV_{avg}	T_{avg}	%DEV_{avg}	T_{avg}
Skorin-Kapov instances					
sko42	15812	**0.000(10)**	1.80	**0.000**	21.43
sko49	23386	**0.007(9)**	5.73	0.039	50.91
sko56	34458	0.033	9.58	**0.027**	58.02
sko64	48498	**0.051(1)**	15.13	0.078	60.99
sko72	66256	**0.111**	15.71	0.250	79.22
sko81	90998	**0.156**	11.16	0.278	68.06
sk90	115534	**0.146**	12.93	0.473	57.84
sko100a	152002	**0.164**	12.21	0.340	70.86
sko100b	153890	**0.221**	17.21	0.408	53.54
sko100c	147862	**0.113**	19.19	0.543	77.63
sko100d	149576	**0.227**	19.51	0.517	88.14
sko100e	149150	**0.182**	11.20	0.460	88.33
sko100f	149036	**0.263**	14.55	0.542	84.19
Average		**0.162**	12.76	0.304	66.09
Taillard instances					
tai20a	703482	**0.000(10)**	0.15	**0.000**	0.56
tai25a	1167256	0.041(9)	0.75	**0.000**	0.55
tai30a	1818146	0.181	1.60	**0.000**	2.04
tai40a	3139370	0.851	8.35	**0.219**	28.01
tai50a	4938796	1.324	11.84	**0.514**	35.43
tai60a	7208572	1.458	9.82	**0.657**	49.24
tai80a	13499184	1.524	17.87	**0.730**	65.05
tai100a	21052466	1.409	19.60	**0.729**	54.36
Average		1.125	8.75	**0.356**	29.41

5   Concluding Remarks

The power of a structural search method relies on its combined characteristics of (i) simplicity, (ii) efficiency, (iii) robustness, (iv) effectiveness, (v) efficacy, and (vi) scalability. Distinguishing these six independent characteristics is aimed at indicating that an effective search may be inefficient, a robust search may be ineffective, or a robust and effective search may lack the property of simplicity, etc.

The number of structural searches in the literature, like the Newton-Raphson method, which has all these characteristics, is very limited. With the hope of developing similar searches, these structural searches can be used as a template for the development of other search techniques.

For tackling the QAP, we used the concept employed in the Newton-Raphson method as a basis for developing a new search procedure called PASS. We showed that the PASS is (i) simple in the sense of being easy to implement and understand (ii) efficient, in the sense that it provides fast solutions for a wide spectrum of benchmark instances, (iii) effective, in the sense that it produces high quality solutions for an assortment of instances, and (iv) scalable, in the sense it maintains solution quality through forcing its linear assignment technique for extra refinement of the solution of a large problem instance.

The conceptual difference over the existing approaches is that the PASS has the unique features of (i) converting the relocation matrix to the positive matrix, (ii) identifying cycles by using the linear assignment procedure which receives the positive matrix as its input, (iii) selecting the best cycle and applying it to the solution, and (iv) removing the selected cycle through modifying reduced costs.

The lessons learned are that (i) the serial programming cannot highly exploit the parallel capabilities of the PASS, and (ii) the employed local search should spend extra time to revise low quality solutions, with low quality initial solutions forcing intensive local searches. Towards resolving these issues and adding further efficacy to the procedure, the following research directions are proposed.

First, because of the scalability of the procedure, different threads of the same module can be programmed to cooperatively optimize different parts of the same solution. In this proposed multi-thread environment, the threads can communicate with one another through a shared memory, and each thread can ignore those pieces of the solution manipulated by other threads, focusing on optimizing only those pieces of solution assigned to it.

Second, a genetic algorithm can be incorporated into the module to identify the beneficial parts of the solutions obtained by the procedure and to combine such parts in producing solutions with higher quality. Here a genetic algorithm is seen from a perspective of a schemata-processing mechanism. In this view, crossover operators identify useful building blocks in different solutions produced by the PASS and combine these building blocks to integrate different judgments made in arranging facilities in different parts of the program under different circumstances.

Third, since the same with the Newton Raphson method, the effect of an initial solution on the performance of the procedure is significant, the employed construction method can be further improved through enhancing its selection mechanism. For this purpose, the employed construction method can employ an improved estimation function to better identify the promising components. In this way, by successively adding proper facilities to a partial layout, the construction method can create a layout with higher quality, albeit in longer time, and this may increase the quality of the overall solution.

References

1. Ahuja, R., Jha, K., Orlin, J., Sharma, D.: Very large-scale neighborhood search for the quadratic assignment problem. INFORMS J. Comput. **19**(4), 646–657 (2007)
2. Anstreicher, K., Brixius, N., Goux, J., Linderoth, J.: Solving large quadratic assignment problems on computational grids. Math. Program. **91**(3), 563–588 (2002)
3. Battiti, R., Tecchiolli, G.: Simulated annealing and tabu search in the long run: a comparison on QAP tasks. Comput. Math. Appl. **28**(6), 1–8 (1994)
4. Baykasoglu, A.: A meta-heuristic algorithm to solve quadratic assignment formulations of cell formation problems without presetting number of cells. J. Intell. Manuf. **15**(6), 753–759 (2004)
5. Burkard, R.E., Karisch, S.E., Rendl, F.: QAPLIB - a quadratic assignment problem library. J. Glob. Optim. **10**(4), 391–403 (1997)
6. Colorni, A., Dorigo, M., Maffioli, F., Maniezzo, V., Righini, G., Trubian, M.: Heuristics from nature for hard combinatorial optimization problems. Int. Trans. Oper. Res. **3**(1), 1–21 (1996)
7. Connolly, D.T.: An improved annealing scheme for the QAP. Eur. J. Oper. Res. **46**(1), 93–100 (1990)
8. Crainic, T.G., Cun, B.L., Roucairol, C.: Parallel branch-and-bound algorithms. In: Zomaya, A.Y., Talbi, E. (eds.) Parallel Combinatorial Optimization Chap. 1, pp. 1–28. Wiley, Hoboken (2006). https://doi.org/10.1002/9780470053928.ch1
9. Demirel, N., Toksar, M.: Optimization of the quadratic assignment problem using an ant colony algorithm. Appl. Math. Comput. **183**(1), 427–435 (2006)
10. Drezner, Z.: Heuristic algorithms for the solution of the quadratic assignment problem. J. Appl. Math. Decis. Sci. **6**, 163–173 (2002)
11. Drezner, Z.: Compounded genetic algorithms for the quadratic assignment problem. Oper. Res. Lett. **33**(5), 475–480 (2005)
12. Drezner, Z.: A new genetic algorithm for the quadratic assignment problem. INFORMS J. Comput. **15**(3), 320–330 (2003)
13. Drezner, Z.: Extensive experiments with hybrid genetic algorithms for the solution of the quadratic assignment problem. Comput. Oper. Res. **35**(3), 717–736 (2008)
14. Gambardella, L., Taillard, E., Dorigo, M.: Ant colonies for the quadratic assignment problem. J. Oper. Res. Soc. **50**(2), 167–176 (1999)
15. James, T., Rego, C., Glover, F.: Multistart Tabu search and diversification strategies for the quadratic assignment problem. IEEE Trans. Syst. Man Cybern. Part A: Syst. Hum. **39**(3), 579–596 (2009)
16. Kelley, C.T.: Solving Nonlinear Equations with Newton's Method. SIAM (2003)
17. Koopmans, T., Beckmann, M.: Assignment problems and the location of economic activities. Econometrica: J. Econometr. Soc. **25**(1), 53–76 (1957)
18. Kuhn, H.: The Hungarian method for the assignment problem. Naval Res. Logist. **2**, 83–97 (1952)
19. Lee, Y., Orlin, J.: Quickmatch: a very fast algorithm for the assignment problem. Report, Sloan School of Management, Massachusetts Institute of Technology (Report number: WP# 3547–93) (1993)
20. Loiola, E.M., De Abreu, N.M.M., Boaventura-Netto, P.O., Hahn, P., Querido, T.: A survey for the quadratic assignment problem. Eur. J. Oper. Res. **176**(2), 657–690 (2007)
21. Maniezzo, V., Colorni, A.: The ant system applied to the quadratic assignment problem. IEEE Trans. Knowl. Data Eng. **11**(5), 769–778 (1999)

22. Matai, R., Singh, S., Mittal, M.: A non-greedy systematic neighbourhood search heuristic for solving facility layout problem. Int. J. Adv. Manuf. Technol. **68**, 1665–1675 (2013)
23. Mavridou, T., Pardalos, P.: Simulated annealing and genetic algorithms for the facility layout problem: a survey. Comput. Optim. Appl. **7**(1), 111–126 (1997)
24. Misevicius, A.: A tabu search algorithm for the quadratic assignment problem. Comput. Optim. Appl. **30**(1), 95–111 (2005)
25. Oliveira, C.A.S., Pardalos, P.M., Resende, M.G.C.: GRASP with path-relinking for the quadratic assignment problem. In: Ribeiro, C.C., Martins, S.L. (eds.) WEA 2004. LNCS, vol. 3059, pp. 356–368. Springer, Heidelberg (2004). https://doi.org/10.1007/978-3-540-24838-5_27
26. Rego, C., James, T., Glover, F.: An ejection chain algorithm for the quadratic assignment problem. Networks **56**(3), 188–206 (2010)
27. Sahni, S., Gonzalez, T.: P-complete approximation problems. J. ACM (JACM) **23**(3), 555–565 (1976)
28. Silva, R., Resende, M., Pardalos, P., Mateus, G., De Tomi, G.: Grasp with path-relinking for facility layout. In: Goldengorin, B., Kalyagin, V., Pardalos, P. (eds.) Models, Algorithms, and Technologies for Network Analysis, vol. 59, pp. 175–190. Springer, New York (2013). https://doi.org/10.1007/978-1-4614-8588-9_11
29. Solimanpur, M., Vrat, P., Shankar, R.: Ant colony optimization algorithm to the inter-cell layout problem in cellular manufacturing. Eur. J. Oper. Res. **157**(3), 592–606 (2004)
30. Taillard, E.: Robust taboo search for the quadratic assignment problem. Parallel Comput. **17**(4), 443–455 (1991)
31. Taillard, E.: FANT: fast ant system. Technical report, Istituto Dalle Molle Di Studi Sull Intelligenza Artificiale (1998)
32. Talbi, E., Roux, O., Fonlupt, C., Robillard, D.: Parallel ant colonies for the quadratic assignment problem. Future Gener. Comput. Syst. **17**(4), 441–449 (2001)
33. Voß, S.: Solving quadratic assignment problems using the reverse elimination method. In: Nash, S.G., Sofer, A., Stewart, W.R., Wasil, E.A. (eds.) The Impact of Emerging Technologies on Computer Science and Operations Research, pp. 281–296. Kluwer, Dordrecht (1995). https://doi.org/10.1007/978-1-4615-2223-2_14
34. Votaw, D.F., Orden, A.: The personnel assignment problem. In: Symposium on Linear Inequalities and Programming, pp. 155–163 (1952)
35. Zhang, H., Beltran-Royo, C., Constantino, M.: Effective formulation reductions for the quadratic assignment problem. Comput. Oper. Res. **37**(11), 2007–2016 (2010)
36. Zhang, H., Beltran-Royo, C., Ma, L.: Solving the quadratic assignment problem by means of general purpose mixed integer linear programming solvers. Ann. Oper. Res. **207**, 261–278 (2010)

Application of MILP to Strategic Sourcing of High-Cost Medical Devices and Supplies

Parimal Kulkarni[1,2] and L. Douglas Smith[1(✉)]

[1] University of Missouri-St. Louis,
One University Boulevard, St. Louis, MO 63121, USA
ldsmith@umsl.edu
[2] BJC Healthcare, 4901 Forest Park Avenue, St. Louis, MO 63108, USA

Abstract. We discuss the application of a mixed integer linear programming (MILP) model to assist in the development of procurement strategies for high-cost medical devices and supplies in one of the largest nonprofit health-care organization in the USA. The MILP model seeks to reduce the costs of providing necessary supplies by qualifying the organization for price discounts through volume purchasing commitments, while maintaining diversity in the supply base, adhering to physicians' preferences for specific products, and considering ratings given to suppliers on several dimensions on vendor scorecards. With results from multiple optimization scenarios, tradeoffs among procurement costs, requirements for diversity in the supply base, and flexibility of physicians in allowing substitute devices are explored in depth. Also revealed are potential consequences of stipulating formal vendor scorecard requirements when negotiating future contracts.

1 Introduction

Healthcare providers are under constant pressure to contain costs in a sector of the USA economy that now exceeds 18% of GDP (Papinicolas et al. 2018). Medical supplies constitute the second-largest category of cost (after medical personnel) for healthcare organizations (Belliveau 2017). Costs connected with high-value medical devices alone have been estimated to result in $5 billion in waste annually in the USA (Dichiaria 2016). Healthcare IT systems are not generally geared to facilitate consideration of alternative sourcing strategies, and physicians' strong preference for medical products with which they are most familiar can complicate strategic sourcing initiatives (Belliveau 2017). Physicians' disinclination to consider alternative devices can increase purchasing costs and reduce flexibility in responding to disruptions in the healthcare supply chain. Analytical tools are required to help healthcare organizations explore alternative procurement strategies and determine the costs of various constraints that may be imposed on procurement decisions. The model presented in this paper is designed to meet this need.

For three decades, large organizations have employed sourcing strategies such as corporate buying, supply base optimization, total supply chain cost minimization, and

© Springer Nature Switzerland AG 2018
R. Cerulli et al. (Eds.): ICCL 2018, LNCS 11184, pp. 461–474, 2018.
https://doi.org/10.1007/978-3-030-00898-7_31

goods and services value maximization in efforts to drive down procurement costs (Anderson and Katz 1998). Summaries of research on supplier selection models are provided by Chai et al. (2013) and Kannan et al. (2013). These authors found that the most frequently used quantitative technique for supplier selection was the Analytical Hierarchical process (AHP), followed by linear programming (LP) and Technique for Order of Preference by Similarity to Ideal Solution (TOPSIS). Research has focused mainly on total cost as the criterion in supplier selection but there is some acknowledgement of tradeoffs between diversification of the supply base for risk reduction, on one hand, and concentration of business with vendors that offer products and services at lowest cost, on the other hand. Pan (1989) proposed a linear programming model to optimally identify the number of suppliers to use and quantities to purchase from each supplier. The objective was to minimize procurement costs, where product demand was deterministic and supply from any source was unlimited. Parlar and Wang (1993) studied strategic sourcing (single source vs multiple sources) by comparing the costs of single versus dual sourcing for a firm assuming that the overall objective was to minimize purchasing and inventory costs. They used an EOQ and newsboy-based ordering policy and demonstrated that dual sourcing was preferred when supplier yield was a random variable. Burke et al. (2009) extended the Parlar and Wang (1993) study by implementing an optimization approach to determine whether a product should be single sourced or sourced from multiple suppliers depending on demand uncertainty. The study concluded that single sourcing was as effective strategy when supplier capacities were large relative to the product demand and when the firm did not obtain diversification benefits. Quantity discounts are important considerations in contract negotiations. Kamali et al. (2011) incorporated quantity discount policies in tandem with joint cost reduction while developing a model for supplier selection. They proposed a MILP model to minimize total system cost while considering various criteria such as quality and delivery. The authors found that solutions were highly sensitive to the changes in demand, production rates, variable costs, defective rates, and late delivery rates. Bui et al (2001) contend that contract negotiations with suppliers should consider qualitative attributes such as product quality, speed, reputation, after sales service, etc. Variants of balanced scorecards proposed by Kaplan and Norton (1992) have been proposed for vendor evaluation and selection (Galankashi et al. 2016). Karsak and Dursun (2015) emphasize the importance of employing group decision-making practices when developing procurement strategies. They tested a fuzzy-logic supplier selection model that considers three supplier selection attributes: cost, quality, and delivery.

In our MILP model, we adapt key ideas from these studies to address specific concerns for medical practice. The model supports the development of data-driven procurement strategies with consideration of physicians' historical preference for specific products, flexibility in the use of alternative products which may be less costly, desired diversification in the supply base, and standards for vendor performance on several dimensions.

2   The Business Setting

Our business setting involves the procurement of expensive medical devices and supplies for cardiac surgery. In this paper, we concentrate on products for cardiac rhythm management. This device category was suggested by clinicians for our initial work in this area because related expenditures are substantial and physicians typically have high preference for particular brands and models in this category. The products (with individual SKUs) are grouped into nine categories as summarized in Table 1. The data were collected from all hospitals in the healthcare system over the course of 12 months. Products within a category may substitute for one another to a designated degree. Limits may be imposed on the total percentage of medical procedures for which substitute products are used, and further limits may be imposed on the extent to which one particular product may be substituted by another (referred to as pairwise substitution constraints). Table 2 illustrates a situation where there is asymmetric pairwise substitutability of one vendor's product for another's. Vendor 1's product may substitute for vendor 4's product but not vice versa. Similar asymmetry exists for vendors 3 and 4.

Table 1. Number of unique SKUs (by category and vendor)

Product Category	Vendor 1	Vendor 2	Vendor 3	Vendor 4	Total
Cardiac pacing leads or electrodes (HV)	7	8	10	10	35
Cardiac pacing leads or electrodes (LV)	4	7	11	5	27
Cardiac pacing leads or electrodes (PM)	5	7	21	13	46
Cardiac resynchronization therapy defibrillator	5	9	5	4	23
Cardiac resynchronization therapy pacemaker	2	4	3	2	11
Implantable Cardioverter Defibrillator Dual Chamber	5	7	4	5	21
Implantable Cardioverter Defibrillator Single Chamber	5	7	3	7	22
Pacemaker-Dual	4	4	7	4	19
Pacemaker-Single	4	3	4	2	13
Total	**41**	**56**	**68**	**52**	**217**

3   Introducing Balanced Scorecards as Procurement Constraints

The healthcare organization had undertaken an initiative for incorporating vendor scorecards formally into the procurement process. Part of our analysis was therefore directed at determining the potential effects of such considerations on future

Table 2. Example of asymmetric pairwise substitution in a product category

Column Vendor's Product May Substitute for Row Vendor's Product
if cell value = 1

	V1	V2	V3	V4
V1	1	0	1	0
V2	0	1	1	1
V3	1	1	1	1
V4	1	1	0	1

procurement plans. Conceptually, information from vendor scorecards may be introduced in constraints which qualify vendors for supplying particular products, in measurements of the extent to which procurement objectives are achieved, or in both. In preliminary discussions with clinical managers and supply-chain managers, it was readily apparent that mixing costs and scorecard measures in a consolidated objective function was impracticable and reaching consensus on a hierarchical set of goals would be challenging. In this pioneering effort at incorporating vendor scorecards into procurement decisions by the organization, we therefore chose to incorporate scorecard ratings by allowing procurement managers to stipulate the minimum weighted average score that a vendor must achieve on each performance dimension in each product category.

Ratings from subjective scorecards are known to be subject to two forms of bias. The first, called common-methods bias or common-method variance, is attributable to "carry-over" or "halo" effects from one dimension to another (Peterson and Wilson 1992). The effect of this is to induce correlations in the ratings given to performance on the different dimensions. This may occur, in this case, as a rater attempts to maintain consistency with an overall impression of the vendor's performance, or as the rater considers performance on a new dimension after giving a rating on other dimensions and finds his or her judgments tempered by that context. It may also occur as a recent experience, possibly when some unrelated issue disposes the rater favorably or unfavorably toward the vendor. Techniques used for mitigating these effects in mass surveys (such as changing the order in which dimensions of performance are considered) require much larger sample sizes than occur in this supply-chain setting where few raters are involved. Testing the possible effects of common methods variance was therefore part of due diligence in testing our model. The need for this was underscored by the tendency of several procurement managers to assign similar ratings to a vendor on all dimensions for different product groups when they worked with draft instruments for the vendor scorecards.

The second type of bias is related to a rater's general tendency to give high or low scores. The effect of this bias is to shift scores upward or downward but without affecting the correlations among the scores (Lance et al. 2010). Our experiments addressed this type of bias as well.

There are various methods that one might employ to inject these types of bias into hypothetical scores for vendor scorecards and, of course, various ways of selecting scores to be used in the scenarios for the deterministic optimizing model. The method we employed involved simulating the sequence with which scores would be assigned and emulating the scoring procedure with a mathematical model designed to generate scores with hypothesized correlation structures and hypothesized tendencies of raters to give high or low scores.

We started with subjective estimates of the extent to which the scores on each dimension might be correlated with scores on other dimensions and established a hierarchical influence by using the sum of the absolute correlations for a variable to establish its order in the hierarchy. Product price was the variable that was judged to be most highly correlated with other dimensions. The first rating (price), also provided an objective anchor for the raters' subjective assessments on the first dimension (price). Subsequent dimensions, in order, were quality, innovation, operational excellence, delivery and service.

Hypothetical scores were generated to exhibit hypothesized correlation structures by having a modal (most likely) score on a dimension reflect systematic carry-over from modal scores on previous dimensions and adding a random component for the dimension under consideration. By choosing relevant weights, "common-methods bias" was thus induced to the desired degree for chosen scenarios. We represented the distribution of possible scores to be assigned by a rater as triangular in form (Fig. 1) with parameters equal to the lowest possible score, the modal score generated as above, and the highest possible score. Raters' general inclination to give higher or lower scores was then incorporated by stipulating a percentile of the distribution of possible scores. The scenarios used in this paper are based on 25th percentiles for rater tendencies toward lower scores and 80th percentiles for rater tendencies toward higher scores.

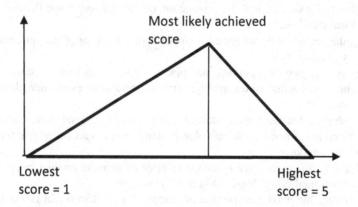

Fig. 1. Triangle distribution for hypothetical vendor scores

4   The MILP Model

We use the following sets when generating variables and constraints:

- P, the set of individual product SKUs p under consideration
- C, the set of product categories c

- V, the set of vendors v
- CV(v), the set of product categories offered by vendor v
- VC(c), the set of vendors offering products in category c
- PCV(c,v), the set of products offered in category c by vendor v
- SC(c), the set of products in category c
- SP(p), the set of possible products that may serve as a substitute for product p
- PS(p), the set of possible products for which p can substitute
- P{p,s(p)}, the set of products p and potential substitutes s (possible combinations of p and s)

Model parameters include:

- $demc_c$, the expected demand for products in category c
- $demp_p c_c v_v$, the expected demand for product p as the preferred product in category c from vendor v
- $befdiscp_p c_c v_v$, the unit price of product p from vendor v before quantity discount is applied
- $bulkvolc_c v_v$, the total unit volume of purchases in category c required to qualify for bulk-buy discount from vendor v
- M, a large number used to relax constraints
- $pricescop_p c_c v_v$, the price score of vendor v for product p within category c
- $maxpricescoc_c$, the highest possible price score granted for items in category c
- $lbpricec_c$, the lower bound on index representing achievement of price goal for products used in category c, across all vendors
- (similar parameters defined for other dimensions used in vendor scorecards (qual for quality, serv for service, inno for innovation, del for delivery reliability and opex for operational excellence)
- $penpc_c$, the penalty assigned for each unit of deficiency in aggregate price points for products in category c
- $penqc_c$ $pensc_c$, $penopc_c$, $pendc_c$, and $penic_c$, similar penalties for deficiencies in aggregate scores achieved for quality, service, operational excellence, delivery and innovation.
- $maxmktshrc_c v_v$, the maximum allowed percentage of demand (which may be set either by count of items or dollar value of items) in product category c that can be covered by vendor v
- $pctrebatec_c v_v$, the percentage reduction in price of items in product category v that occurs when qualifying threshold is met for vendor v
- $maxpctsubp_p$, the maximum portion of demand for product p that is substitutable
- $maxpctsubc_c$, the maximum portion of demand in a product category that may be met by substitute products
- $maxpairp_p s_s$, the maximum portion of demand for product p that may be satisfied by substitute product s
- sif, a small inflation factor applied to product cost when products are used as substitutes instead of as preferred products (to avoid alternative solutions with same costs using substitutes when use of the preferred product would not violate constraints).

Decision variables include:

- $tqp_p c_c v_v$, the total quantity of product p in category c from vendor v to be purchased as either the preferred product or substitute for other products
- $qp_p c_c v_v$, the quantity of product p in category c from vendor v to be purchased as the preferred product
- $subp_p s_s$, the quantity of product p purchased as a substitute for product s
- $subp_s s_p$, the quantity of product s purchased as a substitute for product p
- $qtdisc_c v_v$, a binary variable indicating whether purchases from vendor v in category c are sufficient to qualify for a quantity discount on purchases in that product category
- $rebatec_c v_v$, the magnitude of rebate earned for products in category c from vendor
- $aggpricedefc_c$, the aggregate scorecard deficiency for price
- $aggservdefc_c$, $aggqualdefc_c$, $aggopdefc_c$, $aggdeldefc_c$, and $aggindefc_c$, aggregate scorecard deficiencies for service, quality, operational excellence, delivery and innovation, respectively.

The objective (1) is to minimize the total cost of products purchased (as preferred and substitutes) minus the rebate earned, plus penalties for scorecard deficiencies.

$$
\begin{aligned}
&\sum_{v \in V} \left(\sum_{c \in CV\{v\}} \sum_{p \in PCV\{c,v\}} \left(qp_p c_c v_v * befdiscp_p c_c v_v \right) \right. \\
&+ \sum_{p \in SC\{c\}} \sum_{s \in SP\{p\}} \left(subp_p s_s * befdiscp_p c_c v_v * sif \right) - \left. \left(\sum_{c \in CV\{v\}} rebatec_c v_v \right) \right) \\
&+ \sum_{v \in V} \left(aggpricedefc_c * penpc_c \right) + \sum_{v \in V} \left(aggservdefc_c * pensc_c \right) \\
&+ \sum_{v \in V} \left(aggqualdefc_c * penqc_c \right) + \sum_{v \in V} \left(aggopdefc_c * penopc_c \right) \\
&+ \sum_{v \in V} \left(aggdeldefc_c * pendc_c \right) + \sum_{v \in V} \left(aggindefc_c * penic_c \right)
\end{aligned}
$$

$$(1)$$

In our model testing, if penalties for failure to meet the scorecard requirements emerge in the solution, we simply report that not all the weighted scorecard requirements constraints are able to be satisfied, rather than using the MILP to adjust the procurement plan to resolve the conflicting constraints.

Constraints are imposed as follows. In reading the subscripts, remember that each p (with unique SKU) has an associated c and v.

Sufficient numbers of items in each product category must be purchased to perform the expected number of medical procedures.

$$
\sum_{v \in VC\{v\}} \left(\sum_{p \in PCV\{c,v\}} \left(tqp_p c_c v_v \right) \right) \geq demc_c \text{ for each c} \tag{2}
$$

Purchases of individual products or their substitutes must match total demand registered at the product level.

$$
\sum_{s \in \{SP(p)\}} subp_s s_p + qp_p c_c v_v \geq demp_p c_c v_v \text{ for each p} \tag{3}
$$

Nonsubstitutable demand must be covered by products purchased as "preferred" products.

$$qp_p c_c v_v >= demp_p c_c v_v * (1 - \text{maxpctsubp}_p/100) \text{ for each p} \tag{4}$$

Total purchases of a product are the sum of those purchased as preferred products plus those purchased as substitutes for others.

$$tqp_p c_c v_v = qp_p c_c v_v + \sum_{s \in \{PS(p)\}} subs_s p_p \text{ for each p} \tag{5}$$

The total purchases of products as substitutes in a product category must adhere to the limits allowed.

$$\sum_{p \in sc(c)} \sum_{s \in SP(p)} subp_s s_p \leq \text{maxpctsubc}_c/100 * demc_c \text{ for each c} \tag{6}$$

The total purchases of products to substitute for a particular product must adhere to the limits allowed.

$$\sum_{s \in SP(p)} subp_s s_p \leq \text{maxpctsubp}_p/100 * demp_p c_c v_v \text{ for each p} \in SP(p) \tag{7}$$

Substitution of individual products for a particular product must adhere to the pairwise substitution limitations (with $\text{maxpairp}_p s_s = 0$ if substitution of product s for product p is not allowed and 1 otherwise).

$$subp_s s_p \leq \text{maxpairp}_p s_s * demp_p c_c v_v \text{ for each p and s} \in SP(p) \tag{8}$$

Eligibility for a quantity discount requires sufficient purchases from the vendor in the product category.

$$qtdisc_c v_v \leq \left(\sum_{p \in PCV\{c,v\}} tqp_p c_c v_v \right) / bulkvolc_c v_v \text{ for each v, and c} \in CV(v) \tag{9}$$

Rebates require eligibility for the quantity discount.

$$rebatec_c v_v \leq M * qtdisc_c v_v \text{ for each v, and c} \in CV(v) \tag{10}$$

Rebate amount is dictated by the discount terms.

$$rebatec_c v_v \leq \left(\sum_{p \in PCV\{c,v\}} befdiscp_p c_c v_v * \text{pctrebatec}_c v_v/100 * tqp_p c_c v_v \right)$$
$$\text{for each v and c} \in CV(v) \tag{11}$$

The market share for each vendor in a product category must not exceed the maximum allowed. Stating market share in dollar terms, the constraint set becomes:

$$\sum_{p \in PCV\{c,v\}} \left(tqp_p c_c v_v * befdiscp_p c_c v_v\right) \leq \text{maxmktshrc}_c v_v/100* \left(\sum_{p \in PCV\{c,v\}} \sum_{v \in V} \left(tqp_p c_c v_v * befdiscp_p c_c v_v\right)\right) \text{ for each v and c} \tag{12}$$

The weighted average score on price for all purchases from each vendor must meet the stipulated criterion (designated proportion of the maximum possible rating). This set of constraints is linearized as:

$$\sum_{c \in CV\{c\}} \sum_{p \in PCV\{c,v\}} \left(pricescop_p c_c v_v * tqp_p c_c v_v\right) - ((maxpricescoc_c * lbpricec_c) * \sum_{c \in CV\{c\}} \sum_{p \in PCV\{c,v\}} \left(tqp_p c_c v_v\right)) + \text{aggpricdefc}_c \geq 0 \text{ for each v} \tag{13}$$

Five additional sets of constraints are similarly imposed to ensure that the weighted average score for all purchases from each vendor meet the minimum criteria on each of the other five dimensions. (14–18).

5   Data for the Planning Scenarios

Pricing and usage data were extracted from the most recent year's purchases for hospitals in the organization's healthcare network. They were used as indicative of the number of "preferred" products to be required for medical procedures. Terms for future quantity discounts were based on general industry practice. Hypothetical vendor scorecards were set at averages from 100 replications of the behavioral scoring model set for high and low common methods bias and tendencies toward high and low ratings. Table 3 illustrates, for example, hypothetical vendor scores generated with high common methods bias and raters' tendency to give high scores. Table 4 illustrates hypothetical vendor scores generated with low common methods bias and raters' tendency to give low scores.

The first use of the model was to explore the potential savings that could accrue under different degrees of physicians' flexibility in the use of substitute devices and supplies. For this analysis, we relaxed constraints on scorecard requirements and generated solutions with different degrees of allowed product substitutability with various limits on vendors' market share for items in a product category (thus jointly considering the competing goals of satisfying physician preference, on one hand, and ensuring diversification as a risk-mitigation strategy, on the other hand). In this, we effectively assume that all vendors would meet the performance standards on each performance dimension and that physicians would accept the lower-cost products for their procedures in a product category up to the stipulated level, while conforming to the pairwise substitution limitations.

Table 3. Vendor scores with high common methods bias and tendency toward high scores

alpha1 = .98, alpha2 = .95, alpha3 = .92, alpha4 = .90, alpha5 = .87, alpha6 = .78, cumprob = .8

Variable	N	Mean	Std Dev	Sum	Minimum	Maximum
Price-ratio	3600	0.787	0.081	2832	0.513	1.000
qualsco	3600	3.531	0.023	12710	3.443	3.601
innosco	3600	3.537	0.031	12732	3.436	3.624
opexsco	3600	3.539	0.035	12742	3.431	3.636
delsco	3600	3.547	0.045	12770	3.423	3.662
servsco	3600	3.566	0.073	12837	3.401	3.727
pricesco	3600	3.421	0.083	12316	3.211	3.731

Pearson Correlation Coefficients, N = 3600 (all correlations significant at < 0.0001)

	price ratio	qual sco	inno sco	opex sco	del sco	serv sco	price sco
Price-ratio	1.000	0.735	0.550	0.451	0.350	0.235	−0.996
qualsco	0.735	1.000	0.398	0.340	0.249	0.177	−0.739
innosco	0.550	0.398	1.000	0.258	0.190	0.122	−0.552
opexsco	0.451	0.340	0.258	1.000	0.152	0.127	−0.454
delsco	0.350	0.249	0.190	0.152	1.000	0.077	−0.349
servsco	0.235	0.177	0.122	0.127	0.077	1.000	−0.234
pricesco	−0.996	−0.739	−0.552	−0.454	−0.349	−0.234	1.000

Table 4. Vendor Scores with Low Common Methods Bias and Tendency toward Low Scores

alpha1 = .98, alpha2 = .7, alpha3 = .5, alpha4 = .2, alpha5 = .1, alpha6 = .1, cumprob = .25

Variable	N	Mean	Std Dev	Sum	Min	Max
Price-ratio	3600	0.787	0.081	2832	0.513	1.000
qualsco	3600	2.518	0.118	9065	2.207	2.767
innosco	3600	2.479	0.2	8925	2.023	2.838
opexsco	3600	2.394	0.335	8618	1.728	2.926
delsco	3600	2.383	0.388	8577	1.645	2.965
servsco	3600	2.378	0.383	8562	1.638	2.961
pricesco	3600	1.938	0.147	6976	1.536	2.41

Pearson Correlation Coefficients, N = 3600 (*correlations significant < 0.0001)

	price ratio	qual sco	inno sco	opex sco	del sco	serv sco	price sco
Price-ratio	1.000	0.239*	0.114*	0.020	0.009	0.048	−0.995*
qualsco	0.239*	1.000	0.010	0.015	−0.016	0.019	−0.240*
innosco	0.114*	0.010	1.000	0.010	0.001	−0.001	−0.111*
opexsco	0.020	0.015	0.010	1.000	−0.008	0.028	−0.019
delsco	0.009	−0.016	0.001	−0.008	1.000	−0.003	−0.009
servsco	0.048	0.019	−0.001	0.028	−0.003	1.000	−0.048
pricesco	−0.995*	−0.240*	−0.11*	−0.019	−0.009	−0.048	1.000

6 MILP Solution Process

The MILP model was developed using the Statistical Analysis System (SAS) for data integration. We create the model in sparse-data format for solution either by SAS Proc OPTMILP or by IBM's ILOG-CPLEX (exporting the problem file in extended MPSX format and parsing the CPLEX solution file to convert the solution into the same format as generated by SAS Proc OPTMILP).

The formulation, in this case, contained 3,787 variables (36 binary for indicating whether quantity discounts are achieved) and 7,157 constraints. We selected a relative objective gap of 0.0001 for terminating the search for better integer-feasible solutions. Solution times for an individual scenario vary somewhat according to constraint parameters, but solutions for a particular set of constraints (including report generation using SAS) are typically derived within two minutes with an Intel 64-bit i5-3230m processor at 2.6 GHz and 8 GB of core memory.

7 Sample Results

In this section, we illustrate the type of information derived from the MILP solutions. Table 5 shows the percentage cost savings under several combinations of product substitutability and allowed market share. It is evident that restricting market share to 50% in a product category would have no impact on potential savings, but restricting it to 30% would mute savings somewhat and require considerable flexibility in using alternative devices or supplies for medical procedures.

Table 5. Potential Percent Cost Savings from Physician Flexibility under Different Market Share Constraints

Percent substitutability allowed	Maximum share 100%	Maximum share 75%	Maximum share 50%	Maximum share 30%
20	8	8	8	Infeasible
40	11	11	11	Infeasible
60	15	15	15	13
80	19	19	19	16
100	23	22	22	17

Figure 2 shows the impact of forcing diversification by restricting the dollar market share for each vendor in each of the product categories. Table 6 shows the shift of business for products in Category 1 to more expensive vendors that would be required to comply with changing the maximum market share market share restrictions. Overall savings of 23% are possible if a maximum market share of 50% is tolerated. They are reduced substantially to 17% if market shares are limited to 30% in each product category.

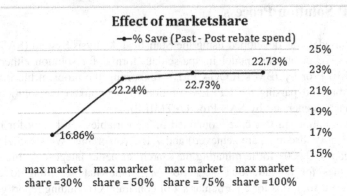

Fig. 2. Potential savings under different market share restrictions

Table 6. Example of Shifts in Market Share for Products in One Category (optimization output to input with 100 pct product level substitutability allowed, 100 pct category level substitutability allowed, scorecard minimum indices = 35 pct)

Max market share allowed: 100%								
Vendor	Price index	Historic usage	Total qty	Pref qty	Hist unit share	Total unit share	Hist dollar share	Current dollar share
1	0.8734	15	42	15	11.4	31.8	9.5	29.3
2	0.9915	23	90	23	17.4	68.2	16.6	70.7
3	1.0248	27	0	0	20.5	0	20.1	0
4	1.1063	67	0	0	50.8	0	53.8	0
		132	132	38	100	100	100	100
Max market share allowed: 30%								
Vendor	Price index	Historic usage	Total qty	Pref qty	Hist unit share	Total unit share	Hist dollar share	Current dollar share
1	0.8734	15	42	15	11.4	31.8	9.5	28.6
2	0.9915	23	39	15	17.4	29.8	16.6	30
3	1.0248	27	38	0	20.5	28.4	20.1	30
4	1.1063	67	13	13	50.8	10	53.8	11.4
		132	132	38	100	100	100	100

Next we illustrate the interactions revealed among market share restrictions, flexibility in allowing substitute devices to be used for medical procedures and possible bias in vendor ratings. Figure 3 shows potential procurement savings if vendor ratings tend to be low; Fig. 4 shows potential procurement savings if vendor ratings tend to be high.

Our investigation of the effects of common methods bias led us to conclude that the tendency of raters to provide high or low scores relative to the norms chosen for the

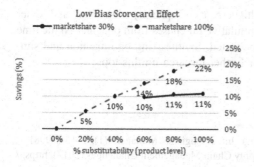

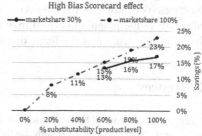

Fig. 3. Potential savings with different market-share restrictions if vendor scorecard ratings are low

Fig. 4. Potential savings with different market-share restrictions if vendor score-card ratings are high

scorecard constraints in the model were more important in shaping procurement strategies than were the correlation structures induced in the ratings.

8 Conclusion

The MILP model was helpful in revealing the impact of various constraints and contractual arrangements in procurement of high-cost medical devices and supplies. With tight constraints on allowed market share, little flexibility to use substitute products, and raters' tendency to assign low scores to vendors, cost-saving opportunities are axiomatically lower, and the likelihood of encountering conflicting constraints increases. With the MILP model, we were able to quantify the potential magnitude of these effects.

The model may also be employed to test the consequences of negotiating contracts with different time frames. There is some systematic seasonal variation in medical procedures which affects product usage. For cardiac rhythm management, total usage of devices in some product categories differed by more than 25% from quarter to quarter, while usage of products in other categories varied less than 5% from quarter to quarter. If qualification for quarterly quantity discounts are based simply on purchasing one-fourth of the quantity that would qualify for discounts based on annual volumes, the number of vendors engaged and share of business given to each can change from quarter to quarter. This calls for building solutions under quarterly planning assumptions and comparing the results (purchases from vendors and related costs) with solutions based on consistent assumptions for annual negotiations.

Finally, we must acknowledge the need to consider other issues that are encountered when implementing procurement plans in the face of normal variation in the conduct of medical procedures and in the event of supply-chain disruptions. Product usage may deviate from expectations and cumulative expenditures with vendors may not progress as expected toward the negotiated amounts that qualify the organization for quantity discounts. This calls for a complementary simulation model in which stochastic variation in medical procedures may be imposed and rules for adjusting purchasing plans to encourage the achievement of quantity discounts and other objectives (depending on deviations of cumulative purchases from plan and on the time remaining in the planning period) may be tested. Synthesizing optimization and simulation in this fashion is the focus of our further research on this topic.

References

Anderson, M.G., Katz, P.B.: Strategic sourcing. Int. J. Logistics Manage. **9**(1), 1–13 (1998)

Belliveau, J.: 3 Most Common Healthcare Supply Chain Management Challenges (2017). https://revcycleintelligence.com/news/3-most-common-healthcare-supply-chain-management-challenges. Accessed 11 May 2017

Bui, T., Yen, J., Hu, J., Sankaran, S.: A multi-attribute negotiation support system with market signaling for electronic markets. Group Decis. Negot. **10**(7), 515–537 (2001)

Burke, G.J., Carrillo, J.E., Vakharia, A.J.: Sourcing decisions with stochastic supplier reliability and stochastic demand. Prod. Oper. Manage. **18**(4), 475–484 (2009)

Chai, J., Liu, J.N., Ngai, E.W.: Application of decision-making techniques in supplier selection: a systematic review of literature. Expert Syst. Appl. **40**(10), 3872–3885 (2013)

DiChiara, J. (2016), Why Executives are Demanding Supply Chain Management Value. Retrieved May 11, 2017, from http://revcycleintelligence.com/news/why-executives-are-demanding-supply-chain-management-value

Galankashi, M.R., Helmi, S.A., Hashemzahi, P.: Supplier selection in automobile industry: a mixed balanced scorecard–fuzzy AHP approach. Alexandria Eng. J. **55**(1), 93–100 (2016)

Kamali, A., Ghomi, S.F., Jolai, F.: A multi-objective quantity discount and joint optimization model for coordination of a single-buyer multi-vendor supply chain. Comput. Math Appl. **72**(8), 3251–3279 (2011)

Kannan, D., Khodaverdi, R., Olfat, L., Jafarian, A., Diabat, A.: Integrated fuzzy multi criteria decision making method and multi-objective programming approach for supplier selection and order allocation in a green supply chain. J. Clean. Prod. **47**, 355–377 (2013)

Kaplan, R., Norton, D.: The balanced scorecard: measures that drive performance. Harvard Bus. Rev. **70**(1), 71–79 (1992)

Karsak, E.E., Dursun, M.: An integrated fuzzy MCDM approach for supplier evaluation and selection. Comput. Ind. Eng. **82**, 82–93 (2015)

Lance, C.E., Dawson, B., Birkelbach, D., Hoffman, B.J.: Method effects, measurement error, and substantive conclusions. Organ. Res. Methods **13**(3), 435–455 (2010)

Pan, A.C.: Allocation of order quantity among suppliers. J. Supply Chain Manage. **25**(3), 36–39 (1989)

Parlar, M., Wang, D.: Diversification under yield randomness in inventory models. Eur. J. Oper. Res. **77**(1), 52–74 (1993)

Papanicolas, I., Woskie, L.R., Jha, A.K.: Health care spending in the United States and other high-income countries. JAMA **319**(10), 1024–1039 (2018)

Peterson, R.A., Wilson, W.R.: Measuring customer satisfaction: fact and artifact. J. Acad. Mark. Sci. **20**(1), 61 (1992)

Location of Electric Vehicle Charging Stations Under Uncertainty on the Driving Range

Mouna Kchaou Boujelben[1(✉)] and Celine Gicquel[2]

[1] College of Business and Economics, UAE University,
P.O. Box 15551, Al Ain, United Arab Emirates
mouna.kchaou@yahoo.fr
[2] LRI, Université Paris-Saclay, 91405 Orsay, France

Abstract. We study the problem of locating electric vehicle (EV) charging stations on road networks. We consider that the driving range, i.e. the maximum distance that a fully charged EV can travel before its battery runs empty, is subject to uncertainty and seek to maximize the expected coverage of the recharging demand. We first propose a new mixed-integer linear programming formulation for this stochastic optimization problem and compare it with a previously published one. We then develop a tabu search heuristic procedure to solve large-size instances of the problem. Our numerical experiments show that the new formulation leads to a better performance than the existing one and that the tabu search heuristic provides good quality solutions within short computation times.

Keywords: Flow refueling location problem
Electric vehicle charging station network design · Stochastic driving range
Mixed-integer linear programming · Tabu search

1 Introduction and Related Literature

In order to reduce CO2 emissions and build a more sustainable future for the next generations, governments are encouraging drivers to adopt electric vehicles. Electric Vehicles (EVs) use battery-powered electric motors and are thus considered "green" transportation modes. However, the main limitation of EVs is that their range, i.e. the maximum distance that a fully charged EV can travel before its battery runs empty, is much smaller than the one of a conventional gasoline vehicle. This implies that, for a long-distance trip, an EV driver will have to stop multiple times at charging stations to refuel the EV battery before arriving to his final destination. The charging stations thus have to be easily accessible and the charging time should be limited to a few minutes. Fast charging technology exists in some countries such as the USA [1] but it requires high installation and operating costs. Therefore, the locations of fast charging stations should be optimized in a way to cover as much recharging demand as possible within the available investment budget.

Unlike classical facility location problems which rely on a node-based representation of the demand, the recharging demand in this problem is represented as a set of origin-destination trips. The reason is that in many cases, drivers do not specifically

© Springer Nature Switzerland AG 2018
R. Cerulli et al. (Eds.): ICCL 2018, LNCS 11184, pp. 475–486, 2018.
https://doi.org/10.1007/978-3-030-00898-7_32

travel to a station to recharge their vehicle but rather recharge it while on their way to another destination. For EVs, this occurs in particular during long-distance trips exceeding the vehicle range. As defined in [2], an EV trip is said to be "covered" or "refueled" if EV drivers can travel from their point of origin to their point of destination and back without running out of fuel. This means that there should be an adequate number of charging stations on the shortest path between the origin and the destination of the trip and that these stations should be carefully located so that the distance between two consecutive stations does not exceed the EV range. The corresponding optimization problem is called the flow refueling location problem (FRLP) and was first introduced in [2]. In the FRLP, the objective is to select the best locations for charging stations so as to maximize coverage of the EV trips on the road network.

Since its first introduction in 2005, different extensions of the problem have been proposed in order to study more realistic assumptions such as capacitated charging stations [3] or deviations from the shortest paths [4]. Other works have focused on developing efficient algorithmic approaches for solving the resulting large-size mixed-integer linear programs (see e.g. [5]). However, most of these works assume that the driving range is deterministically known. This seems a rather strong assumption as in practice, the actual range might be subject to high variations due among others to the traffic conditions, the outside temperature, the use of in-car heater or air conditioning and the age of the battery. Neglecting range uncertainties when planning an EV charging infrastructure might lead to suboptimal solutions as the actual demand coverage provided by the location decisions might be misestimated. It is thus necessary to explicitly take into account in the location model the stochastic nature of the vehicle range.

To the best of our knowledge, the uncertainty of the driving range has been introduced in only two recent works [6, 7]. With a stochastic driving range, it might not be possible to ensure that a trip will be refueled for all possible realizations of the range. Hence, the problem modeling involves the calculation of the coverage probability of each trip as the joint probability that each portion of the trip comprised between two opened charging stations will be shorter than the driving range. De Vries and Duijzer [7] assume that, for a given trip, the realization of the driving range will be the same on each portion of the trip and exploit this assumption to ease the computation of the joint probability of coverage. They introduce two stochastic programming based models. The first one defines the coverage probability of each trip as a decision variable and seeks to maximize the expected coverage of EV trips in the network. The second one is a chance constrained model in which the coverage probability of each covered trip is required to stay above a minimum value. In both cases, the problem is reformulated as a MILP and solved with a mathematical programming solver. Lee and Han [6] also seek to maximize the expected coverage of EV trips but they assume that, for a given trip, the random variables representing the stochastic driving range realization on each trip portion are totally independent of one another. This assumption enables them to compute the joint probability of trip coverage as the product of the reachability probability of each portion of the trip. The problem is then formulated as a mixed-integer non-linear program and solved using a Branch-and-Price approach.

This paper is mostly related to the work of De Vries and Duijzer [7] as we use the same assumption on the driving range realization. The numerical results provided in [7] show that solving the expected flow refueling location problem (EFRLP) under uncertainties of the driving range with a state-of-the-art MILP solver becomes difficult for large size instances of the problem. We thus propose in this paper a new MILP formulation for the EFRLP. This new formulation enables us to divide by two the average computation time needed to obtain guaranteed optimal solutions of the problem. However, even with the proposed formulation, the computation time needed to solve the largest instances is around 7 h. Hence, we also develop a tabu search heuristic approach in order to be able to provide good quality solutions of the problem with reduced computation times.

The remainder of this paper is organized as follows. In Sect. 2, we present the problem and the two formulations. Section 3 describes the tabu search procedure. We discuss in Sect. 4 the numerical results before concluding in Sect. 5.

2 Expected Flow Refueling Location Problem (EFRLP)

2.1 Problem Description

We now describe the problem under study in more detail. We first define the basic elements of the modeling approach: the road network, the recharging demand and the stations. We then discuss the notion of probability of coverage of a trip. Finally, we introduce the concept of cycle segment, which is used in the problem modeling to link the station location decisions and the flow coverage decisions.

We consider a road network $G(N, A)$, where N denotes a set of nodes and A denotes a set of arcs linking these nodes. The recharging demand to be satisfied by the stations is modeled as a set of trips denoted by Q. Each trip $q \in Q$ is described by its origin O_q, its destination D_q and the number f_q of vehicles traveling along it. Drivers belonging to trip q are assumed to follow the shortest path between O_q and D_q. All EVs have a limited range R. The problem is that the distance traveled by a driver on trip q is likely to exceed the range R. In this case, multiple stops in charging stations are required to recharge the EV battery. These stations should be located at nodes carefully selected on the shortest path from O_q to D_q.

In a deterministic range setting, the coverage of a trip is binary: either the distance between each pair of consecutive stations on the corresponding path does not exceed the range R and the trip is covered, or there is at least one pair of stations which are at a distance from one another larger than R and the trip is not covered. However, with a stochastic driving range, it might not be possible anymore to ensure that a trip will be refueled for all possible realizations of the range, even if we open many stations on the corresponding path. In the problem modeling, we will thus consider the probability of coverage of each trip q, z_q, which will be defined as the probability that EV drivers can travel a round trip along the corresponding path without running out of fuel. The optimization problem consists in identifying the best locations to build a predetermined number of stations p so as to maximize the expected coverage of EV trips.

The expected coverage is computed as the sum, over all trips $q \in Q$, of the product of the number of vehicles traveling along the trip f_q by the coverage probability z_q.

In order to model round trips between origin-destination pairs, De Vries and Duijzer [7] introduced the concept of cycle segment: a cycle segment $[k, l]$ is defined as the sequence of consecutive arcs to be crossed (in the outward and/or return direction) when traveling between node k and node l on a trip q. Let N_q (resp. A_q) denote the set of nodes (resp. arcs) belonging to the shortest path between nodes O_q and D_q. The cycle segment $[k, l]$ is defined as follows:

1- If $k = O_q$, $[k, l] = l \to O_q \to l$, i.e. $[k, l]$ corresponds to a round trip from l to O_q to l.

2- If $l = D_q$, $[k, l] = k \to D_q \to k$, i.e. $[k, l]$ corresponds to a round trip from k to D_q to k.

3- $k \in N_q \backslash O_q$ and $l \in N_q \backslash D_q$ then $[k, l] = k \to l$.

Therefore, the length $t_q(k, l)$ of segment $[k, l]$ is computed as follows:

- In cases 1 and 2, $t_q(k, l) = 2 * \sum_{(m,n) \in A_q^{k,l}} d(m, n)$ where $A_q^{k,l}$ is the set of consecutive arcs on trip q that are visited when travelling from node k to node l and $d(m, n)$ is the length of arc $(m, n) \in A$.
- In case 3, $t_q(k, l) = \sum_{(m,n) \in A_q^{k,l}} d(m, n)$.

In what follows, we discuss two MILP formulations for the EFRLP, using the notations summarized in Table 1.

Table 1. Model notation.

Parameter	Description
$d(i, j)$	Length of arc $(i, j) \in A$
N_q	Set of nodes situated along trip q
N_{lq}^b	Set of nodes situated along trip q before node l when traveling from Oq to Dq
N_{kq}^a	Set of nodes situated along trip q after node k whentraveling from Oq to Dq
A_q	Set of arcs on trip q
A_q^{kl}	Set of consecutive arcs on trip q visited when travelling from node k to node l
$t_q(k, l)$	Length of a cycle segment $[k, l]$ on trip q

2.2 Mathematical Formulation (EFRLM1)

Formulation EFRLM1 has been proposed in De Vries and Duijzer [7] for the expected flow refueling location problem under uncertainty on the driving range. We recall it in what follows in order to make the paper self-content.

This formulation relies on the concept of cycle segment. More precisely, for each trip $q \in Q$, the round trip from O_q to D_q and back is decomposed into a sequence of sub-trips between charging stations where the EV successively stops to get refueled. Each of these sub-trips corresponds to one of the cycle segments defined above. Thus, in formulation EFRLM1, a feasible sequence of cycle segments or sub-trips, complying with the stations opening decisions, is built for each trip q. The coverage probability of a trip q is then computed as the joint probability that all cycle segments used by the drivers to carry out round trips on q have a length smaller than the range R.

The formulation uses three sets of decision variables:

- $z_q \in [0,1]$ is the probability that trip q is covered,
- $x_k \in \{0,1\}$ with x_k equal to 1 if a station is opened in node k and 0 otherwise,
- $v_{kl}^q \in \{0,1\}$ with v_{kl}^q is equal to 1 if $[k,l]$ is a cycle segment used on trip q and 0 otherwise.

Before providing the MILP formulation, we explain how z_q can be computed using a set of linear constraints involving variables v_{kl}^q.

$$z_q = \mathcal{P}\left(R \geq t_q(k,l)v_{kl}^q \quad \forall k \in N_q \backslash D_q, \forall l \in N_{kq}^a\right) \tag{1}$$

$$= \mathcal{P}\left(R \geq max_{(k,l)}t_q(k,l)v_{kl}^q\right) \tag{2}$$

$$= 1 - G\left(max_{(k,l)}t_q(k,l)v_{kl}^q\right) \tag{3}$$

$$= 1 - max_{(k,l)}\left(G\left(t_q(k,l)v_{kl}^q\right)\right) \tag{4}$$

$$= 1 - max_{(k,l)}\left(G\left(t_q(k,l)\right)v_{kl}^q\right) \tag{5}$$

$$= 1 - max_{(k,l)}\left(g_{kl}^q v_{kl}^q\right) \tag{6}$$

Equality (1) defines z_q as the probability that the stochastic range R is greater than the length of all cycle segments used to travel along q. Equality (2) makes use of the assumption that the range realization is the same in all the network and computes z_q as the probability that the range is greater than the length of the longest cycle segment used to travel along trip q. In (3), the probability is expressed using the cumulative density function G of the random range R. (4) holds because G is a non-decreasing function. Equality (5) is obtained thanks to the fact that v_{kl}^q is a binary variable and $G(0) = 0$. Finally, in (6), we use g_{kl}^q defined as $G\left(t_q(k,l)\right)$.

In the MILP formulation, we can thus use the following inequalities to compute z_q: $z_q \leq 1 - g_{kl}^q v_{kl}^q \ \forall k \in N_q \backslash D_q, \forall l \in N_{kq}^a$.

Moreover, for a given node k, at most one variable v_{kl}^q can take the value 1. Therefore the constraints can be written as $z_q \leq 1 - \sum_{l \in N_{kq}^a} g_{kl}^q v_{kl}^q \ \forall k \in N_q \backslash D_q$.

This leads to the following formulation denoted EFRLM1:

$$\text{maximize} \sum_{q \in Q} f_q z_q \tag{7}$$

$$s.t. \sum_{l \in N_q} v_{O_q l}^q = 1 \quad \forall q \in Q \tag{8}$$

$$\sum_{k \in N_q} v_{kD_q}^q = 1 \quad \forall q \in Q \tag{9}$$

$$\sum_{l \in N_{kq}^a} v_{kl}^q = x_k \quad \forall q \in Q, k \in N_q \backslash \{O_q, D_q\} \tag{10}$$

$$\sum_{l \in N_{kq}^b} v_{lk}^q = x_k \quad \forall q \in Q, k \in N_q \backslash \{O_q, D_q\} \tag{11}$$

$$z_q \leq 1 - \sum_{l \in N_{kq}^a} g_{kl}^q v_{kl}^q \quad \forall q \in Q, \forall k \in N_q \backslash D_q \tag{12}$$

$$\sum_{j \in N} x_j = p \tag{13}$$

$$x_j \in \{0, 1\} \quad \forall j \in N \tag{14}$$

$$z_q \in [0, 1] \quad \forall q \in Q \tag{15}$$

$$v_{kl}^q \in [0, 1] \quad \forall q \in Q, l \in N_q, k \in N_{lq}^b \tag{16}$$

The objective is to maximize the expected coverage of EV trips. Constraints (8) and (9) state that there is exactly one cycle segment starting at the origin and one cycle segment arriving at the destination of each trip q. Constraints (10) impose that if there is a station at node k on trip q, there must be exactly one cycle segment starting at node k. Similarly, constraints (11) impose that if there is a station at node l on trip q, there must be exactly one cycle segment arriving at node l. Constraints (12) define variables z_q using the inequalities previously explained. Finally, constraint (13) sets the number of charging stations that must be opened to a predetermined number p. Note how variables v_{kl}^q have been defined as continuous variables in $[0, 1]$. Namely, De Vries and Duijzer [7] showed that the binary constraints on v_{kl}^q can be relaxed without changing the optimal solution of the problem.

2.3   Mathematical Formulation (EFRLM2)

We now discuss a new formulation, denoted EFRLM2, for the expected flow refueling location problem under uncertainty on the driving range. This formulation relies on the idea that, if a trip q is covered, each node $l \in N_q$ should be reachable by the EV after refueling at an opened charging station situated before l on the shortest path between O_q and D_q. In other words, if a trip q is covered, each node $l \in N_q$ should be assigned to a single station $k \in N_{lq}^b$ in charge of serving it. If a trip q is not covered, no node-station assignment is required.

Similarly to formulation EFRLM1, the new formulation uses binary location variables x_k and continuous coverage probability decisions z_q. We also introduce a new set of binary assignment decision variables w_{kl}^q defined as follows. For each trip q, each node $l \in N_q \backslash O_q$ and each node $k \in N_{lq}^b$:

- For $k \neq O_q$: $w_{kl}^q = 1$ if the EV battery is recharged at the station located at node k to refuel the driver trip at least up to node l, 0 otherwise.
- For $k = O_q$: $w_{kl}^q = 1$ if the refueling of the sub-trip $l \to O_q \to l$ is ensured by a station located at node l or after on the corresponding path, 0 otherwise.

With this problem modeling, the coverage probability z_q is defined as the joint probability that all nodes belonging to $N_q \backslash O_q$ are assigned to a station located at a distance shorter than the driving range. Thanks to the same reasoning as the one used for formulation EFRLM1 (see Eqs. (1)–(6)), z_q can be computed through a set of linear constraints of the form $z_q \leq 1 - \sum_{k \in N_{lq}^b} g_{kl}^q w_{kl}^q \quad \forall l \in N_q \backslash O_q$.

This leads to the following MILP formulation denoted EFRLM2.

$$\text{maximize} \sum_{q \in Q} f_q z_q \tag{17}$$

$$\text{s.t.} \sum_{k \in N_{lq}^b} w_{kl}^q \geq z_q \ \forall q \in Q, l \in N_q \backslash O_q \tag{18}$$

$$w_{kl}^q \leq x_k \ \forall q \in Q, k \in N_q \backslash O_q, l \in N_{kq}^a \tag{19}$$

$$\sum_{j \in N} x_j = p \tag{20}$$

$$z_q \leq 1 - \sum_{k \in N_{lq}^b} g_{kl}^q w_{kl}^q \ \forall q \in Q, l \in N_q \backslash O_q \tag{21}$$

$$x_j \in \{0, 1\} \ \forall j \in N \tag{22}$$

$$z_q \in [0, 1] \ \forall q \in Q \tag{23}$$

$$w_{kl}^q \in \{0, 1\} \ \forall q \in Q, l \in N_q, k \in N_{lq}^b \tag{24}$$

The objective is to maximize the expected coverage of EV trips in the network. Constraints (18) state that if a trip q has a strictly positive probability of coverage ($z_q > 0$), there must exist, for each node $l \in N_q \backslash O_q$, a node k situated before l on trip q where the vehicle is refueled in order to travel up to node l. Constraints (19) define the relationship between variables w and variables x: if there is no station opened at node k, the EV battery cannot be recharged at node k. Constraint (20) limits the number of charging stations that must be opened to a predetermined number p representing the limited investment budget, while constraints (21) define variables z_q as explained above.

3 Solving the Problem Using a Tabu Search Procedure

Problem (17)–(24) is a mixed-integer linear program which can be solved to optimality with a mathematical programming solver for small to medium size networks. For large size networks, the computation time required to get an optimal solution might become prohibitively long. Therefore, we propose a tabu search procedure in order to obtain good quality solutions in shorter computation times. In the following, we explain the main principles used in building an initial solution for the heuristic algorithm and evaluating the objective function for a given solution. Then we present the tabu search algorithm used to solve the expected flow refueling location problem.

1. Building an initial solution: the objective here is to obtain a set of p stations that provide a good coverage of the network. We first start by sorting the trips of the network in the decreasing order of EV numbers f_q. Then, we iterate over the trips to cover first the trips with the highest numbers of EVs. To cover a trip q, we start by building a station at the origin node O_q, then, by using a greedy approach, we build the minimum possible number of stations on q that ensure that the probability of coverage of the trip (i.e. the probability that the length of each segment used on the trip be smaller than the range R) is higher than a minimum probability, set as input parameter to the heuristic. The greedy algorithm for a trip q consists of multiple iterations that set each a starting point S that contains the last station opened (S set to O_q in the first iteration) and consider one by one the nodes located after S on the shortest path from O_q to D_q to place another station T. For each of these candidate nodes, the value of the probability of coverage of segment $[S, T]$ is computed. This value decreases when we get further from S, until it becomes lower than the minimum required. There, the algorithm stops to locate a station (at the node right before the last visited), which will serve as a starting point in the following iteration. If the destination node is reached while the probability value is still higher than the minimum required, no other station is located on q. Note that any station opened on a given trip q is taken into account in later trips by using an array of opened stations that is updated each time a station is opened. We stop iterating over the trips when the number p of opened stations is reached. At the end of the procedure, we evaluate the objective function for the obtained solution, by checking the probability of coverage of all the trips of the network. The stations opened to cover the trips with the highest EV numbers might belong to other trips in the network, and thus these trips might also be covered.

2. Evaluating the objective function for a given solution: in order to evaluate the objective function for a given feasible solution (set of p stations opened), we iterate over the trips of the network to check their probability of coverage and calculate the expected coverage. To compute the probability of coverage of a trip q, we proceed as follows: for each cycle segment $[k, l]$ of q, we first calculate the probability of coverage $(1 - g_{kl}^q)$ defined in subsection 2.3. Then, we calculate z_q the probability of coverage of trip q as the minimum of the probabilities of coverage of all its cycle segments. Finally, we update the objective value by adding $f_q \, z_q$.

The tabu search procedure is described in Algorithm 1. It starts by building an initial solution using the greedy approach described above. Then, two steps are carried out at each iteration of the algorithm. Step1 consists of selecting a station to be opened. The station should not be tabu (not recently closed) and should lead to the highest expected coverage among all possible openings. Step2 consists of selecting a station to be closed. This station should not be tabu (not recently opened) and should lead to the highest expected coverage among all possible closings. Here, we also consider an aspiration criterion which consists in accepting to close a station that is tabu but leads to improving the best objective value. Note that, due to these two steps, we use two tabu lists: one for the recently opened stations and one for the recently closed stations.

Algorithm 1. tabu search procedure

Build an initial solution;
Initialize the current best solution to the initial solution;
Initialize the current best objective value to the initial solution objective value;

while the maximum number of iterations without improvement is not reached **do**

  Step1: Select a station s_1 to be opened among the $|N|$-p closed stations, such as s_1 is not in the tabu closing list and opening s_1 leads to the highest expected coverage among all possible openings;
  if no stations can be opened (all openings are tabu) **then**
    Break, end tabu search
  else
    Add s_1 to the tabu opening list;
  end
  Step2: Select a station s_2 to be closed among the p + 1 opened stations such that closing s_2 leads to the highest expected coverage among all possible closings (we consider that closing a station s_2 is possible if s_2 is not in the tabu opening list or if s_2 belongs to the tabu opening list but leads to improving the current best objective value);
  if no stations can be closed **then**
    Break, end tabu search
  else
    Add s_2 to the tabu closing list;
    if closing s_2 leads to a better expected coverage than the current best objective value **then**
      Update the current best solution and objective value;
      Set the number of iterations without improvement to 0
    else
      Increment the number of iterations without improvement
    end
  end
end

4 Numerical Experiments

The first objective of our experiments is to compare the numerical performance of the new formulation EFRLM2 with the formulation EFRLM1 recently proposed by De Vries and Duijzer in [7] for the expected flow refueling location problem. The second objective is to analyze the quality of the solutions provided by the tabu search heuristic approach described in Sect. 3. In the following, we first describe how the test instances have been generated. We then present the numerical results for the two formulations as well as for the tabu search algorithm.

4.1 Test Instances

The road networks used in the tests are randomly generated following a procedure similar to the one proposed in [8]. We start by generating $|N|$ nodes in $[1, 1000]^2$ according to a uniform distribution. We evaluate the traveling distance between each pair of nodes of the network using the Euclidian distance. We then apply Kruskal algorithm to determine the minimum spanning tree of size $|N| - 1$. All arcs belonging to this spanning tree are added to the set A. We also select $|N|$ additional arcs to be added to A: these arcs are the shortest potential arcs not yet added to A and such that the degree of each node stays below four. We then randomly select M pairs of origin-destination nodes among the nodes belonging to N. This leads to a set of $|Q| = \frac{M(M-1)}{2}$ trips to be covered. The shortest path (N_q, A_q) corresponding to trip q is determined using Dijkstra algorithm. The population at each origin/destination node is generated randomly between 0 and 1000000 and the number f_q of EVs on each trip q is deter-mined using the gravity model [9]: $f_q = \frac{P_q^O P_q^D}{d_q^2}$, where P_q^O and P_q^D are the populations at the origin and destination nodes, respectively and d_q is the length of the shortest path between the origin O_q and the destination D_q.

We consider two different instance sizes: ($|N| = 100$, $M = 50$) and ($|N| = 200$, $M = 100$) and randomly generate 5 instances for each size. The number of stations p is varied in $\{10, 15, 20, 25\}$, leading to a total of 40 instances. The random range R is represented using a Gamma distribution, with a shape parameter of 50 and a scale parameter of 5.

We employ the C++ language to implement the model and the commercial solver CPLEX 12.6.2 to solve it. All tests are carried out on a PC with Intel i5-3210M Core 2 Duo (2.50 GHz) with 8 GB of RAM. The CPU time of CPLEX solver is limited to 10 h. The tabu search procedure is run with the following settings: the tabu list size is set to 5 and the maximum allowed number of iterations without improvement of the objective value is set to 10.

4.2 Numerical Results

The computational results are provided in Table 2. Each line corresponds to the average value over the 5 corresponding instances. The first two columns display the computation time needed to solve the problem to optimality with the mathematical programming solver CPLEX using either formulation EFRLM1 or formulation EFRLM2. In case no guaranteed optimal solution could be found before the time limit of 10 h is reached, we use the time limit to compute the average computation time. These results show that the new formulation performs better than the existing one as it leads to a two-fold decrease of the average computation time. Moreover, CPLEX failed at finding the optimal solution within the time limit for 16 out of 40 instances with formulation EFRLM1 while it failed for only 3 out of 40 instances with formulation EFRLM2. In the cases where the optimal solutions were not obtained with EFRLM1, the optimality gap went up to 13%.

Table 2. Average performance of the three methods (5 replications).

Instance	1	2	3	4	5	6
	EFRLM2 CPU (s)	EFRLM1 CPU (s)	Nb. B&B nodes EFRLM2	Nb. B&B nodes EFRLM1	TABU CPU (s)	TABU Gap (%)
N100M50p10	165	583	0	484	9	0.2
N100M50p15	487	1595	163	2594	15	0.2
N100M50p20	597	4815	66	2491476	21	0.3
N100M50p25	609	11008	263	8835	24	0.1
N200M100p10	8212	32775	88	645	58	0.8
N200M100p15	17243	36000	126	629	90	0.7
N200M100p20	21237	36000	149	7	102	0.8
N200M100p25	26385	36000	174	93	184	0.6
Average	**9367**	**19847**	**129**	**313095**	**62.6**	**0.5**

Note that formulation EFRLM2 is larger than formulation EFRLM1. Namely, both formulations have the same average number of variables: 170,350 for instances ($|N| = 100$, $M = 50$) and 1,058,379 for instances ($|N| = 200$, $M = 100$). However, the average number of constraints used in EFRLM1 is 52,204 for instances ($|N| = 100$, $M = 50$) and 269,766 for instances ($|N| = 200$, $M = 100$) whereas the average number of constraints used in EFRLM2 is 186,427 for instances ($|N| = 100$, $M = 50$) and 1,143,152 for instances ($|N| = 200$, $M = 100$). Hence, the computation time decrease obtained while using formulation EFRLM2 cannot be explained by a reduction of the time needed to solve the linear relaxation of the problem at each node of the Branch & Bound tree.

A possible explanation might be the fact that formulation EFRLM2 is stronger than formulation EFRLM1. Namely, in formulation EFRLM2, variables z_q are linked to variables w_{kl}^q by two sets of constraints whereas in formulation EFRLM1, variables z_q are linked to variables v_{kl}^q by a single set of constraints. This might contribute in strengthening the linear relaxation bound used at each node of the Branch & Bound tree, thus leading to a decrease in the total number of Branch & Bound nodes explored by the algorithm before closing the optimality gap. Columns 3 and 4 of Table 2 show the average number of nodes explored using EFRLM1 and EFRLM2. On average, the number of nodes is only 129 with EFRLM2 while it is 313095 for EFRLM1.

Columns 5 and 6 of Table 2 report the results obtained with the tabu search heuristic. We provide the average running time of the heuristic as well as the average gap between the best solution found by the tabu search heuristic and the optimal solution. We note that the heuristic performs very well as it provides good quality solutions (tabu gap at 0.5% on average) in short computation times (1 min on average). This shows that the tabu search approach could be a good alternative, especially for large size instances, which are difficult to solve to optimality using a MILP solver.

5 Conclusion

In this paper, we studied the problem of locating EV charging stations under uncertain driving range. We first proposed a new MILP formulation for the problem. Based on randomly generated networks, we compared the new formulation to an existing formulation proposed in a recent paper, with regard to their numerical performance. The new formulation showed significant improvement in performance. Namely, the average computation time needed to solve the problem to optimality was divided by two when using the proposed formulation. However, the CPU time for some large instances remained high (around 7–9 h). We thus developed a tabu search heuristic approach to solve these instances. Our results show that it consistently provides good quality solutions for the problem within short computation times.

In our model, we assumed that for any realization of the random conditions in the road network, the value of the driving range is the same for all trips. An interesting direction for further research would be to relax this assumption and to study the more realistic case where the driving range realization is different on each cycle segment of a trip.

References

1. Tweed, K.: Fast charging key to electric vehicle adoption. www.greentechmedia.com. Accessed 18 Mar 2018
2. Kuby, M., Lim, S.: The flow-refueling location problem for alternative-fuel vehicles. Socio-Econ. Plan. Sci. **39**, 125–145 (2005)
3. Upchurch, C., Kuby, M., Lim, S.: A model for location of capacitated alternative-fuel stations. Geogr. Anal. **41**(1), 85–106 (2009)
4. Kim, J.G., Kuby, M.: The deviation-flow refueling location model for optimizing a network of refueling stations. Int. J. Hydrogen Energy **37**, 5406–5420 (2012)
5. Lim, S., Kuby, M.: Heuristic algorithms for siting alternative-fuel stations using the flow-refueling location model. Eur. J. Oper. Res. **204**, 51–61 (2010)
6. Lee, C., Han, J.: Benders-and-price approach for electric vehicle charging station location problem under probabilistic travel range. Transp. Res. Part B **106**, 130–152 (2017)
7. De Vries, H., Duijzer, E.: Incorporating driving range variability in network design for refueling facilities. Omega **69**, 102–114 (2017)
8. Capar, I., Kuby, M., Leon, V.J., Tsai, Y.-J.: An arc-cover path-cover formulation and strategic analysis of alternative-fuel station locations. Eur. J. Oper. Res. **227**, 142–151 (2013)
9. Fotheringham, A., O'Kelly, M.: Spatial Interaction Models: Formulations and Applications. Kluwer Academic Publishers, Dordrecht (1989)

Author Index

Printed in the United States
By Bookmasters